AF379710

RESEARCH METHODOLOGY
FOR BIOLOGICAL SCIENCES

MJP
PUBLISHERS

RESEARCH METHODOLOGY
FOR BIOLOGICAL SCIENCES

N GURUMANI

MJP
PUBLISHERS

Chennai   Trichy   Tirunelveli   NewDelhi

Honour Copyright
&
Exclude Piracy

This book is protected by copyright. Reproduction of any part in any form including photocopying shall not be done except with authorization from the publisher.

MJP
PUBLISHERS

ISBN 81-8094-016-0   **MJP PUBLISHERS**

All rights reserved   No.44, Nallathambi Street

Triplicane, Chennai 600 005

MJP 014   © Publishers, 2020

Publisher : C. Janarthanan

This book has been published in good faith that the work of the author is original. All efforts have been taken to make the material error-free. However, the author and publisher disclaim responsibility for any inadvertent errors.

PREFACE

Research, which used to be carried out only at the university level, that is, in the university departments, and in special institutes founded for a specific purpose, is now being pursued in almost all post-graduate teaching institutions and, of late, in many distance education institutes. Most of these institutes of higher learning have research programmes leading to M. Phil. and Ph. D. degrees, which invariably include a course on "Research Methodology". In addition, faculty members of the "PG and Research" Departments of many affiliated colleges have minor and major research projects sponsored by various agencies like UGC, CSIR, DST, ICMR, etc. Further, in many of the post-graduate courses students are expected, as a part of the curriculum, to carry out a "project work" and submit the "project report". The aim of such project work is to expose the students to the concept of research at an early stage and to motivate them to take up research as their career.   of the project work, the post-graduate students can opt for a theory paper on research methodology. With the increasing tendency at the higher education level to lay stress on training the students for research, it is imperative that the teachers at this level take up the challenge of providing quality guidance to the students to learn all the basics of research methodology.

Research, as we are fully aware, is an investigation in a very narrow field of biology or any other branch of science. And every field of specialization has its own methodology. However there are several basic things which are common to all the fields in all branches. For instance, literature survey, experimental design, collection of data, analysis of data, and writing and preparing the research report are a few such common aspects. It is very essential that all students who are keen to take up research as their vocation are thoroughly familiar with such basics. Such a familiarity would greatly help them not only in proper planning and execution of their research work but also in the preparation of the report and, more importantly, in getting it published in a reputed journal or getting it approved for the award of a research degree.

Apart from the basics of the research methodology, there are several laboratory methods widely adopted in the different fields of biological research. Microscopy, chromatography, electrophoresis, etc. are a few of such methods. All research workers may not be required to use all the methods. One who is studying the behaviour of a wild mammal may not require the use of, say, confocal laser scanning microscope. However, knowledge of the principle and applications of as many of such methods as possible is desirable, for it would enable one to understand and appreciate the published works involving the use of such methods and techniques, and may even elicit an interest to adopt one or more of them for one's own research.

Thus, the main objective of this book is to provide, as much as possible, the information on the basics of the research methodology and on the principles and applications of some important laboratory methods. Accordingly, the first few chapters (Chapters 1–8) deal with the fundamental aspects of research, research report and experimental designs. The subsequent chapters (Chapters 9–15) discuss the principles and applications of the laboratory methods. All explanations in these chapters have been written with minimal technical details so that even students who may not have background knowledge of biophysics will be able to understand them. As far as possible, all technical terms, even trivial ones, have been defined and/or explained in simple language. These chapters do not describe the handling of the instruments associated with these methods and do not provide detailed procedures and protocols of the various techniques.

Two chapters, one on intellectual property rights (Chapters 16) and the other on laboratory safety (Chapter 17) have been written to provide general information common to many branches of biology and to create an awareness of these aspects among students. A number of appendixes have been added to accommodate those matters which are useful but could not be included in the various chapters. The 'reference' section contains, apart from the list of books referred, addresses of websites that were consulted, and found to be immense sources of information for almost all aspects discussed in this book. The glossary section includes many terms other than what are found in the text. Some of these terms have been described in some detail with an intention to provide as much information as possible.

The Publishers and I are fully aware that there is lot of scope for adding more topics in this book. For instance, research methodology relating to the fields of taxonomy, ecology, toxicology, radiobiology, etc. certainly deserve a place in a book of this nature. It is sincerely hoped that subsequent editions will endeavour to include chapters on such topics.

I feel that research students should be a class above the rest. They should strive to acquire as much theoretical and practical knowledge as possible during their academic research programme, and when they come out of it they should prove to be better teachers and researchers capable of undertaking independent applied research projects. They should not be merely persons of learning but persons worthy of emulation by their students and other research workers. A research degree is not merely for a small piece of work done but also for the broad and deep knowledge gained on all aspects relating to research in general.

I wish to express my gratitude to the Director and the Editorial team of MJP Publishers for their encouragement, for their patient bearing with all the delay on my part, and for their meticulous effort in bringing out this book. I thank all my friends, well-wishers and students who constantly egged on me to take up this project. I welcome critical comments and suggestions about this edition from all, especially the teachers and students.

N GURUMANI
gurumani_n@msn.com

CONTENTS

12. CHROMATOGRAPHY — 275

FIGURES

TABLES

BOXES

ABBREVIATIONS

AFM	Atomic Force Microscope
ANOVA	Analysis of Variance
ANS	anilinonaphthalene sulphonate
AOD	Acoustic Optical Deflector
APD	Avalanche Photo Diode
ASA/BS	American Standard Association/British System
BAC	bisacrylylcystamine
BCA	bicinchonic acid
BME	b-mercaptoethanol
BSA	bovine serum albumin
BSC	Biological Safety Cabinet
CBD	Convention on Biological Diversity
CBE	Council of Biology Editors
CCD	Charge Coupled Device
CEC	Capillary electrochromatography
CGE	Capillary gel electrophoresis
CIEF	Capillary isoelectric focusing
CITP	Capillary isotachophoresis
CMO	common main objective
CPCSEA	Committee for the Purpose of Control and Supervision of Experiments on Animals
CSIR	Council for Scientific and Industrial Research
CTAB	cetyltrimethylammonium bromide
CZE	Capillary zone electrophoresis
DAPI	diamino-2-phenylindol
DATD	diallyltartardiamide
DGGE	denaturing gradient gel electrophoresis
DIC	Differential Interference Contrast
DIN	Deutsche Industrie Norm—German Industry Standard
DST	Department of Science and Technology

DTT	dithiothreitol
EEO	Eelectroendosmosis
EFM	Electrostatic Force Microscope
EOF	electroosmotic flow
FAT	fluorescent antibody technique
FCA	Freund's complete adjuvant
FID	flame ionization detector
GATT	General Agreement on Trade and Tariff
GDS	Gel documentation systems
GLC	Gas-liquid chromatography
GMOs	genetically modified organisms
GPC	Gel Permeation chromatography
GSC	Gas-solid chromatography
HPLC	High performance liquid chromatography
HPTLC	High performance Thin layer chromatography
IAEC	Institutional Animal Ethics Committee
IBW	Instrumental Bandwidth
IEC	Ion Exchange Chromatography
IEF	Isoelectric focusing
IMRAD	Introduction, Material and Methods, Results and Discussion
INSDOC	Indian National Scientific Documentation Centre
IPR	Intellectual Property Right
ISO	International Standard Organization
LCD	Liquid Crystal Display
LSCM	laser scanning confocal microscope
LSD	Latin Square Design
MBA	methylenebisacrylamide
MECC	Micellar electrokinetic capillary chromatography
MFM	Magnetic Force Microscope
NSOM	Near-field scanning optical microscope
OD	optical density
OTC	open tubular column
PAS	(periodic acid Schiff) staining
PCA	Prevention of Cruelty to Animals Act
PCT	patent cooperation treaty
PDA	piperazine diacrylate

PIC	Prior inform consent
PMT	photomultiplier tube
POP	printing-out-paper
PSTM	Photon Scanning Tunnelling Microscope
PZT	Lead Zirconium Titanate
RCF	Relative Centrifugal Force
SCM	Scanning Capacitance Microscope
SCOT	support coated open tubular
SCPM	Scanning Chemical Potential Microscope
SDS	Sodium Dodecyl Sulphate
SDS-PAGE	sodium dodecyl sulphate Polyacrylamide Gel
SEM	scanning electron microscope
SFC	Supercritical Fluid Chromatography
SFM	Scanning Force Microscopy
SHFM	Shear Force Microscope
SICM	Scanning Ion Conductance Microscope
SLR camera	Single-lens-reflex camera
SNOM	Scanning Near Field Optical Microscope
SPM	Scanning Probe Microscope
SRBC	Sheep red blood cells
SThM	Scanning Thermal Microscope
STM	Scanning Tunnelling Microscope
TEM	transmission electron microscope
TEMED	tetramethylethylenediamine
TGGE	temperature gradient gel electrophoresis
TLC	Thin layer chromatography
TRIPS	Trade Related Aspects of Intellectual Property Right
UGC	University Grants Commission
USIS	United States Information Science
WCOT	wall coated open tubular
WHO	World Health Organization
WIPO	World Intellectual Properties Organization
WTO	World Trade Organization

CHAPTER ONE

WHAT IS RESEARCH?

REFLECTION, SCIENCE AND RESEARCH

The concept of *research* is closely associated with the processes of *reflection* and *science*. Therefore, it would be useful if we understand the meaning of, and the difference and relationship between the terms *reflection* and *science*, before venturing into the definition of *research*.

Reflection

Reflection is a planned, serious thinking about one's own environment. The evolution of an organism itself is probably the result of *reflective thinking* by at least a few in a habitat. Whether "lower" organisms such as bacteria, invertebrates, etc. possess the ability to "think" as "higher" organisms such as humans do, is a difficult question to answer. At least we can presume that all those organisms, lower or higher, which are "capable" of "reflective thinking", do possess a genotype that is different to some degree from that of the rest of the kind. Probably, such an innate genetic difference enabled them to get selected by the process of "natural selection", and perpetuate a progeny that is more suitable for a changed or changing environment. It may be noted that a "changing" environment always creates a "need" for its inhabitants also to change. When the product of reflective thinking by a member or members of a habitat helps to fulfill or overcome a "need", that product is a "discovery" and it is utilized by all the other members of the community, in general, and passed on to future generations. The discovery may be a changed genotype that would express as a more useful and adaptive phenotype, a skill or a tool. Such discovery may be refined, modified or, even, discarded—again, by reflective thinking.

Let us elaborate the above point in order to understand that the ability of a few in a community for reflective thinking, i.e., to think in an orderly manner has been influential in the "cultural" evolution of man.

1. The occurrence of a difficult situation, might be as simple as hunger or thirst or may be one that poses a threat to the very existence of the community such as a war, outbreak of a disease, and natural calamities like flood, fire, drought, etc.

2. The felt difficulty or need may have simple solutions such as feeding, drinking water, etc. These solutions for such felt needs are "tradition or habit'. But these simple solutions may lead to new difficult situations such as what and how to feed or drink in order to keep the community healthier and fit. The problem

situations like whether the food or the drink must be processed before consumption, what are the different ways of processing, what are the advantages or disadvantages of each of these ways, etc. might crop up infinitely.

On the other hand, the difficult situation may not have an immediate solution, and it may require serious reflective thinking on the part of at least a few of those affected by the situation. The outcome of such reflective thinking might be agriculture, aquaculture, cloning, genetic engineering and so on. It may be, in general, the development of any one of the fields of science, including medicine and warfare.

3.   Any product of reflective thinking is put to trial in a community to be tested, explained, improved and accepted as a "tradition or habit". A new product may be introduced into the community as a result of reflective thinking by some other member of the community, to meet the same difficult situation. This new product would undergo the same procedure before being accepted or rejected. It may replace the first one or coexist with it. It must also be realized that many products of reflective thinking, which are unexplained or irrational, may continue to exist if they are "useful" in some way or the other to the members of a society. For example, blind beliefs in religious practices, many of which are unexplained and irrational, continue to be followed because they are found to be "mentally or physically" useful to those who practise them.

Science

Science is organized knowledge. And it involves observation, identification, description, experimental investigation, and theoretical explanation of any phenomenon that occurs in nature. In other words, the method of science is an organized way of thinking, again for the well-being and betterment of a society. In general, the aim of any good science is to advance the frontiers of human knowledge, i.e., to add valid information to what is already known. What we must realize is that what we know in biology or any other field of science is as little as a fistful and what we do not know is as big as the earth.

Research

Research is a method to attain the goal of science, i.e., to expand the frontiers of knowledge, and it involves the process of organized reflective thinking by the researcher. Research may be carried out in any field of arts and science, wherever there is need and scope to expand the horizon of knowledge. Almost all the industries have their own Research and Development (R & D) departments to improve the quality and, thereby, the sales of their products.

Research in biology has wider implications on the human society because it encompasses numerous interrelated fields from Agriculture to Zoology. More recently, the research activities in the fields of biotechnology, genetic engineering, bioinformatics, etc. have stressed the need for proper training of researchers not only in basic technical aspects but also in the ethical aspects.

BASIC AND APPLIED RESEARCH

On the basis of its objective a research may be classified as basic and applied. The main objective of a *basic research*, also referred to as *pure research*, is expansion of the existing knowledge. It does not envisage its results to have any social or economic relevance. A problem for basic research is selected from any source and tackled with appropriately planned scientific methodology.

Though the results of a basic research may not have any immediate social and economic value, they may prove to be highly significant and application-oriented at a future date. For example, the results of the basic researches carried out by Charles Darwin and G. J. Mendel, which led to the proposition of the theory of evolution and the theory of inheritance, respectively, are proving, long after their initial submission, to be of immense social and economic importance.

Most of the biological researches carried out in the university and college departments and research institutes specializing in a field, are basic in nature. They are more of training programmes for students aspiring to take up research as their career. Many of these programs expect the trainees not only to carry out a research project in a systematic way but also to present their findings in a forum and to publish them in a scientific journal. On the basis of their performance in the prescribed course work and on the report of the independent evaluation of their research thesis or dissertation, the students are awarded a research degree such as M. Phil. and Ph. D.

Applied research, on the other hand, has its aim as "practical application" of the results, such as cure of a disease, increased crop production, development of a biological weapon, etc. A problem for applied research is selected from a going concern, and is analysed and investigated with well-designed scientific methods. If the result of such an applied research is applied successfully to the solution of the problem, it might be patented and exploited economically. Generally, R&D departments of industries such as pharmaceutical and biotechnological industries are involved in applied research, or they sponsor such researches to be carried out by the universities and other academic research organizations. Government funding organizations such as CSIR, DST, etc. evince keen interest in promoting applied research in educational institutions.

ESSENTIAL STEPS IN RESEARCH

Whether it is basic research or applied research, the general steps necessary to characterize a biological research as a creditable one are the following:

1. Selection of the subject, field, topic and specific problem of the research.

2. General survey of the field to understand the problem of the research.

3. Specific survey of the pertinent literature (already existing published knowledge) and development of a bibliography.

4. Definition of the problem, including differentiating, defining and classifying its components.

5. Determining the parameters required to be studied towards the solution of the problem.

6. Choosing the methodology to study the parameters.

7. Standardizing the methodology and testing its suitability for the specific problem.

8. Designing the experiment, field study or survey with appropriate statistical background.

9. Collecting the data and information.

10. Systematic classification, tabulation, presentation, analysis and interpretation of the collected data.

11. Reporting all the above in the form of a thesis, dissertation or paper.

We shall discuss in the next few chapters the principles of the basic methods involved in biological research.

Chapter Two

Literature Collection

Review of literature in research refers to extensive, exhaustive, and systematic survey of publications relevant to the selected field of investigation. The process of literature review begins even before the stage of defining the research topic or problem and continues till the publication of the research report.

NEED FOR REVIEW OF LITERATURE

A thorough examination of the literature, i.e., the articles in science Journals, review articles, books, monographs and other writings, which deal with a particular subject, is essential for several reasons some of which are:

1. To assess the level of theory and research that have been developed in the field of study, and thus to find what is already known and what remains to be investigated in the specific field of study.

2. To understand the definitions of the established concepts and variables in the chosen field.

3. To identify and adopt the research design, analytical methods, scales, instruments, data analysis, etc.

4. To become fully aware of all the difficulties encountered by other workers, and to thus avoid waste of time and money in the proposed research.

5. To learn how to write a research report.

All the above reasons become more evident when we prepare the final research report, especially while (a) justifying the choice of research topic and methodology and (b) discussing our results and drawing valid inferences from our results.

REVIEW PROCESS AND BIBLIOGRAPHY

The literature review is a continuous process, which involves

1. location of current (about 5 years) pertinent publications

2. summarizing and recording the contents of each publication, and, most importantly

3. comparing and commenting on the merits and weaknesses of the different elements such as theoretical perspective, definitions,

research designs, methodology, instrumentation, data analysis and inferences.

To achieve a satisfactory review of literature we have to prepare a *bibliography*, which is a reading list and a list of works, consulted at different stages of the research. Normally, we begin with a *working bibliography,* one that includes a wide range of works pertaining not only to the specific study but also to those that are required to have a strong grasp of broad background information.

All the works listed in a working bibliography may not require serious reading, as they may not be directly related to our specific work. Therefore, we shortlist the working bibliography to a *pertinent bibliography* that includes only those works which are directly concerned with the field of our study and therefore require serious reading. We may not use all the items of the pertinent bibliography for the preparation of the report, especially if the report is meant for publication in a reputed Journal. All such works that we actually consulted and used (cited) in the final report constitute the *selected bibliography*. Many scholarly works often include an *annotated bibliography*, which is a brief list of references, each entry of which is followed by a note of comment on the content and usefulness of the reference.

RESEARCH READING

The time at our disposal for carrying out any research project is limited and specified, especially in academic institutions. Therefore, we have to make effective use of our time spent in the library. One useful practice is to increase our speed of reading. The following are general guidelines for increasing the speed of reading:

1. To begin with we must force ourselves to read faster. We may be bothered at first about not grasping the matter we are reading. Ignoring this botheration we must continue with the fast, concentrated-reading practice and in short course of time, it will become a habit, especially for youngsters.

2. While reading, we should make as few stops or eye "fixations" in each line as possible. While reading, our eyes move not continuously but by jerks and pauses. Our eyes take quick views or snapshots of successive portions of the printed sentence and then join them together to get the meaning. You can notice that we do not read while our eyes move, but only when it stops.

Therefore, we should not read by syllables or even by words, but by groups of words, phrases, and sentences. A poor reader has an average of 16 eye-stops per line of a given length of sentence, which may be about three lines of this page. This average can be decreased to 6 by a twenty-minute practice period each day for twenty days. Students can practice reading a newspaper line with only one or two eye-fixations. In general we should not fixate on the first word in a line but somewhere inside the beginning of the line; likewise we should make our last eye-stop not at the last word of the line but somewhere inside.

3. We should practice to keep on reading forward, that is, we should not break reading forward to wander back in regressive movements to read something we have missed. This is important especially when we are forcing ourselves to read more rapidly. One of the reasons of regression is the wrong hitting of the line when the eye sweeps back from the end of the preceding line. With a few trials we can establish the most effective way to hit the lines of each type of subject matter we read, and thus avoid unnecessary repeated reading of the same portions of a sentence.

4. We should try and establish a regular rhythm of eye-movement suitable to the length of line and subject matter of an article we read. We should feel comfortable when we swing into each line with this rhythm. When a student has mastered the technique of reading he will feel reading a pleasure and less stressful.

5. It is generally not a good practice to pronounce the words as we read or even allow our lips to move silently, because this slows down our reading pace.

DISCRIMINATIVE READING

As research workers, we should learn to be discriminative in our research reading, that is, we should be able to determine what material should be read to what depth. We need not read everything that we come across with the same speed and care. The main reason for this is the lack of time and furthermore, it is not desirable. Some of the research articles that we examine should be read by title only, or in terms of chapter and section heads, including the preface. Some articles can be read by the title. Some chapters of a book or sections of a paper can be skimmed very rapidly. There are certain papers and books that we should read slowly and carefully, taking notes as we read. The contents of some books and articles are to be mastered, because they are so important and pertinent to our work and will contribute a lot to the planning and designing of our work and to the interpretation of the data we will be collecting.

CONSULTING SOURCE MATERIAL

A valuable starting point for the survey of literature is an authoritative textbook, one that is written by an author who is an authority on the subject. Edited volumes containing chapters written by authorities, review articles published in journals, and annual reviews of specific branches of biology are also useful sources of reference materials. These textbooks and review articles provide lists of references that were consulted and / or a Bibliography which is a list of references consulted and also those not consulted but are relevant and useful. All the information we get from textbooks, review articles and other similar sources constitute secondary literature.

The primary literatures are the original research articles published in reputed journals. A reputed journal is one that accepts research articles for publication only after it has been critically reviewed and evaluated by a panel of authorities on the subject. No other consideration other than the quality and authenticity of the article is involved in the acceptance of a paper for publication in such journals. If a journal has international juries in its review panel, it is described as an international journal. The merit or reputation of a journal increases in proportion to its circulation and indexing by abstracting agencies.

Students must strive, however difficult it might be, to reach the original research articles, pertinent to the field of investigation. There are many sources from where a student can locate these original articles.

1. the department library

2. the college or university library

3. public libraries including those run by other countries such as British Council and USIS

4. personal collections of researchers in the field. In addition, there are also services such as INSDOC, which provide at a nominal cost copies of articles published in research journals.

Many authors of research publications send reprints of their articles on request, either by post or e-mail. Almost all international journals have their own websites through which the articles published in the issues are immediately made available to the subscribers. Many journals provide articles, especially those which appeared in older issues, free of cost, to be downloaded from their websites.

Many research workers have the wrong idea that the information (e.g. results, interpretations and inferences of an experiment) given in a textbook, review article, or in the reviews in a specific research paper, project report or thesis is as good as that given in an original research publication. It must be remembered that such information, however authoritative it might be, is secondary. Somebody else read the information, understood and interpreted it in their own way, and reproduced it in their own words for their own purpose, that is, for example, to discuss a result they obtained in their experiments. It is not unlikely that such reproduced information is one that has been wrongly understood and interpreted.

If a research worker wants to use such information, he/she should reach the original article, read, understand and interpret it before using it for his/her purpose.

Sometimes, it might be necessary to avoid referring to the results of very old works. Such results and generalizations of such results might be used specifying who has reviewed and generalized them.

WORKING BIBLIOGRAPHY

A bibliography is an alphabetic list of all source material to which reference has been made. A working bibliography is one that is prepared by a research student for this purpose. There are different ways of referencing books, journal articles, and other documents and these are dealt with in detail in chapter 3. However, the essential information required for all references includes the following:

1. Author's surname and initials. (If the author is a woman, it is usual to spell out her first name.)
2. The name of the article and/or journal or book.
3. The imprint (place of publication, publisher, and date of publication).

Since a working bibliography is being prepared, it is useful to add three further items of information.

4.   The call number of the book or journal.

5.   The library where the book or journal may be located.

6.   A phrase or sentence indicating the contents.

INDEX CARDS AND REFERENCE CARDS

It is useful to construct a working bibliography by writing each reference on 3" by 5" cards because these are easy to sort alphabetically and store. Such a working bibliography is convenient because new cards can be added, constant reference can be made to existing cards, and additions and corrections to cards can be effected with ease. A typical index card is shown in the Figure 2.1.

The use of card bibliography saves time. If a book or a journal needs to be relocated, no time is lost in searching for the call number. New references can be added to the sequence easily. At the typing stage the bibliography can be typed directly from the information recorded on the cards.

Aaltonen, T. M., E. T., Valtonen, and E. I. Jokinen   1997

Immunoreactivity of Roach, *Rutilus rutilus*, following laboratory exposure to bleached pulp and paper mill effluents.

Ecotoxicology and Environmental Safety, 38: 266–271

Abbreviation: (Ecotoxicol. Environm. Saf.)

Personal collection: NG, Dept. Zool. Pach. Coll.

Figure 2.1   A model index card.

A reference card (Figure 2.2) is a card very similar to an index card, or slightly a larger one, which can be used for taking notes from an article. It can include additional information in addition to the basic requirements listed above.

Lundén, T., S. Miettinen, L. G. Lönnström, E. M. Lilius, and G. Bylund   1999

Effect of florfenicol on the immune response of rainbow trout (*Oncorhynchus mykiss*)

Veterinary Immunology and Immunopathology, 67: 317–325 (Vet. Immunol. Immunopathol.)

Source   Personal reprint collection – NG – Dept. Zool., Pach. Coll.

Abstract   Florfenicol, a drug effective against several bacterial diseases of fish was tested for possible immunomodulatory effects. The aim of the study was to follow the kinetics of the immune response after vaccination with simultaneous oral antibiotic treatment. The fish were immunized with a commercial oil-based divalent (furunculosis/vibrosis) vaccine and were simultaneously given oral antibiotic

P.T.O.

(a) Front side

---

treatment. The specific immune response was monitored by analyzing the levels of specific antibodies with ELISA. As an indicator of the non-specific immune response the phagocytic activity of circulating leucocyte counts and differentials were also monitored. The disease resistance was evaluated by challenge test at the end of the experiment. The results showed that florfenicol did not have any significant effect on antibody production and circulating leucocyte levels but caused a suppression of chemiluminescence response/phagocytic cell, 5-6 weeks after vaccination. The survival after challenge was slightly suppressed by the florfenicol treatment. The RPS-value for the vaccinated group was 98% and for the florfenicol-treated group was 88%.

Notes on any other information: e.g., Design of experiment; ELISA, chemiluminescence analysis, challenge test, RPS – Relative Percentage Survival, Statistical analysis.

(b) Back side

Figure 2.2   A model reference card.

Whenever an entry is made in the index cum reference card, it should be ensured that the content is absolutely accurate. There should not arise a need to go back to the original document to verify the correctness of contents. Great care should be taken to enter spellings of proper names such as those of the species, places, authors, the addresses of the authors, publishers, place of publication, etc., accurately. Likewise, the year of publication, volume number, issue number, page numbers and related details that are necessary for the final preparation of the bibliography should be entered correctly.

Apart from the above-listed essential information, others, that in the opinion of the reader of the article would be useful, may also be entered in the reference card. Such information might be useful at a later stage especially while discussing one's own "observations". If the space in the card is not sufficient, additional cards may be attached (stapled) to the first one. The reference cards, it should be remembered, are only for the use of the research worker and nobody else. They are not official documents to be submitted as evidence for the proficiency of the researcher in the literature of a field of study. Preparation of reference cards is only a method to simplify the procedure for the review of literature. Therefore, what material should be entered in a reference card is at the discretion of the investigator.

Chapter Three

Literature Citation

INTRODUCTION

Research in any field of study requires information on various aspects from different sources. A lot of basic and current information, right from choosing a topic of research to analysing the data collected, are needed. Therefore, when writing a research report, it is essential that we acknowledge all sources that provided us the information that influenced our experiment, conclusions, or interpretation of the data. This we do by including a citation in the body of the manuscript and its corresponding entry in the Reference section of the report. We had already seen in chapter 2 what literature should be cited and what should not be cited in a research report. Now, let us consider the various ways of citing references in the text and the different formats of the Reference section.

DIFFERENT SYSTEMS OF CITING REFERENCES

There are three systems of citing references in the text and the corresponding formats of the Reference section:

1. Name–year or Harvard system

2. Citation-order or citation-sequence system and

3. Alphabet–number system.

The name–year system of referencing is more descriptive and elaborate than the citation-sequence system. In this system, citations in the text of the report consist of author name(s) and source date(s) within parenthesis and the corresponding alphabetically arranged entries can be seen in the reference list. The advantages of the name–year system are:

a) Mention of author names and year of reporting, instead of just numbers, within the text of the report gives at least some immediate relevant information about the reference for the reader.

b) Making alterations in the text citation and list of references at any stage of preparation or publication of the report is easy, that is insertion or deletion of a reference is not a difficult task in this system than in the citation-sequence system.

The main disadvantages of this system are:

a) It consumes more space.

b) Too many citations in the text can be distracting for readers.

c) Formatting the entries in the list of references and organizing the list itself is more complicated than that of the citation-sequence system.

In the citation-sequence system, also referred to as citation-order system, citations in the text of the report are made by way of numbers, which are ordered sequentially throughout the text. These numbers match with and indicate numbered entries in the reference list. Complete bibliographic information is given only in the "References" section of a report. The following are the advantages of the citation-sequence system.

a) Since the text citation includes only numbers, it does not interrupt the flow of the text.

b) This system helps to save space and paper.

The disadvantages are:

a) Readers have to strain to refer to the reference section to know the source of each numeric citation.

b) It is difficult to make alterations (addition or deletion) in the text citations and in the list references, especially when the report is in a later stage of publication.

c) Since their names are not mentioned in the text, authors are deprived of due recognition, which they would have received in the name–year system.

While the name–year system is appropriate for academic research reports such as a theses/dissertations, monographs, books, etc., the citation-sequence is suitable for reports published in journals, provided the journal requires such system.

In addition to these two systems, a third system, which is a modification of the name–year system, is also in vogue. This system, called the alphabet–number system, uses a numbered, alphabetically arranged list of references in the Reference section of a report, and the citation of the references in the text is done by the numbers corresponding to the reference in the list. The use of numbers instead of names and years in the text conserves space and reduces printing cost. The numbered, alphabetically arranged list of references is easy to prepare and to use. This system is more useful for review articles and similar reports.

For journals, especially the biology journals, the style recommended by Council of Biology Editors (CBE) in its publication *Scientific Style and Format* (6th ed.) is very useful. These suggestions are based on the guidelines given in the book *National Library of Medicine Recommended Formats for Bibliographic Citation*. In the following section we shall discuss the "typical" as well as the CBE style.

NAME–YEAR SYSTEM—CITATION IN THE TEXT

Let us understand citation of references in the text by the name–year system with a few examples.

Author(s) Name(s) Mentioned in the Sentence

When a work is reported directly in third person, the year in which the work was published is given in parenthesis immediately following the name(s) of the author(s). There is no comma between the name(s) and the parenthesis. In the following examples, the names of the authors and years have been made bold in order to highlight them. But it is not the practice in print or typing the report.

ONE AUTHOR

Example 1   De Smet (1962) was the first to describe the adrenocortical homolog in *A. calva.*

TWO AUTHORS

Example 2   Youson and Butler (1976) have demonstrated the presence of numerous lipid-filled vacuoles and cytoplasmic organelles which are typical of adrenocortical cells in the mammalian adrenal cortex.

MORE THAN TWO AUTHORS

Example 3   Youson *et al.* (1976) showed that the adrenocortical tissue forms yellow corpuscles which are located near the posterior cardinal and renal veins along the anterior two-thirds of the kidneys.

SIMILAR WORK BY DIFFERENT AUTHORS / REPORTS

Example 4   Hellman and Hellerström (1960), Falkmer and Hellman (1961), Hellerström *et al.* (1964), Hellerström and Asplund (1966) and Östberg *et al.* (1966) have classified the A-cells into argyrophilic A1-cells and non-argyrophilic A2-cells.

Author(s) Name(s) not Mentioned in the Sentence

When a work is reported indirectly, the name or names of the authors and the year of publication are given immediately following the statement, within parenthesis, the name(s) and the year being separated by a comma. In the CBE style the comma between name and year is omitted. Citations of the references should be separated if the statements to which they are attributed are different, as in examples 5 and 6.

> *Example 5*   Yellow corpuscles are formed by loosely arranged cords of cells that demonstrate Δ^5-3 β-hydroxysteroid dehydrogenase activity (Youson *et al.*, 1976) and contains numerous lipid-filled vacuoles and cytoplasmic organelles which are typical of adrenocortical cells in the mammalian adrenal cortex (Youson and Butler, 1976).

> *Example 6*   Hypocalcin isolated from trout SCs (Lefeber *et al.*, 1988) has an N-terminal amino acid sequence which shows substantial similarity to the hypocalcaemic principle of eel (Butkus *et al.*, 1987) and salmon (Wagner *et al.*, 1986).

More than one Source/More than one Work of the Same Author in the Same Year and in Different Years

When we want to attribute a statement to more than one work, the different references are cited together within parenthesis immediately after the statement, in chronological order, separated by semicolon, as in example 7. If any one or more of these works belong to one, two or more groups of authors (more than two authors; same first author; therefore cited in the text with *et al.*), the years of publication of these works should be placed together in chronological order, irrespective of the year of next reference.

> *Example 7*   The D-cells have been speculated to secrete a third pancreatic hormone (Epple, 1963, 1966a, b, 1967a, b, 1968, 1973; Solcia and Sampietro, 1965; Cavellero *et al.*, 1967a, b; Cavellero and Solcia, 1968a, b).

Note in the above example, Epple's name is placed first because the year of publication of his first work in this context is 1963, which is earlier to 1965 of Solcia and Sampietro. However, the other works of Epple belong to years that are later than 1965 or 1967 or, even 1968, the years of the latter references. Yet, these years are placed in chronological order next to 1963. When the same author or group of authors has more than one publication in the same year, these work are distinguished by assigning them suffixes "a", "b" and so on to the specific year. In the example 7, it can be seen that Epple has two publications (again, in this context) each in 1966 and 1967. Therefore, the two works of 1966 are distinguished as 1966a and 1966b. While citing these references together within a parenthesis, the year is given once with the two suffixes, the "a" close to the year followed by "b" separated by a comma. If these works are cited separately in different parentheses, the year should be mentioned with appropriate suffix as 1966a or 1966b. The question of deciding which of the two or more works of the same author should be assigned "a", "b" and so on, can be resolved on the basis of the date of publication of these works. The important thing is, whatever logic we have followed to assign the suffixes, they should be consistent with those we have given to these works in the "Reference" section.

To Cite a Work Cited in Another Work—use of "Cited in"

When we want to cite a work that was referred to in another work, we must specify which work was cited in which work by using the phrase "cited in". Both the works should be included in the List of References.

> *Example 8*   The PP-cells occur as separate elements in the pancreas of *R. catesbiana* (Tomita and Pollock, 1981, cited in Gurumani, 1984).

> *Example 9*   Monacchio's (1964, cited in Epple, 1967) method was adopted with a slight modification in staining time to identify the toluidine blue metachromatic D-cells.

The practice of citing works through second-hand source is common among research students in India. It is always better to avoid it because what we are citing is what another person "understood" of an original work. Whether the person had understood correctly what was written in the original paper is highly doubtful, even more so if the original report is in some language like Chinese and Japanese.

Use of "&" in the Place of "and"

Many journals require the use of "&" in the place of "and" in order to save space.

Example 10   Calcium regulation in fish is dominated by the hypercalcaemic action of prolactin or cortisol (Flik *et al.,* 1986; Flik & Perry, 1988) and the hypocalcaemic action of the hormone from the Stannius corpuscles, hypocalcin (Wendelaar Bonga & Pang, 1986).

According to CBE style a comma may also be used to separate the citations within parenthesis, provided the comma separating author(s) name(s) and the year is replaced by a space, as in example 11. In order to save precious space, some Journals require the use of "&" in the place of "and" in all the citations of references in the text, as in example 10, and in the listing of references at the end of the research article. However, such use of "&" must be avoided in academic dissertations and theses. Use of a full stop at the end of *et al*. is also avoided in some journals. However, the use of it is more appropriate because *al* stands for abbreviation of Latin *alia,* meaning "others", and together with *et* it means "and others". CBE style recommends the use of "and other" in the place of "*et al.*"

Example 11   Although the water economies of a number of species of amphibians have been studied, the majority of these investigations have been interspecific in nature (Smith *et al.* 1998). Such studies have often sought to elucidate adaptive differences among species (Schmid 1965, Ralin and Rogers 1972, Gillis 1979).

NAME–YEAR SYSTEM—LIST OF REFERENCES

The references cited in the text following name–year system are listed in alphabetic order under the "Reference" section of the paper or the thesis. This section is often referred to as "Bibliography", "Cited References", "List of References", "Literature Cited", "References", "References Cited", "Sources Consulted", "Works Cited", etc. The term Bibliography generally refers to a list of references that includes the actual references cited in the text as well as other works that are relevant to the subject discussed. It is more appropriate for use in a review article or in a textbook. In addition, many textbooks give "Annotated Bibliography", which is a list of references cited in the text or sources consulted, with a brief comment (placed immediately below each entry in the list) on the content and usefulness of each reference.

Placement of Reference Section

In a thesis, the Reference section is placed at the end of the text, i.e., immediately after the Summary/Conclusions section, starting on a fresh page. This section usually does not have a chapter number. However, assigning chapter numbers to these sections is not uncommon.

There are no definite rules for listing the literature cited in the text under the Reference section. The system of referencing may vary according to the requirements of the journal to which the report (research paper) is submitted for publication or of the academic institution to which the report (thesis) is submitted for evaluation.

Though there are numerous styles of organizing the entries in the reference section, the basic information required for each entry include:

1. the author(s)

2. the title

3. details such as date, place of publication and publisher in the case of books, and

4. name of the journal, volume number, issue number and inclusive pages, in the case of articles published in a scientific journal. The style of referencing may vary according the Journal specifications, especially with reference to the use of punctuations and the order in which the information is given. However, an academic thesis is expected to follow a typical style giving all the required information in the prescribed format.

Name–year System—Format of the Reference Section

Let us understand the formats of name–year entries in the "Reference" section of a report.

In the following examples of name–year system of referencing, three versions are given—the first one (A) appropriate for a Thesis or Assignment; the second one (B) still followed in many journals; and the third one, a format suggested by CBE for a publication in a journal (C). The main differences are:

1. In a thesis, the entries are typed in double space, whereas in a journal, unless specified otherwise, the entries are typed with single spacing.

2. The font used in a thesis is the same as that used in the text. In a journal a smaller font may be required.

3. An entry in a thesis "Reference" section has the author(s) name (s) placed on the first line of the entry flush with left margin. If there are too many authors, requiring a second line, the second line is typed in single space below and one space indented to the first.

The year of publication is in the second line (i.e., next line to the authors names), slightly (2–3 spaces) indented.

The rest of the information (the title, journal name, volume number, issue number and first page-last page numbers) is given 2–3 spaces away from the year, on the same line and with the same indentations. Altogether there are three indentations, first for author name, second for year and third for article title, journal name etc.

On the other hand, in a journal article, an entry under Reference section begins first author name flush with left margin and runs continuously, the second and other lines slightly (2 spaces) indented to the first. Thus there are only two indentations, the first line and the rest.

The third version is the one recommended by CBE Scientific Style and Format for use in journals. The objective of the CBE version is to reduce as much printing space as possible, and thus conserving paper and cutting down the cost of publication. Accordingly, many of the punctuation marks, spacing, etc. are avoided. Some such conservation measures are:

a) No space nor full stops between the initials of the authors

b) No use of "and" or "&" between the last and the penultimate authors

c) Inclusive page numbers to show the first page number in full and the last page number with only the minimum number of digits required (e.g. 295–300, all the three digits of the end page have to be shown; 1584–6, only one digit to indicate 1586; 5487–91, two digits to indicate 5491).

BOOK

Example 1

(A)

Chester Jones, I.

    1957   *The Adrenal Cortex.* Cambridge University Press, London.

(B)

Chester Jones, I. 1957. *The Adrenal Cortex.* Cambridge University Press, London.

(C)

Chester Jones I. 1957. *The Adrenal Cortex.* London: Cambridge University Press. 340 p.

Example 2

(A)

Gorbman, A., and H. A. Bern

    1962 *A Text Book of Comparative Endocrinology.* John Wiley and Sons Inc., New York.

(B)

Gorbman, A., and H. A. Bern. 1962. *A Text Book of Comparative Endocrinology.* John Wiley and Sons Inc., New York.

(C)

Gorbman A, Bern HA. 1962. *A Text Book of Comparative Endocrinology.* New York: John Wiley and Sons Inc. 265 p.

Example 3

(A)

Prosser, C. L., and F. A. Brown, Jr.

    1961 Comparative Animal Physiology. W. B. Saunders Company, Philadelphia and London.

(B)

Prosser, C. L., and F. A. Brown, Jr. 1961. Comparative Animal Physiology. W. B. Saunders Company, Philadelphia and London.

(C)

Prosser CL, Brown Jr FA. 1961. Comparative Animal Physiology. Philadelphia and London: W. B. Saunders Company.

Notice in the above examples, the last name of the first author is given first with his first and middle names initialized. The last names of the second and other authors are given with initials first or last, as per the requirement of the journal or the academic body to which the report is submitted.

THESIS

Example 1

(A)

Nagalakshmi, C. M.

> 1970   Histophysiological studies on the corpuscles of Stannius and interrenals of the freshwater fish, *Notopterus notopterus* (Pallas). Ph. D. Thesis, Annamalai University.

(B)

Nagalakshmi, C. M. 1970. Histophysiological studies on the corpuscles of Stannius and interrenals of the freshwater fish, *Notopterus notopterus* (Pallas). Ph. D. Thesis, Annamalai University.

(C)

Nagalakshmi CM. 1970. Histophysiological studies on the corpuscles of Stannius and interrenals of the freshwater fish, *Notopterus notopterus* (Pallas) (Ph. D. Thesis), Annamalainagar: Annamalai University. 250 p. Available from: University Library.

As per the CBE guidelines, the format for the reference of a thesis requires details such as place of degree-granting institution, the degree-granting institution, number of pages, availability information and identifying information such as the University Catalogue number.

EDITED BOOK

Example 1

(A)

Martin, A. W. (Ed.).

> 1961   Comparative Physiology of Carbohydrate Metabolism in Heterothermic Animals. University of Washington Press, Washington.

(B)

Martin, A. W. (Ed.). 1961

Comparative Physiology of Carbohydrate Metabolism in Heterothermic Animals. University of Washington Press, Washington.

(C)

Martin AW, editor. 1961. Comparative Physiology of Carbohydrate Metabolism in Heterothermic Animals. Washington: University of Washington Press. 250 p.

Example 2

(A)

Brolin, S. E., B. Hellman, and H. Knutson (Eds.)

> 1964 The Structure and Metabolism of Pancreatic Islets. Pergamon Press, Oxford.

(B)

Brolin, S. E., B. Hellman, and H. Knutson (Eds.). 1964. The Structure and Metabolism of Pancreatic Islets. Pergamon Press, Oxford.

(C)

Brolin SE, Hellman B, Knutson H. editors. 1964 The Structure and Metabolism of Pancreatic Islets. Oxford: Pergamon Press. 420 p.

A CHAPTER IN AN EDITED BOOK

Example 1

(A)

Miller, M. R.

> 1961 Carbohydrate metabolism in amphibians and reptiles. *In* "Comparative Physiology of Carbohydrate Metabolism in Heterothermic Animals" (A. W. Martin, ed). University of Washington Press, Washington, pp. 125–147.

(B)

Miller, M. R. 1961. Carbohydrate metabolism in amphibians and reptiles. *In* "Comparative Physiology of Carbohydrate Metabolism in Heterothermic Animals" (A. W. Martin, ed). University of Washington Press, Washington, pp. 125–147.

(C)

Miller MR. 1961. Carbohydrate metabolism in amphibians and reptiles. In: Martin AW, editor. Comparative Physiology of Carbohydrate Metabolism in

Heterothermic Animals. Washington: University of Washington Press, p 125–47.

Example 2

(A)

Cavellero, C., and E. Solcia

> 1964   Cytologic and cytochemical studies on the pancreatic islets. *In* "The Structure and Metabolism of Pancreatic Islets" (S. E. Brolin, B. Hellman and H. Knutson, eds.). Pergamon Press, Oxford, pp. 83–97.

(B)

Cavellero, C., and E. Solcia 1964. Cytologic and cytochemical studies on the pancreatic islets. *In* "The Structure and Metabolism of Pancreatic Islets" (S. E. Brolin, B. Hellman and H. Knutson, eds.). Pergamon Press, Oxford, pp. 83–97.

(C)

Cavellero C, Solcia E. 1964. Cytologic and cytochemical studies on the pancreatic islets. In: Brolin SE, Hellman B, Knutson H. editors. The Structure and Metabolism of Pancreatic Islets. Oxford: Pergamon Press. p 83–97.

JOURNAL ARTICLES

Example 1

(A)

Aaltonen, T. M., E. T. Valtonen, and E. I. Jokinen

> 1997   Humoral response of roach (*Rutilus rutilus*) to digenean *Rhipidoctyle fennica* infection. Parasitology, 114: 285–291

(B)

Aaltonen, T. M., E. T. Valtonen, and E. I. Jokinen. 1997. Humoral response of roach (*Rutilus rutilus*) to digenean *Rhipidoctyle fennica* infection. Parasitology, 114: 285–291.

(C)

Aaltonen TM, Valtonen ET, Jokinen EI. 1997. Humoral response of roach (*Rutilus rutilus*) to digenean *Rhipidoctyle fennica* infection. Parasitology, 114: 285–291.

According to CBE system, the month/date of publication of the issue may be required to be given along with the year as shown in the following example.

Palani Kumar M, Vairamani M, Prasada Raju N, Charmine Lobo N, Anbumani CP, Girish Kumar, Thangam Menon, Shanmugasundaram S. 2004 June 15. Rapid discrimination between strains of beta haemolytic streptococci by intact cell mass spectrometry. Indian J Med Res 119: 283–8

Example 2

(A)

Falkmer, S.

> 1961   On the morphology and physiology of the endocrine pancreatic tissue of the marine teleost *Cottus scorpius* with special reference to the role of glutathione in the mechanism of alloxan diabetes using a modified nirtroprusside method. Acta endocrinol., 37 (Suppl. 59): 1–122.

(B)

Falkmer, S. 1961. On the morphology and physiology of the endocrine pancreatic tissue of the marine teleost *Cottus scorpius* with special reference to the role of glutathione in the mechanism of alloxan diabetes using a modified nirtroprusside method. Acta endocrinol., 37 (Suppl. 59): 1–122.

(C)

Falkmer S. 1961. On the morphology and physiology of the endocrine pancreatic tissue of the marine teleost *Cottus scorpius* with special reference to the role of glutathione in the mechanism of alloxan diabetes using a modified nirtroprusside method. Acta endocrinol., 37 (59 Suppl): 1–122.

In the above example the article is in a supplement to an issue. If it is in a supplement to the volume, it should be indicated as, for example, J Exp Zool 52 Suppl: 13–9.

ABSTRACT IN A JOURNAL

Example 1

(A)

Lewis, T. L.

> 1974   Blood sugar level in the eel, A. rostrata, under various experimental conditions. Anat. Rec., 178(2): 402. (Abstract)

(B)

Lewis, T. L. 1974.   Blood sugar level in the eel, *A. rostrata*, under various experimental conditions. Anat. Rec., 178(2): 402. (Abstract)

(C)

Lewis TL. 1974   Blood sugar level in the eel, *A. rostrata*, under various experimental conditions (Abstract). Anat Rec 178(2): 402.

ARTICLE FROM A PROCEEDINGS OF A CONFERENCE

(A)

Gingerich, D. A.

> 1984   Pharmacokinetics of drugs used for therapy of the mammary gland. *In* "Proceedings of the 10th Annual Food Animal Medicine Conference", 1984 Sep 25–26. The Ohio State University, Columbus, Ohio. pp. 117–135.

(B)

Gingerich, D. A. 1984. Pharmacokinetics of drugs used for therapy of the mammary gland. *In* "Proceedings of the 10th Annual Food Animal Medicine Conference", 1984 Sep 25–26. The Ohio State University, Columbus, Ohio. pp. 117–135

(C)

Gingerich DA. 1984 Pharmacokinetics of drugs used for therapy of the mammary gland. In: Proceedings of the 10th Annual Food Animal Medicine Conference; 1984 Sep 25–26; Columbus, OH. Columbus: The Ohio State University. p 117–35

ABSTRACT OF A PAPER READ IN A CONFERENCE/SYMPOSIUM

(A)

Sadasivam, V., and N. Gurumani

> 1990   Immunohistochemical localization of pancreatic hormones in the common chunam frog. All India Symposium on Reproductive Biology and Comparative Endocrinology, January 8–11, 1990, M. S. University of Baroda, Baroda. (Abstract)

(B)

Sadasivam, V. and Gurumani N. 1990. Immunohistochemical localization of pancreatic hormones in the common chunam frog. All India Symposium on Reproductive Biology and Comparative Endocrinology, January 8–11, 1990, M. S. University of Baroda, Baroda. (Abstract)

(C)

Sadasivam V, Gurumani N. 1990. Immunohistochemical localization of pancreatic hormones in the common chunam frog (Abstract). In: Proceedings of the All India Symposium on Reproductive Biology and Comparative Endocrinology, January 8–11, 1990, M. S. University of Baroda, Baroda. p 43. Abstract No. 94.

WEBSITE ARTICLES

While listing articles, monographs, books sourced through Internet websites the following information are required:

i)   Last name and initial(s) of the author (followed by, if there are more than one author, the last names and initials of the co-authors

ii) Year of publication

iii) Title of article

iv) Journal title

v)   Volume (issue):Inclusive page numbers

vi) Availability information

vii) Date of accession. Additional and appropriate information such as place of publication, publisher, update details, etc. may also be provided.

Example 1

Bonsor, K. 2004. How DNA Computers Will Work. HSW Media Network—A Convex Company (1998–2004) Available: http://www.howstuffworks.com via the Internet. Accessed 2004 Aug 16.

FORTHCOMING WORK

Example 1

James, G. Effect of synthetic calcitonin on the histology the corpuscles of Stannius of the freshwater catfish *Mystus gulio*. Indian J. Zool. *(In press)*

James, G. Effect of synthetic calcitonin on the histology the corpuscles of Stannius of the freshwater catfish *Mystus gulio*. Indian J. Zool. *(Forthcoming)*.

If details of the volume number, issue number, etc. of the Journal in which the article is being published are known, they may be included in the entry.

Alphabetical and Chronological Arrangement of Entries in the Reference Section

The reference section of a research report should be organized carefully in a logical and consistent format and as per the requirements of the agency to which the report is submitted. Every part of each entry in the list should be thoroughly checked against the original publication. Especially the students who submit a thesis for a research degree should be more careful. A badly prepared thesis, particularly the reference section, would only expose the ignorance and inefficiency of its author.

Some general guidelines

1. Entries are alphabetized according to the first letter of the surname (last name) of the first author.

    Example

Boyd, J. D.	Brown, J. R.
Boyd, W.	Brown, J. W.
Boyde, A.	Brown, R. L.
Boyden, E. A.	Browne, D.
Brown, A. G.	Brownell, W. E.
Brown, D. D.	

2. If an author has a number of papers/reports to his/her credit as single author and as first co-author, the single author works should be listed chronologically first and then the co-authored works, chronologically, without discrimination to the number of co-authors. That is, we have only to decide whether an entry has a single author or more than one author. The alphabetic arrangement of more than one author cases (i.e., same senior author but different co-authors) is decided according to the surname of the second co-author. If they (first and second co-authors) are the same, the surname of the third co-author and so on, should be considered. If two or more works have the same authors in the same order, they should be arranged chronologically. If the date of publication of these works is identical, then they should be arranged in an order in which they were cited in the text. To distinguish between the entries with same authors and date, we must assign suffixes "a", "b", "c" immediately after the year of publication.

Example

Epple, A. 1966a

Epple, A. 1966b

Epple, A. 1967a

Epple, A. 1967b

Epple, A. 1968

Epple, A. 1969

Epple, A., and Farner, D. S. 1967

Epple, A., Jørgensen, C. B., and Rosenkilde, P. 1966

Epple, A., and Lewis, T. L. 1973

Falkmer, S. 1961

Falkmer, S. 1965

Falkmer, S. 1966

Falkmer, S., Grimelius, L., Havu, N., Ljungberg, S., and Unger, R. H. 1965

Falkmer, S., and Hellman, B. 1961

Falkmer, S., Hellman, B., and Voigt, G.E. 1964

Falkmer, S., and Knutson, F. 1963

Falkmer, S., Knutson, F., and Voigt, G. E. 1964a

Falkmer, S., Knutson, F., and Voigt, G. E. 1964b

Falkmer, S., and Matty, A. J. 1966

Falkmer, S., and Wilson, S. 1965

Falkmer, S., and Wilson, S. 1967

Falkmer, S., and Winbladh, L. 1964a

Falkmer, S., and Winbladh, L. 1964b

3. Names starting with Mc and Mac are listed under M assuming the spelling to be Mac. That is, no distinction is made between the two.

Example

MacCabe

McCall

    MacCallum

    McCloskey

4.   Surnames starting with St are treated as Saint

Example

    Saha

    St Helen

    Salmon

5.   Names starting with prefixes are ordered taking into consideration the first letter of the prefix.

Example

    Alexandre

    Al-Lami

    Allanson

    Darazs

    D'Arcy Thompson

    Darlington

    Dearden

    de Beer

    de Castro

    de Duve

    De Fudis

    Deitres

    de Gasperis

    DeGroot

    Del Cerro

    de Sousa

    De Vries, P. A.,

    de Vries, R. A. C.

    De Vries, T. W.

Lecco,

Le Douarin

Le Double

LeDuc

Lee

Obayashi

O'Beirene

Okamota

O'Keefe

Olds

O'Leary

Oliver

Valverde

van Alphen

Van Buren

van Campenhout

Van Hassel

Voneida

CITATION-SEQUENCE SYSTEM

Since this system is mainly used in research articles for publication in journals, the illustrations and formats given hereunder pertain only for such use and not for academic reports such as a Ph. D. thesis.

Citing References in the Text

Citation of references within the text is by numbers, either as superscript or within parenthesis. The reference that appears first in the text is given the number 1, the second, 2 and so on. If any reference appears for the second time, third time and so on, only the number allotted to it in the first instance should be reused. If there are, for example, 25 references in a report, there should be a sequence of numbers 1 to 25, irrespective of the number of times a reference is used.

To cite a single reference

Example   The thymus plays an important role in the development of a fully functional immune system, as demonstrated by thymectomy[1].

In the above example, the superscript number 1 at the end of the statement refers to an entry in the list of references of the report, which has been assigned the same number. Whenever we want to cite this work in this report we have to use the number 1.

To cite more than one reference

Example   Thymus is composed of differentiating lymphoid cells (thymocytes) within a network of epithelial cells and is generally organized in cortical- and medullary-like zones[2–6].

The superscript numbers 2–6 in the above example point to entries 2 to 6 (i.e., 2, 3, 4, 5 and 6) in the reference section.

The superscripts should be placed at the appropriate places to qualify that part of a statement, as in the following example.

Example   The presence of a "cortex" and a "medulla" is not a constant feature in fish[3] and their roles have not been defined[7].

To cite references that are not continuous in the list

Example   New studies[6, 8, 9–12, 15, 16] on the development of immune systems have been carried out following the availability of new monoclonal antibodies specific for leukocyte subpopulations.

To cite a reference within a reference—use of "cited in"

Example   Histochemistry was carried out adopting a modified method of Gomori[21 (cited in 22)].

Use of parenthesis and/or square bracket instead of superscription

A few journals encourage the use of parenthesis to place the citations within parentheses.

Example   The CS plays an important and central role in the mineral metabolism (2, 3, 5–9, 12–17) and in particular, Ca homeostasis (3, 4, 7, 13, 15–17).

While using parenthesis care should be taken to distinguish between numerical citations and any other numerical notes such as sample size, figure number, etc.

Example A sample of (n = 10) fish was subjected stress conditions, as described in an earlier report (10).

To overcome this problem, a few journals have adopted the use of parenthesis and square brackets. The parenthesis is used for any note other than numerical reference citations, while the square bracket is used exclusively for the numerical citations.

Example The earlier hypothesis that pancreatic exocrine and endocrine cells are derived from different cell pools, i.e., the gut endoderm and the neural crest, respectively [63], has been refuted by a number of cell lineage experiments [3, 20, 49] and molecular studies (see below).

Example Rosenberg *et al.* ([70], cited in [71]) developed a model of partial pancreatic duct obstruction in hamsters that provides further evidence of a continuous development from to islet cells.

Example The essential role of transcription factors in pancreatic development was recently demonstrated in investigations using knockout mice lacking certain transcription factors (reviewed in [76]).

Example In the enlarging epithelial buds a treelike ductal system develops, which eventually gives rise to endocrine and acinar cells (Fig. 1) [75].

Format of the Reference Section

The format varies in different journals. The serial number may be required to be given in parenthesis or square brackets. The name of the journal may have to be in italics. The year of publication may also have to be parenthesized. The placement of the year may be immediately after the name(s) of the author(s). The following illustrations are adopted from the CBE style.

BOOK

Format

1. Last name and initial(s), [Followed by last name and initial(s) of co-authors, if any]. Title of book. Place of publication: Publisher; year of publication. Number of pages.

Example

1. Prosser CL, Brown Jr FA. Comparative animal physiology. Philadelphia and London: W. B. Saunders Company; 1961. 450 p.

THESIS

Format

2. Last name and initial(s) of author. Title of dissertation or thesis [Degree. Dissertation or Thesis]. Place of degree awarding institution or University: Degree awarding institution or University; Date/Year of Degree. Number of pages. Availability information. Identifying information/Catalogue number.

Example

2. Nagalakshmi CM. Histophysiological studies on the corpuscles of Stannius and interrenals of the freshwater fish, *Notopterus notopterus* (Pallas) [Ph. D. Thesis]; 1970. Annamalainagar: Annamalai University. 250 p. Available from: University Library.

EDITED BOOK

Format

3. Last name and initial(s) of the editor, [Followed by last name and initials of the co-editors, if any], editor(s). Title of book. Information about Edition (if any). Place of publication: Publisher; Year of publication. Number of pages.

Example 1

3. Martin AW, editor. Comparative physiology of carbohydrate metabolism in heterothermic animals. Washington: University of Washington Press; 1961. 250 p.

Example 2

3. Brolin SE, Hellman B, Knutson H, editors. The structure and metabolism of pancreatic islets. Oxford: Pergamon Press; 1964. 420 p.

A CHAPTER IN AN EDITED BOOK

Format

4. Last name and initial(s) of the author of chapter, [Followed by last name and initials of the co-authors, if any]. Title of chapter. Last name and initial(s)

of editor of the book, [Followed by last name and initial(s) of the co-editors, if any]. Title of book. Edition information. Place of publication: Publisher; Year of publication. Inclusive page numbers.

Example 1

4. Miller MR. Carbohydrate metabolism in amphibians and reptiles. In: Martin AW, editor. Comparative physiology of carbohydrate metabolism in heterothermic animals. Washington: University of Washington Press; 1961. p 125–47.

Example 2

4. Cavellero C, Solcia E. Cytologic and cytochemical studies on the pancreatic islets. In: Brolin SE, Hellman B, Knutson H. editors. The structure and metabolism of pancreatic islets. Oxford: Pergamon Press; 1964. p 83–97.

JOURNAL ARTICLES

Format

5. Last name and initial(s) of the author, [Followed by last name and initial(s) of co-authors, if any]. Article title. Name of the journal [abbreviated according to the National Information Standards Organization] Year [Month and Day, if necessary and available] of publication; Volume number (Issue number): Inclusive page numbers.

Example 1

5. Claydon MA, Davey SN, Edwards-Jones V, Gordon DB.The rapid identification of intact microorganisms using mass spectrometry. Nat Biotechnol 1996; 14: 1584–6.

Example 2

5. Palani Kumar M, Vairamani M, Prasada Raju N, Charmine Lobo N, Anbumani CP, Girish Kumar, Thangam Menon, Shanmugasundaram S. Rapid discrimination between strains of beta haemolytic streptococci by intact cell mass spectrometry. Indian J Med Res 2004 June 5; 119 (2): 283–8.

Example 3

5. Falkmer S. On the morphology and physiology of the endocrine pancreatic tissue of the marine teleost *Cottus scorpius* with special reference to the role of glutathione in the mechanism of alloxan diabetes using a modified nirtroprusside method. Acta endocrinol 1961; 37 (59 Suppl): 1–122.

In the above example the article is in a supplement to an issue. If it is in a supplement to the volume, it should be indicated as, for example, J Exp Zool 52 Suppl: 13–9.

ABSTRACT IN A JOURNAL

Format

6.   Last name and initial(s) of the author, [Followed by last name and initial(s) of co-authors, if any]. Article title (Abstract). Name of the journal [abbreviated according to the National Information Standards Organization] Year [Month and Day, if necessary and available] of publication; Volume number (Issue number): Inclusive page numbers.

Example

6.   Lewis TL. Blood sugar level in the eel, *A. rostrata*, under various experimental conditions (Abstract). Anat Rec 1974; 178(2): 402.

ARTICLE FROM A PROCEEDINGS OF A CONFERENCE

Format

7.   Last name and initial(s) of the author [Followed by last name and initials of the co-authors, if any]. Title of paper. In: Description of proceedings and title of conference; Year, month and days of conference; location of conference. Place of publication: Publisher; Year of publication. Page numbers.

Example

7.   Gingerich DA. Pharmacokinetics of drugs used for therapy of the mammary gland. In: Proceedings of the 10th Annual Food Animal Medicine Conference; 1984 Sep 25–26; Columbus, OH. Columbus: The Ohio State University; 1984. p 117–35

ABSTRACT OF A PAPER READ IN A CONFERENCE/SYMPOSIUM

Format

8.   Last name and initial(s) of the author [Followed by the last name and initials of the co-authors, if any]. Title of abstract [abstract]. In: Title of proceedings/conference/ symposium; Year Month and Days of conference; Location of conference. Place of publication: Publisher; Year of publication. Page numbers. Abstract number [if available].

Example

8. Sadasivam V, Gurumani N. 1990. Immunohistochemical localization of pancreatic hormones in the common chunam frog (Abstract). In: Proceedings of the All India Symposium on Reproductive Biology and Comparative Endocrinology, 1990 January 8–11; Baroda. Baroda: M. S. University of Baroda; 1990. p 43. Abstract No. 94.

FORTHCOMING WORK

Format

9. Last name and initial(s) of the author, [Followed by last name and initials of the co-authors, if any]. Title of the forthcoming document. Abbreviated title of the journal and year of forthcoming publication (if known). *(Forthcoming)* or *(In press)*.

If details of the volume number, issue number, etc. of the Journal in which the article is being published are known, they may be included in the entry.

Example

9. James G. Effect of synthetic calcitonin on the histology the corpuscles of Stannius of the freshwater catfish *Mystus gulio*. Indian J Zool *(In press)*.

9. James G. Effect of synthetic calcitonin on the histology the corpuscles of Stannius of the freshwater catfish *Mystus gulio*. Indian J Zool. *(Forthcoming)*.

WEBSITE ARTICLES

While listing articles, monographs, books sourced through Internet websites, the following information are required:

i)   Last name and initial(s) of the author (followed by, if there are more than one author, the last names and initials of the co-authors

ii) Year of publication

iii) Title of article

iv) Journal title

v)   Volume (issue):Inclusive page numbers

vi) Availability information

vii) Date of accession. Additional and appropriate information such as place of publication, publisher, update details, etc. may also be provided.

Example

Bonsor, K. 2004. How DNA Computers Will Work. HSW Media Network—A Convex Company (1998–2004) Available: http://www.howstuffworks.com via the Internet. Accessed 2004 Aug 16.

ENTIRE WEB PAGE

Format

10. Last Name and Initial(s) of Author ; Title of Webpage [Internet]. Place of Publication: Publisher; Date of Publication [Date of Update/Revision; Date of Citation]. Available from: (Insert URL).

Example

10. British Medical Journal [Internet]. Stanford, CA: Stanford Univ; 2004 July 10; Available from: http://bmj.bmjjournals.com/

ELECTRONIC-JOURNAL ARTICLES

Format

11. Last Name and Initial(s) of Author, [followed by last names and initials of other authors]. Title of article. Abbreviated Journal Title [medium] Year of Publication; Volume (Issue): Inclusive Page Numbers [if available]. Availability Information. Date of Access.

Example

11. Langer WM. "Campesinos" and the crisis of modernization in Latin America. Jour of Pol Ecol [serial online] 1996; 3(1). Available: ttp://www.library.arizona.edu/ej/jpe/volume_3/ ascii-lokeriso.txt via the INTERNET. Accessed 2004 Aug 11.

ELECTRONIC BOOKS (MONOGRAPHS)

Format

12. TITLE OF MONOGRAPH [monograph online]. Place of Publication: Publisher; Year of Publication [Update Information, if applicable]. Availability Information. Date of Access.

Example 1

12. RECOGNITION AND MANAGEMENT OF THE PERIMENOPAUSAL PATIENT IN CLINICAL PRACTICE [monograph online]. University of Medicine and

Dentistry of New Jersey and Robert Wood Johnson Medical School Department of Obstetrics, Gynecology, and Reproductive Services; 1998 May. Available from: Femhealth, http: //peri-menopause.com. Accessed 2004 May 20.

Example 2

13.   [CBE] Council of Biology Editors. 1999 Oct 5. CBE home page. http://www. councilscienceeditors.org/>. Accessed 2004 Oct 7.

ALPHABET–NUMBER SYSTEM

As already indicated, this system makes use of an alphabetically arranged and numbered list of references in the Reference section of research report. These numbers are used to cite the corresponding references in the text of the report. The usage of the numbers is very similar to that of the citation-sequence system, i.e., either as superscripts or within parenthesis, at appropriate places. The format of the entries in the list of references is similar to that of the name–year system with a serial number for each entry.

Example

The most important of these models are duct ligation in adult rats[41, 92], partial (90%) pancreatectomy in rats[10], cellophane wrapping of the pancreatic head in adult hamsters[70, 72], streptozotocin-induced beta cell depletion in newborn rat[8, 10, 19, 92, 96] and beta cell destruction by selective alloxan perfusion in adult mice[90].

The superscript numbers in the above example point to reference entries in the list of References, having those serial numbers as follows, for example:

8.   Bonner-Weir S, Trent DF, Honey RN, Weir GC. 1981 Responses of neonatal rat islets to sterptozotocin: limited B-cell regeneration and hyperglycemia. Diabetes 30: 64–9

9.   Bonner-Weir S, Trent DF, Honey RN, Weir GC. 1983 ....

10.   Bonner-Weir S, Baxter LA, Schuppin GT, Smith FE. 1993 ...

Reference—without Article Title

In the three systems we discussed above, the title of the article referred is an essential component of the entries in the list of references at the end of the report. As we had seen, the formats adopted by many journals strive to minimize the printing cost by conserving space by way of eliminating the periods and spaces wherever possible in the citation and listing of references. A few journals, which are hard pressed for space, have adopted a system of listing reference in which an entry includes only the essential details. It does not give the title of the article, the issue number, and end page number. The citation of the references in the text is either by name–year system or the number (citation-sequence or alphabet–number) system.

Example

1.   Epple A. *Z Zellforsch.* 1961. 53, 731.

2.   Levene C, Feng P. 1964, *Stain Tech.* 1964. 39, 39.

This system of listing the references in not at all suitable for research reports such as an academic thesis or dissertation.

JOURNAL ABBREVIATIONS

The name of the journal is an essential component every entry of journal articles in the reference section. It is given either in full or in an abbreviated form, depending on the requirements of the Journal to which the report is submitted for publication. When given in full, the name might have to be italicized.

Example

Wood, B. P., and Matthews, R. A. 1987. The immune response of the thick-lipped grey mullet, *Chelon labrosus* (Risso, 1826), to metacercarial infection of *Cryptocotyle lingua* (Creplin, 1825). *Journal of Fish Biology,* 31(Suppl. A): 175–183.

When the journal name is required to be given in an abbreviated form every care should be taken to follow the format proposed by the journal in which the reference article has been published. However, the journal to which a report is submitted for publication requires a different and specific format, then the recommended format should be followed. The abbreviations should be consistent in a report. That is, for example, the journal "General and Comparative Endocrinology" should not be abbreviated as "Gen. Comp. Endocr." in one entry and as "Gen. Comp. Endocrinol.," in another.

American National Standards Institute has recommended abbreviations for journal titles and these are widely followed by many journals. For example, the word "Journal" is abbreviated as "J.", "Applied" as "Appl.", "National" as "Natl.", "Quarterly" as "Q.", and "Review" as "Rev." . The abbreviations can be used with or without a period (full stop). All words ending with "ology" (e.g. Bacteriology, Biology, Ecology, Limnology, Physiology, Zoology) are abbreviated at the letter "l" omitting "ogy" (e.g. Bacteriol., Biol., Ecol., Limnol., Physiol., Zool.). One-word journal titles (e.g. Diabetes, Development, Endocrinology, Metabolism, Nature, Parasitology) are not to be abbreviated. A few of these recommended abbreviations are listed in Box 3.1.

BOX 3.1   ABBREVIATIONS FOR WORDS USED IN JOURNAL NAME

Word	Abbreviation	Word	Abbreviation
Abstract(s)	Abstr.	Dental	Dent.
Academy	Acad.	Developmental	Dev.
Acta	Acta	Diseases	Dis.
Advances	Adv.	Drug	Drug (not abbreviated)
Agricultural	Agri.	Ecology	Ecol.
American	Am.	Edition	Ed.
Anales	An.	Entomologia	Entomol.
Analytical	Anal.	Entomologica	Entomol.
Anatomical	Anat.	Entomological	Entomol.
Annalen	Ann.	Entomology	Entomol.
Annales	Ann.	Ethnology	Ethnol.
Annals	Ann.	European	Eur.
Annual	Annu.	Excerpta	(Not abbreviated)
Antibiotic	Antibio.	Experimental	Exp.
Antimicrobial	Antimicrob.	Fauna	(Not abbreviated)
Applied	Appl.	Federal, Federation	Fed.

BOX 3.1   (Continued)

Word	Abbreviation	Word	Abbreviation
Archiv	Arch.	Fish	(Not abbreviated)
Archives	Arch.	Fisheries	Fish.
Archivio	Arch.	Flora	(Not abbreviated)
Association	Assoc.	Folia	(Not abbreviated)
Bacteriological	Bacteriol.	Food	(Not abbreviated)
Bacteriology	Bacteriol.	Forest	For.
Biological	Biol.	Forschung	Forsch.
Biologie	Biol.	Freshwater	(Not abbreviated)
Biology	Biol.	Gazette	Gaz.
Botanical	Bot.	Immunity	Immun.
Botanische	Bot.	Immunology	Immunol.
Botany	Bot.	Indian	(Not abbreviated)
British	Br.	Industrial	Ind.
Bulletin	Bull.	Internal	Intern.
Bureau	Bur.	International	Int.
Canadian	Can.	Japan, Japanese	Jpn.
Cell	(Not abbreviated)	Journal	J.
Cellular	Cell.	Laboratory	Lab.
Central	Cent.	Magazine	Mag.
Chemical	Chem.	Material	Matr.
Chemie	Chem.	Mathematics	Math.
Chemistry	Chem.	Medical	Med.
Clinical	Clin.	Medicine	Med.
Commonwealth	Commw.	Methods	(Not abbreviated)
Comptes	C.	Microbiological	Microbiol.

BOX 3.1   (Continued)

Comparative	Comp.	Microbiology	Microbiol.
Conference	Conf.	Molecular	Mol.
Contributions	Contrib.	Monographs	Monogr.
Current	Curr.	Monthly	Mon.
General	Gen.	Morphology	Morphol.
Genetics	Genet.	National	Natl.
Geographical	Geogr.	Natural, Nature	Nat.
Geological	Geol.	Neurology	Neurol.
Geology	Geol.	Nuclear	Nucl.
History	Hist.	Nutrition	Nutr.
Obstetrical	Obstet.	Station	Stn.
Official	Off.	Studies	Stud.
Organic	Org.	Supplement	Suppl.
Paleontology	Paleontol.	Surgery	Surg.
Pathology	Pathol.	Survey	Sur.
Pharmacology	Pharmacol.	Symposia	Symp.
Philosophical	Philos.	Symposium	Symp.
Physical	Phys.	Systematic	Syst.
Physiology	Physiol.	Technical	Tech.
Pollution	Pollut.	Technology	Tech.
Proceedings	Porc.	Therapeutics	Ther.
Publications	Publ.	Tissue	(Not abbreviated)
Quarterly	Q.	Transactions	Trans.
Rendus	R.	Tropical	Trop.
Report	Rep.	United States	U.S.
Research	Res.	University	Univ.
Review	Rev.	Untersuchung	Unters.

BOX 3.1   (Continued)

Word	Abbreviation	Word	Abbreviation
Revue, Revista	Rev.	Urological	Urol.
Rivista	Riv.	Verhandlungen	Verh.
Royal	R.	Veterinary	Vet.
Scandinavian	Scand.	Virchows	(Not abbreviated)
Science	Sci.	Virology	Virol.
Scientific	Sci.	Vitamin	Vitam.
Series	Ser.	Wissenschaftliche	Wiss.
Service	Serv.	Zeitschrift	Z.
Society	Soc.	Zentrablatt	Zentrabl.
Special	Spec.	Zoologie	Zool.
Zoology	Zool.		

Word	Abbreviation
Stain	(Not abbreviated)
Histology	
Histochemistry	
Microscopy	
Photography	

CHAPTER FOUR

RESEARCH REPORT

INTRODUCTION

A research report is a document prepared at the end of a scientific research project, to communicate the findings to a specific audience. It explains the scientist's motivation for carrying out an experiment, the experimental design, the results obtained, and the meaning, interpretation and significance of the results. The audience generally belongs to the scientific community and may be working in the same or related field. Depending on the importance and utility of the findings, the audience may include students and, even, general public. Since the aim of a research report is to communicate clearly to others, it should be written in a style that is exceedingly unambiguous and concise.

The most important and valuable research report is a paper, i.e., an article published in a reputed journal. This is to emphasize that any research finding not published is not a finding at all. A journal would accept a paper for publication only if it is a first report of original research findings and prepared in a format so that experts in the field can repeat the experiments and validate the findings.

In addition to research articles published in a Journal, there are other forms of reports such as thesis (= dissertation) submitted by students (postgraduate, M. Phil. and Ph. D. courses) to the Universities, oral or poster presentations in conferences and seminars, and project reports submitted to sponsoring agencies. All the research reports have certain basic components, which may or may not be set down as distinct sections with section headings.

COMPONENTS OF A RESEARCH REPORT

A full-length research article generally consists of

i)   a title

ii)   name(s) of the author(s) with the address of the Institution(s) to which the author(s) belonged when the work was carried out

iii)   keywords

iv)   Abstract

v)   Introduction

vi)   Material and Methods

vii)   Results

viii)   Discussion (and conclusions)

viii) Summary

ix) Acknowledgements and

ix) References.

The system of organizing a research paper into **I**ntroduction, **M**aterial and Methods, **R**esults **and D**iscussion is referred to as the IMRAD (or IMRaD) system, the acronym of these four sections. All these components are also found in a thesis. However, a thesis is more than a research article in the sense that it is a treatise of one or more related topics and it generally includes a detailed review of the pertinent literature. The description of the experimental designs and methods and the presentation of original results and their discussion are much more elaborate. Very often, a thesis may have a "General Introduction" to justify the inclusion of several topics in one thesis and a "General Discussion" to bring out the relationship of the results discussed in the different topics.

Preliminary findings of research work can be reported as a "Short Note" or "Short Communication" or "Letter to the Editor" or "Preliminary communication" in a journal. These reports also include the major components of a research article, viz., Introduction, Material and Methods, Results and Discussion, in the same sequence but less briefly and without delineation into sections. Such "notes" are published in order to claim priority in reporting a significant result or a novel method for solving a problem.

Reports for oral or poster presentation in a conference also have the same elements of a research article but are prepared with the economy of time or space available, as the case may be.

Though the IMRAD system is widely followed in the preparation of the research reports in biology, variations may be found. For example, many journals accept articles in which the Results and the Discussion sections have been merged into one section. There are also papers that have the Materials and Methods section and the Results section integrated into an "Experimental" section.

Since all the research reports are meant to communicate the findings to others, we shall discuss in some detail the different elements of a research report, especially how they can be effectively written.

Title

The title of a research report should be clear, concise and adequately descriptive of the contents of the report. It is important for indexing and abstracting purposes. The following general guidelines may be borne in mind while constructing a title for a research report.

- It should be catchy enough to grab the attention of the reader.

- It should be of optimum length, neither too short nor too long.

- The syntax, i.e., the order of words, of the title should be proper so as to convey the correct meaning or to avoid unintended and absurd meaning.

- All words, except the articles and prepositions, of the title should be capitalized.

- Use of complicated grammar and redundancy of words and phrases (e.g. "An investigation of...", "The analysis of...", "Effect of...", "Influence of...", "New method...") must be avoided as far as possible.

- Likewise, it is desirable to avoid the use of abbreviations, chemical formulae, trademark names, jargon, etc. in the title.

A title may be a straightforward one (e.g. Seasonal and Daily Plasma Corticosterone Rhythms in American Toads, *Bufo americanus*) or it may be *series title* having a main title and a numbered subtitle (e.g. Studies on Garden Lizard. VII: Reproductive Behaviour) or a *hanging title* with the main and the subtitles separated by a colon (e.g. Copper and Zinc Exposure of Zebrafish, *Brachydanio rerio* (Hamilton–Buchman): Experimental Listeria Infection). In addition to the full title, a short version of it, referred to as *running title* or *running head*, is printed in journals, at the top of each page of the article.

A few examples to understand how a title can be improved by revision are given in Box 4.1.

BOX 4.1   EXAMPLES OF "TITLE"

Example 1   "Investigations on the Pancreas of Fish"

Unless the above title is the title of a review article on all aspects (morphological, biochemical, physiological, etc.) of the pancreas of all fishes, the title does not tell us what part of the pancreas was studied, pancreas of which fish was studied and what methodology was used. It would be better if the above title reads as follows:

BOX 4.1   (Continued)

"Immunohistochemistry of the Endocrine Pancreas of the Freshwater Catfish *Mystus gulio* (Hamilton) (Bagridae)"

The above modified title is specific in indicating the species of fish used, the specific part of the organ studied and the methodology employed. Further, the terms such as immunohistochemistry, endocrine pancreas, freshwater catfish, and Bagridae will be useful in indexing in the literature database and for searching in a database.

Example 2   "On the Discovery of a New Useful Laboratory Research Method for Isolating and Purifying the Lactose-digesting Enzyme *β*-galactosidase from the Economically Important, Yogurt-producing Bacterial Species *Lactobacillus bulgaricus*"

The title needs a lot of editing. The following revision is concise, specific and informative:

"A New Method for Isolating and Purifying *β*-galactosidase from *Lactobacillus bulgaricus*"

Example 3   "Studies on a Frog"

The above title is too brief. What kinds of studies? What frog? Where? It would read better if revised as "The Reproductive Behaviour of a Population of the Common Frog (*Rana limnocharis*) in the Poondi Reservoir."

Authors and Addresses

In full-length research reports that are published in journals, the names of the authors and their addresses are listed just below the title. Short communications published in a journal carry the names and addresses of the authors at the bottom of the article. Thesis and similar documents have a separate title page that carries the name of the author and the address along with other details such as for what purpose and to which academic or sponsoring body they are submitted and the date of submission.

Generally, the name of an author is written in the sequence of first name, initial of the middle name, and last name (e.g. Elena J. Sandova). If the author has only two names, both the names are spelt out (e.g. Vimal

Shaw). However, much variation could be seen in different journals, such as the use of initials except for the given name, i.e., last name (e.g. S. D. Bradshaw and G. E. Rice). Use of spelt out first and last names is preferable because it helps to avoid confusion about authors having same initials (e.g. A. E. Christopher may mean Abraham E. Christopher, Allen E. Christopher, Andrew E. Christopher, and so on) and it would facilitate more accurate retrieval of information from literature database.

Prefixing of titles (e.g. Dr., Prof., etc.) and suffixing of degrees (e.g. Ph. D., M. D., etc.) to the names are not generally permitted. Mentioning of the designation of the authors (e.g. Director, Head of the Department, Professor, Reader, Lecturer, etc.) has also to be avoided.

If a research report has more than one author, the sequence in which the names of the authors are listed depends on the relative contribution each author has made to the actual research work. The *first author* (= *senior author*, even if he or she is a student working in a laboratory under the guidance of a supervisor or a committee of supervisors) is the one who has chosen the topic of research, formulated the hypothesis, designed and conducted the experiments, collected the data, and prepared the report. The second, third and so on authors (= *co-authors*, even if they happen to be the Director of the laboratory where the work was carried out) are those who have contributed at all or different levels of the work, directly and to an appreciable extent.

For example, in a research project, the principal investigator is the one who is responsible for almost everything right from writing the research proposal to writing the final research report. The Project might have a number of co-investigators, research associates, research assistants, research scholars, etc. All the publications of the results of the project would naturally have the principal investigator as the senior author and those members of the project who contributed significantly (significance has to be decided by the principal investigator) as co-authors.

A publication of a work carried out by the same team, in the same laboratory, along with the Project but not mentioned in the proposal may have different first and co-authors.

A research scholar working under the guidance of the principal investigator of the project may submit a thesis containing results of a part of the work of the project. However, when the same result is published in a journal, the principal investigator has to decide who should be the senior author.

Inclusion of names of persons in the list of authors of a research report, who were in no way directly connected with the proposal and execution of the research work, is not desirable. The practice of incorporating the name of the "head" of the laboratory, department or institution, at the end of the list of authors is not uncommon. There are also instances where names of technicians and those of who help in writing the report are listed as co-authors. The general suggestion in this regard is that only those who were directly involved in the work and, therefore can take the responsibility of defending the contents of the report, should be listed as authors.

The address of the authors, i.e., the place where the work was actually carried out, is given just below the list of the authors. If the authors belong to different institutions, the name of each author and the corresponding address is marked with a superscript like a, b, c ... or 1, 2, 3 .... If one or more of the authors have shifted their place of work, their present address may be given as a footnote. Though door numbers and street names are not required in the address, it should be clear enough for anyone who wishes to contact the author, especially for requesting reprint copies of the research article.

Abstract

All the research articles of a journal carry an abstract, in about 250 words. An abstract is a summary of the information in the research article. It gives the salient features of the main sections (Introduction, Materials and Methods, Results and Discussion) of a paper. The abstracts of research articles of standard journals are published or made available on the Internet by indexing services such as *Biological Abstracts*, *Medline*, etc. Any reader would first go through the abstract and then decide whether to study the entire article or not. Therefore, care must be taken while preparing the abstract. A few suggestions are given below.

- Introduction—general problem—objectives or hypotheses or scope of the particular study—2–3 sentences
- Brief description of the materials used and the methods adopted, without experimental details—1–2 sentences
- Summary of the results—only the most significant results—1–3 sentences
- Statement of the significance of the results, and the conclusions—1–2 sentences

- Should not contain any reference to bibliography, tables, figures, etc.

- Should not contain any information that is not mentioned in the research report.

- Should not contain any abbreviations, unless a long word or phrase is repeatedly used and should be properly introduced.

- Should be typed in a single paragraph.

- Should be written in short and concise sentences, in past (perfect) tense, and in passive voice, as far as possible, because it pertains to a work done.

Deviations to the above suggestions may be seen in different journals. Abstracts written with specific headings (Background, Purpose, Materials and Methods, Results and Conclusions) printed in bold fonts can be found in journals like Indian Journal of Medical Microbiology. Abstracts are also written in paragraphs, for articles that deal with more than one aspect. For example, a paper dealing with the morphology of immune organs in fish may have an abstract with more than one paragraph, one for each organ.

The abstracts used in journals and reproduced without change in indexing media, are referred to as *informative abstracts* because they provide brief information of what is contained in the article that follows. *Descriptive* or *indicative* abstracts contain more details of a work, and such abstracts are useful for publication in conference reports, government or sponsoring agency reports, etc. An example of an abstract in a journal article is analysed in Box 4.2.

BOX 4.2   AN EXAMPLE OF "ABSTRACT"

Ontogeny of the lymphoid organs in the turbot
Scophthalmus maximus: a light and electron
microscope study

F. Pardós, S. Crespo

Laboratori de Biologia Animal, Facutat de Veterinaira,
Universitat Autonomoa de Barcelona, 08193
Bellaterra, Barcelona, Spain

BOX 4.2   (Continued)

Abstract

A histological and ultrastructural study was made of the development of the kidney, thymus and spleen in the turbot *Scophthalmus maximus,* from hatching until the end of metamorphosis. Primordial haemopoietic stem cells are first observed in the pronephric kidney very early after hatch and rapidly differentiate into different cellular types. The spleen develops later, soon becoming rich in blood capillaries, red blood cells and thrombocytes. The thymus is the last lymphoid organ to appear but show a quick development. This organ seems to originate from haemopoietic stem cells migrating from the head region of the kidney. Lymphoid organs become lymphoid in the sequence thymus, kidney and spleen. Although a small number of lymphocytes appear only in the later stages, cellular types involved in non-specific defense mechanisms, such as macrophagic and reticular cells, originate in early stages. These observations suggest that non-specific system may play an important role in the immunocompetence mechanisms of the turbot during early larval development.

Key words   Larval development; Lymphoid organs; *Scophthalmus maximus*

Given above is the Abstract, along with the Title, Authors, Address of the Authors and Key words, of an article published in *Aquaculture 144 (1996),* along with the title, authors and address.

Let us first analyse the Abstract.

The first sentence clearly states the scope of the study including the organism (larval stages of turbot, *Scophthalmus maximus*), the organs (kidney, thymus and spleen), and the methodology (histology and electron microscopy).

The next six sentences give the salient features of the observations made. While the first three of these sentences are allotted to the three organs, one each, the next three are used to highlight the interrelationship between the organs and the type of defense mechanism in the early stages of development.

The last sentence states in clear terms the conclusion arrived at about the nature of defense mechanism during the larval development of turbot.

BOX 4.2   (Continued)

Now, let us comment on the Title, Authors and Key words. Note that the Title, a hanging type, deviates from what was suggested under the section Title. It does not have the words capitalized. The two Authors do not have a conjunction (and) between them. The number of Key words is fewer. Instead of "Larval development", words such as ontogeny, kidney, thymus and spleen would have been more appropriate because the investigation was not about the larval development of turbot but about the ontogeny of the three lymphoid organs.

Summary

In addition to the term "abstract", two other terms viz., "summary" and "synopsis" have been used in research articles published in journals and in other research documents such as a thesis. There is not much difference in the dictionary meanings of "abstract" and "summary". Abstract of a thesis is usually called *summary*. It is given at the end, after the Discussion. It is much more elaborate than an abstract in an article of a journal. Since it deals with more than one chapter of a thesis, it will have to be in numbered paragraphs, sections, etc. running into more than one page.

Synopsis

A concise outline survey of the contents of a thesis or a monograph or a review paper is called a *synopsis*. Research articles published in the past, in the 60's and 70's, used the terms "Summary" and "Synopsis" to sections homologous to "Abstract" of the articles in the present-day journals. Many review articles give the synopsis more in the form of a table of contents with page number indicated against each section.

Universities require from every doctoral student a synopsis of the thesis he or she is writing, at least six months prior to the submission of the actual thesis. The main use of such documents is to get the prior "acceptance" of the examiners to evaluate the thesis. The unfortunate thing about the thesis synopsis system is that it is required to be prepared much earlier, at a time when the contents of a thesis is not completely known even to the candidate. When the final draft of the thesis is approved by the supervisor or the committee of supervisors, many of the findings mentioned in the synopsis may not find a place in the thesis. Many other discrepancies are also likely to creep in, between the synopsis and the thesis. Fortunately the synopsis is not part of the thesis and is not likely to be made public.

Whether it is an abstract, a summary or a synopsis, it is prepared with utmost care. The brevity or the descriptive nature of it depends on the interest of the target audience and the requirements of publishing agencies.

Key words

A few words used in the article, about 3 to 4, which will be useful for indexing services, are given as key words just below the abstract in a journal. The choice of the words should be such that they facilitate searching for the article in some database. The example of abstract discussed in Box 4.3 also shows key words.

Introduction

The first section of a research article paper is its Introduction. Its aim is to introduce the specific subject of research to the reader, to justify the choice of the topic and, if necessary, the specific methodology adopted, and to state clearly the objectives or hypotheses of the investigation. The following are a few general guidelines for writing an Introduction.

1. To begin with, a relatively broad background of the topic of investigation is given. In this process, information from pertinent primary literature and other appropriate technical sources may be cited. This brief review, limited to studies that relate directly to the present study, is intended to provide some sort of orientation to the readers, even to those who may not be very familiar with the topic of the report. It also helps to point out the gaps in the literature.

2. The background scope is progressively narrowed to the specific problem that is being investigated. With the help of citations from primary literature, the need for studying and solving the specific problem is justified.

3. Finally, the specific objectives or hypotheses of the investigation, and the material of investigation are stated clearly but briefly in one or two sentences.

In addition, there may also be a need in the Introduction of an article to justify the choice of a specific methodology employed and/or of the choice of the specific organism. It all depends on the nature and objectives. For example, the structure of spleen of a common laboratory mammal might have been investigated at the light-microscopic level using histological and histochemical methods but not employing immunohistochemical techniques. Therefore, when the results of a immunohistochemical study on the spleen of a laboratory mammal is reported, the Introduction of the report will have to contain justification (in what ways the results of this technique would throw more light on the structure of the spleen than what is already revealed by the other methods) for the choice of the method. On the other hand, different workers might have reported the light-microscopic structure of the spleen of several vertebrates, but none may have described that of an aquatic mammal previously. And, when a report on the structure of the spleen of a species of whale is prepared, justification for such a study on whales (their peculiar habitat, ability to counter infections through wounds, etc.) would be appropriate in the Introduction.

The Introduction may also be utilized to define any specialized terms and/or abbreviations that will be used in the following text. There is also a suggestion that principal results and conclusions might be stated in the Introduction.

The Introduction of a thesis has to be more elaborate. If a thesis deals with the results of more than one aspect of a topic, and all these aspects are presented in different chapters, then there is a need for a separate chapter for a General Introduction to bring out the relationship of the different aspects. Providing the background information on these different aspects will become more complicated and would call for a separate chapter on Review of Literature (See chapter 2). However, the basic composition of an Introduction in all forms of research report remains the same.

The Introduction is written in both present tense and past tense. Established knowledge is expressed in present tense. Results of the previous works are given in past tense. Generally, use of passive voice is practiced in writing Introduction and other sections of a paper. However, many eminent editors of journals encourage use of active voice because it is less verbose, and unambiguous. An example of an Introduction to a research paper and its analysis is given in the Box 4.3.

BOX 4.3   AN EXAMPLE OF "INTRODUCTION"

Turbot (*Scophthalmus maximus*) farming is one of the main activities in European aquaculture. Production of turbot juveniles has increased from around 300 000 units in 1985 to an estimated 3 million in 1992 (Pearson-Le-Ruyet, 1993). However, the larval rearing of this species is still characterized by high though variable mortality rates, which makes the final survival results rather unpredictable. Several studies on turbot larval pathology have been carried out to elucidate the causes of mortality, which have been related to either nutritional or infectious aetiologies (Cousin *et al.*, 1986; Nicolas *et al.*, 1989; Gatesoupe, 1990; Padrós *et al.*, 1993; Padrós and Crespo, 1995). Morphological and physiological events throughout turbot development have been extensively studied (Cousin and Baudin-Laurencin, 1985; Cousin *et al.*, 1987; Munilla-Moran *et al.*, 1989; Munilla-Moran and Stark, 1990; Pardós *et al.*, 1991). In contrast, very little is known about the disease defense mechanisms of this species during early development, and no reports can be found in the literature on the ontogeny of its lymphoid organs. Since knowledge of this is necessary to the understanding of either specific or non-specific immunocompetent mechanisms in the turbot, and might be useful to improve therapeutic and immunoprophylactic measures in the rearing systems, a light and electron microscope study of the lymphoid organs of this species throughout larval development was undertaken. Because the chronological age of the larva does not necessarily indicate its physiological age, the larvae were staged on the basis of morphological features (Al-Maghazachi and Gibson, 1984), and histological and ultrastructural findings were related to these stages.

Let us analyse the above Introduction to an article entitled, "Ontogeny of the lymphoid organs in the turbot *Scophthalmus maximus*: a light and electron microscope study" authored by F. Pardós and S. Crespo and published in *Aquaculture*, 144: 1–16 (1996).

1. Note how the Introduction started with a relatively broad idea (importance of turbot, a species of flat fish, in aquaculture and the problem of unpredictable mortality rate in the rearing of the larvae) and then it reviewed literature to indicate what is known and what is not, thus identified a general "gap" in the literature (disease defense mechanism in larva and ontogeny of lymphoid organs of turbot).

BOX 4.3   (Continued)

2. After having introduced the background to the research, the Introduction justifies the need for the study (to understand the specific or non-specific immunocompetence mechanism) what the findings might contribute (improvement of therapeutic immunoprophylactic measures).

3. The final section of the Introduction stated the specific work carried out (light and electron microscope study of the immune organs of turbot throughout the larval life) and the objective (to relate the structure with the physiological age).

4. Thus, the Introduction looks like an upside-down triangle— broadest topic at the top and progressively narrows to a "point" which is the statement of the specific work and objectives.

5. Also note that the Introduction includes a number of citations to primary literature.

Materials and Methods

We might have indicated in the Title itself what materials we used and what methodology we had adopted, and probably justified their choice, in the Introduction. Now we must give, in the Materials and Methods section, the details of all the materials and methods used and, if necessary, the details of the specific techniques (Box 4.4 discusses the difference between method, technique, procedure and protocol). The reason for providing all the necessary details is to enable any competent worker to repeat our experiment. If the result reported in our report is so significant and valuable, then it should be reproducible and get validated. Further, when a research report is submitted for publication or for evaluation, any reviewer or examiner would like to ascertain that the methodology is appropriate and flawless.

BOX 4.4   PROCEDURE, METHOD, TECHNIQUE, PROTOCOL

Procedure   A set of established forms or methods for conducting a research.

Method   A systematic means and orderly arrangement of parts or steps to accomplish an end.

Technique   The systematic procedure by which a complex or scientific task is accomplished.

Protocol   The plan for a course of medical treatment or for a scientific experiment.

The use of the terms "procedure", "method", "technique" and "protocol" needs a little clarification. The "procedure" is a way of doing something and includes a set of established forms of "methods" for conducting a research. It tells "what" is planned to be done in a research to achieve the stated objective. The "methods" explain "how" the procedure is to be carried out. And it (methods) may include one or more "techniques". The "techniques" describe the selected measuring or observing devices to be employed in each of the methods. The "protocol" gives the step-by-step course of action in each technique.

Let us see an example to distinguish the above terms. The general aim of a research project is to understand the reason for the high larval mortality rate in turbot, a species of flatfish. There are several "procedures" to attain this objective, like the study of larval pathology, physiology, etc. We may choose one of them, viz., the study of the ontogeny (the origin and development of an organ) of lymphoid organs in this fish. The choice of this procedure has its own aim, namely, to understand the existence of specific and non-specific immunocompetence mechanism during the development, i.e., to understand whether the larvae die due to lack of proper defense mechanism in the early stages. Having chosen the "procedure" (ontogeny of lymphoid organs), we must proceed to decide "how" we are going to carry it out. In other words, we have to find "methods" for studying the ontogeny of lymphoid organs of turbot. The methods may include several aspects such as collection of eggs of turbot, rearing of the larvae in the laboratory, staging of the larvae, study of the anatomy, histology, electron microscopy, micrometry,

---

BOX 4.4   (Continued)

photomicrography, statistical analysis and so on. Let us consider the method of histology alone. How many "techniques" have we to use? Apart from the basic microtechniques, we may have to use special histochemical and immunohistochemical staining techniques. Techniques to evaluate the changes in the morphology, cell types, cell populations, etc. in the organs will have to be employed. Let us assume we have made choices of the techniques for each of the methods for carrying out the procedure. Now, we have to describe each technique, if necessary (if it is a new one, or if substantial modifications have been incorporated), including the apparatus (a group of glassware, devices and instruments used in a technique), preparation of the reagents and media at specific concentrations, the duration of each step, etc. in detail. One way of describing it is in the form of a "protocol", a numbered step-by-step scheme. In fact, a protocol is used to present not only a technique but also a procedure and a method. It is very similar to a script of cinema film, which fills in all the details like the time and place of the action, the characters' physical appearances, all the dialogue and action, position of the camera, movements of the camera and even, transition devices between scenes.

Though we have attempted to differentiate the terms procedure, method, technique and protocol, the use of these terms is not always strict in their suggested meanings. We may describe the "procedure" for a histochemical "method". We can also give the "protocol" of our research "technique". However, we should strive to convey the correct and intended meaning and be consistent in the use of these terms.

It is somewhat difficult to decide to what extent details should be given. It will always be useful to follow the "Instructions to Authors" given in a Journal to which a report is to be submitted, or, if the report is a thesis, the Departmental or University guidelines. The following are a few general suggestions for writing Material and Methods.

1. Normally, it should be written in third person and past tense, but use of first person (active voice) is encouraged in many of the Modern Journals.

2. Both Materials and Methods are described together, i.e., no separate subsections are allotted to them. However, several subsections with or without headings, numbered or unnumbered may be created. For example, the Materials and Methods section of a paper on "Cellular Immune Responses in Rainbow Trout (*Onchorhynchus mykiss*): Flow Cytometric Study" has the following subdivisions: Fish, Leucocyte collection, Flow cytometric analysis, Cell subpopulations, Phagocytosis, Oxidative burst, and Nonspecific Cytotoxic Cells.

3. Experimental materials such as animals, plants, and microorganisms should be identified specifying the genus, species and strain. The source of these materials, i.e., the place of collection or purchase should be given. The characteristics of the organisms, viz., age, sex, genetic and physiological status should be provided. How these organisms were reared or maintained in the laboratory (feeding food and water, maintenance of specific photoperiod, ambient temperature, etc.) should be described.

4. If large number of materials were used, presentation of the details in a tabular form would be ideal. For example, suppose eleven species of snakes belonging to six families were used. Details of these species, such as the taxonomy, number of specimens used, number of males and females in each species, whether mature or immature, habitat, etc., can be effectively and concisely tabulated.

5. As for the other materials such as chemicals, media, antigens, antibodies, etc., technical specifications, quantities, source and method of preparation should be given. Unless it is essential, it is better not to mention the trade names. If trade name has to be mentioned, for reasons of critical differences between products, the name of the manufacturer also should be included. If any material (e.g. an antibody) had been gifted by some scientist of some other laboratory, the source information should be provided with the details of its synthesis and purification. If materials like antibodies, media, etc. are used in considerable numbers, use of a tabular form for details is desirable.

6. In reports of research involving fieldwork (e.g. Seasonal Variations in the Hydrobiological Features of the Veeranam Lake in South India), it would be useful to include details regarding the location of the field, the stations from which samples were collected, how samples were preserved and transported to the laboratory, etc.

Depiction of the field of study in the form of an appropriately labelled map may also be considered if it is difficult to describe in words the complicated details of sampling stations.

7. The presentation of the Methods should follow some logical sequence. Related methods (microscopic, biochemical, immunological, etc.) may be grouped together. The sequence may be consistent with that followed in the Results section of the report.

8. Not all methods and techniques need be described in detail. If a method we followed is one published in a standard Journal (one which has wide national and international circulation and indexed by agencies like Biological Abstracts), it is sufficient if literature reference is given. Any technique that has been developed in our laboratory or a standard technique that has been radically modified to suit conditions of our experiment should be described with sufficient detail, along with literature citation if they had been published in a not-so-popular Journal.

9. While describing the techniques, the measurements of the reagents (e.g. quantity in L, mL, µL, g, mg, µg, etc., concentration in percent weight/volume, volume/volume, etc., strength in molarity, normality, etc., dilutions in 1:100, etc.) should be specified.

10. The routine methods used for statistical analysis of the data (e.g. arithmetic mean, standard deviation, *t*-test, ANOVA, etc.) may be mentioned without any description or reference. If any advanced techniques or software packages were used, they should be cited with appropriate reference as to where the details about them could be found.

11. No results should be included in the Materials and Methods section.

12. In a Thesis, the Materials and Methods section could be more elaborate than that of a paper in a journal. The general organization is the same. Details regarding the biology, economic importance, etc. of the animals, plants, microorganisms, etc. might be included. Experimental designs, randomization procedures, collection of samples, preservation of the samples, etc. might be described in detail. Description of the preparation of the reagents, media, etc. may also be included. However description of trivial basics such as washing of the glassware, maintenance of aseptic conditions, sterilization procedures, etc. might be avoided.

Examples of Materials and Methods of articles published in Journals, along with our comments, are given in Box 4.5.

BOX 4.5   EXAMPLES OF "MATERIALS AND METHODS"

Example 1

The following is an extract from Materials and Methods section of paper entitled "Cellular Immune Response in Rainbow Trout *(Oncorhynchus mykiss):* Flow Cytometric Study". Published by Chilmonczyk, S., and Monge, D. *Acta vet. Brno,* 67: 207–213, 1998.

Materials and Methods

Fish

Leucocytes were collected from healthy (control) or experimentally infected rainbow trout (*Oncorhynchus mykiss*–RBT). Fish (5–10 g) were subjected to waterborne Viral Haemorrhagic Septicaemia Virus (VHSV) while PKX parasite (proliferative kidney disease causing) infections (intraperitoneal injection) were performed in 30 g RBT. Control and infected fish were reared in recirculated filtered water thermoregulated at 10 ± 1°C (VHSV infection) or 16 ± 1°C (PKX infections).

Leucocyte collection

Freshly collected blood anticoagulated with heparin was diluted (1:20) with Minimum Essential Medium (MEM). Cells from pronephros, spleen and thymus were obtained by forcing the tissues through a stainless steel mesh in cold serum-free MEM. Then leucocytes were isolated through a Ficoll cushion. The leucocyte-rich interphase was collected, washed in PBS and resuspended in MEM.

Flow cytometric analysis

Flow cytometric analyses were carried out using a standard fluorescence activated cell analyzer (FACScan, Becton Dickinson). For each sample 10 000 to 20 000 individual cells were recorded using a dotplot combination of low angle forward scattered (FSC) and right angle scattered (SSC) laser light. Data were analyzed with the program LYSIS II version 2.0.

Comment   Concisely written with enough and necessary details of the experiment and the methods. Abbreviations properly introduced. However, information on the number of fish used and doses of infective agents would have been useful.

Example 2

The following is an extract from the Materials and Methods section of paper entitled, "Antigen handling in the spleen of the yellow-bellied

BOX 4.5   (Continued)

toad, *Bombina variegata*" published by Jozef Dulak, in *Folia Histochemica et Cytobiologia*, Vol. 32, No. 2, pp. 85–90, 1994.

Animals   Adult specimens of the yellow-bodied toad, *B. variegata* (about 5 g body weight) were collected in Southern Poland in May and June. In the laboratory the animals were kept in plastic boxes with small amount of tap water. The toads were fed by redworms, earthworms and flies. Animals were killed by decapitation and pithed.

Comment   Insufficient information. Were the toads both males and females? At what temperature and photoperiod were they maintained? The toads were fed with redworm, etc. and not by these worms! (MS Word grammar check suggests, "Redworms, earthworms and flies fed the toads". With what?). How were the toads sacrificed? By both decapitation and pithing? Is it possible to pith a decapitated toad?

Antigens   Rabbit horseradish peroxidase-antiperoxidase (PAP)(Sigma, St Louis, USA, P2026), TRITC-labelled anti-rabbit IgG goat antiserum (Sigma, St Louis, USA, T5268) and blue dyed latex beads (0.8 mm in diameter)(Sigma, St Louis, USA, L1398) were used. FITC-Ficoll was kindly provided by the Department of Cell Biology, Division of Histology, Free University, Amsterdam (thanks to the courtesy of Prof. Taede Sminia and Dr. Nico Van Rooijen).

Comment   Unduly very elaborate acknowledgement for the gift of an antigen! FITC-Ficoll antigen could have been added to the previous sentence with the acknowledgement in parenthesis as "(courtesy by Prof. Taede Sminia and Dr. Nico Van Rooijen of Free University, Amsterdam)."

Experiment   The toads were injected into dorsal lymph sacs with a mixture of 50 ml of PAP, 50 ml of FITC-Ficoll and 50 ml of latex beads (about 4×10^8 particles). Amphibian saline (0.6% NaCl) was added to a total volume 200 µl. Control toads were injected with 200 ml of amphibian saline. Three toads were killed 2, 24 and 48 h after injection three toads served as controls.

Comment   What the authors mean here is that they injected the mixture of antigens into the dorsal sac of the toads! How much was injected? A total volume of 200 ml consisting of 50 ml of each of the three antigens in 50 ml of amphibian saline was injected. How many experimental toads and control toads were used? It appears,

---

> **BOX 4.5   (Continued)**
>
> one experimental and one control frogs were sacrificed at each of the post-injection intervals. Is one animal a valid sample? Or, was it three control and three experimental toads at each interval?
>
> The above article was written by a non-native-English speaker, a Polish. And, it was published in a Polish Journal. It is natural that the non-native-English speakers, as most of us in India do, think in their respective mother tongue and write in English. As a result, even a good research work gains a dubious image. Every student aspiring to have their research published in an International, peer-reviewed Journal, or research thesis accepted without any need for revision or correction, should equip himself or herself with a proficiency in scientific writing. This could be achieved only by reading, reading original research articles published in reputed Journals. It is not simply reading. It is reading with critical analysis.

Results

The Results (= Observations) section of a research report is its nucleus. In other words, all the other sections of a report depend on the Results, and not vice versa. We write Introduction, Materials and Methods and Discussion to suit the Results. Therefore, we will have to take extra care while presenting the results of our investigation. The following are a few general guidelines we have to bear in mind while writing the Results.

1. We should present the results of our investigation with accuracy, brevity and clarity, and should avoid discussing the results.

2. While writing the Results section we generally use past tense and passive voice. However, many editors of journals encourage the use of active voice (e.g. Instead of saying, "blood glucose was found to increase significantly following glucagon injection" we can say, "blood glucose increased significantly..."). This section may be divided into paragraphs with headings.

3. To begin each paragraph of this section, we should use a good topical sentence. It may be a brief overall description of the specific method and the inference of the specific result, without any repetition of what has already been given in the Materials and

Methods (e.g. Phagocytosis assays carried out on leucocytes from the main RBT lymphoid organs showed that pronephros was the main phagocytic organ). In the following sentences, we highlight the results, i.e., evidence to support the inference, in the form of data in tables and/or figures.

4. We do not give raw data in this section, but only processed, summarized, discriminatorily selected data, in the form of text, tables and/or figures. If the data are few, they may be presented descriptively in the text. The text is used to describe and highlight the data given in a table or figure, to compare effects of "treatments" and to emphasize the significant findings.

   We should avoid redundancy of results: data shown in tables or figures should not be repeated in the text.

5. If the results of our investigation are descriptive in nature (e.g. Cytoarchitecture of the thymus gland of the adult frog), we should strive to present them with utmost clarity with the use of appropriately and adequately labelled diagrams and photomicrographs. It is also desirable to present descriptions in a table to avoid monotony. For example, the qualitative data on the main ontogenetic events in the lymphoid organs—kidney, thymus and spleen—during the larval stages (1 to 5) of turbot can be summarized in a table. Such a table would have four columns for Stage, Kidney, Thymus and Spleen, and five rows for the five stages. Such a table not only simplifies the presentation of the description but also helps to compare the progress of development in the different organs during the different stages.

6. Whether graphs, diagrams, maps, photomicrographs or any other illustration (electropherogram, chromatogram, etc.), all of them are referred to as "Figures" (always capitalized to distinguish from figures, meaning numbers).

7. All the Tables and Figures should be separately and serially numbered (Fig.1, Fig. 2... and Table 1, Table 2...) and appropriately cited in the text. They are numbered in the same sequence in which they are first cited in the text. They should be placed immediately after and as close as possible to their first citation in the text.

8. While presenting the results of the statistical analysis of the data, we need to give only the summary value and its significance. For example, we can say, "The difference between the mean weights of adrenal glands of control and treated rats was significant

(t: 4.35; $P < 0.05$)". We need not give the details of the calculations and other basic information (e.g. null hypothesis, critical values, rejection rules, degrees of freedom, etc.).

9. We should not include the same data in both a Table and a Figure. It is best to present the data in a table unless there is visual information that can be gained by using a Figure. For instance, a figure is useful for reporting a trend analysis (e.g. seasonal changes in the antibody response to SRBC in the rockfish) using a line graph, or comparing the several treatment levels using bar diagrams with error bars. It is always better to avoid using Figures (graphs) that show too many variables or trends, all in one, because they can be difficult to understand.

10. Each Table and Figure has a legend (or caption) that explains the information that is being presented. A legend is made to stand alone. A Table's legend appears above it, while the legend for a Figure appears below the Figure. If our Table includes the results of a statistical analysis, we must provide the information necessary for the reader to properly evaluate the analysis (the name of the test, the calculated critical value, levels of significance, sample size, etc.).

11. If the Figures used in a research report are photomicrographs, we should select the appropriate enlargement so that what we want to show could be seen. For example, to show the nature of the cytoplasmic granules, nuclear shape, number of nucleoli, etc. in an eosinophil of the blood of frog, choice of a photomicrograph with a final magnification of around 150X will be inappropriate. Such details would require a photomicrograph enlarged not less than 1200 times. Likewise, use of very high magnification photographs to show the histological organization of the intestine of frog will not serve the purpose.

All photographs in a report should contain information regarding their final magnification, at the end of the legend or within each Figure as magnification scale or bar. It is important to realize that magnification of photograph (including the microscope magnification and enlargement of the final photograph is different from the microscope magnification. The former is greater than the latter, because the negative (e.g. 35-mm format) is printed in different sizes using an enlarger (See Chapter 9 for the calculation of magnification of photographs). We shall discuss the construction of Tables and Figures in the coming chapters. Box 4.6 gives examples of Results with tips on the construction of Tables and Graphs.

BOX 4.6   AN EXAMPLE OF "RESULTS"

The following is an extract from the Results of a paper, "Effects of X-irradiation and thymectomy on the immune response of the marine teleost, *Sebasticus marmoratus*" published in *Developmental and Comparative Immunology*, 10: 519–527, 1986.

Effect of X-irradiation on the humoral and cellular immune response

In non-thymectomized and irradiated fish, injected with SRBC one week after irradiation, antibody production was completely suppressed (Figure 1). They required twice the time for completion of the rejection process when compared to the non-irradiated control, that is, the rejection time of irradiated and non-irradiated fish were 9.8–16 days and 5.2–6.7 days, respectively, following scale allografting three days after irradiation (Table 1). However, at the third month after irradiation, recovery from injury was observed to some extent in both antibody production and allograft rejection (Figure 2, Table 1).

Comment   Note how clear and concise this Results section is written. Note also how Table 1 and Figure 1 and Figure 2 were cited. While "Figure" is abbreviated as Fig., Table is not abbreviated. Even if a Figure is cited in a running sentence, it is written as Fig., as in "Fig. 1 shows the humoral response of rockfish to SRBC in each month".

A few things to remember while constructing a Table are

1. Table number.
2. A title at the top of the table, sufficiently clear, concise, and self explanatory without need to refer to the text.
3. Column and row headings and subheadings, if necessary.
4. Double-spaced lines.
5. Columns of numbers that are carefully aligned by their decimal points.
6. Zeros placed before the decimal points of numbers less than 1 (e.g. 0.1, not .1).
7. Precisely placed horizontal lines (all tables should have at least 3 horizontal lines).

BOX 4.6   (Continued)

8. Only summary-level data, such as means ± standard deviations.

9. Additional information (statistical methods used, statistical significance, etc.) in the footnote.

A few things to remember while constructing a Graph

1. Horizontal axes = independent variable, vertical axes = dependent variable

2. Axes clearly labelled

3. The legend is understandable without reference to the text

4. The Figure caption ("Figure 1. Seasonal changes in ....") is placed below the figure

5. All Figure elements (lines, bars, symbols, etc.) are in black or gray-scale (not in colour).

Discussion

The Discussion section of a research report is the most challenging section to write. In a good discussion we try to bring out the principles, relationships and generalizations shown by the results of our investigation. To write a meaningful discussion we must be thoroughly familiar with the pertinent literature, up-to-date and have good biological insight and mastery of relevant ideas. Let us not forget that this section is not meant for repeating the results but to interpret them and discuss them in the light of what is already known and, importantly, to state the significance of and conclusions arrived at out of our results. Let us consider a few guidelines that will help us to write the Discussion.

1. First, we must emphasize our key findings and their interpretations. This does not mean simple recapitulation of the results. If necessary we can cite Tables and Figures that were given under Results.

2. Though generally we deal with Results in the Discussion section, often we may have to discuss the methodology we had adopted. We might have fabricated a new apparatus to measure the rate of respiration in an air-breathing fish. Or, we might have adopted a

histochemical method, originally designed for mammalian tissues, for fish tissues. If our attempt is the first of its kind, then we may have to discuss the validity and suitability of the method we used. Such a discussion must precede that of the results obtained using that method.

3. We must compare our results and interpretations to other studies citing references from the primary literature, i.e., show how they agree or contrast with the previously published work. If our observations are deviating to a considerable extent, we try to explain it. In doing so, we are not trying to disprove the previous findings but only pointing out to the existence of a different result observed under the stated conditions of our study.

4. We must try to justify how our results have contributed to our knowledge of the specific organism or system. We can discuss the theoretical implications and any possible practical applications of our results.

5. Any unexpected results or problems that we might have encountered during our investigation must find a place in the discussion, because they may be important. We must try to provide possible explanation. We can even point out the absence of information in the literature on such or related observation.

6. When we discuss the results obtained in one organism (e.g. a species of fish), it will not be inappropriate if we use references of published works on other distantly related organisms (e.g. amphibians, reptiles, birds or even mammals) especially if there is a dearth of pertinent literature on that organism.

7. Finally, we must state the conclusions arrived at as clearly as possible. Each of the conclusions should be accompanied with a summary of the evidence we have obtained. We may also suggest future course of research in the specific organism or even in a wide variety of organisms, and mention the possible yield of such works.

8. Use of headings for paragraphs dealing with different aspects (e.g. different animals, glands, or parameters), would be useful to overcome the problem of discontinuity between paragraphs.

An example of Discussion is given, along with our comments, in Box 4.7.

BOX 4.7   AN EXAMPLE OF "DISCUSSION"

The following is an extract of the Discussion of the research article, "Humoral response of roach (*Rutilus rutilus*) to digenean *Rhipidocotyle fennica* infection" published in *Parasitology*, 114:285–291, 1997.

Rhipidocotyle fennica-specific antibodies were found in wild roach in lakes where the fish were infected by the parasite. Previously uninfected roach produced antibodies when immunized with homogenized cercariae or infected with cercariae emerging from infected clams also in aquaria. These findings are parallel to previous studies on humoral antibody response against metazoan parasites. Antibodies have been reported in rainbow trout (*Salmo gairdneri*) against a digenean trematode *Diplostomum spathaceum* (Bortz *et al.*, 1984; Whyte *et al.*, 1987), in plaice (*Pleuronectes platessa* L) against *Rhiphidocotyle johnstonei* and in plaice and grey mullet (*Chelon labrosus*) against *Cryptocotyle lingua* (Correl, 1977; Wood and Mathew, 1987).

Roach infected with *R. fennica* were found in 5 or 6 lakes studied and also anti-*R. fennica* antibody–positive fish were found in these lakes. The emergence of *R. fennica* cercariae begins in mid-July in Lake Saravesi (Taskinen *et al.*, 1994). Thus, the *R. fennica*-specific antibodies in fish sera already present in early June must originate from infections in the course of the previous year. This is in accordance with other studies, where antibodies have been demonstrated to persist for a long period. For example Thuvander *et al.* (1987) found antibodies as long as 46 weeks after *Vibrio anguillarum* vaccination. The increased antibody levels in September are most probably due to fresh infections during the late summer.

Low levels of antibody against *R. fennica* were detected in the sera of 4 wild roach (total n = 41) from Lake Peurunka, where *A. piscinalis* does not occur. This was an unexpected finding that probably results from cross-reactions with serum antibodies against other parasites of microbes. The antibody concentrations in roach from Lake Peurunka were an order of magnitude lower than those of fish in the other lakes and they were the background level of the assay.

Anti-*R. fennica* antibodies were found in both experimental groups A and B. Strong antibody response was noted in groups A and B. Strong antibody response was noted in group A which was infected with living cercariae. Antibody levels found in this group were similar to those measured in wild roach from lakes containing clams infected with *R. fennica*. In contrast to

BOX 4.7   (Continued)

group A, the response in group B was very weak. The percentage of roach responding to immunization with homogenized parasite was only 27% compared to 63% in group A. In addition, serum antibodies were present only in low concentrations.

The weak response in immunization cannot be explained by a small amount of antigen. Williams and Hoole (1992) induced antibody response in roach by using 100 µg of protein from homogenized worm *Lingula intestinalis* and in our experiment fish were given 150 µg of protein from cercariae. The timing of sampling (30 days post-immunization) cannot be the reason for low level response in immunization because our earlier studies showed that high levels of specific antibodies can be determined in serum of roach 28 days after i.p. immunization (Aaltonen *et al.*, 1994). Further, a weak response is not likely to be due to the poor immunogenicity of the antigen because there was a strong antibody response after the infection with living cercariae (group A). The route of immunization, i.p. injection compared with infection by living cercariae, is the most probable explanation for the difference in the antibody development.

When infecting a host, cercariae must penetrate the skin. During penetration the cercariae contact with immune cells in the blood and in the skin resulting in the synthesis of antibodies. Antibodies against pathogenic organisms are found in serum and also in the mucus of fish (Ingram, 1980; Peleteiro and Richards, 1985). It is not possible to say whether the protection against *R. fennica* is due only to antibodies. Other defence mechanisms, such as cell-mediated immunity or non-specific humoral factors, may have been activated. For example activation of trout macrophages increase larvicidal activity for diplostomules *in vitro* if the larvae are opsonized with immune serum (Whyte, Chappell and Secombes, 1989). Also, other serum components like complement or lysozyme have been found to have effects against the protection of *D. spathaceum* (Whyte, Chappell and Secombes, 1990). The role of these other factors should not be ignored as potential resistance mechanisms against *R. fennica* and they need to be studied.

In conclusion, we found that antibodies against *R. fennica* do develop in roach. Specific antibodies were detected in blood of wild, naturally infected fish and this finding was verified by performing experimental infections in aquaria. The specificity of the antibody depends on the route of immunization, i.e., via intraperitoneal injection or natural infection by living cercariae. Our results suggest that previous infection may confer some protection against the parasite.

BOX 4.7   (Continued)

Comments

- *1st paragraph*   Important and significant findings of the investigation are presented. The similarity of findings with those of other works indicated.

- *2nd paragraph*   Explains/accounts for a significant result citing information from the literature.

- *3rd paragraph*   Considers an unexpected result and offers a possible explanation.

- *4th and 5th paragraphs*   Explains the probable cause for another finding in the experiment. Considers and rules out several causes, with supporting evidence from literature. Settles for one highly probable reason, viz., route of immunization, and provides evidences.

- *5th paragraph*   Stresses the importance of other possible factors in the defence mechanism of fish and the need for further studies.

- *6th paragraph*   Summarizes highlighting the conclusion arrived from the findings of the investigation.

Acknowledgements

This section is meant to express the researcher's gratitude to (1) those who provided significant technical help whether in their own laboratory or in other laboratories; (2) the sources, individuals, laboratories, companies, etc. which helped the researcher with special equipment, chemicals, cultures, specimens, etc.; and (3) the agencies (UGC, CSIR, etc.) that sponsored the research programme, and provided scholarships or fellowships to the authors.

A few examples of the expression of acknowledgement in a research report are given in the Box 4.8.

BOX 4.8   A FEW "EXPRESSIONS OF ACKNOWLEDGEMENT"

- We/I greatly appreciate receiving help from xxx for ...

- We/I wish to thank xxx for.....

- We/I gratefully acknowledge the support (financial, technical, etc.) of xxx.

- We would like to thank xxx for ....

- This research was financed by ......

- The authors thank xxx for ...

- The author is grateful to xxx for ...

- This study/work/research/investigation was supported by grants from...

- Particular gratitude is expressed to xxx for...

- The authors would like to thank xxx for ....

- xxx made useful contributions ....

- xxx provided technical help in ...

- xxx made available .... (laboratory, etc. facilities).

- We are indebted to xxx for ....

- xxx is acknowledged for ...

- I/We express our cordial thanks to xxx for ...

- We express our appreciation to xxx for ...

- This study was made possible, thanks to the courtesy and guidance provided by xxx.

- The authors thank xxx for his encouragement and advice/helpful suggestions/valuable comments.

- The authors express their appreciation to xxx for his/her technical assistance.

- Thanks are also extended to xxx for ....

- We are also grateful to xxx for ...

- Thanks are due to xxx who kindly provided us with .... (materials).

- My/our thanks are due to xxx for ....

BOX 4.8   (Continued)

- We express our special thanks to xxx for ....

- This work was carried out under YYY award to xxx from ... (funding agency).

- Especial thanks are due to xxx for....

- I am most grateful to xxx for ...

- Expert technical assistance was provided by xxx.

- The author was in receipt of a YYY scholarship/post-doctoral fellowship.

- The work was supported financially by ....

An academic thesis may have more elaborate Acknowledgements, and it may include, besides those listed above, Head of the Institutions, Head of the Department, and Research Guide and Supervisor. Acknowledging parents, friends and other well-wishers, a common feature in theses of Indian Universities, is not desirable. The reason is, a thesis in our Universities is more like an examination answer script; it is submitted for evaluation by a Board of Examiners; and it may be accepted, recommended for drastic revision and resubmission or rejected outright. We do not acknowledge our parents, relatives, and friends or, even, God in an answer script. All these "thanksgiving" can wait till the thesis has been accepted and acclaimed as one deserving to be published as a book.

General Introduction and General Discussion

A research report, especially an academic thesis, usually comprises several chapters dealing with different aspects of a particular topic. In such cases it would be very useful to have a General Introduction at the beginning, to introduce the interrelationship of the different aspects. Likewise, a General Discussion at the end would serve to establish the interrelationship introduced in the General Introduction section. These sections, if included, are treated as the first and the penultimate chapters, respectively. The last chapter is, of course, Summary and Conclusions.

Summary and Conclusions

This section of the report is perhaps the one that is read by many. Therefore, it must be prepared with utmost care. It gives the gist of the entire report including the materials used, methods adopted, observations made, the inferences drawn and the conclusions arrived at. However, this section has no subdivisions or subheadings like a regular chapter does. It is simply organized into paragraphs, which may be numbered (Arabic), and each of which highlights specific observations and conclusions. If necessary, the paragraphs may be grouped into numbered (Roman) sections, corresponding to the chapters.

Appendixes

This section presents supplementary material, which is helpful but not essential to be included in the text. Materials, which if included in the text might hinder the free flow of presentation, are relegated to an appendix. Interested readers can always refer to such matters. Raw data, detailed protocols of procedures, questionnaires, statistical tables, glossary of terms, and extensive quotations are a few such matters.

If there is only one appendix in a report, it is referred to as APPENDIX. If there are more than one appendix, each one is identified with an English alphabet and a descriptive title (e.g. APPENDIX A IMMUNOHISTO-CHEMICAL PROTOCOLS).

Each appendix should be introduced in the text, at the earliest point where the matter presented in the appendix is relevant to the discussion.

Example

The details of the immunohistochemical procedures adopted are presented in Appendix A.

The placement of the appendix or appendixes (appendices) should be logically before the Reference section, because, an appendix may contain one or more references cited in it. All appendixes should be listed together with page numbers in the Table of Contents.

References

The References or Literature Cited or Cited References section of a research report is a list of all references cited in the report (in Introduction,

Review of the Literature, Discussion, etc.). Let us make the difference between "literature citation" (or reference citation) and "Literature Cited" (or Reference Cited). Citation is acknowledging in the text the source of information we have used while writing the different sections of our report. Literature Cited is the section of the report that gives the details of these sources. References appropriate to cite when writing a scientific research report include primary literature and several other categories of technical literature. What is a primary literature? Research articles, which report the results of original research, published in peer-reviewed scientific journals constitute the *primary literature*. A peer-reviewed journal is one in which the manuscript submitted for publication is critically evaluated by peers, i.e., experienced scientists. If, in the opinion of the peers, the manuscript is not worthy of publication, it may be rejected. As many as 50 percent of the manuscripts are rejected by such peer-reviewed, "International" journals. Even those manuscripts that have been accepted for publication will have been revised and resubmitted for final approval and printing.

A research paper accepted for publication in a peer-reviewed journal may be cited and listed in the References, giving the name of the journal and indicating that it is *"In press"*.

If the report we are preparing is meant for publication in a peer-reviewed journal, citation of secondary literature and other sources of information may or may not be accepted. Theses, project reports, proceedings of a conference, textbooks, unpublished data, etc. constitute the *secondary literature*. However, research reports published by government research agencies, technical books, and edited books consisting of series of articles, each written by one or more experts may be cited as reference, especially in research reports like the academic theses.

Generally, sources of information on the Internet are not cited as reference in a research report, because they are not peer-reviewed and are highly transient. However, online peer reviewed science journals and recognized scientific databases are used for citation as sources of information.

Newspapers and popular magazines (e.g. Natural History, National Geographic, Discover, Frontline, Outlook, etc.) are definitely inappropriate sources of information for citing in a research report.

The format of the Reference section considerably varies somewhat among journals and organizations, and according to the style of citing references in the text. Therefore, we have to be sure to determine what format is appropriate for our manuscript and follow it carefully and consistently. We had already discussed in detail the various systems of citing literature and organizing the References section of a research report in chapter 3.

An ideal way to master the technique of writing scientific reports is by critically reading papers published in standard international journals. The more we read, the more we become familiar with the various styles of writing, especially the Discussion part of a report. As already indicated, writing a thesis is not much different from writing a paper for publication. A thesis can be more elaborate. The candidate who submits the thesis has to demonstrate his/her broad and deep knowledge in the field of investigation. Therefore, the Review of Literature section of a thesis can be much more elaborate than in a paper. Likewise, the candidate has to be meticulous in the methods he has adopted for the investigation. And therefore, details of the techniques, any modification introduced, etc. may have to presented in a thesis.

PLAGIARISM

The native language of most of the research workers in India is one other than English. Many of us have very limited control over English which is a dominant language in research publications. However, we depend to a large extent on a literature largely written in English. In order to incorporate the information given in the literature into our research report, we have to first understand what is written in the original source and then present it, again in English, in our own version, a process generally referred to as paraphrasing. Proper praphrasing requires thorough mastery over that language, which, we should admit, is lacking in most of us. If we reproduce verbatim what is reported in the literature, not because we had not understood it but for the fear of wrongly paraphrasing it, we would be blamed of committing plagiarism.

Let us understand what is plagiarism, before considering what we can do to avoid it while writing a research report.

Plagiarism is (i) reproducing materails from other workers' reports, verbatim, as our own or (ii) paraphrasing materials from another report without giving due credit to the author of the original report, i.e., without citing the source or without obtaining permission.

We must also have a clear idea of what is meant by "materials" of other reports. All facts, figures, statistics that are not common knowledge, particular theories or thoughts that have been put forward by another person and any specific data that is not general knowledge are all "materials" that should be properly acknowledged, if used in our report. The "material" may also include original ideas and creative expressions in the form of sentences, phrases, innovative terminology, formatting, charts, graphs, diagrams, websites, computer programs, etc.

In other words, the use of anything from any source (books, magazines, newspapers, websites, textbooks, etc.) in our report, if not given proper credit, runs the risk of being branded as plagiarism.

Since we had already discussed (chapter 3) in detail how to cite Reference (i.e., the primary way of giving credit to the source), we shall now consider a few precautions to avoid plagiarism.

1. The best way is to learn to use our own words while expressing the materials from other sources for our purpose, that is, learning to paraphrase.

Paraphrasing is expressing a written or spoken matter of another person in our own words. This will be possible only if we understand clearly what has been written or spoken. If not, the paraphrase may convey a different, wrong meaning.

Superficial or partial paraphrasing, i.e., change of a few words with synonyms or voice of the sentence may not be sufficient to avoid plagiarism. Exact wordings or the sentence structure as found in the source should not find a place in a paraphrase.

2. If we feel that certain expressions should be reproduced in its original form, then it should be done in proper way with due credit to the original source.

If the expression is a word or a phrase or a short sentence, it can be incorporated into our text with in quotation marks. If it is a passage of several sentences, it can be given either as an italicized paragraph or as a paragraph within quotation marks.

Materials of common knowledge need not be acknowledged. But we must be careful in deciding what is "common knowledge". Any undocumented standard information which we come across in several sources (e.g. textbooks), which our target audience already know and which can be easily retrieved from a database are some of the "common knowledge" categories. But, we should not forget that even the "common knowledge" should be in our own words. A few suggestions for expressing a "material" in our own words are:

i)   substituting all words other than generic ones, with appropriate synonyms

ii)   changing the voice, active to passive and vice versa

iii) altering the structure of a sentence

iv) reducing the clauses to phrases

v)   modifying the parts of speech

CHAPTER FIVE

RESEARCH REPORT—TABLES

NEED FOR TABLE

A Table, if properly constructed and placed, would be very useful for presenting a variety of information in a research report. It can be used to present a review of literature, materials, methods, or, needless to say, even the results of a research report.

Presentation of results reported in a literature about a specific aspect in different species in the form of a Table would eliminate the monotony of repeating the same matter several times. Further, highlighting the similarities and differences of the results obtained by various workers in different species is made much easier and effective. Examples of use of Tables for presenting reviews, materials, methods and results have been given in Boxes 5.1, 5.2, 5.3 and 5.4, respectively. We shall discuss in detail the construction and use of Tables for the presentation of numerical data, especially in a research report.

A Table should be used only when there is a need for it. Any information that can be presented in textual form in a few sentences need not be converted into a Table. For example, to report that a sample of 15 *Catla catla* was collected at the landing station of the lake, every month for one year, we need no Table. Only when we have sufficient data that may require about half-a-page in textual format, we should resort to use of a Table. The same data should never be presented in both text and Table. However, data presented in a Table can be highlighted in the text.

USE OF "TABULAR" FORM

Sometimes, it is useful to give information in a tabular form (i.e., not in a typical Table format), as in the following example.

The number of fish collected, the sex and the body-weight range of the two species of fish are as follows:

	Male	*Female*
Catla catla	15–25 g	30–65 g
	($n = 10$)	($n = 12$)
Cirrhina rheba	15–40 g	60–120 g
	($n = 16$)	($n = 15$)

However, if the information requires more than 5 to 6 lines in the "tabular" form, then it is desirable to have a typical Table to present it.

INTRODUCTION AND PLACEMENT OF A TABLE

Every Table in a research report should first be introduced in the text. The Table should be placed as immediately as possible next to its first introduction (mention) in the text.

Example   The relative proportions of heavily carbon-laden macrophages in the different compartments of the spleen after different time periods are summarized in Table 1.

Table 1   Time kinetics of the distribution of carbon-laden macrophages in the spleen[1].

Spleenic regions	5 min	1 h	3 h	6 h	12 h	24 h	36 h
Marginal sinus	++	++	+	+	+	+	+
Marginal zone	±	+	+	++	+	±	±
Marginal-zone bridging channel	-	+	±	+	+	±	±
Lymphocyte corona	-	±[2]	+	+	+	±	±
Germinal center	-	-	-	±	+	±	±
Outer PALS	-	±	±	+	+	±	±
Inner PALS	-	-	±	+	±	±	
Red pulp cord	+	+	+	++	+[3]	+[3]	+
Red pulp sinus	±	+	±	+	±	++	±

[1] –, no macrophages found; ±, macrophages occasionally found; +, few macrophages constantly found; ++, many macrophages constantly found. [2]Macrophages found only close to the marginal sinus. [3]Macrophages showing localized accumulation.

In the above example, the Table is introduced, i.e., mentioned in the text, as "…summarized in Table 1." If the report is an article in a journal, the placement of the Table is decided by the editor depending on the layout. As we are familiar, most of the journals are published in two columns per page. If the width of the Table is such that it requires the entire width of a page, its placement may be either at the top or at the bottom of a page. If it is in the top, it may so happen that the introduction of the Table in text may appear below the Table. If the width of the Table requires only one column, it is mostly placed immediately below or very close to its introduction in the text in the same page or, at the most, next page.

In a thesis, which is always typed in single column, the placement of a Table depends on its size. A Table that occupies less than half a page in length may be placed in the same page in which it is introduced, provided enough space is available, or at the beginning of the next page.

When it is placed in the page of its introduction, it is better to complete the paragraph of its mention before inserting the Table. That is, a paragraph should not be broken for the sake of placing a Table.

An alternate suggestion, a better one, is to place a Table, whatever be its length or width, in a separate page, immediately following the page of its mention in the text. If necessary, two small Tables, consecutively numbered, may be accommodated on a single page.

More than one Table may be introduced in a page in numerical order, and all these Tables must be inserted in the following consecutive pages in the same numerical order.

FORMAT OF A TABLE

Any introductory textbook of Statistics would give basic guidelines for constructing a Table. Here, we shall confine to a few general suggestions that should be borne in mind while drawing a Table for a research report.

A typical Table has:

1. Table number
2. Title or Caption
3. Headings for the columns, also referred to as boxheads
4. Labels for the rows, i.e., the stub headings
5. The field or the cells, i.e., the intersections of columns and rows, which accommodate the actual data.

Numbering of Tables

All the Tables in a report should be serially numbered for easy identification. The numerals are either arabic or roman. Thus we can have Table 1, Table 2 and so on or Table I, Table II, and so on. Use of arabic numerals is more common. The word Table when it refers to a specific Table in the report should always be spelt as Table with the first letter capitalized.

There is another way of numbering a Table. The entire word can be capitalized as TABLE followed by the number, as TABLE 1, TABLE 2, and so on. However, while citing these Tables in the text they are typed only as Table 1, Table 2 and so on.

Tables can also be numbered for every chapter separately. These Table numbers include chapter number also. For example the Tables in Chapter 10 of a report can be numbered as Table 10.1, Table 10.2, ... Table 10.10 and so on. This method is useful in theses, manuals, textbooks and other such report. It cannot be used for research articles in Journals.

Whatever style is followed for numbering the Tables, it should be consistent throughout the entire report. A model blank Table is shown in Figure 5.1.

Table No.   Title of the table

(Head note)

Stub heading	Caption			Total
	Column heading 1	Column heading 2	Column heading 3	
Stub entry 1	Cell	Cell	Cell	Row total
Stub entry 2	Cell	Cell	Cell	Row total
Stub entry 3	Cell	Cell	Cell	Row total
Total	Column total	Column total	Column total	Grand total

Footnote:

Figure 5.1   A model blank table to show the components of a typical table.

Title of the Table

Every Table in a research report should have a title that is always placed above the Table. A Table title should be brief, clear, complete, and unique.

The title of a Table

- should not be a complete sentence

- should not be divided into two or more clauses or sentences
- should not contain unnecessary words
- should not give any background information
- should not describe the results shown in the Table
- should not use, as far as possible, abbreviations
- should not begin with phrases such as "Table showing ..."

Stub

The stub of a Table is the column at the extreme left of a Table (or the one next to the column for "serial number". It contains items of subject matter about which data are given in columns to its right. The items in the stub should be logically arranged so that the purpose for which the Table is given is served. For example, if we want to show the areas (km^2) of the different states and union territories of India, we can list the different states in alphabetic order starting from Andhra Pradesh and ending with Tamil Nadu, for example. On the other hand, if our intention were to show the wide variations in the areas of the different states, then it would be useful if the states were arranged according to their areas, either in ascending or descending order.

Box Headings

Each column, including the stub column, has a heading that is referred to as box heading or column heading. The box heading of the stub indicates the nature of the items for which data are given in the neighbouring columns. It would be useful if the stub box heading were derived from title of the Table. For example, if the title of the Table is: "Table 2. Area and population of the different states and union territories of India", the stub heading may read as "States and Union Territories of India". The headings of the other columns on the right of the stub may be Area in km^2 and Population in $\times 10^6$.

Numbering of the columns is useful when a Table is so large that it requires several pages to accommodate.

Units of Measurements

It is very important that the units in which the data given in a Table were measured are indicated at appropriate places. Usually, the units of measurements are shown in abbreviated form in the appropriate column heading. If possible, these units may also be given in the title in expanded form.

Footnotes

Essential information that could not be accommodated in the body of the Table are given in footnotes that appear beneath the Table. Footnotes are generally used for three purposes:

1. *Source notes*   It is used to indicate the source of the data shown in the Table: Bibliographic information for the source from which the data were obtained (of course, with permission) is given with citation of author(s) and year. If the Table is reproduced entirely, the footnote should read as: SOURCE(S): Government of India (2004). Sometimes, the source is acknowledged as: Reproduced with permission from: Smith (2001). If the data is modified, i.e., the style is changed or the data is rearranged, extended or deleted, the source is acknowledged as: ADAPTED FROM: Berkley (1997).

2. *General notes*   These footnotes pertain to the entire Table. For example, each value given in a Table is Mean $\pm$ Standard Deviation of 10 measurements. This information may be given as: NOTE(S): Each value, Mean $\pm$ SD of 10 measurements. This general note would eliminate mention of this information anywhere else, either in the title or in the body of the Table.

3. *Specific notes*   Necessary comments on specific items of Tables are incorporated in a Table by way of specific notes. These notes are indicated by some symbol as superscript to the specific item. All English letters, in lower case (e.g. a, b), Arabic numerals (e.g. 1, 2) and non-sense symbols (e.g. *, §,) can be used as superscripts to the items in the Table that require explanation. The use of asterisk mark in single or in groups of two or three (*, **, ***) to indicate statistical inference is common. If this symbol is used for this purpose in a Table in a report, it is better not to use it for any other note in any Table of the report.

If all the three types of footnotes are required for a Table, they should always appear in the same sequence, i.e., source, general and specific.

BOX 5.1   PRESENTATION OF "REVIEW" IN A TABLE

In teleosts the spleen is usually smaller than in holocephalans and elasmobranchs (Table 1).

Table 1   The weight of the spleen of fishes in percent of the body weight (ranges)

Species	Spleen weight	References
Holocephali Rabbit fish (*Chimaera monstrosa*)	0.42–1.82	Fänge and Sundell (1969)
Elasmobranchii Nurse shark (*Ginglymostoma cirratum*)	0.21–0.36	Fänge and Mattisson (1981)
Dogfish (*Scylorhinus canicula*)	0.22–0.63	Fange and Pulsford (1982)
Various species	0.10–0.80	Fänge (1982)
Shark (*Mustelus manazo*)	2.65–4.00	Murata (1980)
Teleostei Bass (*Lateolabrax japonicus*)	0.05–0.14	Murata (1980)
Mullet (*Mugil cephalus*)	0.12–0.34	Murata (1983)
Salmon (*Salmo salar*)	0.04–0.86	Miescher–Ruesch (1979)
Various species of marine teleosts	0.02–0.18	Fänge and Nilsson (1985)

BOX 5.2   PRESENTATION OF "MATERIALS" IN A TABLE

Table 1   Fish studied from 6 lakes in Central Finland at 2 sampling times

Lake	Type of Lake	Occurrence of *Anodonta piscinalis**	Fish (June/September)		
			n	Mean weight (g)	Mean length (mm)
Saravesi	Eutrophic	+++	21/18	27/31	148/147
Kuusvesi	Oligotrophic	+	18/20	27/34	140/149
Vatia	Eutrophic, Polluted	+	19/20	36/34	162/219
Leppavesi	Eutrophic	++	21/21	23/25	143/141
Ahveninen	Eutrophic	+	14/16	29/36	149/157
Perunka	Oligotrophic	–	21/20	34/61	156/195

* Occurrence of *A. piscinalis*: many (+++), moderate (++), few (+), none (-). The occurrence of *A. piscinalis* was monitored during several years using bottom dredge, by diving or by enumeration of glochidia on fish.

BOX 5.3   PRESENTATION OF "METHODS" IN A TABLE

Four groups of fish (10 fish/group) were immunized as shown in Table 1.

Table 1   Immunization schedule

Group	Antigen	Amount of inoculum (ml)	Route of injection
1. Experimental, exposed	SRBC	1.0	ip
2. Control, unexposed	SRBC	1.0	ip
3. Experimental, exposed	AH + FCA	0.5	im
4. Control, unexposed	AH + FCA	0.5	im

SRBC = Sheep red blood cells, AH = *Aeromonas hydrophila,* FCA = Freund's complete adjuvant, ip = intraperitoneal, im = intramuscular

BOX 5.4   PRESENTATION OF "RESULTS" IN A TABLE

These results are summarized in Table 1.

Table 1   Relationship between the stage, length and lymphoid development in the trout embryo/fry

Stage	Age in days	Length in cm	State of lymphoid development
26	8 days pre-hatch	1.0	
28	3 days pre-hatch	1.2	First appearance of thymic anlagen and of haemopoietic anlagen in kidney.
30	Day of hatching	1.5	Active lymphopoiesis in the thymus.
32	3 days post-hatch	1.7	Small lymphocytes in thymus. First appearance of spleen analgen.
33	5 days post-hatch	1.8	Small lymphocytes appear in extra thymic tissue and in the kidney.
34	6 days post-hatch	1.9	Thymus fully differentiated. Kidney fully lymphocytic. First appearance of a very few small lymphocytes in spleen.
36	13 days post-hatch	2.2	First appearance of occasional leucocytes in the gut epithelium.
	25 days post-hatch	2.8	Thymus and kidney are the major lymphoid organs present. Spleen mainly elytroid with no division into red and white pulp.

Chapter Six

RESEARCH REPORT—FIGURES

INTRODUCTION

A research report may require illustrations such as photographs, charts, diagrams, drawings, maps, graphs, etc. as supportive evidence for the methods used, observations made and inferences drawn. The term Figure (abbreviated as Fig., Figs.; always begins with capital first letter in order to be differentiated from figure meaning number) is commonly used to refer to any illustration. Sometimes, the term Plate is used for photographs arranged in a separate sheet. It is very convenient to use the term Figure to refer to any type of illustration in a report instead of identifying them as Plate, Graph, Diagram, etc.

WHEN TO USE FIGURES

There is no general rule to decide when to use an illustration. It is true that illustrations are more explanatory than words. But we cannot just insert Figures into a report without rationale. There is an undesirable tendency among the research students to have too many unnecessary Figures simply for the sake of increasing the bulk of the report. Redundancy in the presentation of data may be tolerated, to a certain extent, in academic research reports, but certainly not in Journal publications. The main reason is, again, the cost of printing and the need to conserve space.

If the data had already been presented in a table, it need not be represented in the form of a graph or diagram, unless there is a need to demonstrate certain trend, correlation, significant variations, etc., which are complicated to be perceived from the tabulated data.

Illustrations should not be created to represent trivial and insignificant data. For example, we may collect data on biochemical parameters such as the serum levels of calcium, magnesium, potassium, protein, glucose, and cholesterol from two groups of subjects, say normal adult males and females of a particular age group. If there is no significant difference in these parameters between the two groups there is absolutely no need for a diagram (e.g. bar diagram) to show the absence of difference in these parameters. A statement in one or two sentences is sufficient. We can very well imagine how meaningless it would be if we construct a full-page bar diagram for each of the above parameters.

INTRODUCTION AND PLACEMENT OF FIGURES

Every Figure in a research report should first be introduced in the text. The Figure should be placed as close as possible to its first mention in the text. If the report is an article in a Journal, the placement of the Figure depends on the size of the Figure, whether it can be reduced to one column width, or requires two column width and half-a-page height or would occupy a whole page.

In a thesis, Figures are generally organized in separate sheets. If the Figures are photographs, each sheet may be referred to as a Plate. Since graphs, maps, diagrams, etc. are also referred to as Figures, it would be better to avoid the use of the term Plate in the report, i.e., for citing the Figures in the text or in the caption. However, we will use the term Plate to refer to a sheet that contains only Figures and no text. With the availability of high-quality (photo-quality) photocopying facilities, it is now possible to have even photographs copied on sheets as thin as normal bond paper. Therefore, it is possible to insert the Figures on pages immediately next to the page on which they are introduced in the text. Care must be taken to place only those Figures that have been referred to in the previous page. A Plate may contain even eight Figures. If only some of the Figures (e.g. seven) had been introduced in the previous page, it is wrong to insert the Plate in the next page because we would be breaching the rule of not to place Figure before being introduced in the text. In the above example, we are placing the eighth Figure without its prior introduction in the text. We may have to shift this Figure to a Plate to be placed elsewhere.

To overcome this problem, all the Figures can be placed at the end of the report. We need not worry to split the Plates to place the Figures at appropriate places after their mention in the text.

Insertion of Plates of Figures at the end the Results section or at the end of each chapter is also possible. There are also research journals that publish articles with Plates of Figures (mostly photographs) at the end of each article.

NUMBERING OF FIGURES

As in the case of Tables, every Figure in a research report should bear a number. Figures, whether they are graphs, diagrams, maps or photographs, are serially numbered in the same order in which they are cited in the text. The most common and convenient practice is to use Arabic numerals for numbering.

The number (without any prefix of Figure or Fig.) should be placed within and in one corner of the illustration. The placement of the Figure number should be consistent throughout a report. Usually, the right bottom corner is used for numbering.

The number should be distinct from any other numerals in the illustration and stand out clearly against the background. Therefore, black against lighter background and white against darker background are useful.

If each Figure is placed in isolation with its caption immediately beneath it, there is no need for numbering the illustration.

The word Figure is abbreviated as Fig., both in the caption of the Figure and while citing the Figure in the text.

CAPTION OF FIGURE

While the title of a Table is placed above the Table, the caption of a Figure is generally placed below the Figure. However, in research reports like an academic thesis where separate pages are used for Figures, the caption may be shifted to the left-hand page facing the Figures.

A caption should be brief, clear, complete, and unique. Unnecessary words like "Figure showing", "Graph showing", etc. must be avoided. However, introductory words like "Electron micrograph of ...", "Freeze-fracture replica of ...", "Low-power scanning electron micrograph showing ...", "Photomicrograph of thin section of ...", "Diagrammatic ventral view of ...", "Western blot analysis of ...", "Scatter diagram showing the relationship ...". "Proportion of ..." and so on could be used.

The caption of a Figure may also include relevant explanatory information about abbreviations used for labelling, distinguishing legend key markers in a graph, stains and fixatives used and so on. Let us consider an example.

Example

Figure 1. The proportion of roach infected by *Rhiphidocotyle fennica* in the lake studied: Lake Saravesi (Sa), Lake Kuusvesi (Ku), Lake Leppävesi (Le), Lake Ahveninen (Ah), and Lake Peurunka (Pe). Classification of infection: Strong infection was > 30 cercariae/tail, weak infection was ≤ 30 cercariae/tail and not infected: no cercariae in the tail. Numbers of fish and descriptions of lakes are given in the Table 1.

PREPARATION OF STATISTICAL DIAGRAMS

We are not going to describe how a graph is constructed. In yesteryears graphs or other statistical diagrams were drawn in graph sheet, traced onto an appropriate white sheet such as a Bristol board, and then photographed. Now, construction of a statistical diagram has been made very simple by computer technology. Any standard Word processing software package (e.g. MS Office) would include tools to prepare statistical graphs and diagrams. The inkjet and laser printers aid in the production of high-quality computer-generated illustrations.

We shall, however, list some important considerations, which we should remember while preparing illustrations for research reports.

1. The font type and the font size of the labels and the size of the symbols should be carefully chosen to be clearly readable and visible after final reduction to a size required for printing or incorporation into a thesis.

2. Each illustration should be as simple as possible. Addition of too much of information in one Figure (e.g. too many curves in a graph) would only create confusion.

3. The choice of scale for graphical illustrations, in both the axes, should be appropriate. Too much of empty space in any part of a graph should be avoided.

4. The key to the symbols and shades should be given, as far as possible within the illustration itself. If this is not possible, the key may be included in the caption of the Figure.

5. Only standard symbols such as open circle (O), triangle (△), and square (□), and closed circle (●), triangle (▲) and square (■) should be used. For example, if a graph has only one curve, the symbol should be open circle, and if there are two or more curves the other symbols should be used in the same sequence.

6. The connecting line should not go through the symbol.

7. To distinguish between the curves in a graph, we can use either different symbols, as shown above, or different types of connecting lines (solid, dashed, etc.). But we should not use different types of symbols and different types of connecting lines in a graph.

Statistical diagrams generated by computers use different shades of colours to distinguish the different components such as the bars, connecting lines of a curve, sectors of a pie, and so on. The cost of making copies of these illustrations by colour photocopying is not very high. Therefore, colour illustrations are commonly being used in theses and similar reports. However, journals do not accept colour illustrations in articles for publication because of the high production cost.

PREPARATION OF PHOTOGRAPHS AND MICROPHOTOGRAPHS

Photographs and microphotographs are essential illustrations in many research reports. For example, any report of light and electron microscopic investigations depend largely on microphotographs (i.e., photographs taken through a light or electron microscope) to provide evidence for the observations made. Therefore every care must be taken in the preparation of these illustrations.

While an academic thesis generally does not impose any restriction on the number of photographs, Journals accept only essential photographs. Many Journals even charge the author of the article, the cost of printing photographs. The cost of printing colour photographs is much higher than that of black-and-white pictures. However, colour photography and microphotography is more common in theses and such reports, because of low cost of prints and colour photocopying. In fact, preparation of black-and-white prints is becoming more costly because of the non-availability of negative films and studios willing to take up the processing work.

The following are a few points to remember while preparing photographs for a research report.

1. *Print paper*   The prints should be only on high quality glossy paper. It should not be on mat paper.

2. *Layout*   The size of the prints, after cropping (trimming off unnecessary parts), should be such that it fits the width and length of the column or page of the Journal. If the photographs are prepared as a Plate used in both Journals and thesis-like documents, the dimensions of the cropped prints should be matching appropriately with those of the neighbouring photographs. Whatever be the number of photographs in a Plate and whatever be the size of the individual photographs, the overall layout should fit the outer prescribed margin (Figure 6.1).

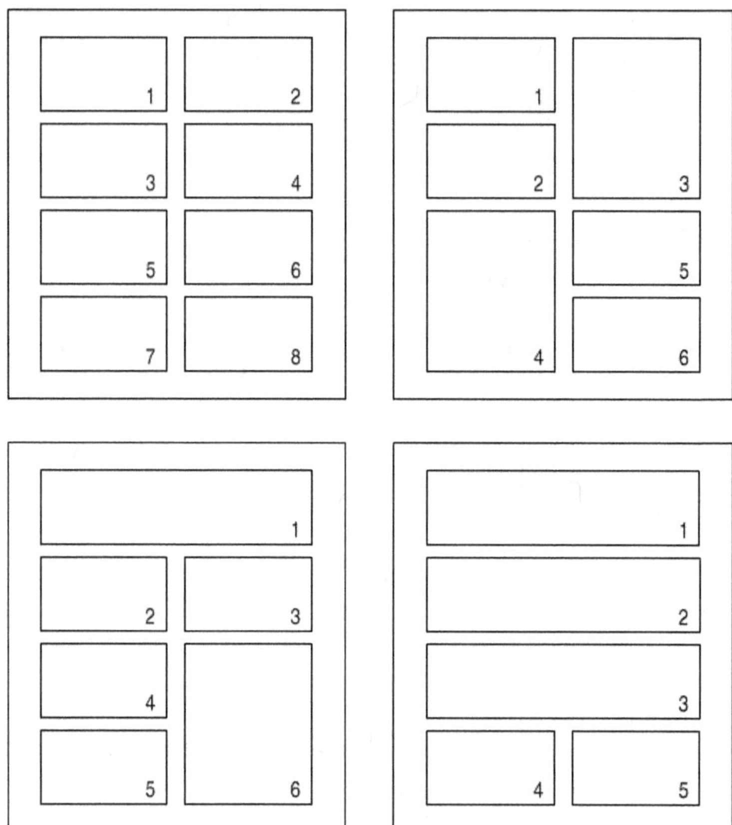

Figure 6.1   A few examples of layout for photographs.

3. *Framing* Framing of the photograph in order to highlight the feature that we want to show to the reader is very important. The photograph should have been taken at the appropriate magnification of the microscope and/or the final print should have been made at the desired enlargement. It would be useless if a microphotograph taken under lower power magnification (e.g. $10\times$ Objective and $10\times$ Eyepiece) of a microscope to show the morphological and cytological details of a type of WBC because, the print of such a micrograph would show only hundreds of RBCs with very few WBCs, as dots. A higher magnification ($100\times$ and $10\times$), focusing on one or two specific types of WBCs would yield a better result. Enlargement of specific portions of the negative, containing those specific types of cells might also help to show the features that we want to highlight.

4. *Labelling*   If a photograph or a microphotograph is meant to show more than one feature (e.g. several types of WBCs in one frame) appropriate labelling is necessary. We can use alphabets and /or symbols like different types of arrows and arrowheads. These labels should be of appropriate size (neither too large nor too small) and should stand out clearly against the background. Same alphabet or group of alphabets should not be used to label different features in different illustrations of a report. For example, the label "A" should not be used to label both A-cells in one illustration and antenna in another, of the same report.

   The labelling should be parallel to the top margin of the photograph, i.e., it should not be radiating in different directions. However, the pointer line or the arrow, if necessary, may be placed in a convenient direction in order to avoid masking other details in the photograph.

   The key to the labels should be given in the caption of the photograph. The expansions of the abbreviations used as labels in photographs that are grouped into Plates, may be given commonly for each Plate, i.e., instead of including the expansions in each caption.

5. *Magnification*   Magnification of each of the photographs, especially the microphotographs, should be clearly indicated in the caption. The magnification of microphotograph is not the same as the magnification of the microscope (objective power $\times$ eyepiece power) at which the photograph was taken, because the printing involves enlargement of the negative. One way of calculating the exact magnification of the final print of a micrograph is to use the actual measurement (obtained using standardized ocular micrometer) of a dimension (e.g. diameter of RBC in microns) of an object shown in the microphotograph and the measurement of the same dimension of the same object in the final print. For example, if the diameter of an RBC measures 10 μm and the same measures 10 mm (= 10 $\times$ 1000 μm = 10,000 μm) in the final print, the magnification of the microphotograph can be calculated as 10000 $\div$ 10 = 1000. It is indicated as $\times$1000, i.e., the microphotograph has a magnification of 1000 times the original size. Suppose we photograph a snake that measures 150 cm in length. In the final print, the image of the snake may measure 15 cm. The magnification of this final print is 15 $\div$ 150 = 0.1. It is indicated as $\times$0.1. We can easily understand that if the magnification of photograph is more than unity the image in the photograph is larger than the original

object. If the magnification is less than unity, the size of the image is smaller than the original object.

The difficulty here is that we have to calculate the magnification of every microphotograph individually because the enlargement of the negative may not be constant for all.

An alternative way is to give an approximate magnification. All final prints of photographs taken at a particular microscopic magnification are assumed to have the same magnification, and the magnification calculated for one is assigned to all. Such an approximate magnification is specifically indicated as ×1000 ca. The abbreviation "ca." stands for "*circa*" meaning "about" or "approximately".

Yet another method to show the magnification is to place a micrometer marker directly on the microphotograph. A micrometer marker is a line, distinctly visible against the background, of known length (in micrometres). For example, we can draw a 10-mm line in one corner of the microphotograph and indicate, on the print itself, just above or below the line, that it represents 10 μm. This method is useful for microphotographs submitted for publication in Journals. While printing the photographs in a Journal, they are invariably reduced in size to be accommodated to the column width or the page width. Therefore, the calculated magnification may not be appropriate for the final print in the Journal. A micrometer marker, if given on each microphotograph, would also get reduced in the final print and thus be indicative of the true magnification.

While photographing larger objects, placement of a scale with clear markings alongside the object would help to indicate the actual size of the object in magnification of the final print.

LINE DIAGRAMS

Hand-drawn illustrations using black ink (India ink) are very useful in the field of descriptive biology like morphology and anatomy. For example, a hand-drawing of the digestive system of an insect, showing the relative positions of the various organs, would be more descriptive than a photograph. Camera Lucida diagrams are useful in depicting microscopic histomorphological organization of organs. All such hand-drawn illustrations must have the necessary details used to make a photograph or a microphotograph. If any of them were not drawn to scale, it should be specified in its caption.

CHAPTER SEVEN

RESEARCH REPORT—
FORMATTING AND TYPING

INTRODUCTION

The final stage in reporting a research work is the preparation of the manuscript to be submitted to the agency which will consider it for publication or for evaluation for the award of a degree. A badly prepared manuscript will not be and should not be accepted for consideration. It is the responsibility of the research worker to take every care, from the quality of the paper used for typing the report to the mode of dispatching the report to the agency. All the guidelines and regulations prescribed by the agency (e.g. Journal or University or College) should be followed scrupulously. The research supervisors or guides in the departments should also share the responsibility of report preparation. The young research workers must be encouraged to master the basics of the Thesis and Paper preparations.

All the requirements for the formatting and typing of a research report should conform to those specified by the target agencies. Usually, the Universities or Colleges provide guidelines for the preparation of theses. Journals specify the format of the manuscripts to be submitted, in every issue under "Instructions to the authors".

PAPER

Only white bond paper should be used. There are variations in the dimensions for bond papers: (1) 225 × 285 mm (8.9 × 11.2 inches), thesis bond; (2) 210 × 297 mm (8.27 × 11.69 inches), A4 bond; (3) 212 × 297 mm, ISO A4 bond. The selection of the paper for typing a research report depends on the requirement stipulated by the target agency. In general, for an academic thesis, bond paper of size 225 × 285 mm is used, and for manuscripts to be submitted for a journal, A4 or ISO A4 is used. In addition, bond paper of 8.5 × 11 inches (215.9 × 279.4 mm) dimension is permitted in many academic institutions.

In most of the computer printers (inkjet or laser) the thesis bond may not be accommodated. An easy way out is to set the Page in Custom Size with a width of 225 mm and a length of 285 mm. Such custom-sized pages can be printed in the usual A4 papers for which most of the PC printers are designed, taking care to set the Print properties for A4. These printouts can be photocopied onto the thesis bond papers with appropriate margins.

MARGINS

The ideal typing margins for academic theses (typed in thesis bond) are: 4 cm (1.57 inches) on the Left (binding edge) and 3 cm (1.2 inches) on all the other three sides (top, bottom and right).

The typing margins for reports typed in A4 bonds are: 3.81 cm (1.5 inches) on the left and 2.5 cm (1 inch) on all the other sides.

These margins should be maintained for all pages including those of Figures and Tables.

Pages carrying major headings such as Acknowledgement, Chapter, Table of Contents, References, etc. should have a top margin of two inches (5 cm).

PARAGRAPH INDENTATIONS

Generally in reports like a thesis, each paragraph starts with an indentation of five to ten spaces. It must be uniform throughout a report. The indentation facilities provided by the software packages can also be manipulated to suit our requirements. Variations may be permitted by different research Journals. For example, a paragraph beginning may not have any indentation at all, the demarcation being an extra space between paragraphs. Whatever be the pattern, it should be uniform throughout the report.

WIDOW AND ORPHAN LINES

We should avoid typing single lines of a paragraph at the top (widow line) and bottom (orphan line) of a page. If a paragraph has to be divided at the bottom of a page, we must make sure that at least two lines appear at the bottom and carry at least two lines to the top of the next page. If there is no room for a complete heading and at least two lines of text at the bottom of a page, we must begin the new subdivision on the next page.

SPACING

Spacing refers to both vertical and horizontal spacing. Vertical spacing pertains to the number of typed lines that will occupy a page. The space between lines (single, one-and-a-half and double-line spacing) makes a manuscript appear either cluttered or uncluttered. The horizontal spacing determines the space between letters, which is necessary for easy reading (e.g. WA-normal; WA-condensed; WA-expanded). Most of the software packages (e.g. MS Word) provide default settings for both vertical and horizontal spacing that is pleasing to the eye. We can easily manipulate these settings to suit our requirements.

The entire text, including the preliminary pages (those of Certificate, Acknowledgement, Contents, etc.) and References, of an academic thesis should be typed with double-line spacing.

However, single-line spacing may be used in a thesis for the following cases:

1. Titles of Tables and Captions of Figures (Plates)
2. Footnotes of Tables
3. Tables, especially those presenting textual matter
4. In the "Contents", when an entry requires more than one line

Some academic institutions may permit use of 1.5 line spacing for typing References, double-line spacing between entries.

The spacing of the different sections of a research article may vary. However, while preparing the manuscript for submission, it should be typed with double-line spacing, because editing, corrections, modifications, etc. of the text requires sufficient spacing and margins. The spacing in the final product will be decided by the editor and publisher of the Journal.

ALIGNMENT

The text of a research report is typed either left aligned or justified. In justified alignment, the length of the lines, irrespective of the number of words in each line, are same. As a result the text appears flush with both left and the right margin. Whatever alignment is followed, it should be followed consistently throughout the report.

HYPHENATION

Hyphen is a small horizontal line (-), which is used while typing or printing a document, to join two or more elements of compound word, as in "two-third," "one-and-a-half," etc. Such a hyphen is called a link-hyphen. A hyphen is also used as a break-hyphen to indicate the split of a word at the end of a line. Though there is lot of difference of opinion especially in the usage of break-hyphen, we must follow, while typing a research document, the established conventions. One of the ways is to refer to a standard dictionary. Different dictionaries may suggest different breaking points (syllables) for a word. Therefore the easiest way to decide on use of break-hyphen is to leave it to the Computer word processor. However, it might be useful if we bear the following in mind regarding hyphenation.

We should hyphenate

1. Modifiers composed of a number and unit (e.g. a five-kilometre stretch of highway).

2. Modifiers involving comparative and superlative forms (e.g. better-qualified candidate).

3. Fractions written in full when used as adjectives (e.g. two-third majority, one-and-a-half kilogram).

4. Words in which prefixes and proper nouns are joined (e.g. mid-July, pro-Canadian, pre-Christian).

We should not hyphenate

1. Modified adjective phrases: short-lived compounds, but very short-lived compounds.

2. In a range, if the word "from" is used, the word "to" must also appear: "from 7 to 24 hours" not "from 7–15 min." "A period of 7–24 hours" is correct.

   • Words that are always hyphenated should be split only at the hyphen.

   • We should avoid hyphenating the end words of more than three lines in a row.

   • Compound, unhyphenated words should be broken only between the roots (e.g. thunderstorm, micro-physical).

FONTS

The conventional font type is the Times New Roman. Other types may be used only if permitted by the target agency. A font size of 12 should be used. A slightly smaller font may be used in a Table in order to accommodate it within the specified margins of a page. Unreadable font sizes in Tables and in labels of Figures should be avoided. The font size of the title of a Table or the caption of Figure should be maintained at 12, i.e., it should not be reduced to match the contents of the Table or the Figure.

PAGINATION

Every page of a research report should be assigned a page number. In a manuscript meant for submission to a Journal, all pages must bear the page number.

In reports like an academic thesis, there are two separate series of page numbers. The preliminary pages of a thesis (i.e., pages preceding the main text; e.g. pages of Title, Certificate, Acknowledgement, Contents, List of Tables, List of Figures, List of Abbreviations, etc.) are numbered using lower case roman numerals (e.g. i, ii, iii, and so on). The title page is given the number "i" but not typed. The numbers of the succeeding preliminary pages are placed at the bottom centre, without any punctuation, one inch from the bottom of the page.

Starting from the first page of the text, the first page of the first chapter (e.g. GENERAL INTRODUCTION), all the pages are numbered consecutively throughout the report, including the Summary, References and Appendixes, with arabic numerals. Use of letter suffixes to page numbers (i.e., supplementary page numbering; e.g. 7a, 7b, etc.) is not allowed. The page number of a text page is placed at the top right corner at the intersection of one inch from top and right edges. This placement of page number applies even to pages containing tables typed in landscape (horizontal) orientation.

The pages containing captions of Figures are not given page numbers. The plates containing photographs or graphs are also not assigned any page numbers. But, pages containing Figures or Tables with text above and below are assigned numbers.

FORMAT OF A THESIS

An academic research report (e.g. Thesis, Dissertation, Project Report) is a bound volume submitted by an individual investigator to a University for the purpose of evaluation and, if favourably adjudicated, for the award of a degree like M. Phil., Ph. D., etc. It is more like an examination answer script. Unlike in many Universities in western countries, where a thesis is bound only after the approval of a committee of examiners, our Universities require submission of bound thesis, which is placed before a committee of Examiners for evaluation and adjudication.

The organization of a thesis is different from that of research paper published in a Journal. A thesis usually consists of

- Title page
- Certificate (by the Supervisor and Guide) page
- Declaration (by the Candidate) page (if required)
- Acknowledgement page
- Abstract (if required)
- Table of Contents
- List of Tables (if required)
- List of Figures (if required and if the Figures are spread throughout the text)
- List of Abbreviations (if required)
- Chapters ...
- Summary
- References
- Figures (if not already given chapter-wise or spread throughout; specify Figure numbers, as 1–120, for example)
- Appendixes

There may be variations in the organization outlined above, according to the requirements of the different universities, institutions, departments and research schools.

Title Page

Academic institutions prescribe their own format for the title page of a thesis or dissertation. The directions should be followed in total especially with regard to matters of content, margin and spacing. Generally, the following information will be required to be provided for a thesis.

1. Title of the report (thesis/dissertation/project report).

2. Degree for which the report is submitted, specifying the subject and whether it is partial fulfillment or otherwise.

3. Name of the candidate, if necessary, with degrees already obtained.

4. Name of the Department /Institute where the work was carried out (optional).

5. Name of the University to which the thesis is being submitted.

6. Date (month and year) of submission of thesis.

A sample title page of a Ph. D. thesis is shown in Figure 7.1. The title of the thesis is typed in all capitals, about 4 cm from the top and centered. Any scientific nomenclature is italicized. If the title requires more than one line, it should be typed in an inverted pyramid model, with double-line spacing.

The other information are given below the title, centered and balanced to occupy the page leaving a space of about 4 cm below the date of submission.

The title page is the first of the preliminary pages and therefore has the page number "i", but it is not typed.

Certificate Page

The certificate page of a thesis is the second preliminary page and, therefore, carries the page number "ii". The certificate is issued by the guide/supervisor to the effect that the research work reported in the thesis is the original work carried out by the candidate under his / her supervision, and that the work has not formed the basis for the award of any other degree or diploma by any other institute. The title CERTIFICATE is typed centred, 2 inches from the top edge. The text of the certificate is typed with double-line spacing. Every care should be taken to type the name of the candidate and the title of the report without any error. A sample certificate page is shown in Figure 7.2.

Declaration Page

The third preliminary page with a page number "iii", carries the declaration by the candidate to an effect similar to that of the certificate issued by the guide. It is also typed in the same format with care as those of the certificate page. A sample declaration page is shown in Figure 7.3.

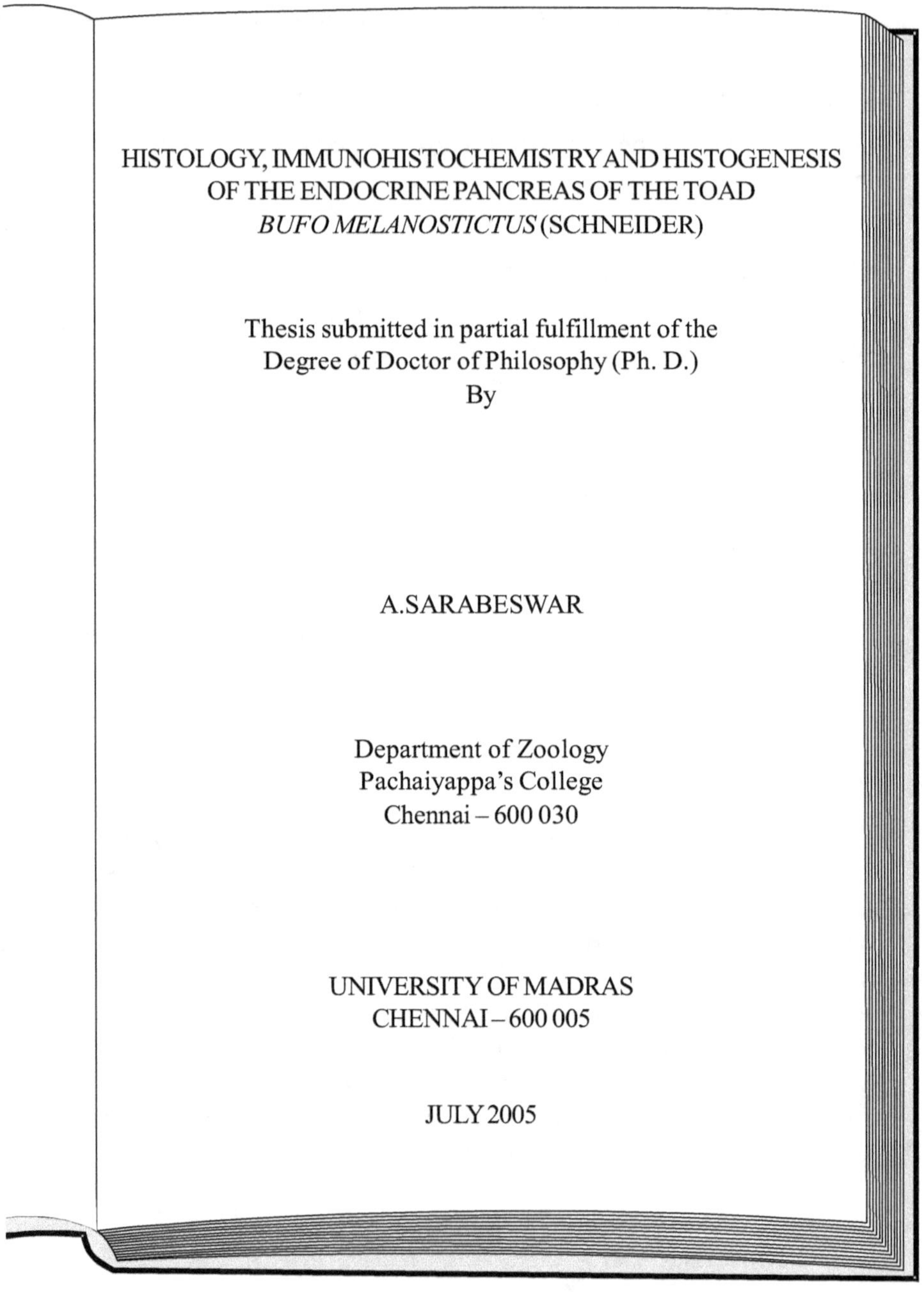

HISTOLOGY, IMMUNOHISTOCHEMISTRY AND HISTOGENESIS
OF THE ENDOCRINE PANCREAS OF THE TOAD
BUFO MELANOSTICTUS (SCHNEIDER)

Thesis submitted in partial fulfillment of the
Degree of Doctor of Philosophy (Ph. D.)
By

A.SARABESWAR

Department of Zoology
Pachaiyappa's College
Chennai – 600 030

UNIVERSITY OF MADRAS
CHENNAI – 600 005

JULY 2005

Figure 7.1   A sample title page of a thesis.

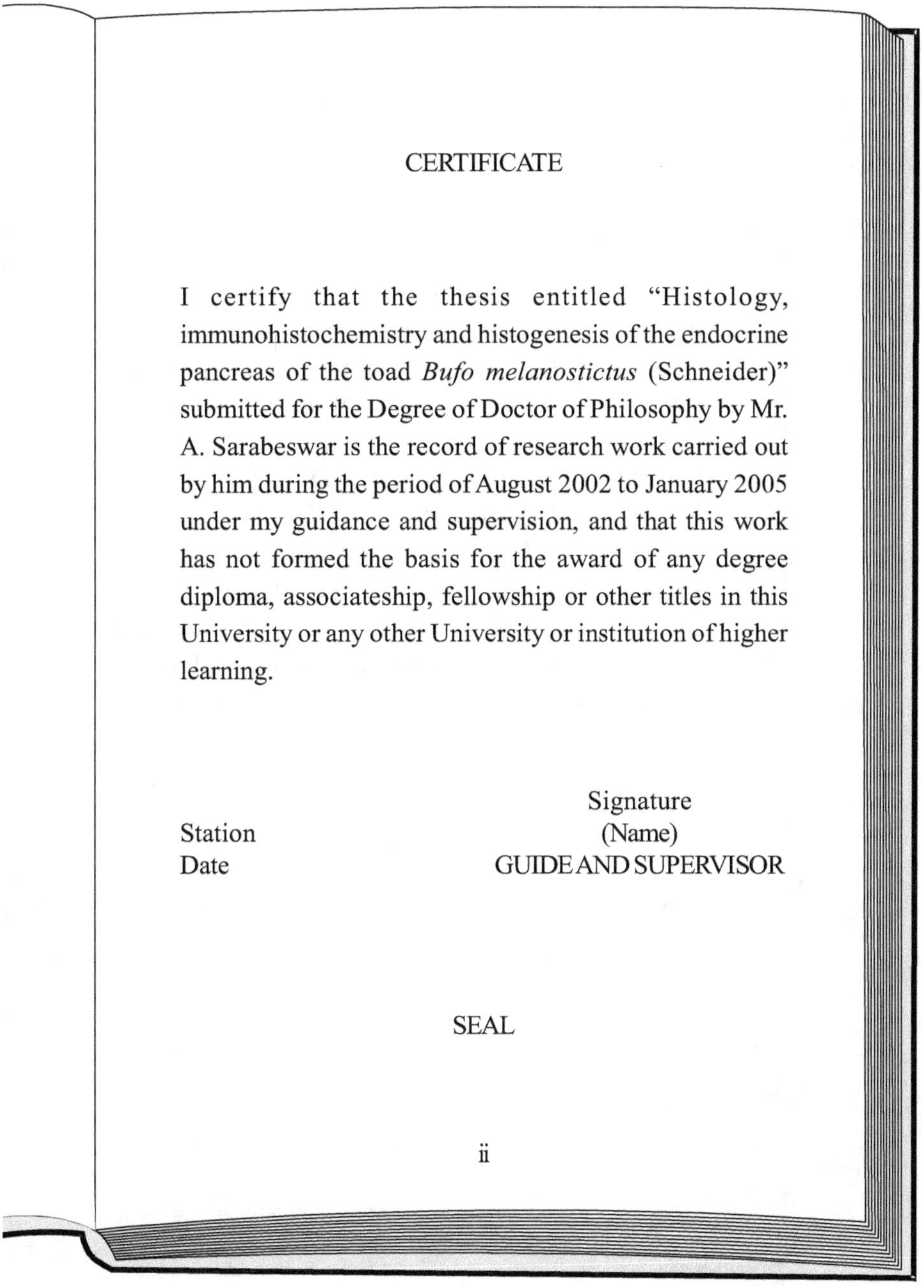

CERTIFICATE

I certify that the thesis entitled "Histology, immunohistochemistry and histogenesis of the endocrine pancreas of the toad *Bufo melanostictus* (Schneider)" submitted for the Degree of Doctor of Philosophy by Mr. A. Sarabeswar is the record of research work carried out by him during the period of August 2002 to January 2005 under my guidance and supervision, and that this work has not formed the basis for the award of any degree diploma, associateship, fellowship or other titles in this University or any other University or institution of higher learning.

Signature
Station   (Name)
Date   GUIDE AND SUPERVISOR

SEAL

ii

Figure 7.2   A sample certificate page of a thesis.

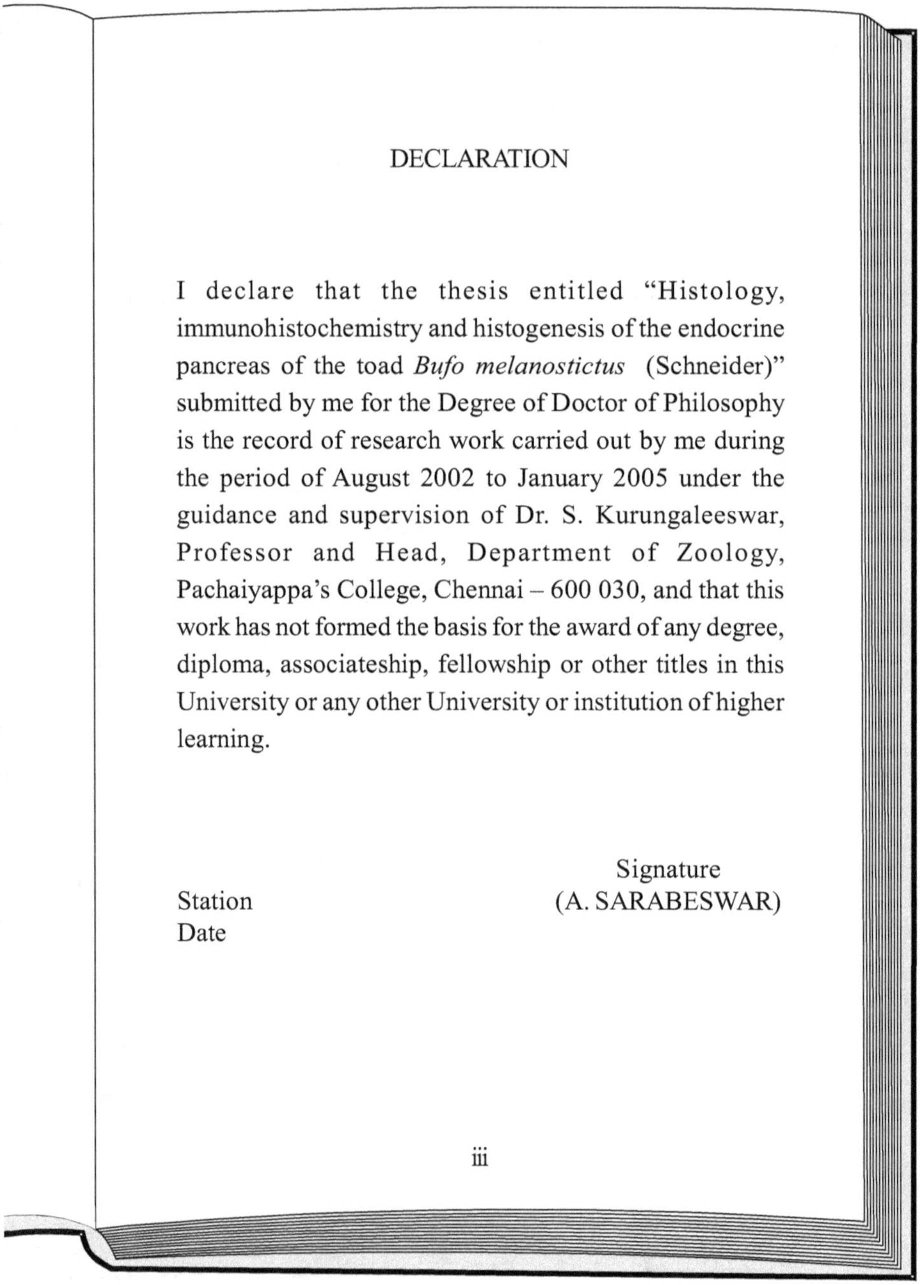

DECLARATION

I declare that the thesis entitled "Histology, immunohistochemistry and histogenesis of the endocrine pancreas of the toad *Bufo melanostictus* (Schneider)" submitted by me for the Degree of Doctor of Philosophy is the record of research work carried out by me during the period of August 2002 to January 2005 under the guidance and supervision of Dr. S. Kurungaleeswar, Professor and Head, Department of Zoology, Pachaiyappa's College, Chennai – 600 030, and that this work has not formed the basis for the award of any degree, diploma, associateship, fellowship or other titles in this University or any other University or institution of higher learning.

Signature
Station   (A. SARABESWAR)
Date

iii

Figure 7.3   A sample declaration page of a thesis.

Acknowledgement Page

This page is also formatted and typed on the same pattern as those of the Certificate and Declaration pages. Its page number is "iv". An example of an acknowledgement page is given in Figure 7.4.

Table of Contents

There are different ways of constructing the Table of Contents of a research report. The main objective of Table of Contents is to help the reader to locate specific sections of the report as easily as possible. Therefore, it should be constructed as a true reflection of what the report contains.

The title of the Table of Contents (TABLE OF CONTENTS or CONTENTS) is typed, centred, 2 inches (about 5 cm) from the top edge.

As already noted, the Table of Contents is also part of the preliminary pages and, therefore, its pages are also numbered in roman numerals, in continuation with the previous pages.

Generally, in a Table of Contents, the preliminary pages that are placed before it (certificate page, declaration page, acknowledgement page, etc.) are not listed in it, but those that are placed after it are listed indicating their page numbers in roman numerals.

If there are no preliminary pages after the Table of Contents, the listing begins with the first chapter. Headings of all sections of the report including those of Chapters, Summary, Appendixes, References, and Figures should be listed. Since, no page numbers are assigned to plates of Figures, only the figure numbers (e.g. Figures 1–120) are given.

In addition to chapter headings, subheadings, to a certain extent (e.g. up to fourth level), may also be listed in the Table of Contents. Subheadings used in Appendixes are not listed in the Table of Contents.

Wording, spelling, capitalization, and punctuation of the headings listed in the Table of Contents should be identical to what is found in the text of the report. Indentations in the Table of Contents should correspond to the levels of subdivisions followed in the text.

The following guidelines may be borne in mind while constructing a Table of Contents.

- Chapter titles are aligned on the periods following the chapter number.

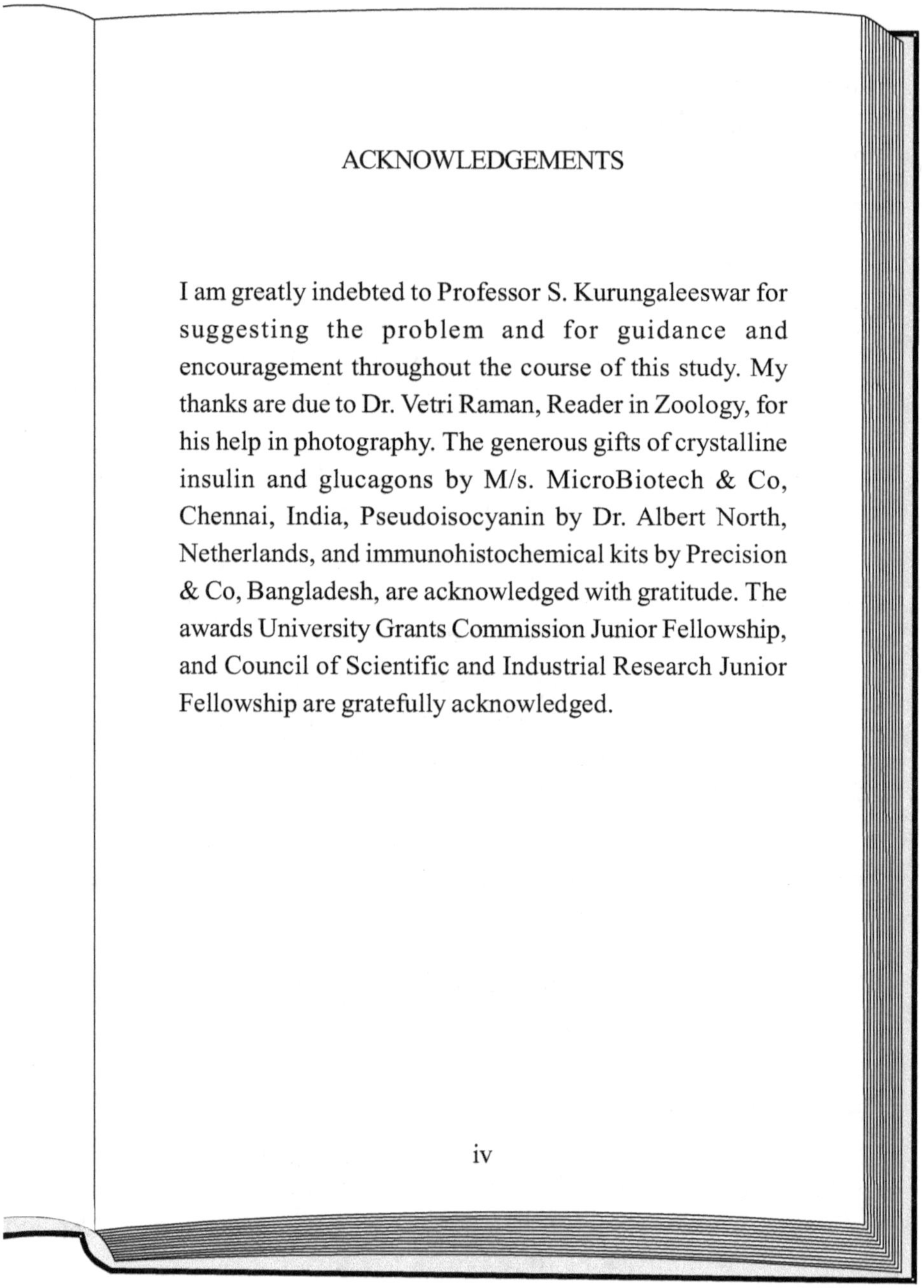

ACKNOWLEDGEMENTS

I am greatly indebted to Professor S. Kurungaleeswar for suggesting the problem and for guidance and encouragement throughout the course of this study. My thanks are due to Dr. Vetri Raman, Reader in Zoology, for his help in photography. The generous gifts of crystalline insulin and glucagons by M/s. MicroBiotech & Co, Chennai, India, Pseudoisocyanin by Dr. Albert North, Netherlands, and immunohistochemical kits by Precision & Co, Bangladesh, are acknowledged with gratitude. The awards University Grants Commission Junior Fellowship, and Council of Scientific and Industrial Research Junior Fellowship are gratefully acknowledged.

iv

Figure 7.4   A sample acknowledgement page of a thesis.

- Subheadings are indented in increments of tabs from the chapter titles.

- Double-or one-and-a-half-line spacing should be used between entries.

- In a very long Table of Contents, single-line spacing may be used between lesser levels of subheadings.

- Entries that require two or more lines should be single-line spaced within the listing.

- The word Chapter is not repeated each time.

- Subheadings should not be underlined, even if they are in the text.

- Entries should not overlap into the space of the page numbers.

- The last digits of page numbers should be aligned at the right margin.

- Leader lines are the spaced periods leading from the end of each entry to its page number. They are optional.

- Page references should be double-checked.

Examples of Table of Contents are given in Figure 7.5 and Figure 7.6.

Abstract Page

This is the first preliminary page to be placed after the Table of Contents and, therefore, listed first. The format of this page is the same as the acknowledgement page. An example of abstract page is shown in Figure 7.7. This page is to be included only if the target agency requires it. It bears a page number, in lower case roman numeral, in continuation with the last page of the Table of Contents.

List of Tables

If a research report, such as a thesis, contains at least five tables, a List of Tables may be included in the document. The List, counted as a preliminary page, is placed next to the Abstract page. If the document has no abstract, the List will be next to the Table of Contents, and numbered accordingly in lower case roman numeral.

A List of Tables gives the table numbers, captions and page numbers for all tables, including those in the appendixes. A few guidelines to remember while constructing a List of Tables are as follows.

CONTENTS

v

Figure 7.5   A sample Table of Contents.

(Continues)

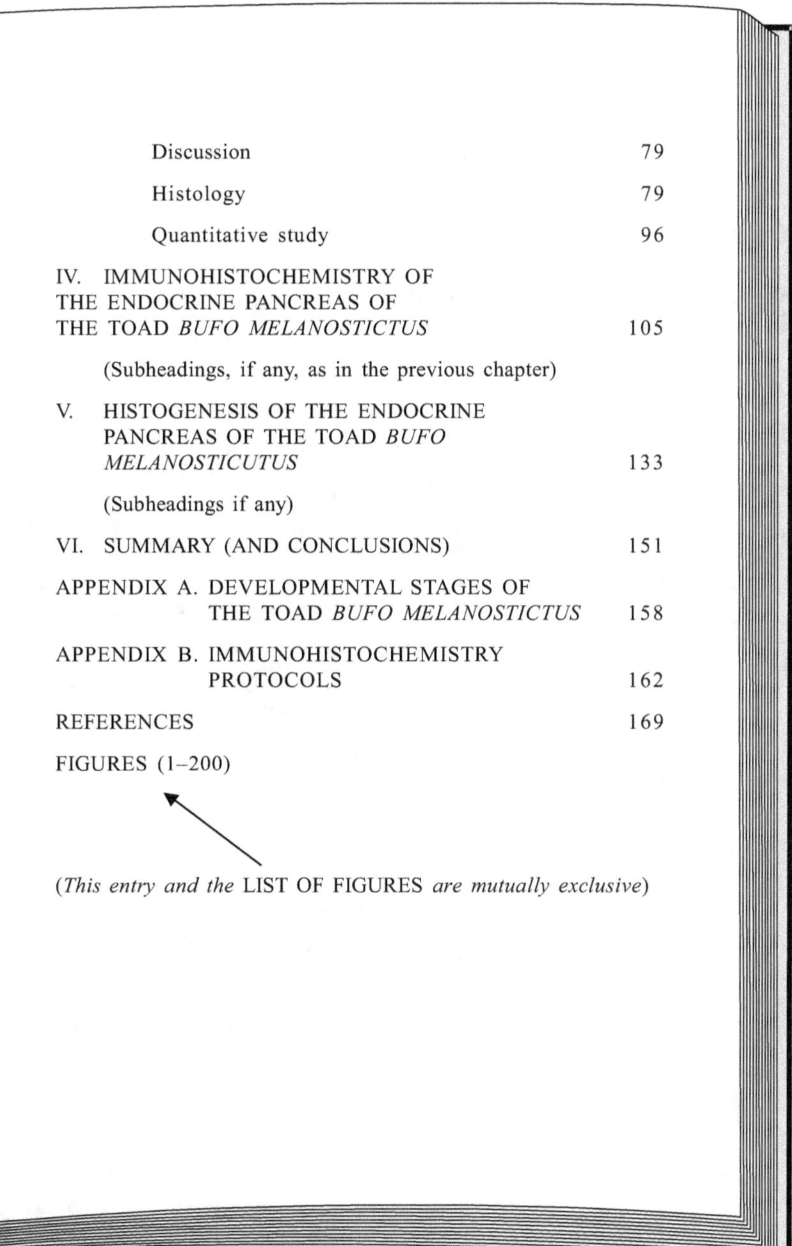

(*This entry and the* LIST OF FIGURES *are mutually exclusive*)

Figure 7.5 A sample Table of Contents.

CONTENTS

V

Figure 7.6   A sample decimal system Table of Contents.

(Continues)

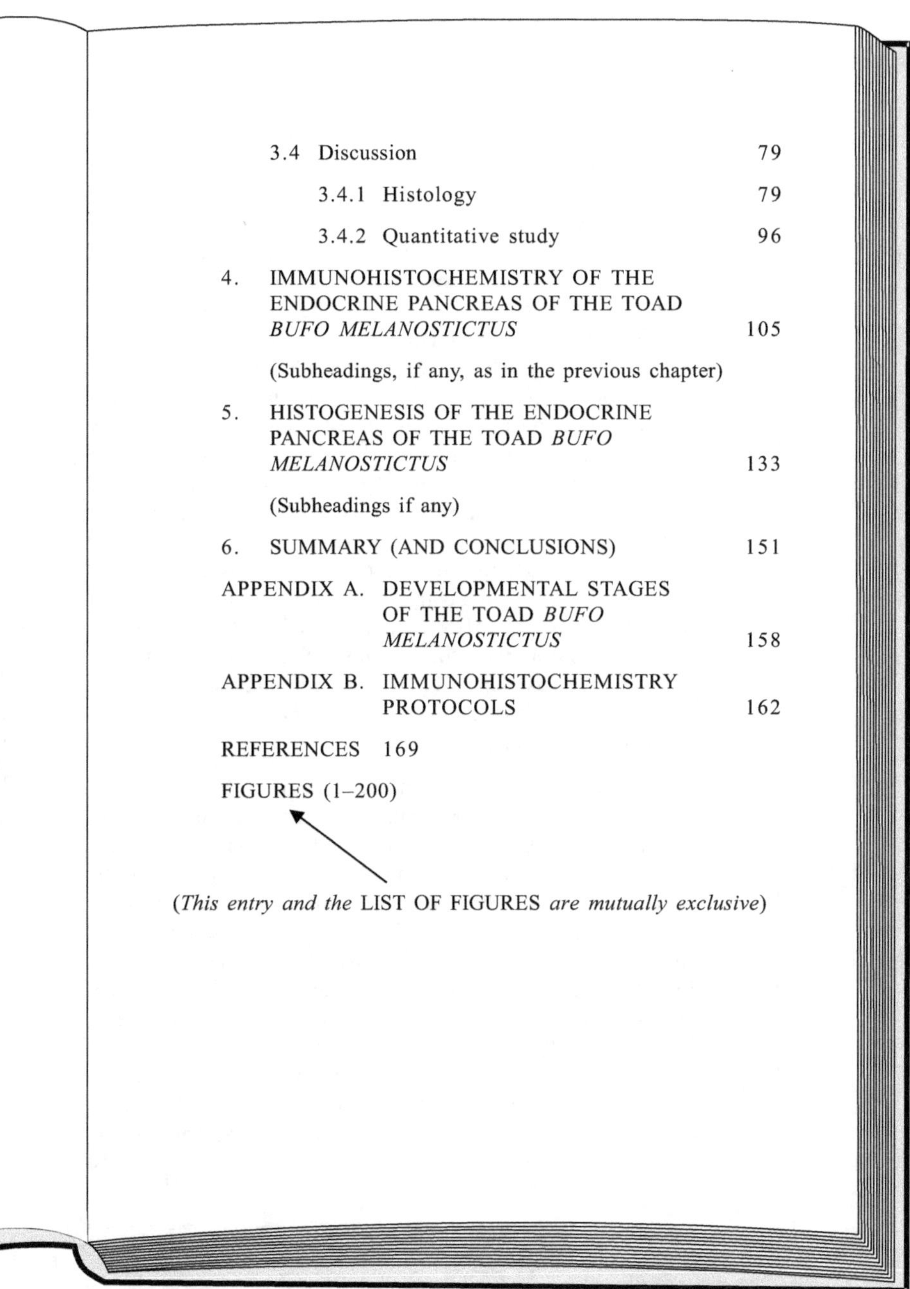

Figure 7.6   A sample decimal system Table of Contents.

ABSTRACT

The influence of biodiversity on ecosystem processes has been the subject of considerable research effort and debate among plant and animal ecologists but the ecosystem consequences of microbial diversity are largely unknown. In this investigation the hypothesis that soil microbial diversity affects ecosystem function was tested by evaluating the effect of denitrifier community composition on nitrous oxide (N_2O) production. A soil enzyme assay to evaluate the effect of oxygen concentration and pH on the activity of denitrification enzymes responsible for the production and consumption of N_2O was employed. By controlling, or providing in non-limiting amounts, all known environmental regulators of denitrifier N_2O production and consumption, conditions were created in which the only variable contributing to differences in denitrification rate and the relative rate of N_2O production in soils from two fields in southwest Michigan was denitrifier community composition. It was found that the denitrification enzymes of the denitrifying communities from the two fields differed substantially in their sensitivities to oxygen and pH, indicating that the denitrifying communities in these two soils respond to environmental regulators differently.

Denitrifying bacteria were also isolated from these same soil samples and, for 31 representative isolates, the sensitivity of their *nos* enzymes, which catalyze the reduction of N_2O to N_2, to low levels of oxygen, was measured. Cluster analysis of the cellular fatty acid profiles of 93 denitrifying bacteria isolated from the agricultural field and 63 from the successional field showed 27 denitrifying taxa with only 12 common to both soils. In addition, it was found that there is substantial diversity in the degree of sensitivity of the isolates' *nos* enzymes to oxygen, indicating that the taxonomic diversity present among denitrifiers in these two soils is functionally significant. These results demonstrate a clear potential for differences in denitrifier community composition to affect differences in N_2O production among ecosystems, independent of direct environmental controls.

Figure 7.7   A sample abstract of a thesis.

- The heading (TABLES or LIST OF TABLES) is typed in all capitals, centred, leaving a 2-inch (5 cm) margin from the top of a new page.

- Table numbers are aligned on the periods.

- Double or one-and-a-half-line spacing is used between table titles.

- Single-line space is used within titles, if it has more than one line, and the lines are aligned on the first word.

- The text of the title, the wording and capitalization, should be the same as that given for the table in the text. However, the following are exceptions.

  □ The term Table (as in Table 1) is not repeated with each title in the list.

  □ If a title has a more than one complete sentence, only the first sentence is given in the list.

  □ Any matter given within parenthesis in the title may be omitted in the list. And it should be adopted for the all the table titles.

Examples of List of Tables are given in Figure 7.8 and Figure 7.9.

List of Figures

As we had already seen, the Figures (charts, diagrams, graphs, maps, photographs, etc.) may be given all through a report or collectively at the end of the report, after the Reference section. A List of Figures may be included only if the Figures have been inserted at appropriate places, spread throughout the report, like the Tables, and only if there are more than five Figures in the research report. The list is placed in a page next to that of the List of Tables, and the page is given the next consecutive number in roman numerals. The title LIST OF FIGURES is typed in all capitals, no punctuation, centred, two inches below the top edge.

The List of Figures gives the numbers, legends and page numbers of all figures. We should list captions and legends exactly as they are found in the text. The entries are formatted as those of the tables in the List of Tables. Long legends, which have more than three lines, may be shortened either by using the first sentence of the legend or by giving a clear, concise summary of the legend.

The page numbers and leader lines are as in the Table of Contents.

Examples of List of Figures are given in Figures 7.10 and 7.11.

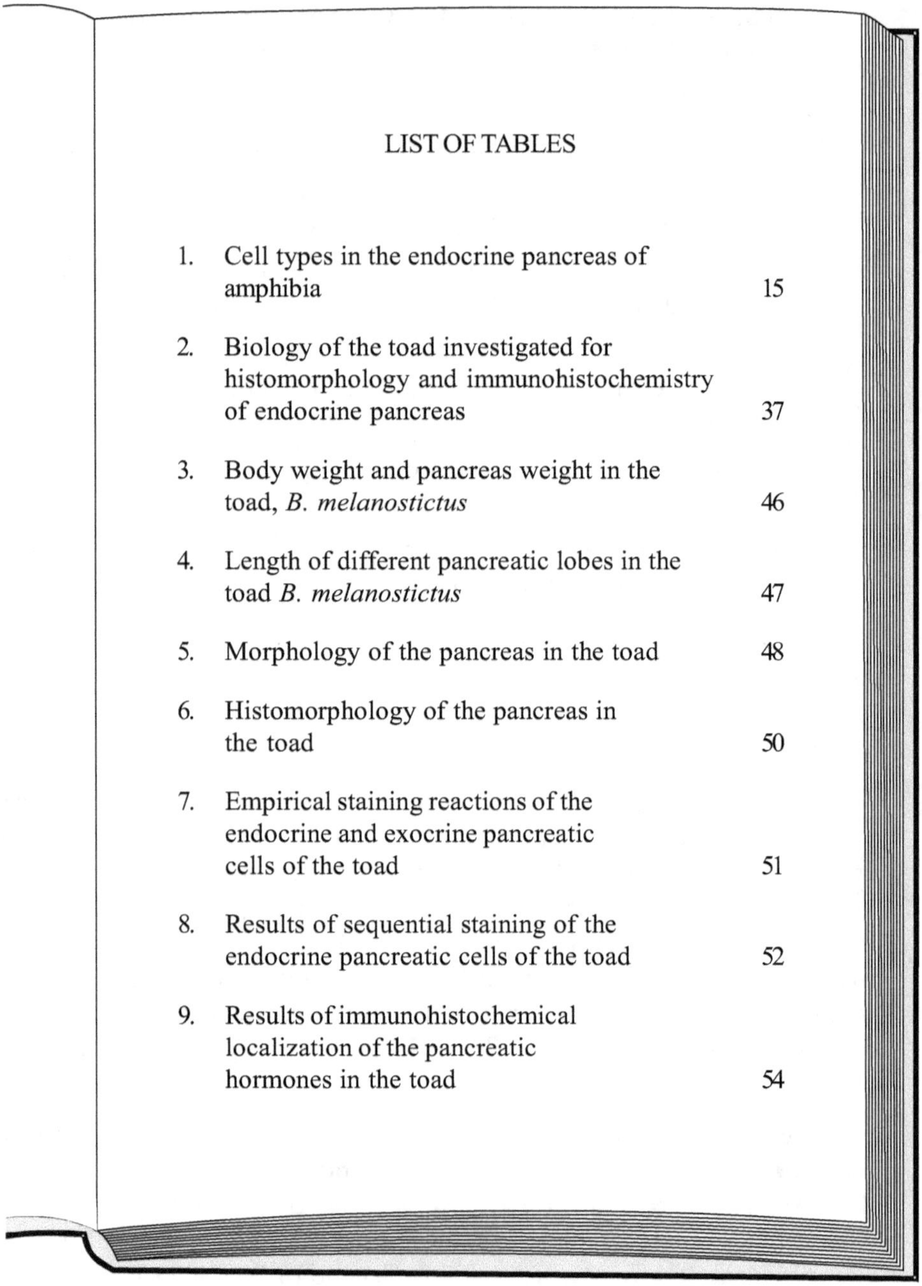

LIST OF TABLES

1. Cell types in the endocrine pancreas of amphibia 15
2. Biology of the toad investigated for histomorphology and immunohistochemistry of endocrine pancreas 37
3. Body weight and pancreas weight in the toad, *B. melanostictus* 46
4. Length of different pancreatic lobes in the toad *B. melanostictus* 47
5. Morphology of the pancreas in the toad 48
6. Histomorphology of the pancreas in the toad 50
7. Empirical staining reactions of the endocrine and exocrine pancreatic cells of the toad 51
8. Results of sequential staining of the endocrine pancreatic cells of the toad 52
9. Results of immunohistochemical localization of the pancreatic hormones in the toad 54

Figure 7.8 An example of List of Tables—continuous numbering.

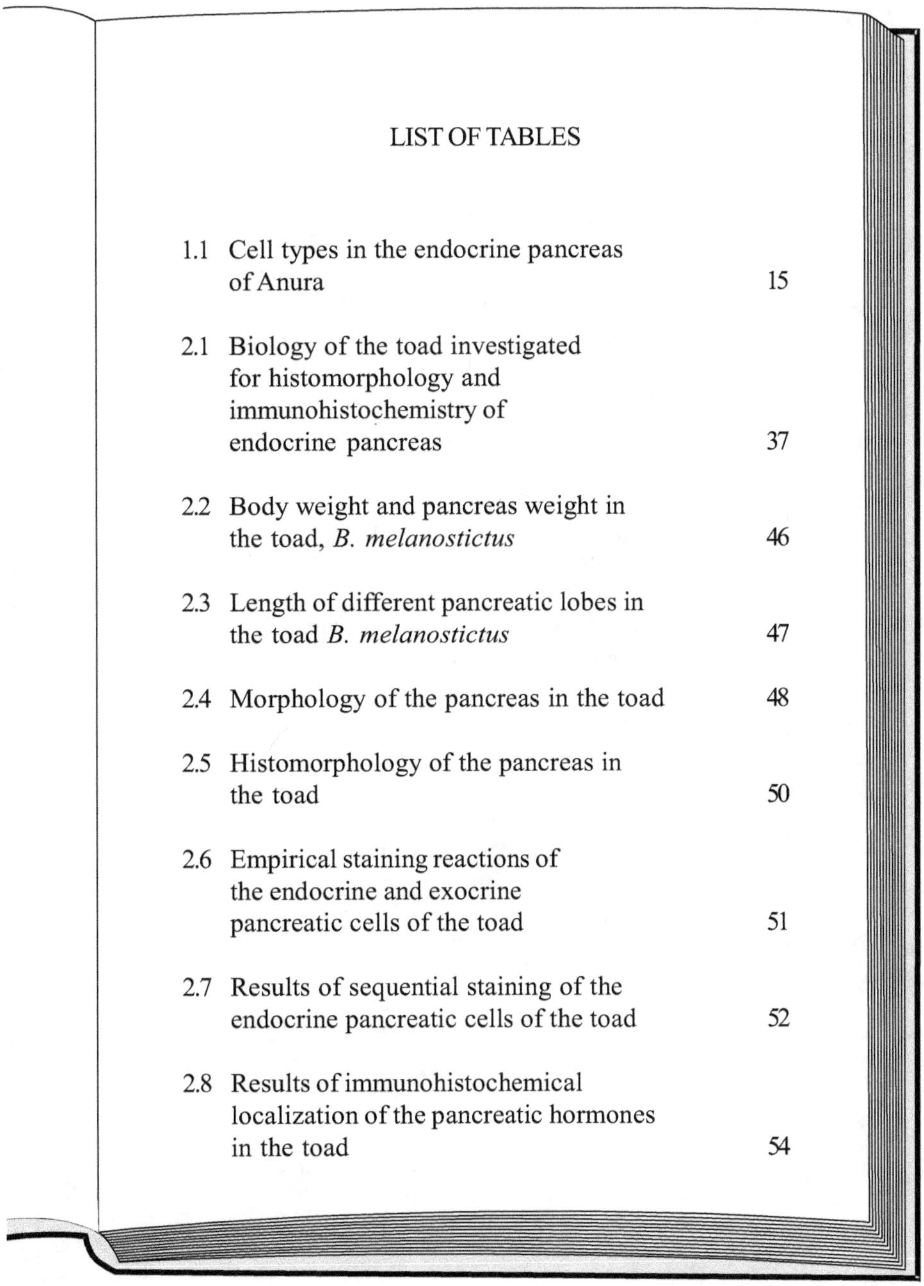

LIST OF TABLES

Figure 7.9   An example of List of Tables—decimal numbering.

LIST OF FIGURES

1. Pancreas volume (PV; Mean ± SD) and endocrine pancreas volume (EPV; Mean ± SD) in the three species of Anura   43

2. Percent (Mean ± SD) endocrine pancreas volume (EPV) in the three species of Anura   45

3. Islet cell population (percent in total population; Mean ± SD) in the three species of Anura   61

4. Endocrine pancreas volume (Mean ± SD) during the larval development of the toad   97

5. Distribution of insulin-immunoreactive cells in the endocrine pancreas of larval stage I of the toad   105

6. Distribution glucagons-immunoreactive cells in the endocrine pancreas of larval stage I of the toad   107

Figure 7.10 An example of List of Figures—continuous numbering.

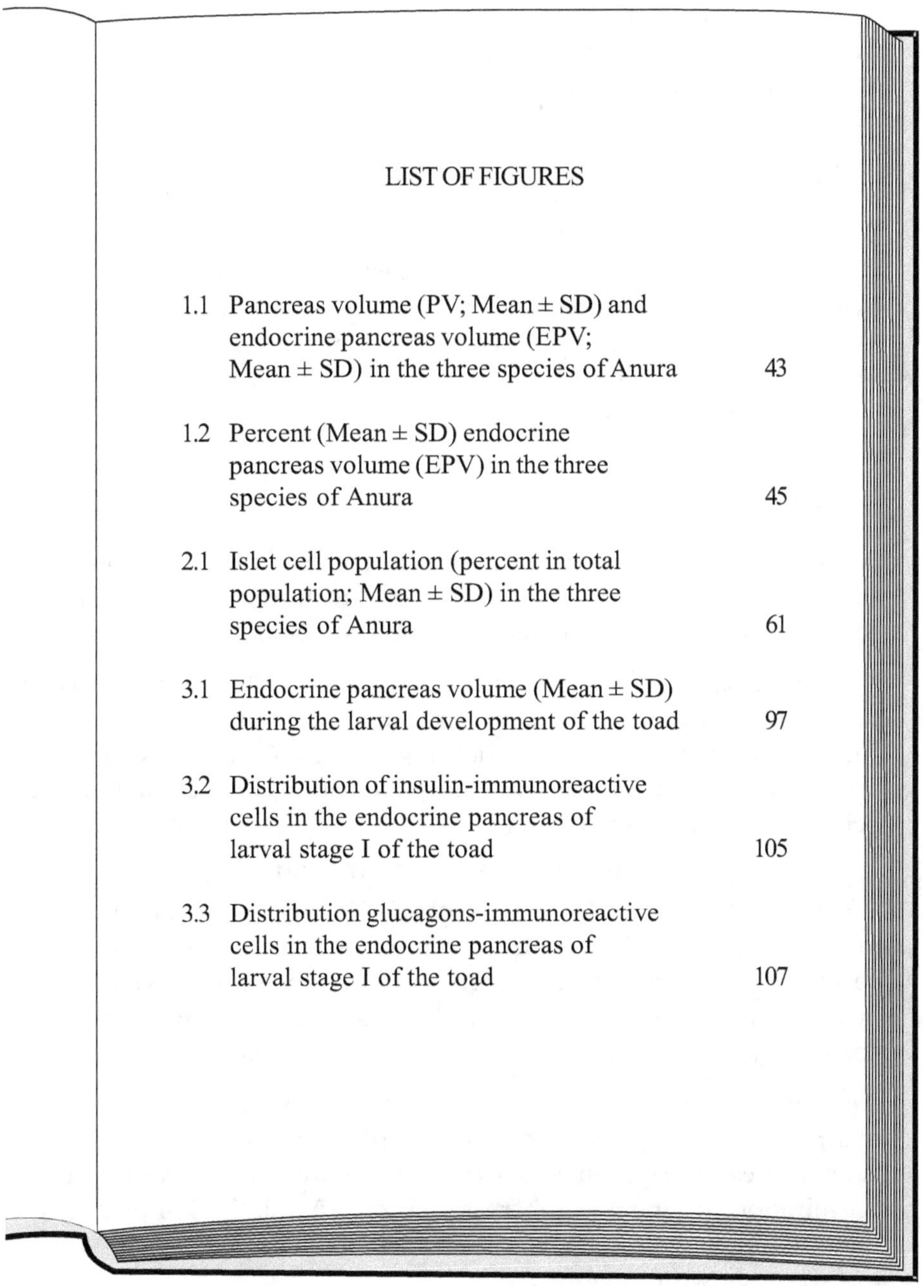

LIST OF FIGURES

Figure 7.11   An example of List of Figures—decimal numbering.

The List of Tables and the List of Figures can be presented in a single page, if feasible, with triple space between them. But they should never be combined with the Table of Contents.

List of Abbreviations and Symbols

This list is the last of the preliminary pages. Entries in this list are alphabetized first by abbreviation and then by symbol. Single-line spacing is used when two or more lines are needed per explanation, and double-line spacing (or one-and-a-half-line spacing) between entries. The terms included in this list must be defined when they are first mentioned in the main text of the report.

Chapters

A research report such as a thesis may contain more than one chapter, such as General Introduction, Review of Literature, Chapters dealing with specific investigations, General Discussion, and Summary and Conclusions.

Each chapter is given a number and a title. The number may be in capital roman numeral or arabic numeral, if decimal system is followed. If roman number is assigned, the heading (e.g. CHAPTER I, CHAPTER II, etc.) is typed in all capitals with no punctuation, centred and two inches below the top edge of the page.

The title (e.g. HISTOLOGY OF THE ENDOCRINE PANCREAS OF THE TOAD *BUFO MELANOSTICTUS*) is typed double space below the heading in all capital letter, centred, with no period at the end. If the chapter title is more than four inches in length, we must divide it into two or more lines, arranging them centred in an inverted pyramid model. The chapter title should not be in bold or in a larger font size than the text.

Subdivisions of chapter   Further, each chapter may be divided into a number of sections and subsections in order to make both presentation and reading easy. A research paper in a Journal is also more like a thesis chapter in its organization, having a number of divisions. A subdivision of a chapter is indicated by its heading. The number of subdivisions may vary from two to five, as shown in Figure 7.12.

The hierarchy of subheadings is as follows:

1. centred, underlined;
2. centred, not underlined;

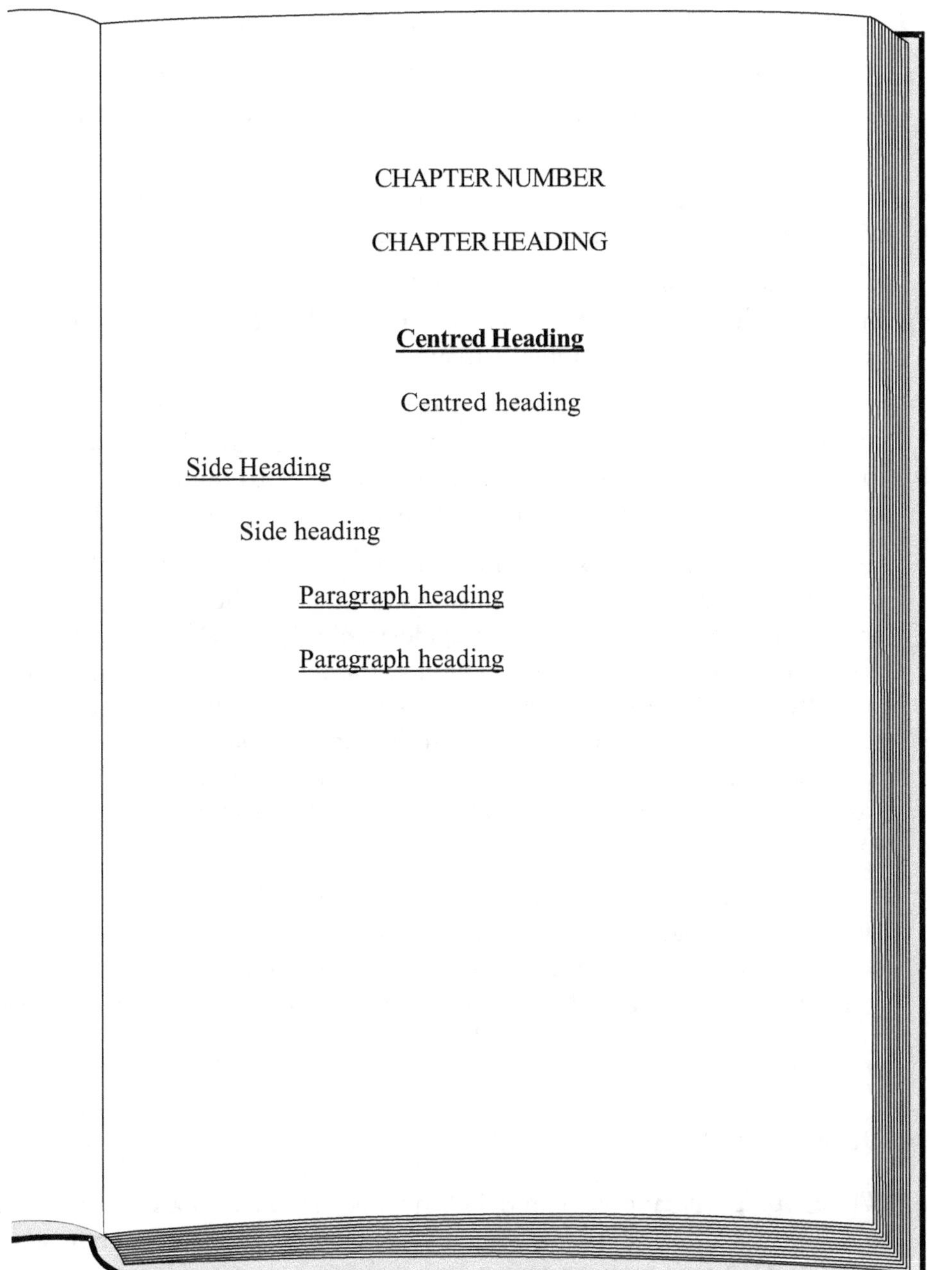

Figure 7.12   Subdivisions of chapter at five different levels.

3. side margin, underlined (at left margin—no tab);

4. side margin, not underlined (at left margin—no tab);

5. paragraph (one tab), underlined and end with a period (no extra space before this level).

A subheading should be in the same size and style of font as the text as well as not being in bold, italics, or all capital letters.

It is not necessary that the number of subheadings should be the same in all the chapters of a thesis. For example, Chapter I may use two levels of subheading while Chapter II may require four levels.

A much easier way of subdividing a chapter is to use the "decimal" system. The Chapter is given an Arabic number, and all the subdivisions in the chapter are given additional numbers, separated by a period, as required for the level of subdivision, as illustrated in Figure 7.13.

In the "decimal" system, any number of subdivision levels may be accommodated. However, having subdivisions higher than fourth level (having five digits) is generally not desirable. Since all the headings are distinctly numbered, they are aligned at the left margin, that is, there are no centred headings. Further, there is no need to underline or bolden these headings. The text matter, followed by two-line spacing under each heading, begins as a paragraph with set indentations. Four-line spaces (two double-line spaces) separate the end of a paragraph and the next subheading.

Each chapter begins on a new page. It is also advisable to start sections with centred heading or first level division (e.g. 1.2 Materials and Methods, 3.4 Discussion) on a new page. The page number of the first page of a chapter may or may not be typed, but those of all other pages, even if they begin on a new page, must be typed.

Quotations

While typing quotations, material eliminated from a quotation is indicated by ellipses (3 periods); material inserted by the author is enclosed in brackets. Direct quotations of four lines (not sentences) or less are enclosed in quotation marks and embedded into the text lines. Direct quotations of five or more lines are indented, blocked and single-spaced with the quotation marks removed, and the period placed after the text.

1. CHAPTER HEADING
*(The number may be 1.0; Typed in all
capitals, centred)*

1.1   First Level Subdivision (Chapter 1, Subsection 1)
*(Typed in lower case with initials in capital, aligned to
left margin)*
*Paragraph begins with an indentation of a tab space
(between 5 and 10).*

1.1.1   Second level (Chapter 1, Subsection 1, Subsection 1)
*(Lower case; no capitalization of all intials; aligned to
left margin or one tab space)*
Paragraph begins with an indentation of a tab space.

1.1.2   Second level (Chapter 1, Subsection 1, Subsection 2)
Paragraph begins with an indentation of a tab space.

1.1.2.1 Third level (Chapter 1, Subsection 1, Subsection 2,
Subsection 1)
*(Lower case; no capitalization of all initials; aligned to
left margin or two tab spaces)*
Paragraph begins with an indentation of a tab space.

1.1.2.2 Third level
Paragraph begins with an indentation of a tab space.

1.1.2.3 Third level
Paragraph begins with an indentation of a tab space.

1.2   First Level Subdivision
Paragraph begins with an indentation of a tab space.
(Further subdivisions, if required, as in 1.1.2)

1.3   First Level Subdivision
Paragraph begins with an indentation of a tab space.
(Further subdivisions, if required, as in 1.1.2)

Figure 7.13   Subdivisions of a chapter—decimal system.

Table

When a small table (occupying less than half-a-page) is typed along with the text, it is separated from the text by a triple-line spacing above and below. Table number, if given separately (as TABLE I or TABLE 1 or TABLE 3.1) should be typed, all capital, centred, three-line spaces below the margin or three-line spaces below the text. The title is typed with two-line spacing below the table number.

If the table number is given along with the title (as, for example, Table 1. Changes in … or Table I. Changes in… or Table 3.1 Changes in…) the term table is not capitalized. The title is centred and, if it requires more than one line, single-line spaced in an inverted pyramid model. The table is typed two spaces below the title.

The entries in the stub and box heads are centred. The entries in the stub are left aligned. The numerical entries in the boxes are right aligned. Other types of entries, such as text (NA) or symbols (dash, plus, asterisk, etc.), are centred.

When numerical values are given as Mean ± SD, a space is given on either side of the ± symbol.

The left and right sides of a table are not generally closed, probably because of the difficulty of typing a vertical line in a conventional typewriter. In computer-generated tables, however, not only closing the sides is not a problem, insertion of lines delineating the columns is easy. The bottom of the table is closed by a single horizontal line. Footnotes, if any, are typed two-line spaces below the bottom line of the table, two letter spaces indented from the left margin of the table, left aligned, and in single-line space within a note. Double-line space is allowed between notes.

Two small tables can be typed, if necessary, in a single page if there is at least a triple space between them.

If a table cannot be accommodated in the normal vertical (portrait) orientation, it can be typed in a horizontal (landscape) orientation. The head of the table is always along the binding edge of the page. The typing of the table number, title, footnotes, etc. is in the same format as for the vertical table. The page number of the page that contains a horizontal table is, however, in its usual position, i.e., top right corner, as in other pages.

A large table that cannot be accommodated in a single page, even after sufficient reduction in the font size, may be carried over to more than one page. The continuation page bears only the table number as TABLE 3

(Continued). Footnotes of such tables are typed on the last page of the table. The column headings may be repeated in every page of the table. However, if the columns are numbered on the first page of the table, it will be enough if these numbers alone are mentioned in the subsequent pages.

Summary

The formatting of the summary (or summary and conclusions) section of a research report is similar to any other chapter. However, it is not subdivided into sections like a chapter is, but organized as paragraphs. The paragraphs may be numbered (arabic) as if to enumerate the findings and inferences. If necessary, the paragraphs may be grouped (e.g. using roman numbers) into sections corresponding to the chapters. This section is typed double-line spaced; the pages are numbered in arabic numerals as any other page of the main text.

The summary section (i.e., summary of all the chapters) of a thesis may or may not be given a chapter number. If given, it should be on the same pattern as in the other chapters.

Appendixes

No chapter number is given for an appendix. Instead, it is identified by English alphabet (e.g. APPENDIX A, APPENDIX B). If there are more than one appendix, each of them is also given a descriptive title.

Each appendix begins in a new page. The heading (e.g. APPENDIX A) is typed in all capitals, centred, two inches from the top edge of the page. The descriptive title (IMMUNOHISTOCHEMICAL PROTOCOLS) is typed three-line spaces below the heading, in all capitals and centred. The materials presented in the appendixes are typed double-line spaced. However, depending on the nature of the material, the spacing may be reduced.

The pages of appendixes are numbered in continuation to SUMMARY section.

References

We discussed elsewhere the citation of the references, i.e., the sources of information we used in the report, in the text as well as in the References section. As seen therein, the formatting of the references may differ in different Journals and in different academic bodies. For an academic thesis, it is advisable to format them in a standard form, i.e., without omitting any of the punctuation marks or spaces. When a part of the thesis is prepared for publication, necessary modifications can be incorporated according to the requirements of the Journal.

The reference section of a thesis is generally not assigned a chapter number, especially if appendixes are inserted before it. If there are no appendixes and if a number is assigned, it should be continuation of that of the summary section.

The title REFERENCE (or any other equivalent title such as LITERATURE CITED, BIBLIOGRAPHY, LIST OF REFERENCES, etc.) is typed in all capitals, centred, two inch below the top edge. The first entry begins two-line spaces below aligned to the left margin. If the entries are alphabetized (name and year system), they are not numbered. All the entries are typed with double-line spacing. If permitted, one-and-a-half-line spacing may be used within an entry, with double-line spacing between entries.

The pages of the REFERENCE section are numbered in continuation with those of the APPENDIXES section, or, if there are no appendixes, with those of the SUMMARY section.

Figures

Illustrations such as flow charts, statistical diagrams and graphs, photographs and microphotographs, etc., generally referred to as Figures, are presented in a research report in one of the following two ways: (1) They may be inserted at appropriate places in the text, thus spread out throughout the report. (2) They may be collectively presented at the end of the report (e.g. thesis) in several pages. A page may contain one or several Figures, serially numbered using Arabic numerals.

When a Figure is placed along with text (i.e., the Figure occupying one-half, either upper or lower, of the page and the text, the other half), the Figure number is typed in all capitals, centred, double-line spaced below the Figure (Figure 7.14). The caption is typed two-line spaces below the Figure number, centred, single-line spaced, inverted pyramid model, if more than one line is required.

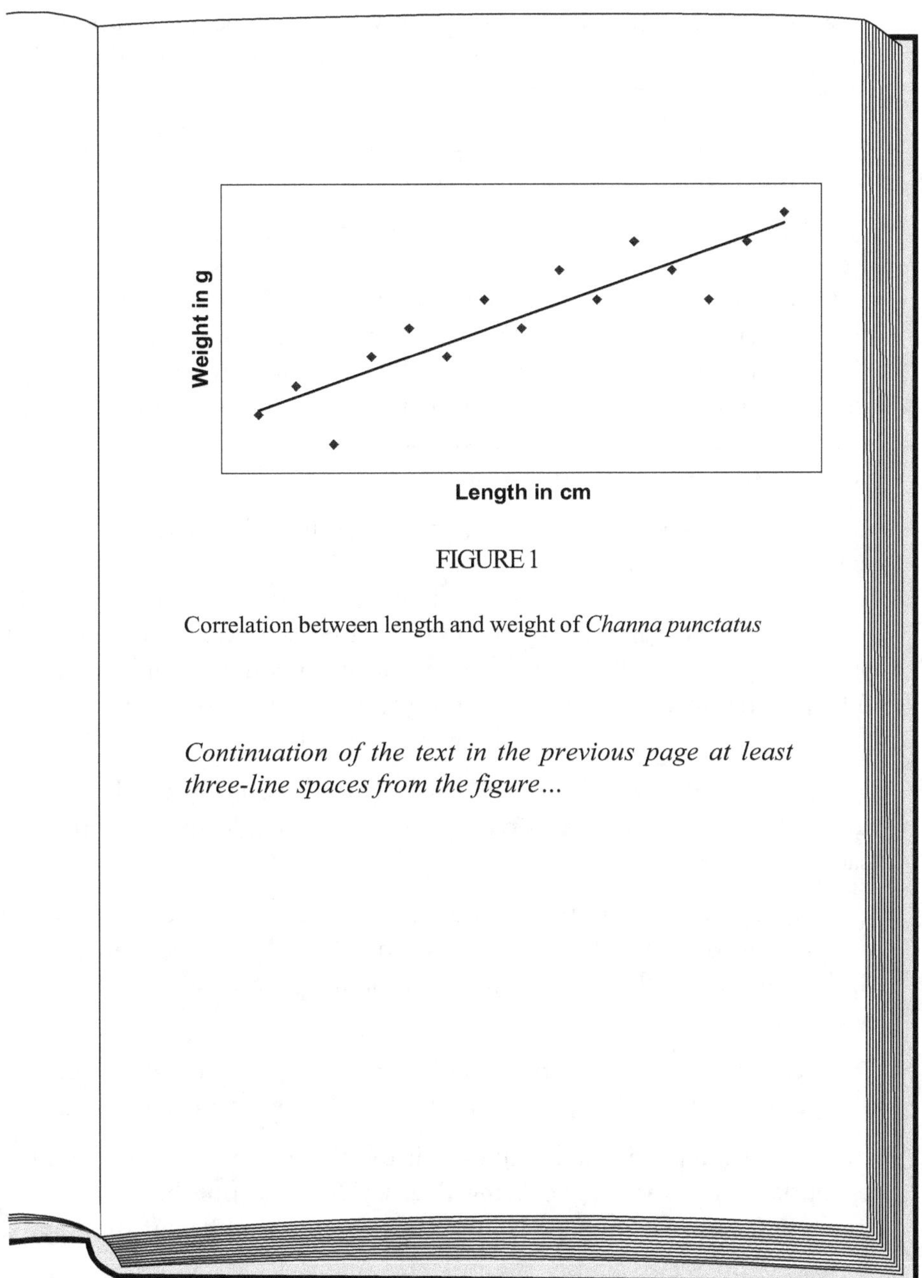

FIGURE 1

Correlation between length and weight of *Channa punctatus*

Continuation of the text in the previous page at least three-line spaces from the figure...

Figure 7.14   An example of a Figure inserted at the top of a text page.

A much simpler and space-saving way is to type the caption as, for example, "Fig. 1. Correlation between length and weight of *Channa punctatus*." An example is shown in Figure 7.15.

If the caption, with number, is given immediately below the Figure, the Figure is not separately numbered.

When a Figure is given separated from the text, i.e., in a separate sheet, often referred to as a Plate, its caption is given below it only if there is sufficient space. A Plate containing one to three Figures might be able to accommodate their captions beneath them on the same page itself. But if a Plate carries more Figures, as in most of the Biology theses, the Figures are numbered (as already discussed, it is convenient to number all the Figures in a report continuously without assigning any Plate number), and their captions are given on the facing page of the Plate.

A separate sheet (other than an inter-leaf, if any) is inserted in front of each Plate. The front side of this sheet is not used for typing any text except the page number in the usual position (top right). The backside of this sheet, i.e., the page facing the Plate is used for typing the caption(s) of the Figure(s) in the Plate. This side of the sheet is not numbered. The left and right margins of this page are the reverse of those of a usual page, i.e., 1.5" on the right (binding edge) and 1" on the left.

The content of a caption page are balanced vertically and horizontally. The caption of each Figure is typed left aligned, single-spaced with double space in between captions.

A common heading for all the Figures may also be given to include matters common to all the Figures, thus avoiding repetition in every caption. An example of captions on a facing page of a Plate is given in Figure 7.16.

Even when a Figure is presented in a Plate in a landscape orientation, its caption on the facing page should be in the vertical orientation.

A list of Figures is made out only if the Figures are given distributed at appropriate places throughout the thesis. The page number typed on the front side of the facing page is given as the page number of the Figures.

When all the Figures of a report are presented at the end, after the REFERENCES, it is enough if we just include an entry in the Table of Contents as FIGURES (1–204). No page number is typed on the front side of the facing pages and, therefore, no page number is indicated for this entry in the Table of Contents.

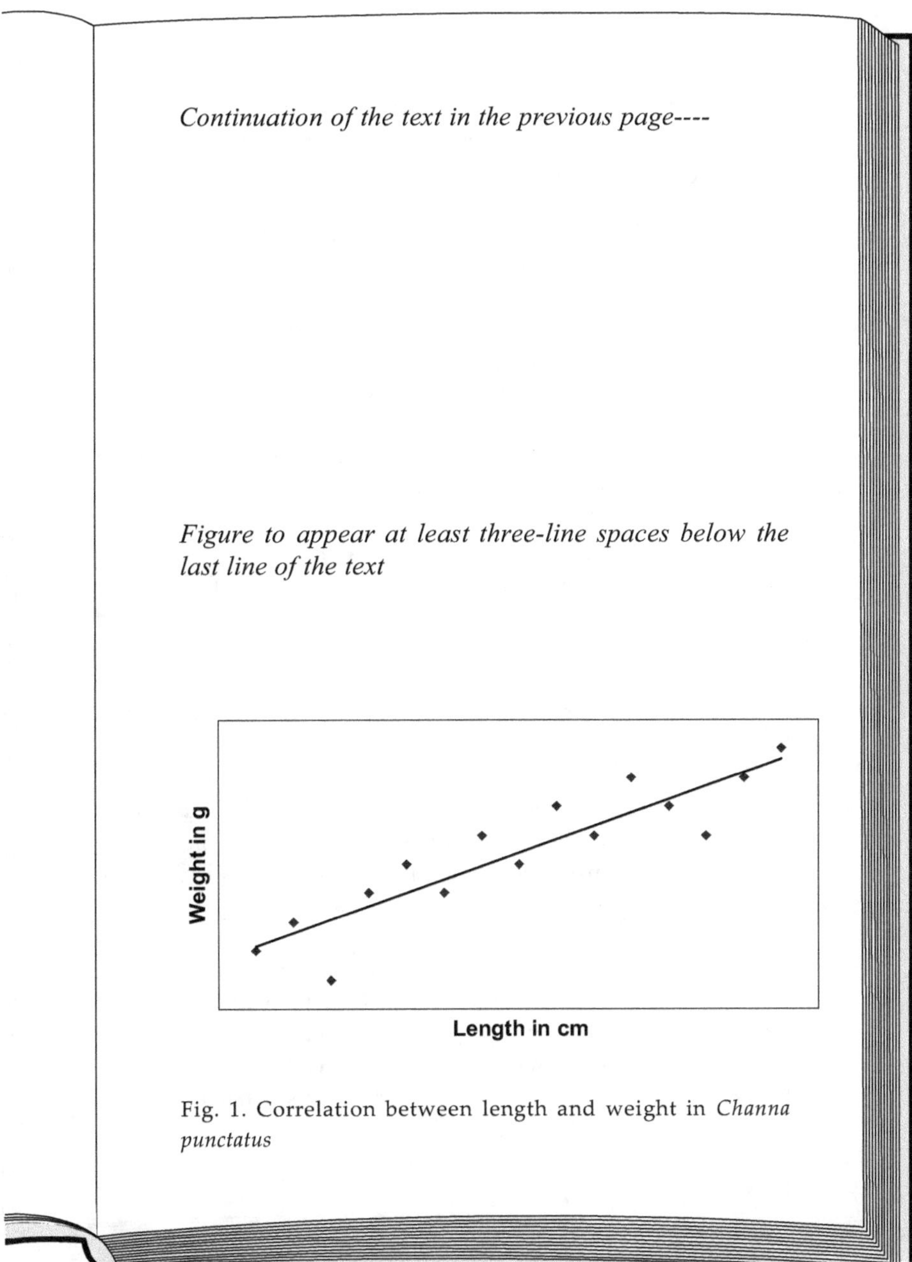

Fig. 1. Correlation between length and weight in *Channa punctatus*

Figure 7.15   An example of a Figure inserted at the bottom of text page.

Figs. 73–78. Photomicrographs of the cross sections
of the embryonic stages of *B. melanostictus*.
Bouin, 4 mm, Antiinsulin, PAP

Fig. 73. Stage 22 showing the gut, and the yolk sac filled with yolk. × 60

Fig. 74. Enlarged view of the gut shown in Fig. 7.3, displaying its position in relation to the yolk mass. × 180

Fig. 75. Stage 23 showing the differentation of the gut lying above the yolk sac. × 60

Fig. 76. Stage 24 Showing dorsal and ventral pancreatic buds arising from the distal end of the foregut. The buds are solid and filled with yolk platelets. × 60

Fig. 77. Stage 25 showing rapidly growing dorsal bud and its fusion with the ventral bud. × 60

Fig. 78. Enlarged view of fused pancreatic buds shown in Fig. 77, having primitive tubules and lacking distinction between exocrine and endocrine cells. × 160

DB – dorsal bud	NT – neural tube
FG – foregut	PT – pancreatic tubule
GT – gut	VB – ventral bud
IN – intestine	YK – yolk
LI – liver	YP – yolk platelet
NC – notochord	

Figure 7.16 Example of captions of Figures in a plate, on the facing page.

CHAPTER EIGHT

Experimental Designs

INTRODUCTION

Biological research includes not only observation of natural events and their description, but also giving explanation for those events. It is also concerned with finding solutions to problems of the natural world. In these endeavours biologists adopt scientific methods.

Generally, an investigation may be a case study, a cross-sectional study or a longitudinal study. A *case study* is observation and description of one or a few occurrences of an event. The event may occur anywhere in the world, among humans, other animals and plants, in the atmosphere, freshwater bodies, oceans and so on. For example, the event may be a human subject with an abnormality, which may be physical, physiological or psychological. The natural corollary of a case study is the proposition of a hypothesis to explain the observed event.

A *cross-sectional study* is the study of a defined population by observing samples collected from that population. In such studies, the variables are not manipulated. A possible correlation, association, etc. among the variables are investigated. Such studies, however, do not provide any evidence to suggest that a particular variable is the cause of a given event.

A *longitudinal study* is an experimental investigation. We manipulate the variables and factors in experimental units and observe the effects. Such a study helps us to provide tangible evidence to demonstrate the causative agent of a particular event.

Let us consider an example to understand the above basics of a research study. Suppose a physician observes a few persons with oral cancer and also notes that all these persons have tobacco-chewing habit. He would report the description of the cases of occurrence of oral cancer in these persons, and even might hypothesize that tobacco-chewing is the causative agent. However, there is no proof that this habit actually caused oral cancer. There may be other reports of occurrence of oral cancer in persons who do not have this habit. This report is simply a case study.

The next step is cross-sectional study of a population of persons who have the tobacco-chewing habit. Samples from such populations may be screened for the occurrence of oral cancer, symptoms of oral cancer, several biochemical parameters, etc. Simultaneously samples from a "control" population of those who do not have this habit, may be screened for the same factors. A comparison of the incidence of oral cancer and other factors would certainly help to ascertain the association among the factors studied. Establishment of correlation among various variables is also possible.

However, such a cross-sectional study does not provide evidence that tobacco-chewing causes oral cancer. The results simply add strength to the hypothesis.

Only a longitudinal study involving well-designed experiments would provide substantial evidence to demonstrate that tobacco-chewing is the cause of oral cancer. It may not be possible to conduct the experiment on human subjects. But, animal models such as monkeys, which can be induced into tobacco-chewing habit, can be used (if permitted by CPCSEA).

OBSERVATION

The basic requirement for any biological research is observation. Observation of what has happened and is happening in nature leads to an attempt to explain the event, i.e., answering questions such as what, how and, if possible, why. Very often, the explanation is in the form of a hypothesis.

Let us consider an example. We observe large-scale mortality of fish in a lake. Naturally, we are concerned about the happening of this event and would like to have an explanation for it. As biologists, we would test the quality of the water, especially for possible pollution. We analyse samples of water from the lake for all its physico-chemical properties and finally find out that the water in the lake has unusually very high level of cadmium, a heavy metal. We suspect that the high level of the cadmium is the cause of death of fish. We hypothesize that cadmium at high levels is toxic to fish. This is only one of the several possible hypotheses. Several other factors such as bacterial, viral, or any other parasitic infection or a combination of factors might have caused the mortality.

HYPOTHESIS AND NULL-HYPOTHESIS

A hypothesis is put forth as a solution to a problem or as an explanation for an observed event. A hypothesis is a statement making a prediction that an event will occur under stated, specific conditions.

Examples

1. *Prolonged exposure of men to lowered oxygen pressure will cause increased haemoglobin level in them.*

This hypothesis might have arisen as a result of the observation of higher haemoglobin levels in men who are living at high altitudes for several years. It is common knowledge that with increase in the altitude from the mean sea level there is decrease in the atmospheric oxygen pressure.

2. *Spirulina* promotes growth in fish.

3. Cadmium inhibits growth in plants.

4. Ginger has antimicrobial activity.

5. *Lantana camara* has cytotoxic activity.

6. Eucalyptus oil has antibacterial activity.

A hypothesis should be one that can be tested. It is very difficult to "prove" a hypothesis beyond any doubt. All that we can do is to conduct well-designed, controlled experiments to collect evidence in the form of valid data to support our hypothesis. The data that we collect from the experiments are subjected to statistical analysis, which would test a null form of our hypothesis, the null-hypothesis. The result of statistical test of significance calculates the probability for occurrence of the null-hypothesis. If the probability is very low, i.e., lower than the Level of Significance, we reject the null-hypothesis. Rejection of null-hypothesis leads to the suggestion that we have collected evidence to support our hypothesis, that is, in other words, we cannot reject our hypothesis. But we should not forget that the "evidence" we have obtained from an experiment using a few samples may not be sufficient to "prove" our hypothesis beyond doubt.

On the other hand, if the statistical test gives us a probability of the occurrence of the null-hypothesis equal to or greater than the Level of Significance, then we cannot reject the null-hypothesis. Failure to reject null-hypothesis leads to the suggestion that the evidence we collected from our experiment is not sufficient to support our hypothesis. In other words, we have to reject our hypothesis, at least for the time being, until we are able to provide sufficient evidence.

A null-hypothesis is generally a hypothesis of no difference, hence the name null-hypothesis. For example, our null-hypothesis may state, in words or symbolically:

1. No significant difference between the observed mean and the population mean, $H_0 : \overline{X} - \mu = 0$.

2. No significant difference between the two observed means. $H_0 : \overline{X}_1 - \overline{X}_2 = 0, \; H_0 : \mu_1 - \mu_2 = 0$.

3. No significant difference among the observed means, $H_0 : \mu_1 = \mu_2 = \mu_3 = \mu_4$.

4. No significant difference between the observed correlation coefficient and population correlation coefficient, $H_0 : r - \rho = 0$.

5. No significant difference among the observed variances.

The advantage of using a null-hypothesis is that it is a testable hypothesis with only two specific options to decide about it: Reject it or Fail to reject it. Therefore, our decision about our hypothesis also becomes more specific. Since our ultimate intention is to "prove" our hypothesis, we have to produce evidence to reject the null-hypothesis, which is easier than providing evidences to "prove" it.

When we say a hypothesis or a null-hypothesis should be testable, that does not mean those that cannot be tested are totally false. Untestable hypotheses may be true or false. Similarly testable hypotheses may be true or false. What we, experimental biologists, are concerned is only about testable hypotheses.

BASIC PRINCIPLES OF EXPERIMENTS

While observing what has happened or is happening in nature is the basis for any hypothesis, experiment is the basic scientific method of testing the hypothesis that tries to explain an event. In a way it is also observation, observation of events (results) that occur under controlled conditions. The purpose of such observations is to demonstrate or "prove" the functions, the relationships, the causes, the effects, etc. of one or more factors that exist in nature.

Generally in any biological experiment, either in the field or in the laboratory, we try to manipulate (i.e., change by removing or adding or altering) one or more factors in a sample obtained from a defined population, and observe the resulting changes in specific parameters. On the basis of these observations we infer about the population. If our inferences based on the results of the experiment are to be valid, we must lay out all aspects of the experiments before the commencement of the experiment, i.e., we must design our experiment to the minutest detail. The designing varies depending on the specific field of biology, which may be ecology, toxicology, ethology, genetics, immunology, molecular biology and so on. It will not be possible to go into the details of designing experiments in each field, and therefore we shall discuss only the general aspects, which apply to all or most of the designs.

The following are the three important basic steps we have to take at the beginning of any biological experiment.

1. *Aim*   We must define and record precisely the objective of the experiment that we are proposing to carry out.

2. *Plan*   We must write down in detail the strategy we are going to adopt in the conduct of the experiment. The planning includes definition of the population, sampling procedure, sample size, determination of dosage, mode of treatment, control, randomization, methodology for the evaluation of the parameters, collection and analysis of the data, etc.

3. *Procedure*   We must clearly state in writing the experimental procedure, i.e., how we are going to perform the experiment in practice. For example, how to collect the sample of fish, water, air, etc. should be specified. How to transport the sample from the field to the laboratory? How to sacrifice the experimental animals? All physical activities, operational details and requirements (glassware, apparatus, instruments, chemicals, etc.) expected during an experiment should be thought of in advance. Even the time, the specific time to start and end an experiment, the time for lunch-break, etc. should be decided.

Before we venture into the different experimental designs it would be useful if we understand the meaning and usage of certain common terminology we come across in any experimental design.

Experimental Unit and Sampling Unit

We must clearly understand the distinction between experimental unit and the sampling unit. The experimental unit refers to the material used, such as a laboratory animal, cage, culture plate, agriculture field, a plot in the garden, a pot in a greenhouse, etc. A treatment is applied to an experimental unit. A sampling unit may be the experimental unit itself or it may be a fraction of it. For instance, in order to test the toxic effects of a heavy metal on fish, we may have two aquaria one treated with a heavy metal and the other serving as control. Each aquarium may have a number of fish, say, fifty. At specified time intervals we may sample fish from each tank for various analyses such as haematological, histological and biochemical. In this experimental set-up, the experimental unit is the aquarium and not the fish in it. The fish are the sampling units, a part of the experimental unit.

Experimental Error

Another important aspect we have to bear in mind while designing experiments is the experimental error or residual variance. Experimental error refers to the variations among observations made on the experimental units, which were treated alike. For example, we may inject a specific dosage of hormone into each of a number of mice. Though the treatment is "similar" to all the experimental units, the response to the treatment (e.g. level of glucose in the serum) may not be the same. The variation among the observations is the experimental error.

There are two important sources of experimental errors. They are (a) the natural variability among the experimental materials and (b) the lack of consistency in or during the conduct of the experiment. The natural or inherent variability in experimental materials may be due to genetic background, age, sex, health, physiology, immunology, parasitemia, etc. Absence of uniformity in the actual conduct of the experiment may be due to errors in recording, instrumentation, chemicals and reagents used, etc. When we conduct experiments with living things we should realize that the experimental units continue to undergo developmental changes and such changes may also cause errors.

Discrimination

Any experiment that we design should be able to provide an answer to a specific hypothesis. It should not yield results that can explain more than one hypothesis. In other words, experiments should be capable of discriminating between different hypotheses.

Replication

As we are familiar, no two living things, even those that are genetically identical, are similar. There is a large degree of variability in gene expression patterns in particular tissues, even within supposedly genetically identical individuals. In addition, differences can be expected as a consequence of time of day, age, and physiological state such as sex and stage of the estrous cycle, and as a response to disease.

Biological variables are highly unpredictable. Even those variables that are considered to be constant are only regulated by some homeostatic mechanism, to remain within a wide but acceptable range of values. Therefore, in any experiment an inference drawn on the basis of data collected from a single observation will be invalid. For example, to test a hypothesis that bitter-gourd lowers blood sugar level in man, we may conduct an experiment using one human subject. We can prepare an aqueous extract of known quantity of bitter-gourd and administer orally the extract to a diabetic patient after measuring the initial blood sugar level. After a lapse of half-an-hour the blood sugar level of the patient may show a lower value. Can we infer that bitter-gourd has the potency to lower blood sugar level? Certainly not! Apart from so many other pertinent questions that can be raised against the above "experiment", the one that asks "is the difference between the initial and final blood sugar levels statistically significant" can never be answered. A sample of n = 1 is not a sample at all. Suppose, we had a second diabetic volunteer to whom also the bitter-gourd extract was administered. The blood sugar level at the end of the experiment might be slightly higher than the initial level, or the two levels, initial and final, might be the same.

Thus the response to a treatment may vary in different individual units of a sample. This natural variance can be accounted for only if the treatment is replicated, i.e., the same treatment is administered to different sampling units. How many times a treatment should be replicated, i.e., what should be the size of the sample? The answer to this question depends on the expected variance of the data, before and after treatment. It also depends on the accuracy we desire to have in our final prediction. Higher the accuracy, i.e., narrower range of prediction, larger is the size of the sample. If the number of replicates is very small, our inferences will be inconclusive and invalid. On the other hand, too many replicates may yield diminishing returns.

Generalization

The aim of a biological investigation is to make an inference about the population from the sample, i.e., estimation of population parameters using sample statistics. In other words, we attempt to generalize what we observed in a sample, very often, a small sample. We must be guarded while making any generalization. What we observe in a few fish from a laboratory stock may not apply *in toto* to wild fish in a lake. Likewise, we must be very cautious while using observations made on laboratory-bred guinea pigs, for example, for extrapolation to other mammals, especially to humans.

Controls

All biological experiments should have appropriate controls. When a sample unit receives a "treatment" by one of several methods, it is possible that the effect observed might be due to factors other than the one that the treatment contained. Therefore, it is obligatory on our part to demonstrate that the effect is in fact due to the treatment factor and not due to any others including the method of treatment itself (e.g. handling of the experimental animals, medium in which the treatment factor is prepared, etc.). This we can do by incorporating a proper control in our experiments.

Suppose an experiment is designed to find the efficacy of a plant extract, which we suspect has the potency to reduce blood-glucose level in rabbits. The experimental condition would be to administer rabbits with this extract and to find whether the treatment had an effect on the blood-glucose level of the experimental rabbits. In order to ascertain that the effect observed in the experimental rabbits is in fact due to the extract and not due to other factors such as the medium in which the extract was prepared, the technique of administering the extract, etc., we must have a control. Rabbits that are similar (genetically and physiologically) to the experimental rabbits, are administered with placebo, which has everything except the plant extract (e.g. if the extract had been prepared in physiological saline, the placebo would be the physiological saline minus the plant extract; it would have undergone all the steps of the extraction process including adjustment of pH, temperature, etc.). Such a control is often referred to as *negative control*, because of the removal of basic factor that is being tested. Comparison of the effects of the two treatments, experimental and control, would help us judge the efficacy of the plant extract in lowering blood glucose levels in rabbits.

A *positive control* is one in which the factor being tested is added in the placebo. Suppose, for example, we wish to know the effect of removal of pituitary gland on the blood-glucose level of a species of fish. The experimental fish are those that were hypophysectomized. We can have two types of controls for such an experiment. In one control, we can have fish that were sham operated, to eliminate the effect of the operation procedure itself. Another control is hypophysectomized fish injected with extract of pituitary or implanted with pituitary. The second category is a positive control in the sense that it has something added that is missing in the experimental animal.

The negative and positive controls, also referred to as baseline controls, are basically opposite of experimental units, with reference to only one specific condition while all the other conditions are the same. Normal, healthy, untreated animals, plants, etc. may also serve as baseline controls.

Another type of control, referred to as *known standard control*, is used to control the possible experimental errors. Suppose, we are testing immunohistochemically the presence of a hormone, pancreatic polypeptide, in the pancreas of fish. We might be using antiserum raised in rabbits against avian pancreatic polypeptide. It is possible we might get negative results. Our inference might be that the pancreas of fish is lacking pancreatic polypeptide. But to ascertain that the procedure we followed was correct and that the antiserum we obtained was potent enough to cross-react with the polypeptide, we must have a control test run in the pancreas of a bird and/or a mammal, which are known for certain to give positive results.

When we talk about histochemistry or immunohistochemistry, the need for controls is more obvious. For example, when we use a histochemical test to localize a particular chemical substance in a tissue (e.g. carbohydrate, proteins, lipids, nucleic acids, enzymes, hormones, etc.), we need to make sure that the substance stained is really the substance intended to be stained by the method we adopted. A simple control for such a situation is a negative control. We can remove the particular chemical substance (e.g. removal of carbohydrates by enzyme digestion) from the tissue and then apply the histochemical test. Comparison of the experimental and control tissues would ascertain the validity of the method applied.

Blind controls are useful in many experiments, especially when we know what results are expected to occur. For example, in an experiment designed to test the effectiveness of growth promoting hormone, the expected result is obvious. There is always the possibility of "falsifying' data on our part if we have a prior knowledge of the likely results. This problem can be overcome by the "blind" design in which a person (a technician, perhaps) who is not informed which plant received which treatment is entrusted with the task of evaluation of the results. Thus the bias toward an "expected" result can be avoided. *Double blind controls* can also be used, in which not only the person who evaluates the results is blind as to what treatment is given to which subject, but also the subjects are blind to the nature of the treatment they receive.

Randomization

Ensuring that every experimental unit gets equal chance of being assigned to a treatment, in all aspects including space and time, is randomization. In other words, all treatments have an equal chance of being assigned to each unit in the experiment. Randomization begins from the collection of the sample and continues to the end of the experiment.

An experiment may contain two or more groups receiving different treatments. The allotment of sampling units to the different groups should be random. For example, we may procure 20 mice with known and similar genetic background, for an experiment to test the effectiveness of a drug to reduce heartbeat rate. We then divide the 20 mice into two groups of ten each, one for drug treatment and one for control (placebo) treatment. How are we to select the first ten and second ten? "Catching" blindfold the mice from the cage may not be really a random selection of a group. It can be claimed that the mice that allow themselves to be caught earlier than the others are in some way less active than those that do not allow themselves to be caught. An easy way out is to assign numbers to each of the mice and pick the numbers by lottery. The first ten numbers would form one group, and the remaining ten, the second group. Which group will receive which treatment should also be at random, may be by the toss of a coin.

Within a group, say the experimental group, which of the ten mice will get first injected with the drug? Is there any difference between the first one to be injected and the last one to be injected? It is possible that the investigator who injects the drug or the placebo gets tired with the passage of time and therefore, the amount of test material injected, handling of the animal, etc. are different for those treated last when compared to those treated first. If the experimental procedure is more complicated than a simple intramuscular injection, such as one involving a surgery, the difference due to elapse of time may be more pronounced. This necessitates randomization in time.

Randomization in time pertains not only to the order of treatment at the beginning of the experiment but also to the order of observations or measurements to be taken during the course and at the end of the experiment.

Randomization is also necessary wherever there is variation due to space. For example, when we arrange experimental cages, aquarium tanks, pots, etc. in the laboratory, their position may make a lot of difference with reference to light, aeration and like factors. A pot placed close to a window might receive more light and air than those that are kept away from the window. Unless we have the facility to maintain artificially a uniform lighting, aeration, humidity, temperature, etc., variations are bound to arise due to space and influence the result of the experiment.

Randomization is more important especially in field experiments mainly because of the heterogeneity of soil. While soils in adjacent areas are more homogeneous, those far apart are more heterogeneous.

Randomization is becoming more and more important and complex in the field of molecular biology where microarray experiments are routinely conducted to ascertain the functions of genes in relation to different environmental factors.

We shall consider more about randomization when we discuss the different experimental designs, later in this chapter.

Measurement

Very often, the results of an experiment require to be measured using some kind of scale, which may be an ordinary metric scale, a digital balance or a sophisticated spectrophotometer. Whatever instrument we use, we must be concerned with the accuracy and precision of our measurements. *Accuracy* is closeness of measured value to the true value. Ironically, the true value of a measurement is never known and, therefore, we would never be able to say whether our measurement is accurate or not. For example, suppose we measure the length of a fish as 4.3 cm. This value is obtained by using a scale that could give values up to a tenth of a centimetre. If we use a still finer scale, one with vernier facility, we might be able to get measurements up to a hundredth or even a thousandth of a centimetre. But how can we be certain that the value we obtained is accurate? Suppose we know the true value of the length of the fish as 4.25 cm. Any measurement that is close (e.g. 4.3 cm) to this true value can be described as more accurate when compared to a value that is far away (e.g. 4. 0 cm or 4.5 cm).

Precision is the reproducibility of a measurement or numerical result. Suppose, we measure using a digital balance the weight of a fish and get a result as 10.6 g. We weigh the same fish again in the same balance and get a result as 10.5 g. A third measurement may give a result of 10.9 g. Our results are not precise. And suppose we use another balance and our results are something like 10.6 g, 10.5 g, and 10.5 g. Now we can say our results are more precise than the ones obtained using the first balance. We can describe the second balance as a precision instrument because it yields reproducible measurement values. Precision is practical and more valuable than accuracy.

A Few Common Experimental Designs

All experimental designs require a thorough understanding of the basic principles of statistics, especially those of inferential statistics. At the end of an experiment we collect data from the sample, analyse it, and infer about the population from which the sample was obtained. If we know beforehand what statistical test should be applied to the data, we can design the experiment accordingly. That is why it is suggested to all experimental biologists to think about the statistical tests to be used before the start of the experiment. Sir R. A. Fisher (1890–1962), whose contribution to Design of Experiment in Biology is well known, said in his Presidential address to the First Indian Statistical Congress in 1938, "to consult the statistician after an experiment is finished is often merely to ask him to conduct a *post mortem* examination. He can perhaps say what the experiment died of".

The following examples of experimental designs are based on statistical principles. However, we are not going to consider the detailed statistical procedures of each design but only the principle behind and the possible applications.

One-group design   A random sample is collected from the population and its statistics (e.g. mean) is compared with the population parameter (e.g. μ). For example, a sample of persons having the habit of consuming alcohol may be collected and examined for their blood cholesterol level. It may be tested whether the mean blood cholesterol level of this sample is significantly different from mean population level. An example of statistical analysis of data of a one-group design is shown in Box 8.1.

BOX 8.1   ONE-GROUP EXPERIMENTAL DESIGN

Large Sample statistics vs. Population parameter

Testing the significance of the difference between the sample mean and population mean

Procedure

1. Statement of the problem—Hypothesis

2. Null-hypothesis: $H_o: \overline{X} - \mu = 0$

3. Level of Significance, 0.05 or 0.01

4. Sampling distribution of sample means. Computation of the Standard Error of the mean either from the population SD (σ) or sample SD (S), as

$$\sigma_{\overline{X}} = \frac{\sigma}{\sqrt{n}} \qquad \text{or} \qquad S_{\overline{X}} = \frac{S}{\sqrt{n-1}}$$

5. Location of the sample statistics in the sampling distribution, as

$$Z = \frac{(\overline{X} - \mu)}{S_{\overline{X}}}$$

6. Decision regarding the rejection of the H_o

Minimum Z required for rejecting the H_o

Level of Significance	One-tailed test	Two-tailed test
0.05	1.64	1.96
0.01	2.32	2.58

If the calculated Z > the table Z, reject the H_o.

7. Inference: Difference between $\overline{X}$ and μ is significant if H_o is rejected, and not significant if H_o is failed to be rejected. The given hypothesis is discussed based on the above decision

---

BOX 8.1   (Continued)

Example   A certain random sample of 100 men from a hill-tribal village gave a mean height of 167 cm with a SD of 5 cm. Discuss the suggestion that the men of this tribal village do not form a part of the Dravidian race whose mean height is 170 cm.

Step 1   Statement of the hypothesis: Mean height of the tribal men ($\overline{X}$) is significantly different from the population mean (μ).

Step 2 H_o: $\overline{X} - \mu = 0$; LS: 0.05, two-tailed test (because no direction of test is specified).

Step 3   Assumption of a sampling distribution of $\overline{X}$'s of samples from the population and computation of the SE of the mean as

$$S_{\overline{X}} = \frac{S}{\sqrt{n-1}} = \frac{5}{\sqrt{100-1}} = \frac{5}{9.95} = 0.5$$

Step 4   Location of the observed mean in the sampling distribution in terms of the Z-score as

$$Z = \frac{(\overline{X} - \mu)}{S_{\overline{X}}} = \frac{167-170}{0.5} = \frac{3}{0.5} = 6$$

Step 5   Decision about the H_o: Since the calculated Z (6) > the table Z (1.96, at 0.05 LS, two-tailed test), we reject the H_o: $\overline{X} - \mu = 0$; That is, there is a significant difference between the observed mean and the population mean, with a significance of P<0.05. The calculated Z is greater than even 2.58 (Z value required to reject H_o at 0.01 LS, two-tailed test). Therefore, we can claim a greater significance of the difference, P<0.01.

Step 6   Inference: Since the null-hypothesis is rejected, the suggestion that the men of the hill-tribal village whose mean height is 167 cm do not form a part of the Dravidian race whose mean height is 170 cm cannot be rejected.

Two-group design   Samples may be obtained from two different populations, and their statistics ($\overline{X}_1$ and $\overline{X}_2$) may be used to compare the population parameters (μ_1 and μ_2) for a possible difference. Or, a sample may be randomly divided into two and allotted to "experimental" and "control" groups, in order to assess the possible effect of the "experimental treatment". Statistical analysis data of a two-group design, for large samples and for small samples is explained with example in Box 8.2 and Box 8.3, respectively.

BOX 8.2   TWO-GROUP EXPERIMENTAL DESIGN (LARGE SAMPLES)

Comparison of means of two large samples

Example   Samples of one-year-old adult male *Tilapia mossambica* were collected one from each of two geographically isolated lakes, and their body lengths were measured to the nearest millimetre. From the data below, determine whether there is statistically significant difference between males of the two populations in terms of body length.

$$\overline{X}_1 = 74 \qquad \overline{X}_2 = 78$$
$$S_1^2 = 225 \qquad S_2^2 = 169$$
$$n_1 = 42 \qquad n_2 = 56$$

Step 1   Statement of the problem: To test whether there is significant difference between $\overline{X}_1$ and $\overline{X}_2$

H_o: $\mu_1 - \mu_2 = 0$ (i.e., the mean of the population from which the sample 1 was drawn is not different from the mean of the population from which the sample 2 was drawn).

LS: 0.05 / Two-tailed test

Step 2   Sampling distribution of difference between means, $\overline{X}_1 - \overline{X}_2$ with a mean of 0, and a SE of difference between means

---

BOX 8.2   (Continued)

$$S_{\overline{X}_1 - \overline{X}_2} = \sqrt{\frac{S_1^2}{n_1 - 1} + \frac{S_2^2}{n_2 - 1}}$$

$$= \sqrt{\frac{225}{41} + \frac{169}{55}} = \sqrt{5.49 + 3.07} = \sqrt{8.56} = 2.93$$

Step 3   Location of the observed difference between means in the sampling distribution in terms of Z score, as

$$Z = \frac{(\overline{X}_1 - \overline{X}_2) - (\mu_1 - \mu_2)}{S_{\overline{X}_1 - \overline{X}_2}} = \frac{(74 - 78) - 0}{2.93} = \frac{4}{2.93} = 1.365$$

Step 4   Decision about H_o: Minimum Z required to reject H_o at 0.05 LS two-tailed test is 1.96. The calculated Z (1.365) < table Z (1.96). Therefore, we fail to reject the H_o: $\mu_1 - \mu_2 = 0$.

Step 5   Inference: There is no statistically significant difference between the mean lengths of the two geographically isolated populations of one-year-old male *Tilapia mossambica*.

BOX 8.3   TWO-GROUP EXPERIMENTAL
DESIGN (SMALL SAMPLES)

Student's t-test

Comparison of means of two small samples (Analysis of uncorrelated groups)

1.   Data: Group 1: n_1, $\overline{X}_1$ and S_1; Group 2: n_2, $\overline{X}_2$ and S_2

2. H_o: $\mu_1 - \mu_2 = 0$ (no significant difference between the means of the two populations from which the samples n_1 and n_2 were obtained, respectively).

BOX 8.3   (Continued)

3. Sampling distribution of differences between means. Standard Error of difference between means of two uncorrelated groups is obtained using pooled variance $(S_p{}^2)$

$$S_p^2 = \frac{n_1 S_1^2 + n_2 S_2^2}{n_1 + n_2}$$

SE of difference between means is calculated as

$$SE = \sqrt{\frac{S_P^2}{n_1 - 1} + \frac{S_P^2}{n_2 - 1}}$$

4. Calculation of t-value as, $t = \overline{X}_1 - \overline{X}_2 / SE$

5. Table t at specified LS and $n_1 + n_2 - 2$ degrees of freedom (one- or two-tailed test)

6. *Decision*   If calculated t > table t, reject H_o

7. *Inference*   Based on the decision the given H is discussed.

Example   Two horticultural plots were divided each into six equal sub-plots. Organic fertilizer is added to Plot 1 and chemical fertilizer is added to Plot 2. The yield of fruits from Plot 1 and Plot 2, in kg/sub-plot, is given below. Can we say the yield due to organic fertilizer is higher than that due to chemical fertilizer?

Plot 1	6.2	5.7	6.5	6.0	6.3	5.8
Plot 2	5.6	5.9	5.6	5.7	5.8	5.7

Step 1   Using a hand-held electronic calculator we can easily obtain the following statistics for the above data.

Plot 1	Plot 2
$n_1 = 6$	$n_2 = 6$
$\overline{X}_1 = 6.08$	$\overline{X}_2 = 5.716$
$S_1 = 0.279$	$S_2 = 0.116$

Step 2 $H_o: \mu_1 - \mu_2 = 0$; LS $= 0.05$ (one-tail);
$DF = 6 + 6 - 2 = 10$

BOX 8.3   (Continued)

Step 3   Assumption of a sampling distribution of difference between means, with a mean $\mu_1 - \mu_2 = 0$ and SE of difference between means computed as follows.

$$\text{Pooled variance } S_P^2 = \frac{n_1 S_1^2 + n_2 S_2^2}{n_1 + n_2} = \frac{6(0.279)^2 + 6(0.106)^2}{6 + 6}$$

$$= \frac{6(0.08) + 6(0.01)}{12} = \frac{0.48 + 0.06}{12}$$

$$+ \frac{0.54}{12} = 0.045$$

$$SE = \sqrt{\frac{S_P^2}{n_1 - 1} + \frac{S_P^2}{n_2 - 1}} = \sqrt{\frac{0.045}{5} + \frac{0.045}{5}}$$

$$= \sqrt{0.009 + 0.009} = \sqrt{0.018} = 0.134$$

Step 4   Location of the observed difference between means in the sampling distribution in terms of t, calculated a follows.

$$t = (\overline{X}_1 - \overline{X}_2)/SE = (6.08 - 5.716)/0.134 = 0.364/0.134 = 2.72$$

Step 5   Decision about the H_o. Table $t_{0.05 \text{ LS (one-tail) /10 DF}} = 1.812$.

Since the calculated $t >$ the table t, H_o: $\mu_1 - \mu_2 = 0$ is rejected. There is significant difference between the means of the two plots.

Step 6   *Inference*   The yield due to organic fertilizer (Plot 1) is significantly higher than that due to chemical fertilizer (Plot 2).

Matched-pair data analysis design   In this design only one group of sample is used. The data may be collected "before" and "after" an experimental treatment. Or, data may be collected after the control treatment and, again, after the experimental treatment. Thus, in this design, each sample unit serves for both control and experimental treatments. The pairs of data obtained are matched, i.e., the difference between each pair is obtained and tested whether the mean difference is significant. An example of a matched-pair data analysis is shown in Box 8.4.

BOX 8.4   MATCHED-PAIR DATA ANALYSIS DESIGN

1. Same subject yields pairs of data. e.g. (1) Values obtained before and after treatment; (2) Values obtained after control treatment and after experimental treatment; (3) Values obtained at two different periods—now and after a gap of a day, a month, a year, etc.

2. Significance of the mean difference $(\overline{D})$ is tested using t-test.

3. H_o: $\overline{D} - \mu_D = 0$; $n - 1$ degrees of freedom (n = pairs of data); Sampling distribution of mean differences ($\overline{D}$'s) with a population mean of $\mu_D = 0$.

4. Computation:

   [1] D = matched difference between n pairs of values—if the "after" value is greater than "before" value the difference is shown as +; if the "after" value is less than the "before" value, the difference is –. This + or – sign should be taken into account while calculating ΣD and $(D - \overline{D})$

   [2] ΣD

   [3] Mean difference, $\overline{D} = \Sigma D / n$

   [4] $(D - \overline{D})$, $(D - \overline{D})^2$ and $\Sigma(D - \overline{D})^2$

   [5] SD $S = \sqrt{\Sigma(D - \overline{D})^2 / n}$

   [6] SE of mean difference, $S_{\overline{D}} = S / \sqrt{n - 1}$

5. $t = \overline{D} - \mu_D / S_{\overline{D}}$, this is, $t = D - 0 / S_{\overline{D}}$

6. Table t at specific LS and $n - 1$ degrees of freedom (one-tail, if direction specified; two-tailed, if not)

7. *Decision*   If calculated t > table t, reject H_o

8. *Inference*   Based on the decision, the given H is discussed.

Example   A pharmaceutical company develops a drug which, it claims, increases haemoglobin content in aged people. The haemoglobin content (g/100 ml) of 10 subjects is measured before and after administration of the drug. On the basis of the following data, determine whether the company's claim is valid.

BOX 8.4   (Continued)

Subject	1	2	3	4	5	6	7	8	9	10
Before	10	9	11	12	8	7	12	18	10	9
After	12	11	13	14	9	10	12	14	11	12

Step 1 H_o: ; 0.05 LS one-tailed test with $10 - 1 = 9$ DF.

Step 2   Computation of and

$$\overline{D} = \Sigma D / n \text{ and } SD = S = \sqrt{\Sigma(D - \overline{D})^2 / n}$$

Before	After	D	$D - \overline{D}$	$(D - \overline{D})^2$
10	12	+2	0	0
9	11	+2	0	0
11	13	+2	0	0
12	14	+2	0	0
8	9	+1	−1	1
7	10	+3	1	1
12	12	0	−2	4
10	14	+4	2	4
10	11	+1	−1	1
9	12	+3	1	1
		20		12

$$\overline{D} = \Sigma D / n = {}^{20}\!\!/_{10} = 2\, g/100\, mg/100\ ml$$

$$S = \sqrt{\Sigma(D - \overline{D})^2 / n} = \sqrt{12/10} = \sqrt{1.2} = 1.1\ mg/100\ ml$$

Step 3   Assumption of sampling distribution of $\overline{D}$ with a mean of $\mu_D = 0$ and SE computed as

$$\text{SE of mean difference} = \frac{SD}{\sqrt{n-1}} = \frac{1.1}{\sqrt{10-1}} = \frac{1.1}{3} = 0.37$$

Step 4   Location of the observed mean difference in the sampling distribution in terms of t as

$$t = (\overline{D} - \mu_D)/SE = \frac{2}{0.37} = 5.41$$

BOX 8.4   (Continued)

Step 5   Decision about the H_o: $\overline{D} - \mu_D = 0$
Table t at 0.05 LS (one-tail) and 9 DF = 1.833
The calculated t (5.42) > table t; ∴ Reject H_o.

Step 6   The sample mean difference is significant. The drug is effective in increasing the haemoglobin content in aged people. The claim of the company is valid.

Multiple-group design   Application of ANOVA would enable a variety of multiple group designs. A few of the common designs are one-way classification design, Randomized Block Design, Factorial design or two-way classification design, Latin square design, Split-plot or nested design. Application of analysis of covariance is more useful in the design of experiments because it would enable the incorporation of correlation and regression between variables into ANOVA.

One-way classification design   This is relatively a simple and common design in which two or more groups are compared. More specifically this design is described as *one-way classification, completely random design with fixed effects*. It is referred to as one-way classification because only one factor is under study in each experiment. The factor may be studied at different levels, i.e., in a number of groups. In other words, one factor is studied in k groups.

Let us consider an example to illustrate the above. Suppose we are interested in studying the effect of temperature on bacterial growth. In such a study the factor is the growth of bacteria measured in terms of number of colonies/plate. The different temperatures such as 25°, 30°, 35°, 40°, and 45° at which the bacteria are cultured are the levels, i.e., 5 groups ($k = 5$). At each temperature we may have a number of replicates, i.e., culture plates, which refer to the sample size in each level or group. The number of replicates in different groups may be equal or unequal. The samples of the k groups are independent of one another. The allotment of a culture plate is random. Hence the design is *completely random design*. On the other hand, we decide the number of levels. Though we could have selected any number of temperatures, say, from 0° to 100°, we choose only 5 specific temperatures. Even these temperatures are not selected randomly. That is why this design of experiment is described to have *fixed effects*, meaning that the levels of factor are specifically selected by the investigator because of their significance.

In the above experimental model, the factor, growth of bacteria (number of colonies/plate), may be studied in a number of bacterial species at a particular temperature, say, 37°. Here the levels of the factor are the number of species. Each species is cultured in replicates at 37°.

Whether it is "one species–different temperatures" or "different species–one temperature," the basic assumption for applying ANOVA is that the samples of each level come from a "population" having that characteristic. The sample of bacteria cultured at 25°, 30°, 35°, 40°, and 45° are assumed to have come from the respective populations of bacteria cultured at 25°, 30°, 35°, 40°, and 45°. Similarly, samples of k number of species of bacteria are supposed to have come from their respective population.

The null-hypothesis to be tested in the one-way classification, completely random design with fixed effects is $\mu_1 = \mu_2 = \mu_3 = ... = \mu_k$, represented by $\overline{X}_1, \overline{X}_2, \overline{X}_3...\overline{X}_k$ from $n_1, n_2, n_3 ... \mu_k$ samples. The procedure of ANOVA is described in Box 8.5.

BOX 8.5 ONE-WAY CLASSIFICATION DESIGN—ANOVA

Example The following data represent the gain in weight (in kg) of a species of edible fish cultured in 4 diet formulations (D1, D2, D3 and D4) for a period of 3 months. Analyse these data for significant difference among the diet formulations in terms of gain in weight.

D1	D2	D3	D4
4	8	5	1
5	7	7	4
1	9	8	1
3	6	6	3
2	10	9	1

BOX 8.5   (Continued)

Step 1   Computation of the basic statistics

	D1	D2	D3	D4	Grand
n	5	5	5	5	20
ΣX	15	40	35	10	100
$\overline{X}$	3	8	7	2	5
ΣX^2	55	330	255	28	668
$(\Sigma X)^2$	225	1600	1225	100	10000
$(\Sigma X)^2/n$	45	320	245	20	500

Step 2   Computation of SS (sum of squares)

$$SS_{total} = \Sigma X^2 - \frac{(\Sigma X)}{N}$$

$$= 668 - \frac{(100)^2}{20} = 668 - \frac{10000}{20} = 668 - 500 = 168$$

$$SS_{between} = \frac{(\Sigma X_{D1})^2}{n_{D1}} + \frac{(\Sigma X_{D2})^2}{n_{D2}} + \frac{(\Sigma X_{D3})^2}{n_{D3}} + \frac{(\Sigma X_{D4})^2}{n_{D4}} - \frac{(\Sigma X)^2}{N}$$

$$= \frac{(15)^2}{5} + \frac{(40)^2}{5} + \frac{(35)^2}{5} + \frac{(10)^2}{5} - \frac{(100)^2}{20}$$

$$= \frac{225}{20} + \frac{1600}{5} + \frac{1225}{5} + \frac{100}{5} - 500$$

$$= 45 + 320 + 245 + 20 - 500 = 630 - 500 = 130$$

$$SS_{within} = SS_{total} - SS_{between} = 168 - 130 = 38$$

Step 3   ANOVA table

BOX 8.5   (Continued)

Source of variation	Degree of Freedom (DF)	Sum of squares (SS)	Mean Squares (MS)	F ratio
Between	3	130	43.33	$F = \dfrac{MS_{between}}{MS_{within}}$
Within (error)	16	38	2.375	$= 43.33/2.375$
				$= 18.24$
Total	19	168		

Step 4   Decision about H_o: $\mu_{D1} = \mu_{D2} = \mu_{D3} = \mu_{D4}$

Table F with 3 and 6 DF at 0.05 LS = 3.238867

Since the calculated F (18.24) > table F (3.238867) we reject the H_o. The means of the four diet groups are not the same. P < 0.05

Step 5   There is significant difference among the diet formulations in terms of weight gain

It is also possible to have *one-way classification, completely random design with random effects*. In the experiments described earlier, we selected the levels of treatment, the temperatures or the number of species and, therefore, the design was described to have fixed effects. Instead, if the levels of treatment are selected at random, the experimental design is said to have *random effects*. For example, if the five temperatures are selected randomly from a wide range of all possible temperatures in which the bacteria can grow, then the bacteria cultured at each of the random temperatures would represent a population of bacteria growing in that temperature. Here, these populations from which the samples were drawn, have themselves been selected randomly from large number of populations. Hence, the design is said to have random effect. The method of handling of the data is the same as described earlier.

Two-way classification design or factorial design   When two factors are involved in an experiment, the design is referred to as two-way classification, completely random design with fixed effects or factorial experiment. In a factorial experimental design, the effect of two or more factors operating on a third factor is investigated. The first two factors may be in two or more levels.

For example, we may be interested in knowing the effects of temperature and pH on bacterial growth. The levels of the factor temperature may be two, say, 30° and 40°, and the levels of the pH may also be two, say, pH 4 and pH 10. Application of ANOVA would enable us to separate and evaluate not only the effects of the two factors, temperature and pH, but also the interaction of these two factors. The procedure for handling the data in a factorial design of experiment is given in Box 8.6.

BOX 8.6   FACTORIAL DESIGN OF EXPERIMENT

Let A and B be two factors affecting a specific parameter in the sample units, A_1 and A_2 be the levels of A, and B_1 and B_2 be the levels of B, at which the effect is studied, n_{A1B1}, n_{A1B2}, n_{A2B1}, and n_{A2B2} be the sample in the k groups, and N be the total of all the samples.

Step 1   The data is organized as follows:

Factorial design table

Factor B↓	Factor A		Sum
	Level A_1	Level A_2	
Level B_1	n_{A1B1} ΣX_{A1B1}	n_{A2B1} ΣX_{A2B1}	ΣX_{B1}
Level B_2	n_{A1B2} ΣX_{A1B2}	n_{A2B2} ΣX_{A2B2}	ΣX_{B2}
Sum	ΣX_{A1}	ΣX_{A2}	

Step 2   Add all the N values of X to get ΣX. Square each X and Compute ΣX^2 (Sum of the squares of all the N values)

Step 3 $$SS_{total} = \Sigma X^2 - \frac{(\Sigma X)^2}{N}$$

$$SS_A = \frac{(\Sigma X_{A1})^2}{n_{A1}} + \frac{(\Sigma X_{A2})}{n_{A2}} - \frac{(\Sigma X)^2}{N}$$

$$SS_B = \frac{(\Sigma X_{B1})^2}{n_{B1}} + \frac{(\Sigma X_{B2})}{n_{B2}} - \frac{(\Sigma X)^2}{N}$$

BOX 8.6   (Continued)

Step 4

$$SS_{\text{interaction total}} = \frac{(\Sigma X_{A1B1})}{n_{A1B1}} + \frac{(\Sigma X_{A1B2})}{n_{A1B2}}$$

$$+ \frac{(\Sigma X_{A2B1})}{n_{A2B1}} + \frac{(\Sigma X_{A2B2})}{n_{A2B2}} - \frac{(\Sigma X)^2}{N}$$

To remove the separate effects of A and B, SS of these factors is subtracted from SS interaction total.

$$SS_{\text{interaction}} = SS_{\text{interaction total}} - (SS_A + SS_B)$$

Step 5 $SS_{\text{Error}} = SS_{\text{total}} - (SS_A + SS_B + SS_{\text{interaction}})$

Step 6   The ANOVA table is constructed as follows.

ANOVA table

Source of variation	DF	SS	MS	F
Factor A	$(2-1)$	SS_A	$SS_A/DF = MS_A$	$F_A = \dfrac{MS_A}{MS_e}$
Factor B	$(2-1)$	SS_B	$SS_B/DF = MS_B$	$F_B = \dfrac{MS_B}{MS_e}$
Interaction of A and B	$(2-1)$	$SS_{\text{interaction}}$	$SS_i/DF = MS_i$	$F_i = \dfrac{MS_i}{MS_e}$
Error (within)	$(N-k)$	SS_{error}	$SS_e/DF = MS_e$	
Total	$(N-1)$	SS_{total}		

Step 7   Referring to the F-table, critical F's required, at the appropriate DF, to reject the following H_o, are obtained.

1. $H_{o1}: \mu_{A1} = \mu_{A2}$

2. $H_{o2}: \mu_{B1} = \mu_{B2}$

3. $H_{o3}: \mu_A = \mu_B$

If the calculated F is greater than the table F, H_o is rejected and the inference is discussed.

Randomized block design   Also referred to as *randomized complete block design (RCBD),* this design is used mainly to reduce the "error" due to natural variations. It involves the arrangement of the experiment in the form of *blocks.* The assignment of the experimental units to each block is *random.* For example, suppose we want to compare the yield of four varieties of paddy, A, B, C and D. A simple randomized block design with 4 replications is to have an agriculture field divided into 4 blocks of equal size, each block having 4 sub-blocks of equal sizes. The four varieties of paddy are grown in all the 4 blocks. Further, the allotment of a paddy variety to each sub-block of a block is *random.* Such a randomization procedure would result in minimum variation within blocks and maximum variation between blocks. It reduces the total "random error" due to variations between blocks. Figure 8.1 illustrates the above randomized block design.

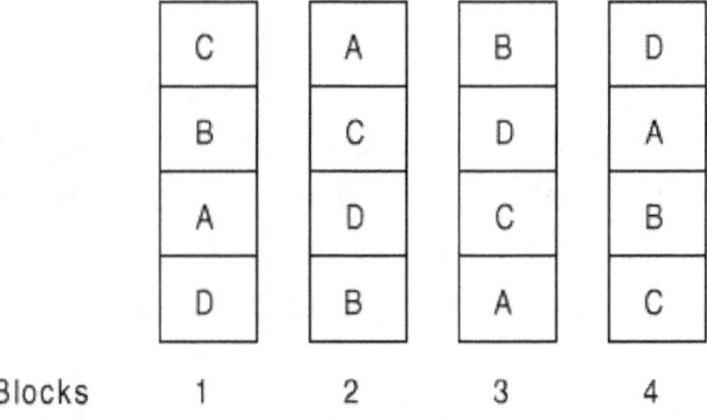

Figure 8.1   Randomized Block Design for 4 varieties of paddy with 4 replicates each.

Besides being commonly used in agricultural research, the randomized complete block design can be adopted in numerous other experimental conditions. If replications of several treatments in an experiment are to be carried out in one day, they (several treatments) will have to be performed over a period of several days. In such a case, the different days are classified as blocks. When the same experiment is replicated in different laboratories, these laboratories become the blocks. Whenever experimental conditions are heterogeneous, these conditions are subdivided, as far as possible, into more homogeneous blocks. The blocks can be age groups, genotypes, environmental factors, etc.

Latin square design (LSD)   This design is a versatile one. When we wish to control the variations in an experiment that is related to rows and columns in the field, LSD is the most appropriate one. This design can easily be adopted in many experimental situations in a greenhouse or a laboratory.

In a field study, LSD basically consists of a row-and-column design for n treatments in n rows and n columns. Each treatment occurs once in each row and once in each column as shown in Figure 8.2.

Columns

	1	2	3	4
I	A	B	C	D
II	C	D	A	B
III	D	C	B	A
IV	B	A	D	C

Rows

Figure 8.2 A simple Latin Square Design involving 4 treatments with 4 replicates.

Let us consider an experimental situation in which Latin square design can be adopted. Suppose we are interested in testing the effectiveness of the extracts of four plants, A, B, C and D (say, herbal formulations, of which one or more could be controls) in controlling blood sugar level in diabetic patients. Testing a single extract requires one whole day. Let us also assume that there are only four diabetic patients (I, II, III, and IV) who have agreed to be the subjects of our experiment, and that they will be available with us on four consecutive Sundays (1, 2, 3, and 4). The experimental unit in this set-up is a "patient-Sunday". If we adopt the Latin Square Design, the rows (as in Figure 8.2) would represent the patients and the columns the consecutive Sundays. On each Sunday all the four extracts are tested. The decision as to which patient would receive which plant extract on which Sunday is made by the random procedure. For example, as per the design shown in Figure 8.2, on the first Sunday, patient II would receive the extract of plant C.

The LSD is more useful in field studies where we would like to control variations in two-dimensional space, i.e., different directions, rows and columns. Treatments (e.g. different fertilizers, pesticides, etc.) are allotted at random within rows and columns, with each treatment once per row and once per column.

The ANOVA of the data collected from an experiment based on LSD is as shown in Box 8.7.

BOX 8.7   ANOVA OF DATA FROM LATIN SQUARE DESIGN

ANOVA table format

Source of variation	Degrees of freedom*	Sums of squares (SS)	Mean square (MS)	F
Rows (R)	R–1	SS_R	$SS_R/(r–1)$	MS_R/MS_E
Columns (C)	R–1	SS_C	$SS_C/(r–1)$	MS_C/MS_E
Treatments (Tr)	R–1	SS_{Tr}	$SS_{Tr}/(r–1)$	MS_{Tr}/MS_E
Error (E)	(r–1)(r–2)	SS_E	$SS_E/((r–1)(r–2))$	
Total (Tot)	$r^2–1$	SS_{Tot}		

*where r = number of treatments, rows, and columns.

Sample ANOVA table

Source of variation	Degrees of freedom	Sums of squares (SS)	Mean square (MS)	F
Rows	3	40.77	13.59	5.91[†]
Columns	3	125.39	41.80	18.16[†]
Treatments	3	160.57	53.52	23.26[†]
Error	6	13.81	2.30	
Total	15	340.54		

[†]Table F at 0.05 Level of Significance and at 3,6 degrees of freedom is 4.76

Split-plot design   This design is useful for two factors, one of which may have to be applied to larger experimental units and the other factor, which requires smaller areas within the larger units. For example, we may be interested in knowing the effects of 5 different fertilizers (A, B, C, D and E) on the growth of 10 genotypically different plants of species (1, 2, 3, 4... 10). An experimental field is divided into 5 five main plots. Each main plot is applied with one fertilizer, referred to as main-plot factor. The allotment of fertilizers to the main plots is by random procedure. Each of the 5 main-plots is subdivided into 10 equal sized subplots. Each sub-plot in a main-plot is planted with a genotypically different plant. The allotment of the different plants to the sub-plots in every main-plot is once again by random procedure.

The split-plot design can be adopted in a variety of experimental situations such as effect of environmental conditions, disease trials and nutrient trials on different plants.

The ANOVA of the data obtained, involves

- comparison of variation between treatments applied to main plots to random variations between main plots, and

- comparison of variation between treatments applied to sub-plots, and variation of the interaction, to random variation between sub-plots.

CHAPTER NINE

Microscopy

INTRODUCTION

Microscope is a common appliance used in all biology laboratories. There are numerous varieties of microscopes serving different purposes, and more sophisticated and powerful ones are being designed and developed. We shall consider the principles and applications of some of the commonly used microscopes.

LIGHT

Since light is one of the crucial requirements for the working of a microscope, let us understand a few of its basics.

Light is an electromagnetic wave. The energy carried by light waves is packaged in discrete particles called *photons* or *quanta*. The propagation of light wave is described by its wave properties. The emission and absorption aspects of light are explained by its particle properties.

Light, as a part of electromagnetic wave, is produced from objects as a result of *thermal radiation*, which is emission of electromagnetic radiation due to thermal motion of the molecules of the object. All objects emit enough visible radiation, i.e., light, at sufficiently high temperatures, and become self-luminous. For example, when we heat a metal loop (routinely used in a microbiology laboratory for streaking microbial cultures) over a Bunsen flame, it glows "red hot". The emission of light by the tungsten filament in an electric bulb or the element of an electric heater is also due to thermal radiation. Sun, which is the most common source of light for earth, emanates light as a result of thermal radiation.

While sun and other distant stars are described as self-luminous because they emit light of their own accord, the common objects around us are not self-luminous. We are able to see these objects because they reflect light falling on them.

Using electrical energy, we can induce certain materials to emit light. For example, when electricity is discharged through ionized gases, light is produced. The mercury-arc lamp producing bluish light and sodium-vapour lamps emanating orange-yellow light, and 'neon' lamps of various colours are a few examples. The fluorescent lamp ('tube-light'), which we commonly use in our household, is a variation of the mercury arc lamp; it uses a material called phosphor to convert the ultraviolet radiation from mercury into visible light.

Laser (acronym, Light Amplification by the Stimulated Emission of Radiation) is another source of light that has very wide applications ranging from CD player to microsurgery. A laser is a device that induces atoms to emit monochromatic light as a very intense, narrow, coherent, parallel beam.

All electromagnetic radiations travel in vacuum at the same speed, i.e., the speed of the light. The speed of the light in vacuum has been estimated to be 299,792,458 m/s (approximately, 3×10^8 ms^{-1}).

Definitions, explanations and illustrations of a few terms relevant to light wave are given in Box 9.1.

BOX 9.1   LIGHT WAVE, WAVELENGTH, AMPLITUDE, AND PHASE

Light is any electromagnetic radiation in the range of infrared to ultraviolet. It has three basic dimensions namely amplitude, frequency and polarization.

A light wave is a disturbance that spreads like the waves on the surface of water body. It has all the structural features of a typical wave as graphically shown in Figure 9.1.

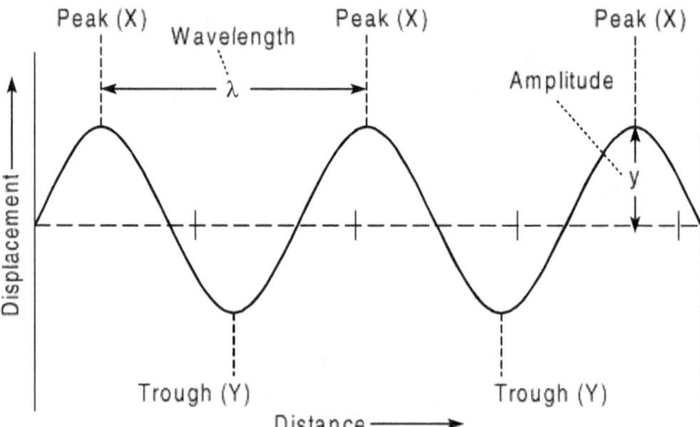

Figure 9.1   Graphical representation of a waveform.

The term *"waveform"* refers to the shape of a graph of the varying quantity (I, e.g. strength of electric or magnetic field) against time

BOX 9.1   (Continued)

or distance. An oscilloscope can be used for the visual demonstration of the shape and properties of a wave.

The *wavelength* of a monochromatic light is the distance between two consecutive peaks. It is denoted by the Greek alphabet *lambda* (λ). Wavelengths of visible light can be measured in metres or in nanometres (nm).

Frequency of a light wave is the number of times a wave repeats per unit time, i.e., the number of "wavelengths" that cross a certain point in space in a given amount of time. It is measured in cycles per second or hertz (Hz). Frequency is directly proportional to the speed of the wave and is inversely proportional to the wavelength.

Since all light waves travel at the same speed, more short waves will cross a point in space, per second, than long waves. That is, shorter waves have a higher frequency than longer waves. The relationship between wavelength, speed, and frequency is expressed by the equation:

$$c = lf$$

Where c is the speed of a light wave in m/sec (3×10^8 m/sec in a vacuum), l is the wavelength in meters, and f is the wave's frequency in Hz. We perceive frequency as the colour of the light.

The *amplitude* (more specifically, peak amplitude) of a light wave is the height of the wave. It is measured from a point midway between a peak and a trough to the peak of the wave (Y in the graph shown in Figure 9.1). This height corresponds to the magnitude of oscillation, to the maximum strength of the electric and magnetic fields and to the number of photons in the light. Amplitude of a light wave pertains to the light's intensity, and we perceive it as brightness.

Polarization is the angle of vibration of the wave and it cannot be normally perceived by human eye.

The term *phase* of a wave describes the position of a feature (peak or trough) of a wave, in relation to another wave. The phase may be measured as a time, distance, a fraction of the wavelength. A *phase shift* is a difference or change in phase. The phase shift describes

BOX 9.1   (Continued)

how far the wave slides, to the left or right, in relation to another wave.

Let us try to understand the concept of phase and phase change with the help of graphical representations of waves shown in Figures 9.2, 9.3, and 9.4.

Consider the two waves A and B in Figure 9.2:

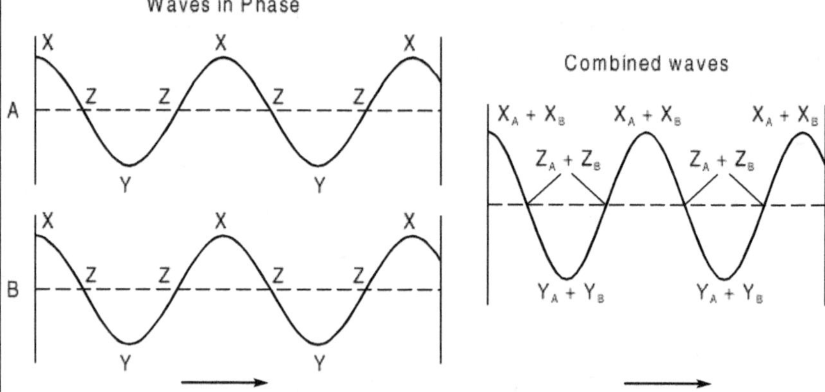

Figure 9.2 Graphical representation of waves in phase.

Both the waves A and B have the same amplitude and the same wavelength. We can see that the positions of the peaks (X), troughs (Y) and zero-crossing points (Z) of both waves all coincide. The *phase difference* of the two waves is thus zero, or, the waves are described to be *in phase*.

Suppose, the two in-phase waves A and B are combined together (we assume them to be two lights shining on the same spot), the result will be a third wave of the same wavelength as A and B, but with twice the amplitude. Doubling of the amplitude will have increased brightening effect. This is known as *constructive interference*.

Now let us consider waves A and C shown in Figure 9.3:

BOX 9.1 (Continued)

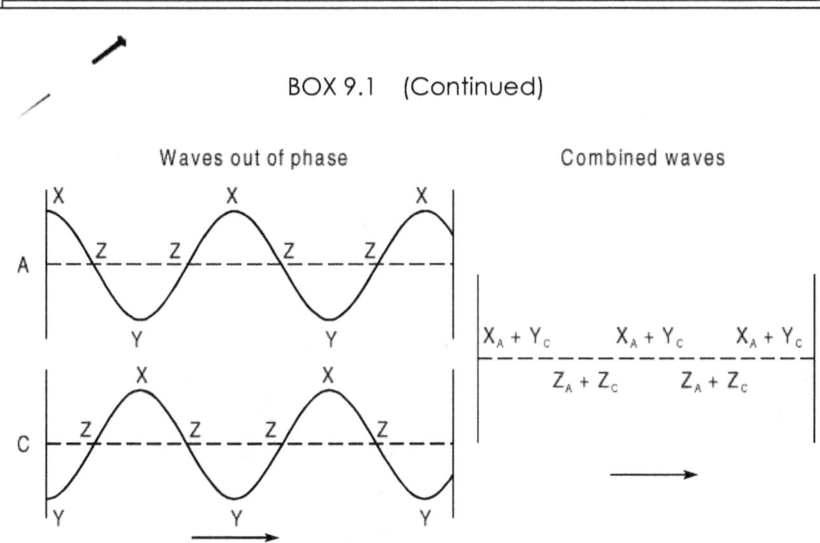

Figure 9.3 Graphical representation of waves out of phase.

Both the waves, A and C, are also of the same amplitude and wavelength. It can also be seen that the zero-crossing points (Z) are coincident between A and C. However, we can notice that the positions of the peaks and troughs are reversed, that is an X on A becomes a Y on C, and vice versa. In this case, the two waves are described to be *out of phase* or *in antiphase*, or the phase difference of the two waves is π radians, or half the wavelength ($\lambda/2$). If we add the two waves A and C, the result is a wave of zero amplitude, and this is referred to as *destructive interference*.

Waves in phase and waves out of phase are illustrated in Figure 9.4.

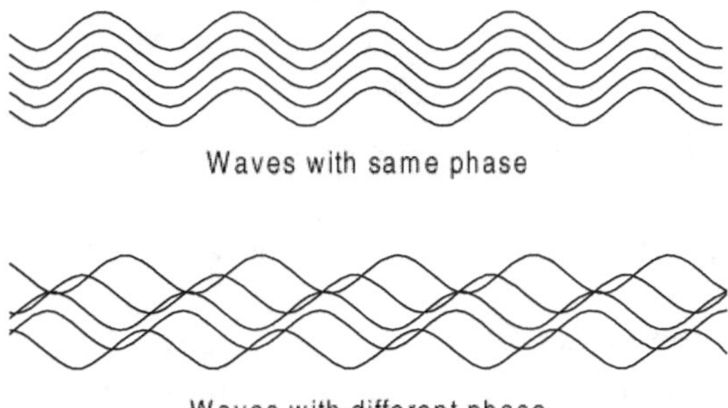

Figure 9.4 Waves travelling in same phase (in phase) and in different phase (out of phase).

Reflection and Refraction

When a beam of light strikes the smooth surface of a non-transparent material (e.g. polished surface of a mirror), it is bounced back, i.e., reflected. When a beam of light strikes a smooth interface separating two transparent media (e.g. air and glass; air and water; glass and water), it is partly reflected and partly transmitted, i.e., refracted, into the second medium. Refreshing our knowledge on reflection and refraction of light would also help us in understanding the principles of microscopy.

A ray is the direction of the path in which the light is propagating. We depict rays in a diagram by lines with arrowheads. A beam is a stream of light energy, which we diagrammatically represent as a number of rays; these rays may be diverging, converging or parallel.

Reflection   Consider Figure 9.5. Let MM´ be a plane surface (e.g. surface of a plane mirror). AO is the *incident ray* representing the direction in which light strikes the plane surface. O is the *point of incidence* and OB is the *reflected ray*. ON is the perpendicular to the reflecting surface at the point of incidence and is called *normal*. The angle, *i*, formed between the incident ray and normal (i.e., angle AON) is called the *angle of incidence*. Likewise, *r* is the *angle of reflection*, which is formed between the reflected ray and normal (i.e., angle BON).

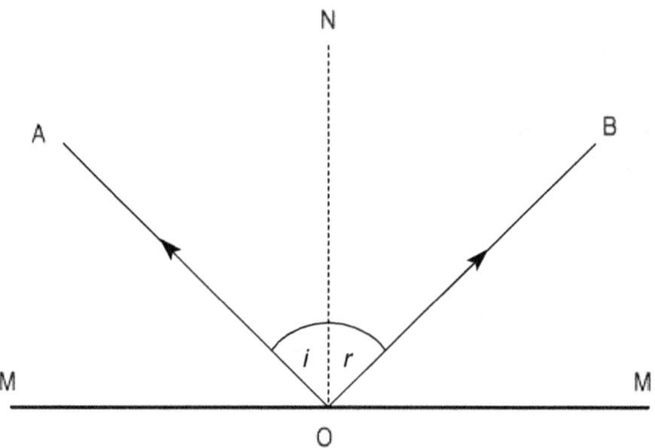

Figure 9.5   Reflection of light from a plane surface.

According to the **laws of reflection of light**, (1) the incident ray, the reflected ray and the normal at the point of incidence all lie in the same plane, and (2) the angle of incidence is equal to the angle of reflection.

Refraction   Consider Figure 9.5. AO is the incident ray striking at the point of incidence O on an air–glass interface. OB is the refracted ray. ON is the normal at O in air and ON′, the normal at O in glass. *i* is the angle of incidence, and *r* the angle of refraction.

We should remember that when a light ray passes from one medium to a more optically dense medium, the ray bends towards the normal, as shown in Figure 9.6, where the light passes from air into the glass. On the other hand, when a ray passes from an optically dense medium into a less dense medium (e.g. from glass or water into air), the ray bends away from the normal. The refraction is due to change in the speed of light when it passes from one medium to another.

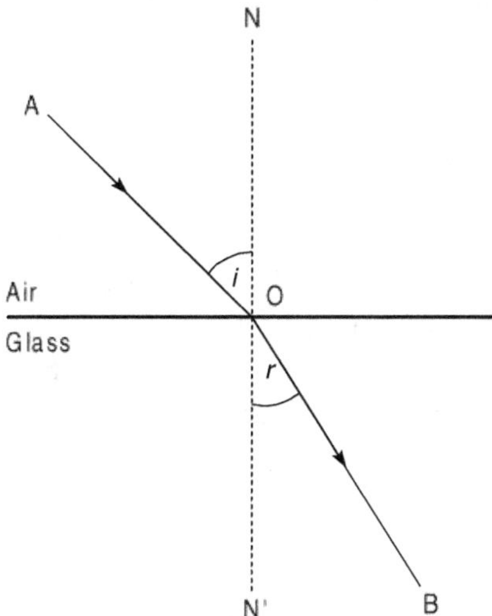

Figure 9.6   Refraction of light at the interface of air and glass.

The **laws of refraction of light** are as follows:

1. The incident ray and the refracted ray are on the opposite sides of the normal at the point of incidence and all the three rays are in the same plane.

2. The ratio of the sine of the angle of incidence (sin *i*) to the sine of angle of refraction (sin *r*) is a constant. The second law is also known as Snell's law.

Refractive index   The constant $\dfrac{\sin i}{\sin r}$ for a ray passing from one medium to another medium is called the refractive index of the second medium with respect to the first. The refractive index (RI) is denoted by *n*. RI can also be expressed as a ratio of the speed of light in vacuum (c) to the speed of light in a given medium (v). Thus,

$$RI = n = \frac{c}{v}.$$

Light travels more slowly in any medium than in vacuum. Therefore, the RI of any medium is always more than 1. The RI of a material and the speed of light in that material are inversely proportional. The greater the RI of a material, the slower the speed of light in that material. The refractive index of water is 1.33, and that of glass, about 1.52–1.80. RI of glasses is useful while making lenses for various optical instruments such as telescope, microscope, camera, etc.

LENS

All light microscopes and many other optical instruments such as a camera make use of lenses. So, let us recall some basics about them. A lens is an optical system with two refracting surfaces. There are two main categories of lenses, converging and diverging. While converging lenses are thickest in the middle, the diverging lenses are thinnest in the middle (Figure 9.7).

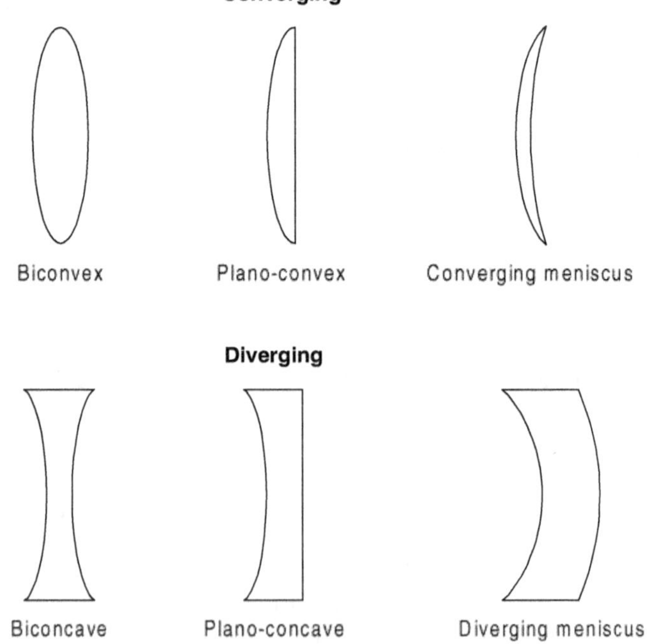

Figure 9.7   Types of converging and diverging lenses.

A simple, thin lens has two spherical surfaces so close that we can ignore the distance between them (thickness of the lens). The *principal axis* of a lens is the line joining the centres of curvature of its surfaces.

The *principal focus* of a converging lens is that point on the principal axis to which all rays originally parallel and close to the axis converge after passing through the lens. When a parallel beam of light, parallel to the principal axis, is incident on a converging lens, the rays after passing through the lens, all converge to a point on the principal axis (Figure 9.8). This point is the principal focus, F.

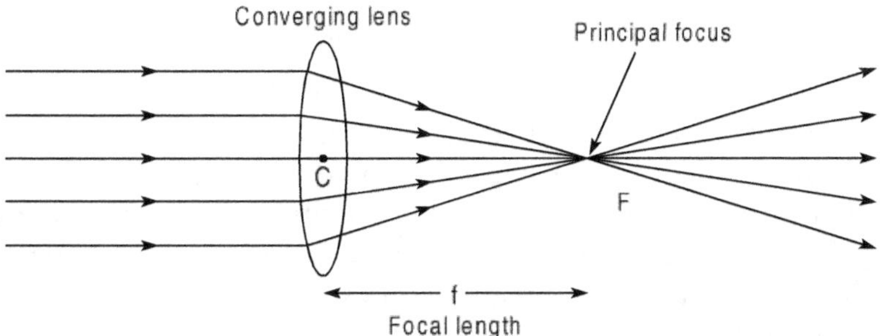

Figure 9.8 Principal focus of a converging lens.

In a diverging lens, a beam of incident rays, parallel to the principal axis, will spread out after passing through the lens, as if diverging from a focus behind the lens (Figure 9.9). The point from which the rays appear to diverge is the principal focus, F, of a diverging lens.

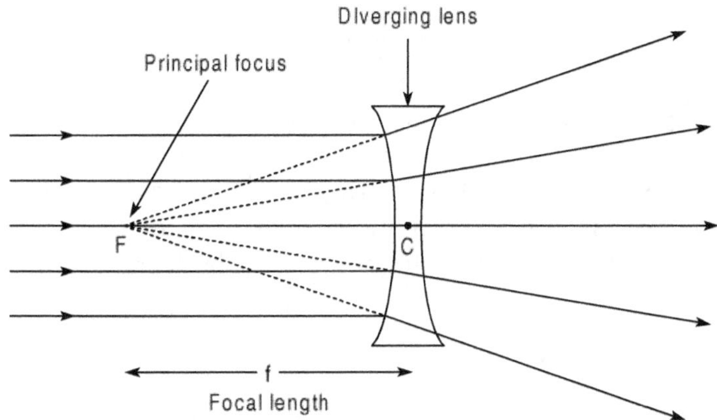

Figure 9.9 Principal focus of a diverging lens.

The principal focus is thus real for a converging lens and virtual for diverging lens.

The central point of the principal axis within the lens is referred to as the *optical centre*. Rays passing through optical centre are not refracted, or the deviation is so negligible that we can ignore it. The *focal length* (f) of a lens is the distance between the optical centre (C) and the principal focus (F) (Figure 9.8 and 9.9). A lens has two principal foci, one on either side. These foci are indicated as F and F´. They are equidistant from the optical centre.

Image Formation by a Lens

The illustration in Figure 9.10 shows the use of a lens as a magnifying glass. When the object (OA) is placed between the lens and the F´, the image (IB) formed is virtual, erect and magnified. The image appears on the same side of the lens where the object is placed. A magnifying glass is described as a simple microscope in contrast to a compound microscope, which is made up of two or more lenses mounted in a tube.

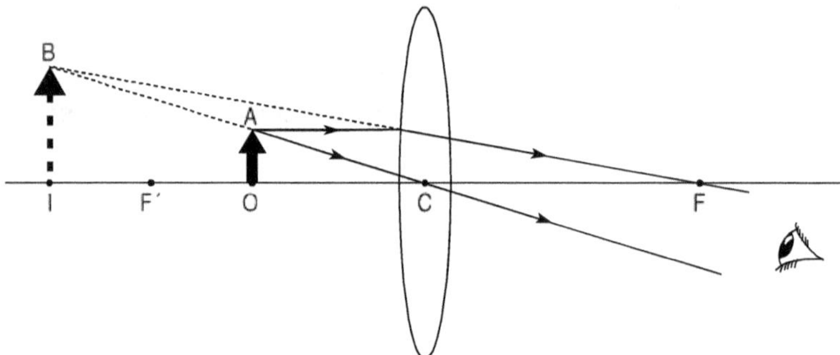

Figure 9.10   Formation of image by a converging lens.

In the case of diverging lens, whatever be the position of the object, the image formed is virtual, erect and smaller than the object. The image appears to be situated between the object and the lens.

Magnification

The size of the image produced by a lens varies according to the position of the object. The ratio of the height of the image to the height of the object is called the *linear magnification, m.*

$$m = \frac{\text{height of image}}{\text{height of object}}$$

Since m depends on the distance of the image from the lens (v) and the distance of the object from the lens (u), m can also be expressed as

$$m = \frac{\text{Distance of the image from the lens}}{\text{Distance of the object from the lens}} = \frac{V}{u}$$

COMPOUND MICROSCOPE—PRINCIPLE

When we require a magnification of an object much greater than what we get with a simple magnifying glass, we have to use a compound microscope. A simplified illustration of the principle of compound microscope is given in Figure 9.11. A compound microscope functions on the principle that an

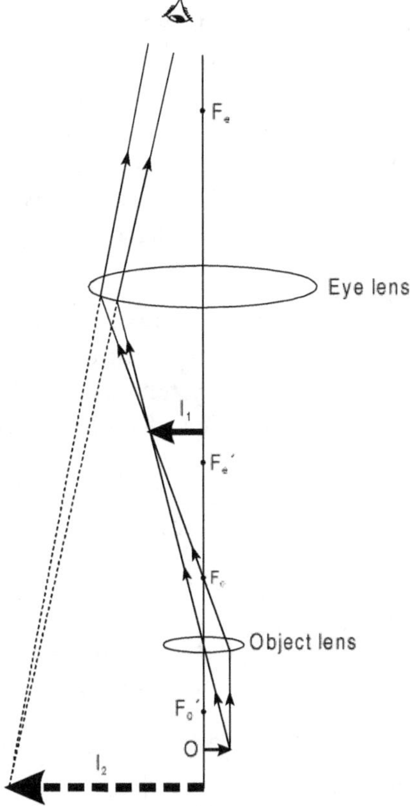

Figure 9.11   Image formation in a compound microscope.

image formed by one lens can serve as the object of second lens. A typical compound microscope employs two lenses of short focal lengths. The first is the *object lens* or *objective* and the second is the *eye lens* or *eyepiece* or *ocular*. Both the lenses are converging lenses. The object O, which we want to view through the microscope, is placed just beyond the first focal point of the objective (F_o'). The objective, which is a converging lens, forms a real, enlarged and inverted image of the object (I_1). This first image, I_1, lies just inside the first focal point (F_e') of the eyepiece. The I_1 now acts as the object of the eyepiece. The eyepiece functions like a simple magnifying glass, and forms a final, virtual and still further enlarged image (I_2) of I_1. This is the image that our eye sees through the eyepiece. The objective and ocular lenses of an actual microscope are highly corrected, coated lenses with several optical elements.

Resolving Power of a Microscope

The *resolving power* of our eye is its ability to see two closely placed objects as two distinct objects. In other words, it is the minimum distance between two objects that we can see as separate objects. Theoretically (on the basis of the diameter of the cones in the retina, the distance between the eye lens and retina) our eye can resolve two objects separated by a distance of at least 20 μm. But practically, due to factors like illumination, contrast and object size, the resolving power of our eye is about 100 μm.

Like that of our eye, the resolving power of a microscope is its ability to differentiate two objects (e.g. two adjacent chromatin granules, flagella, etc). It is determined by the optical system used in it and the wavelength of the light used to illuminate the object. Resolving power of a microscope is expressed as follows.

$$d = \frac{0.5\lambda}{NA}$$

where,

d =   the minimum distance between two objects that reveals them as separate units or resolving power

0.5 =   a constant which depends upon contrast

NA (Numerical Aperture) = $n\sin\theta$

The above relationship (also referred to as Abbè equation, after Ernst Abbè, the German Physicist who largely contributed to the optical theory used in designing a microscope) shows that the resolving power of an objective and, therefore, that of a microscope has an inverse correlation with the wavelength of the illumination. The shorter the wavelength (450–500 nm), the greater is the resolving power.

The *numerical aperture* (NA) of a system of lenses (e.g. objective) is an index of its light gathering capacity and measured in terms of the angle subtended by the optical axis and the outermost ray entering into the objective. However, the value is not expressed in degrees but as a sine value, hence the term "numerical" aperture. NA is given by the value $n \sin \theta$. Theta (θ) is ½ the angle of the cone of light entering the system (Figure 9.12). The illuminating light passing through the condenser and striking the specimen is cone-shaped and therefore the light emerging from the specimen and entering the objective is also cone-shaped. That is, the light that emerges from every minute object in the specimen and entering into the objective of the microscope is cone-shaped. If the angle of this cone of light is narrow, closely packed objects of the specimen will not be distinguished as separate entities, and the resolution is poor or low. On the other hand, if the cone of light has a wide angle, that is, the light spreads out after striking the specimen, the objects in the specimen will be seen as distinct entities and, therefore, the resolution is high.

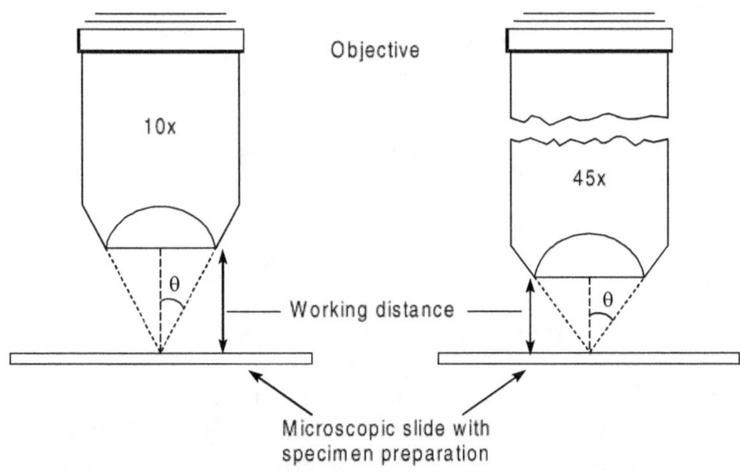

Figure 9.12 Numerical aperture of low power (10×) and high power (45×) objectives.

The angle of the cone of light entering into an objective lens is dependent on the refractive index (n) of the medium through which the light passes before entering into the lens. The refractive index of air is 1.00. The maximum θ (1/2 the angle of the cone of light) is 90° and sin 90 is 1.00. Therefore the maximum numerical aperture for any lens system working in an air medium cannot be greater than $1(n \sin\theta = 1 \times 1 = 1)$.

A practical way to increase the numerical aperture of an objective of microscope greater than 1.00 is to increase the refractive index of the medium. This is achieved by replacing the air space between the specimen and the objective lens with immersion oil that has a larger refractive index than that of air, and equal to that of glass. The higher refractive index of immersion oil enables all the reflected and refracted light rays that previously failed to enter into the objective lens, to do so now (Figure 9.13). As a result, the numerical aperture of the objective is greatly enhanced.

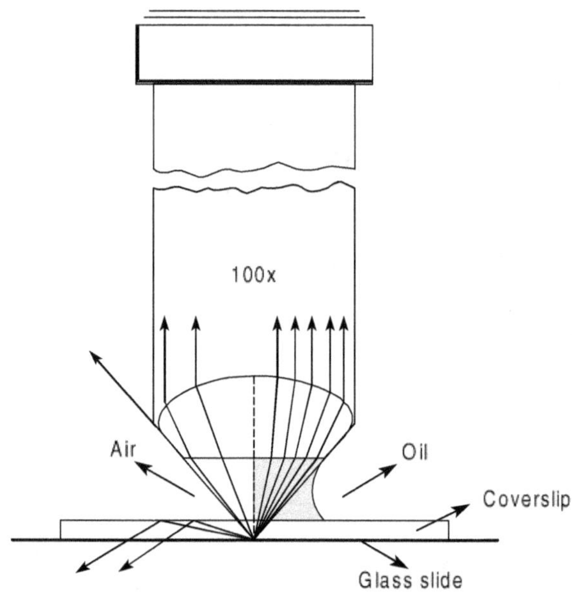

Figure 9.13   Oil immersion objective.

The resolving power of a microscope depends on the numerical aperture of not only the objective but also that of the substage condenser. Taking into account the two numerical apertures the resolving power of a complete microscope can be expressed as follows.

$$d \text{ (of complete microscope)} = \frac{\lambda \text{ (wavelength of the illuminating light)}}{\text{(NA of objective)} + \text{(NA of condenser)}}$$

The NA of a condenser which is generally about 0.9 can be increased by adding a drop of immersion oil between the top of the condenser and the bottom of the specimen slide.

For all our practical purposes the resolving power of a light microscope can be calculated using Abbè equation, as illustrated in the following example.

Example Calculate the maximum theoretical resolving power of compound microscope with (1) 10× objective with 0.25 NA and blue light of 450 nm wavelength; (2) 45× objective with 0.65 NA and blue-green light of 530 nm wavelength; and (3) 100× oil immersion objective with 1.4 NA and 450 nm wavelength light.

1. $$d = \frac{(0.5)(450)}{0.25} = \frac{225}{0.25} = 900 \text{ nm} = 0.9 \ \mu\text{m}$$

2. $$d = \frac{(0.5)(530)}{0.65} = \frac{265}{0.65} = 408 \text{ nm} = 0.4 \ \mu\text{m}$$

3. $$d = \frac{(0.5)(450)}{1.4} = \frac{225}{1.4} = 160 \text{ nm} = 0.16 \ \mu\text{m}$$

When we say that the resolving power of a microscope at a specific magnification, NA and illumination is 0.16 mm, it means that the microscope is capable of distinguishing between two spots separated by a distance of at least 0.16 mm.

Working Distance

A compound microscope generally has three to four objectives (2.5×, 10×, 45×, and 100×). We can notice the change of distance between the objective lens and the specimen slide when a specimen on glass slide is focused under each of these objectives; higher the magnification, shorter the distance. This distance, specifically between the surface of the objective lens and the coverslip or the specimen itself, when the specimen is in sharp focus, is called the *working distance*. Objectives with larger NA and higher resolving power have shorter working distance. For example, a 10× objective with an NA of 0.25 and *d* of 0.9 mm has a working distance of 4–8 mm. A high-power objective of 45× with NA = 0.65 and *d* = 0.35 mm has a working distance of 0.5–0.7 mm. The working distance of an oil immersion objective of 100× with NA = 1.4 and d = 0.18, is very much shorter, about 0.1 mm.

Useful Magnification

The magnification of a microscope (magnification of objective ×
magnification of eyepiece) that increases the size of the smallest resolvable
object sufficient to be clearly visible is referred to as the *useful magnification*.
In most of the light microscopes the largest useful magnification is 1000×
when a 10× eyepiece is used, and 1500× when 15× eyepiece is used along
with 100× oil immersion objective. Use of still higher magnification would
result only in the increase in the size of a blurred image. That is, we will
see a large but unclear image of the specimen.

Illumination

An important factor in determining the resolving power of a microscope
is the proper illumination of the object. Illumination involves a light source,
a mirror, filter and a condenser.

The source of light can be daylight (sunlight) or an electric bulb. The
daylight (preferably diffused light such as reflected sunlight) or artificial
light can be reflected into the microscope with the help of an adjustable
mirror mounted on the base of the microscope. The microscope mirror
has a plane side and a concave side. The plane side, though not good for
optimum illumination, can be used effectively with substage condenser,
for routine work. The concave side of the mirror is useful even without a
condenser. All modern-day microscopes have built-in lamps or provision
for fitting a lamp in the place of the mirror. These sources of illumination
are more flexible that we can easily manipulate the brightness of the light
with the provided rheostat control, or the amount of light with an iris
diaphragm fitted above the bulb. Facilities for inserting desired filters may
also be found in these lamps.

A bulb with coiled tungsten filament emanates uneven light and, as a
result, the image also appears unevenly lighted. Use of filter with ground
glass would help to overcome this problem to a large extent. Lamps with
tungsten bead or ribbon filament give a uniformly bright source of light.
Mercury or zirconium arc lamps can also be used with good effect.

Use of appropriate filters modifies the intensity or wavelength of the
light. Neutral density filters help to adjust the brightness. Colour filters
control the wavelength of the light used. A red or an orange filter helps to
improve the transparency and clarity of dense structures and structures
stained with blue stains. Green filters are useful to enhance the contrast

of histological preparations made with red stains. Filters are very much valuable in black-and-white photomicrography.

Optimum Illumination (Köhler Illumination)

Precise control of the light path before it reaches the specimen is very essential for any sophisticated microscopic work such as photomicrography. Such a control is given by the Köhler illumination (after August Köhler who first described it). Köhler illumination requires a low-voltage (6–10 volts), high-intensity bulb with a small concentrated filament, fitted into a built-in illuminator or into an external lamp condenser. The illuminator should have an iris diaphragm in front of it. The following general guidelines are useful in achieving Köhler illumination for most of the microscopes.

1. *Alignment of light*   If the illuminator lamp is external, it is aligned in such a way that a small, bright spot of light is projected onto the plane surface of the mirror fitted at the base of the microscope. This alignment can be checked by the formation of an image of the lamp filament on a sheet of white paper held over the surface of the mirror.

2. *Alignment of the mirror*   The condenser is raised to the top. The eyepiece is removed and the back lens of the objective is viewed through the tube. The mirror is adjusted so that the back lens of the objective is filled with light.

3. *Centring of light*   The eyepiece is put back into the tube. A slide preparation of a specimen is placed on the stage. The specimen is focused with a 10× objective. The iris diaphragm of the external illuminator lamp is almost completely closed. Now, the light through the specimen appears as a round, unsharp spot (field spot), within the field. This round image is moved to the centre of the field of view by adjusting the mirror.

   The edge of the field stop is focused by slightly lowering the condenser. Now, both the specimen and the field stop are in sharp focus. The iris of the illuminator is opened until the field of view is bright and clear. Adjusting the lamp condenser, a uniformly illuminated field of view can be obtained.

4. *Adjusting the condenser*   If the iris diaphragm of the condenser (aperture stop) is fully open, the image of the specimen appears

very bright. Therefore, we have to close it enough to eliminate unnecessary glare and to make the image appear with sufficient contrast. For stained preparations (e.g. stained sections of tissues for histology), fully opened condenser diaphragm would give satisfactory brightness and contrast. Unstained, fresh specimens may require closure of condenser diaphragm to reduce the brightness and increase the contrast. Adjusting the voltage of the illuminator can also decrease the intensity of the light. Use of appropriate filters (e.g. grey or green) would also help in this regard.

5. If the field of view is only partly illuminated, which might happen especially while using low-power objectives, the front lens of the condenser can be moved out and the condenser diaphragm fully opened. By adjusting the lamp diaphragm, the desired contrast can be achieved.

6. When the microscope has a built-in illuminator, as in all the research microscopes, steps 1 and 2 described above, can be omitted. The centring of the light is done by manipulating the centring screws of the condenser.

Advantages of Köhler illumination   The Köhler system provides optimum and appropriate illumination of the field of observation and it has several advantages.

1. *Low power input*   Though the lamp filament is small, the entire area illuminated by the lamp, i.e., the lamp field stop, has the same luminance as the filament. Therefore, only a low power input is required for the lamp.

2. *Elimination of unnecessary irradiation*   The condenser diaphragm is adjusted so that the area illuminated in the specimen corresponds to the area that is viewed through the eyepiece. As a result, unnecessary irradiation of specimen, especially living organisms, is avoided.

3. *Increasing contrast and depth of focus*   Adjustment of the condenser diaphragm by manipulating its iris helps to eliminate the unnecessary stray light from the specimen area outside the field of view. This increases the contrast and depth of focus.

4. *Irregularities of illumination source ineffective*   Under Köhler method, the light source is at infinity from the specimen, i.e., parallel rays of light impinge on the specimen. Consequently, any irregularity

in the source of light (lamp) has little effect on the specimen plane and, therefore, on the final image viewed.

5. *Brilliant image*   Due to the adjustment of the lamp diaphragm, the field of view through the eyepiece is fully filled with light. Further, light rays that may cause disturbing flare do not reach the inner wall of the tube. As a result, a brilliant image of the specimen is obtained.

COMPOUND MICROSCOPE—INSTRUMENTATION

A typical compound microscope has the following parts:

1. Oculars
2. Objectives
3. Stage
4. Substage condenser
5. Light source

Oculars

Interchangeable oculars (eyepieces) with different magnification powers (e.g. 2.5×, 6×, 10×, 12× and 15×) are provided with a microscope. Each ocular is a combination of lenses set with precise alignment into a cylindrical tube. The ocular is fitted into the upper end of the microscope tube, which is usually 160–180 mm in length. The total magnification of an ocular is engraved on its upper surface.

A simple eyepiece design, referred to as Huygenian eyepiece, has two plano-convex lenses mounted at the two ends of a cylindrical tube. The convex side of both the lenses faces down, toward the objective. The lens through which we look into is the *eye lens* and the one at the other end is the *field lens*.

There are other eyepiece designs such as Ramsden, Kellner, widefield, etc. These designs are modifications of the simple design, having multiple (doublet, triplet) pieces for the eye lens and field lens. Further, these oculars are highly corrected for optical aberrations (Box 9.2) and are *compensating*, i.e., they can work with any power objective. They are also compatible with different types of reticules such as ocular micrometer.

BOX 9.2   OPTICAL ABERRATIONS

When we observe specimens through optical instruments such as simple hand lens, a dissection microscope, a stereomicroscope or a common compound microscope we might see distorted image of the specimen. The various kinds of distortion of the image are due to lens errors. The lens errors are commonly referred to as optical aberrations. The defects in a lens are generally associated with the spherical geometry of the lens surfaces.

In a microscope, there are three primary sources of lens aberrations. The first kind of aberration is due to errors in the orientation of wavefronts and focal planes with respect to microscope optical axis. These errors are therefore called on-axis lens errors, and they cause chromatic and spherical aberrations. The second source of error is referred to as off-axis lens errors, and these errors cause aberrations such as coma, astigmatism and field curvature. The third kind of aberrations, encountered in stereomicroscopes fitted with zoom lens system, is due to geometrical distortion of the lenses. Barrel distortion and pincushion distortion are of this type of aberration. All the optical aberrations in a microscope result in faults in minute details of the specimen, especially while recording the observations microphotographically or digitally.

Chromatic aberration   A common aberration observed in spherical lenses is the chromatic aberration. This aberration occurs because the lens refracts the various colours present in the illuminating white light at different angles according the wavelengths of the colours. For example, red rays are not refracted at the same angle as green or blue rays. The effect is the focal points of these colours vary in the optical axis of the lens (Figure 9.14). The red is focused farther away when compared to green and blue. Blue is focused closest to the lens. Chromatic aberration, also referred to as dispersion, results in the formation of a colour fringe or halo surrounding the image of the specimen. Chromatic aberrations can be reduced to a great extent, or even eliminated, by designing compound lenses that consist of individual elements having different colour properties. Refinement of the optical system with more sophisticated glass formulas and shapes has helped to almost eliminate chromatic aberrations in modern microscopes.

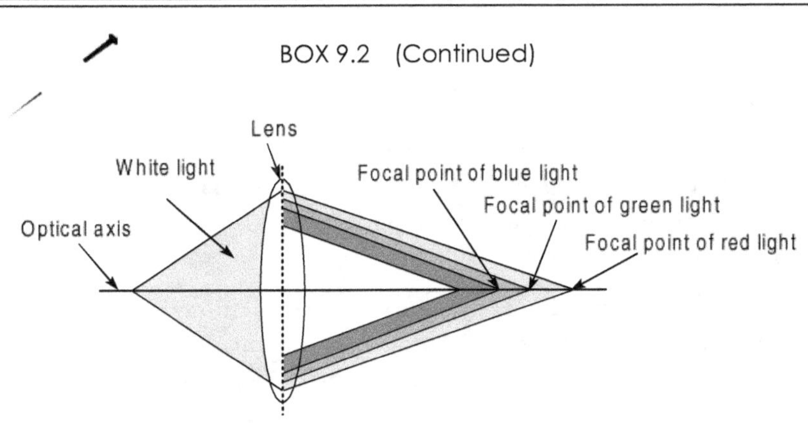

Figure 9.14   Chromatic aberration.

Spherical aberration   All spherical lenses, i.e., those having spherical surface, cause spherical aberration of the image of specimen observed in a microscope. It is due to the inability of a spherical lens to focus light waves passing through different regions, i.e., the thicker central and thinner peripheral regions, in the same plane (Figure 9.15). The result is the absence of a well-defined image plane, the image appearing blurred in the periphery when the central portion is focused and vice versa.

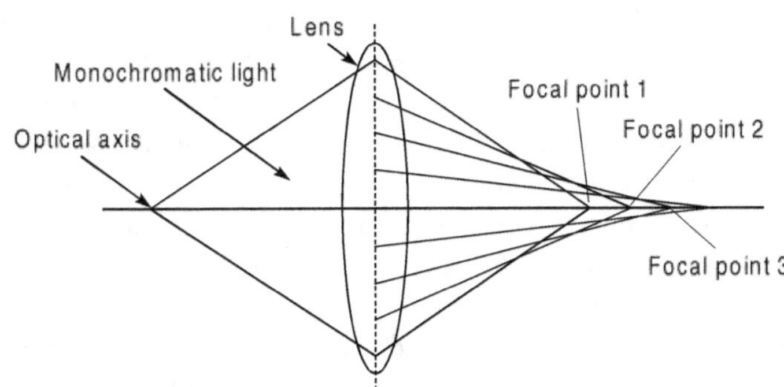

Figure 9.15   Spherical aberration.

Spherical aberration is rectified in a microscope by the use of a lens group which is a combination of positive and negative lens elements with different thicknesses, all cemented together into a single group.

Spherical aberration can be reduced by preventing light from striking the periphery of a spherical lens by closing the diaphragms to the

BOX 9.2   (Continued)

necessary extent. Modern microscopes have objectives that are highly corrected for spherical aberration, using special glass-grinding techniques, improved glass formulations and superior control of optical pathways.

Coma   This aberration is also due to off-axis light rays, and mainly occurs in microscopes with improper alignment. Coma causes an aberration in which the point of focus, especially one at the periphery of the viewfield, exhibits a streak of light that resembles the tail of a comet, hence the name coma. When the angle of incidence of the light rays on different parts of the lens becomes more and more oblique (off-axis), the refraction differences become more pronounced, and result in coma. The degree of coma aberration is directly related to the objective aperture, it being greater for lenses with wider apertures. Therefore, it can be partially corrected by reducing the aperture size to accommodate the diameter of the object field for a given objective and eyepiece combination.

Astigmatism   A microscopic optical aberration in which the off-axis image of the observed spot of specimen appears as a line or ellipse instead of a discrete point, is called astigmatism. Depending on the angle of the off-axis light rays passing through the lens, the line image may be oriented either meridionally or equatorially. Further, the contrast decreases as the distance of the image from the centre increases, and consequently, the details of the spot of the specimen being observed will not be well defined. Astigmatism in microscopes is due to one of the following three reasons: (i) asymmetric lens curvature, a manufacturing defect; (ii) improper mounting of lens in its frame; and (iii) incorrect orientation of the lenses within the objective barrel. Astigmatism aberration can be corrected by proper designing of objectives with precise spacing of individual lens elements, appropriate lens shapes and refractive indexes.

Field curvature   This type of aberration is also referred to as *curvature of field* and is the result of use of lenses with curved surfaces, a common feature in most microscopes. When light rays are focused through a lens with curved surfaces, the image plane will be curved

BOX 9.2 (Continued)

(Figure 9.16). The image can be focused over a certain range in the centre or the periphery of the viewfield. The entire image cannot be simultaneously focused especially onto a flat surface such as a film or digital imagers. Field curvature is corrected by adding corrective lens elements to the objective. Such corrected objectives are called *flat-field* or *plane* or *plano*-objectives. In spite of the corrections, field curvature aberration cannot be totally eliminated. However, the aberration is diminished by the plano-objectives to such a degree that it cannot be easily detected, especially in higher magnifications. The problem of field curvature is more pronounced in stereomicroscopes.

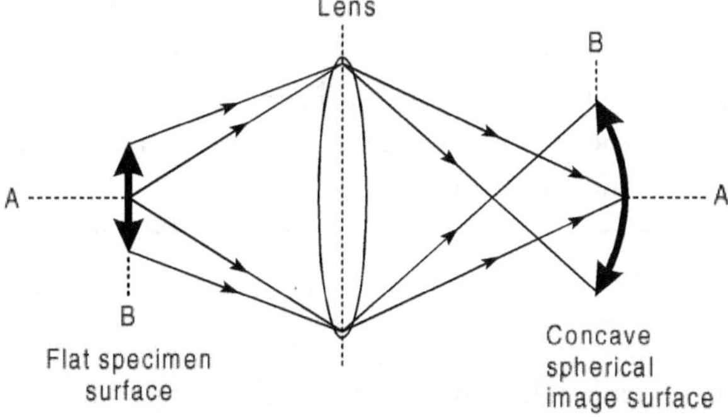

Figure 9.16 Field curvature aberration.

Geometric distortion An aberration involving geometric distortion of the image of the specimen is common in stereomicroscopy. Geometric distortion causes changes in the shape of the image but does not affect the sharpness or colour spectrum. Positive and negative geometric distortions, also called as pincushion and barrel, respectively, are the common aberrations (Figure 9.17) of this type.

Even objectives that have been designed with corrections for other types of aberrations such as chromatic, spherical, coma, astigmatism, would produce images suffering from geometric distortions. We may not be able to detect geometric distortion especially when the aberration is insignificant and the specimen does not have well defined periodic structures. However, we can easily notice this type of distortion of the image when the specimen has straight lines, grids,

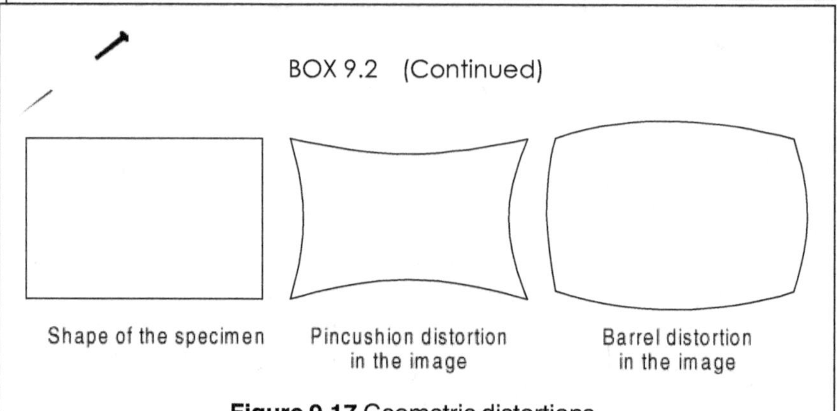

Figure 9.17 Geometric distortions.

squares, rectangles or other regular polygon shaped features. The main cause of this aberration is the complex lens system containing meniscus, concave hemispherical, and thick convex lenses. Such lens systems having variable focal lengths, produce pincushion distortion at long focal lengths and barrel distortion at short focal lengths. This is the reason why stereoscopic zoom microscopes have pronounced geometric distortions.

Objectives

The objectives (so called because of their closest proximity to the object being viewed) are the most crucial components of a light microscope because they form the primary image of the specimen. They determine the quality and magnification of the final image of the object viewed in the microscope. An objective of a modern microscope is a combination of several lenses. The lenses may be organized in singles, doublets or triplets. All the lens groups are carefully oriented, and tightly packed into a tubular brass shell that in turn is encapsulated by the objective barrel. High power objectives (e.g. 100× oil immersion) have hemispherical front lens and a meniscus second lens. This combination works synchronously to capture light rays at high numerical aperture with least spherical aberration. A spring-loaded retractable nose cone assembly protects the front lens elements from damage due to collision with the coverslip or the glass slide.

Parameters of the objective, such as numerical aperture, magnification, optical tube length, and degree of aberration correction are imprinted or engraved on the outer surface of the objective barrel. For example, an objective may have the following inscription.

Plan 40/0.65

160/0.17

It means,

Plan = planachromat (flat-field achromat)

40 = initial magnification of 40×

160 = for use with a microscope with a mechanical tube length of 160 mm (the distance between objective flange and eyepiece seating face)

0.17 = cover glass thickness, 0.17 ± 0.01 mm. Other pertinent information such as type of microscope (UV, polarized light, phase contrast), type of immersion (oil, water, glycerine or other specialized hydrocarbon-based oils), etc. might be included. In addition, coloured rings provided near the knurled ring help to visually recognize the magnification of the objective. White ring indicates 100×; light blue, 40×; yellow, 10×; and brown 2.5×.

The different objectives are fitted to the nosepiece of the microscope. The nosepiece can be rotated to exchange objectives. The different objectives of a modern microscope are *parfocal*, i.e., they are optically and mechanically designed so that their working distances remain constant. In other words, we can focus an object under low-power objective (10×), change the objective to high powers (45× or 100×) and find the object remain focused. This parfocal design is useful especially while working with oil immersion objective.

Substage Condenser

The condenser of a light microscope, as the name suggests, functions to condense the scattering light rays from the source (sunlight or external lamp through mirror or built-in lamp) into a cone of bright light, align and focus onto a tiny spot of the specimen. This illuminated part is the area that is perceived by the objective lens. The setting of the condenser, i.e., its position and its opening size is critical in obtaining optimum illumination. The condenser aperture diaphragm (iris diaphragm) can be adjusted to the desired opening size, which in turn would alter the illumination cone to match the optical properties of the objective.

The position of the condenser from its upper stop can be lowered to attain optimum illumination. Condensers are often equipped with a swing-out front lens, and are useful with low-power objectives.

Specific types of condensers are used with the different purposes such as bright-field, dark-field, phase-contrast, polarization, etc.

Light Source

The source of light for a compound microscope is either the reflected sunlight or a lamp. The lamp can be external or built-in. An adjustable mirror, fitted to the base of the microscope beneath the substage condenser, is used to reflect the light beam from external source, i.e., sunlight or lamp, along the axis of the condenser. Built-in microscope lamps are equipped with a lens system to project a sharp image of the light source into entrance plane of the substage condenser. An iris diaphragm (field diaphragm) positioned in front of this light-gathering lens system can be adjusted to regulate the area of illumination.

TYPES OF LIGHT MICROSCOPES

There are several types of light microscopes designed for specific purposes in biology, geology and related fields. A microscope may be a simple monocular type having only one ocular lens fitted to an upright or inclined tube. Binocular microscopes have two oculars fitted to a wide inclined tube. Finely aligned prisms inside the tube reflect light to both the oculars. Trinocular microscopes are binocular microscopes with a third upright tube to which a microphotographic camera or digital camera can be attached. The digital camera can be connected to a TV monitor or to a PC.

We shall consider the principle and uses of a few of the microscope types that have relevance to biology.

Bright Field Microscope

The common compound microscopes used in biology laboratories and clinical laboratories are referred to as bright field microscopes because the specimen is viewed as dark objects in a bright background. This is in contrast to the dark-field microscope in which the objects appear bright in a dark background. Because of the bright background, the specimen to be viewed must be stained to have sufficient contrast. The principle of a compound microscope has already been explained earlier and shown diagrammatically in Figure 9.15. The uses of bright field microscopes are many. They are commonly used in biology laboratories for routine observations of cytological and histological preparations.

Dark-Field Microscope

A dark-field microscope is just a modification in the illumination of the bright-field microscope to view living, unstained, transparent specimens. In contrast to the bright-field microscope, in a dark-field microscope we see bright objects against a black background. The dark-field (also referred to as dark-ground) illumination involves focusing a hollow cone of light on the specimen so that direct light (unreflected and unrefracted) does not enter into the objective lens, because the objective lies in the hollow dark portion of the cone. However, light reflected and refracted by the specimen will enter the objective and form an image. The clear portions of the specimen will not reflect or refract any light and therefore will appear dark. The medium in which the specimen is mounted also does not reflect or refract light, and therefore it will appear black. The result is the formation of brightly illuminated specimen against a dark background.

Dark-field illumination involves the use of substage condenser that blocks the central portion of a beam of light from the lamp, resulting in the formation of a hollow beam of light. The blocking is effected by *dark-field stop* placed underneath the condenser lens system (Abbé condenser). The spherically concave top lens of the condenser allows the ring of light (hollow beam) to form an inverted cone of light with its apex centred in the specimen plane. If there is no specimen on the stage, and if the numerical aperture of the condenser is greater than that of the objective, the emerging oblique rays will not enter the objective because of the obliquity. In fact, the front lens of the objective would be lying in the hollow portion of the emerging cone. Since no light enters the objective, the field appears dark. However, when a specimen—which is transparent, unstained, non-light-absorbing and mounted in a medium (e.g. water)—is placed on the stage, the oblique rays cross the specimen. Optical discontinuities of the specimen such as the body wall, cell membrane, nucleus, organelles, etc., diffract, reflect and refract the oblique rays of light striking them. Only these refracted and reflected light rays enter the objective to form a bright image of the specimen against a dark background. A simplified light pathway of dark-field microscope is shown in Figure 9.18.

The dark-field objective contains an internal iris diaphragm, which can be adjusted to reduce the numerical aperture of the objective to a value lesser than that of the inverted hollow light cone emitted by the condenser. The *cardioid condenser* is a reflecting dark-field design, which has internal mirrors to project an aberration-free cone of light onto the specimen plane.

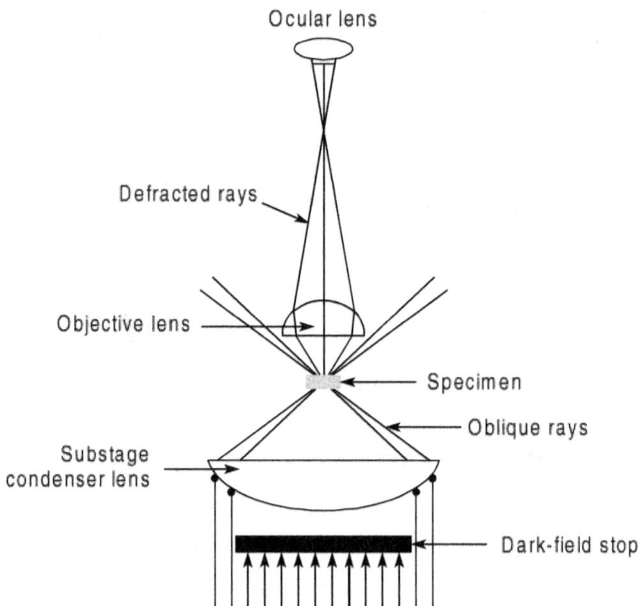

Figure 9.18   Light pathway of a dark-field microscope.

While using dark-field illumination, it is necessary to fill the space between the top lens of the condenser and the bottom of the specimen slide with oil (immersion oil) in order to effect the light to reach the specimen. The oblique hollow cone of light rays emerging from the dark-field condensers cannot emerge from the top parabolic lens without oil because the rays will be totally reflected back into the condenser. In an oiled condenser, where the refractive index of the condenser glass, immersion oil and glass slide are equal, the light is allowed to pass through the specimen, unrefracted by the glass-air interface.

Dark-field microscopy is used in many biological and medical investigations. It is effectively employed at high magnifications to observe and photograph live bacteria and other microorganisms. It is also used to observe cytological details of unstained large eukaryotic cells such as those in tissue culture, bone, fibres, hair, etc. At low-power magnifications, dark-field microscope is useful to observe small insects, many freshwater and marine algae, plankton, etc. Dark-field illumination has been found to be advantageous when used in fluorescence microscopy. Non-biological specimens such as mineral and chemical crystals, colloidal particles, and thin sections of polymers and ceramics can also be effectively viewed with dark-field illumination.

A simplified procedure for dark-field microscopy is given in the Box 9.3.

BOX 9.3   DARK-FIELD MICROSCOPY (OIL IMMERSION)—
SETTING UP PROCEDURE

1. Align the Köhler illumination, using the bright-field condenser.

2. Replace the bright-field condenser with dark-field condenser.

3. Apply a drop of immersion oil on the surface of the top lens of the condenser.

4. Prepare a wet mount of the specimen (e.g. live bacteria, minute aquatic organisms such as paramecium) on a very clean and thin slide (preferably thinner than the usual glass slide) and place it on the stage.

5. Raise the dark-field condenser to its top position so that the top lens (with oil drop) touches the specimen slide.

6. A ring of light can be seen. Adjusting the condenser, the ring is focused to get the smallest spot.

7. Apply a drop of immersion oil on top of the coverslip.

8. Focus the oil-immersion objective on the specimen.

9. Adjust the iris diaphragm of the objective to gain the maximum dark background.

10. Move the slide to choose an ideal field of observation. Adjust condenser focus for maximum brightness of reflections from the specimens.

Phase-Contrast Microscope

Another design of light microscope that is useful in observing live, transparent, unstained microorganisms and cells is the phase-contrast microscope. A phase-contrast microscope uses a special condenser with annular stop and a special objective with phase analyser.

The principle of a phase-contrast microscope is to convert the different degrees of phase retardation of light rays passing through a translucent material into amplitude differences to create an image with increased contrast. Biological specimens such as tissue sections or minute whole organisms are prepared for microscopic observation by fixing, staining and clearing (making transparent). In such materials, the various organelles modify the amplitude and wavelength of the transmitted light by absorption and scattering. The result is the formation of a sharp image of the specimen with appreciable contrast.

The objects that can be viewed in a bright-field microscope are often referred to as *amplitude objects* (or specimens) because their images are formed due to reduction in the amplitude (brightness or intensity) of the illuminating light.

But unstained cells and microorganisms do not produce sufficient modification in the amplitude of the transmitted light to give enough contrast in the image. However, the components of such materials are translucent and possess refractive indices and density different from the surrounding structures and from the mounting medium. These components with different refractive indices cause different degrees of phase retardation in the light passing through them when compared to the light that passes through the other surrounding materials and through the mounting medium. The phase-contrast microscope converts the minute change in the phase of the light into amplitude differences and forms an image with strong contrast. The objects that diffract and shift in phase the illuminating light are called *phase objects* or *specimens*.

The optical pathway of a phase-contrast microscope is shown in the Figure 9.19. The substage condenser of phase-contrast microscope has an *annular stop (condenser annulus)*, which is an opaque circular disk with a thin transparent ring. The annular stop is located at the lower focal plane of the condenser. Illumination produced by the tungsten–halogen lamp is directed through a collector lens and focused on the annular stop. The annular stop converts the cylindrical beam of light into a hollow cone of light. As this cone of light traverses through the specimen, some rays are diffracted to various minute degrees proportional to the variations in the density and refractive indices of the components. These diffracted rays are retarded by about ¼ wavelength. This is referred to as *phase retardation*; the diffracted light rays are described to be *out of phase* with the phase of the non-diffracted light rays.

Both diffracted and non-diffracted rays enter into the special phase-contrast objective. This objective is equipped at its rear focal plane with a *phase plate (phase analyser)*.

A phase plate is a special optical flat disk with a groove called phase ring. The groove is so designed that the phase ring is less thick glass when compared to other areas of the plate. The ring can cause at least a ¼ wavelength difference in phase retardation compared to non-diffracted light.

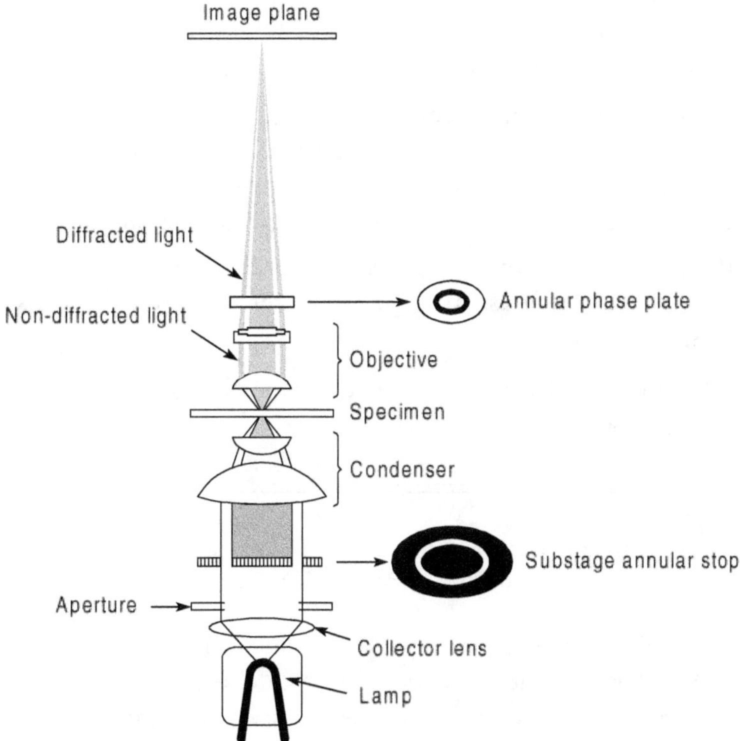

Figure 9.19 Optical pathway of a phase-contrast microscope.

Both the rays, diffracted and phase retarded as well as non-diffracted, take part in the formation of the image. The non-diffracted light rays strike the phase ring, while the diffracted rays pass through the other areas of the phase plate. The phase difference of the two kinds of rays are converted into amplitude difference, which causes either destructive interference (dark contrast) or constructive interference (bright contrast) when the image of the object is formed. A bright background is formed by the non-diffracted light and the unstained specimen appears dark and well-defined. Colour filters can be effectively used to improve the contrast. This type of phase-contrast microscopy, used in most laboratories, is referred to as the *dark-phase-contrast microscopy*.

Phase-contrast microscopy, the principle of which was first described in 1934 by the Dutch Physicist Frits Zernike, has several uses:

1. Study of microbial motility.

2. Cytology and dynamics of live prokaryotes and eukaryotes.

3. Detection of bacterial components such as endospores and inclusion bodies containing poly-β-hydroxybutrate, polymetaphosphate, sulphur or other chemical substances.

4. A more modern phase-contrast microscope coupled with electronic enhancement and post-acquisition image processing enables detection of very small components (subcellular particles) containing, for example, even a few protein molecules.

5. Phase-contrast microscope is effectively used in the observation of materials such as lithographic patterns, fibres, latex dispersion, glass fragments, etc.

A brief and simplified procedure for setting up a phase-contrast microscope is described in Box 9.4.

BOX 9.4   PHASE-CONTRAST MICROSCOPE— SETTING UP PROCEDURE

1. The phase-contrast microscope requires much more brighter light than that for an ordinary bright-field microscope. Any unwanted chromatic aberrations in the extra bright light can be reduced by the use of appropriate green filter.

2. Köhler illumination is set without the annulus in the condenser, i.e., with the normal bright-field condition (Phase condenser provides for both conditions, with or without annulus).

BOX 9.4   (Continued)

3. Phase objective is adjusted to focus on the specimen.

4. Condenser annulus is moved in place.

5. One of the eyepieces (usually the right one) of the binocular head is removed and replaced with the "telescope" eyepiece. A grey ring image of the phase plate of the objective is focused through the "telescope" eyepiece.

6. Through the "telescope" eyepiece the image of the condenser annulus, which appears as a bright ring (or as a part of a ring), is aligned with the objective phase ring, which appears as faint rings on each side of the groove in the phase plate. This alignment can be done by adjusting the condenser using the centring knobs.

   The bright image of the condenser annulus should be completely enclosed by the grey annulus of the phase plate when centring and condenser position are correct.

7. The telescope eyepiece is replaced with the regular eyepiece. Regulating the intensity of the light, the image of the specimen is fine-focused.

Differential Interference Contrast Microscope

A microscopic design very similar to the phase-contrast microscope is that of the *differential interference contrast (DIC)* microscope; a design that is capable of producing increased contrast and three-dimensional effect to the image. The DIC microscope also creates an image by detecting differences in the refractive indices of the different component parts of the specimen.

The light source of a DIC microscope produces a white or monochromatic polarized light with the help of an appropriate filter. A normal light wave vibrates in any direction but always at right angles to the direction of travel. In a *plane-polarized light*, however, there is only one vibration direction. This polarized light passes through a *birefringent prism* called *Wollaston prism*. The function of the prism is to generate two plane-polarized beams. The two beams are at right angles to each other and they fill the condenser. A cone of polarized light (a complex of the two beams) emerging from the condenser illuminates the specimen mounted on a slide. One of the beams, the *object beam*, passes through the specimen while the other beam, the *reference beam* traverses through the clear areas of the slide. The object beam undergoes phase differences in the specimen depending on the varying refractive indices of the different components of the specimen. After passing through the specimen and the clear areas of the slide, the two plane-polarized lights entering the objective are combined when they interfere with each other to form an image. An *analyser-polarizing filter*, located above the objective lens, achieves the interference between the two beams. If white light is used for illumination of the specimen, it is possible, with appropriate manipulation, to create a coloured image of the specimen. The colours are correlated to the amount of the phase change. Photometric measurement of such an image might provide valuable information about the components of the specimen.

Fluorescence Microscope

The usual light microscopy works on the principle that the light transmitted through the different components of a specimen creates an image of the specimen. On the other hand, the *fluorescent microscopy* is based on the principle that certain substances are capable of emitting light if they are excited by radiation and that the emitted light forms an image of the substance. While other light microscopes are used to study the structural details of an organism or cell, a fluorescent microscope is generally used as a probing tool, to ascertain the presence of a specific organism, cell, organelle, or molecule, in the specimen.

Fluorescence   Fluorescence can be defined as luminescence of a material when it is excited. The material may be a whole organism like bacteria, specific cells in a tissue or even molecules in a cell. There are some biological materials (e.g. chlorophyll), which have *primary fluorescence* and therefore *autofluorescence*, i.e., emit light of their own accord without any addition of fluorescent molecules. However, most biological materials do not have autofluorescence property, and therefore need to be linked with some exogenous fluorescent molecules, usually called as *fluorochromes* or *fluorescence dyes*, in order to create the required fluorescent probe.

Only lights of shorter wavelengths (ultraviolet, violet or blue), which are high-energy lights, are capable of exciting specific molecules, and therefore they are described to have an *excitation wavelength* or *spectrum*. When a molecule in the specimen is exposed to such an excitation light, the molecule absorbs the high-energy light. This results in the increase of the molecule's own energy, and now such molecules are "excited". The excited molecules must give out, i.e., emit the absorbed energy in order to return to the normal stable state from the excited state. Much of the absorbed energy is lost internally. Therefore, when the excited molecule releases the absorbed energy, it emits a light of low energy, with longer wavelength, and therefore, the colour of the emitted light (e.g. green, yellow, orange, red) is different from the colour of the excitation light. The emitted light is described to have the *emission wavelength* or *spectrum*. The wavelength and therefore the colour of the excitation light are dependent on the material. Likewise, the wavelength and colour of the emitted light are material-dependent. That is, each molecule requires specific wavelength light for maximum excitation and it emits light of specific wavelength.

Fluorochromes   Fluorochrome stains are acidic and basic dyes that are mostly yellow or orange in colour. They fluoresce intensely when exposed to lights of shorter wavelengths such as ultraviolet, violet, and blue lights. *Fluorescein* is a common fluorochrome used as fluorescence probe for free protein amino groups. It emits green light when struck with blue excitation light. A basic fluorochrome, *acridine–orange* permits discrimination between living and dead cells or microorganisms. A number of basic fluorochromes such as auramine O, acridine yellow, and acriflavine, are used to create probes for nucleic acids, especially, DNA.

An advantage of fluorescence microscopy is that we can attach fluorescent dye molecules to specific components of a sample, so that only those parts can be seen selectively. We can also use more than one type of fluorochrome, a mixture of more than one fluorochrome. By changing the excitation light, we can cause the different fluorochromes to fluoresce one after another, to distinguish two or more different components of specimen.

Types of fluorescence microscopes   There are two basic types of fluorescence microscopes, one with transmitted light illumination (*diascopic fluorescence* or *trans-fluorescence*) and the other with *epi-illumination*. The transmitted light illumination fluorescence microscope might use a bright-field condenser or, more often, a dark-field condenser. In an epi-illumination fluorescence microscope (*episcopic fluorescence* or *epi-fluorescence*), the objective functions as a condenser as well.

Transmitted light fluorescence microscope   The optical pathway of a trans-fluorescence microscope is illustrated in Figure 9.20. The lamp providing illumination for fluorescence microscope must produce a light that is capable of exciting the specific fluorochrome. One type of lamp, a high-pressure mercury-vapour arc lamp can generate UV rays and a very wide spectrum of short wavelength lights. The advantage of this source of illumination is that it can excite any fluorochrome to fluoresce. Another type, a quartz-halogen-tungsten filament lamp produces light in the blue end of the visible spectrum. This source of illumination has the advantage of being very limited to specific fluorochromes.

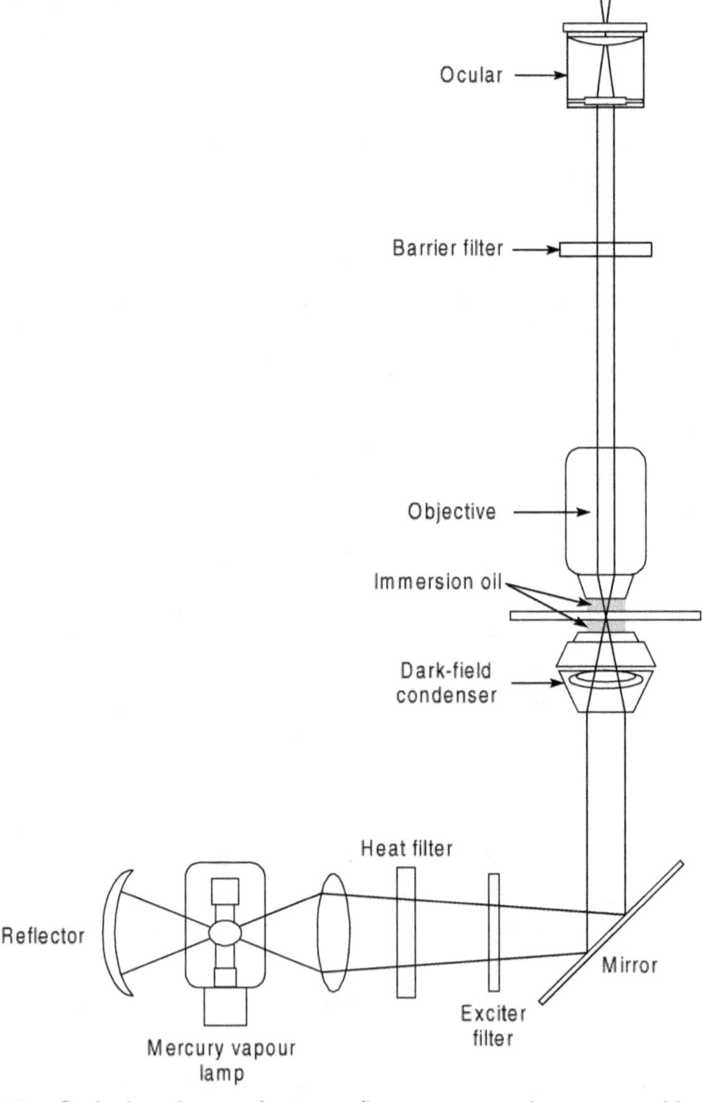

Figure 9. 20   Optical pathway of a trans-fluorescence microscope with oil-immersion dark-field condenser.

There are two filters between the illuminating lamp and the specimen. The first one is a long wave *heat-absorbing filter* and it is more useful when UV light is used. The second filter is an *exciter filter*, which is in fact a combination of filters designed to transmit only wavelengths required for the excitation of a fluorochrome, and to absorb other wavelengths.

In addition to the above-mentioned two filters, there is one more filter placed between the objective and the ocular lens. It is a *blocking* or *barrier filter* designed to absorb any remaining exciting radiation, such as UV radiation, which is harmful to the eye, or blue and violet lights, which would diminish the contrast of the image.

When a bright-field condenser is used to illuminate the specimen with excitation light, it is difficult to separate the excitation light from the fluorescing light because both kinds of light directly enter the objective. Therefore, these bright-field condensers have become obsolete and they have been replaced by condensers high numerical aperture, oil-dark-field.

The dark-field condenser creates a hollow cone of excitation light so that oblique rays of light illuminate the specimen. The fluorochromes in the specimen get excited and very quickly begin to fluoresce. Only the emitted long-wavelength rays gain entry into the objective. Most of the excitation rays do not enter the objective and even those which sneak through the objective will get blocked by the barrier filter. The image of the material formed by the emitted light rays appears as brightly fluorescing regions against a dark background.

Epi-illumination fluorescence microscope   The optical pathway of an epi-illumination microscope is shown in Figure 9.21. In an epi-illumination fluorescence microscope the excitation radiation comes from above the specimen in contrast to the transmitted light fluorescence microscope in which it comes from beneath the specimen through the substage condenser. Further, the illumination passes through the objective before striking on the specimen. Thus, the objective of an epi-illumination fluorescence microscope has a dual function of a condenser to focus illumination of the specimen as well as that of an objective to gather the reflected and emitted lights.

One other essential component required is a *dichroic beam splitter* (*dichromatic beam splitter* or *chromatic beam splitter* or *partial mirror* or *dichroic mirror* or *dichromatic mirror*). This beam splitter is designed to reflect lights of lower wavelengths and to transmit lights of higher wavelengths. The beam splitter therefore can isolate the emitted light from the excitation wavelength. The epi-illumination creates a dark background so that emitted light forms fluorescing images can be clearly seen. The wavelength at which the dichroic beam splitter lets the higher wavelength to pass is set between the excitation and emission of wavelengths of any given fluorochrome. Thus the excitation light will get reflected while the emission light will get transmitted to form the image.

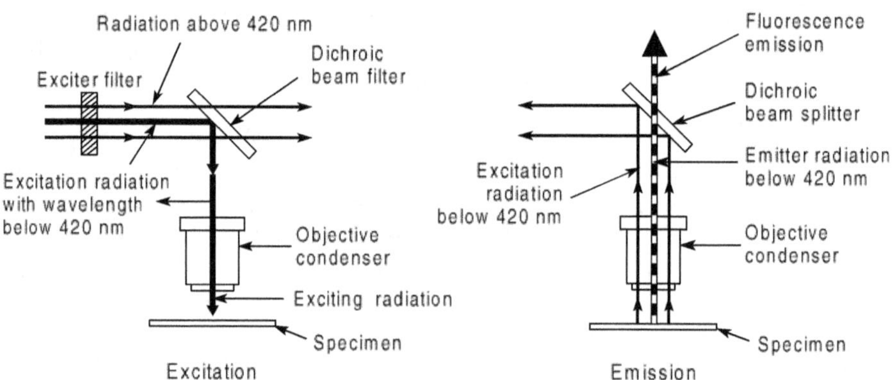

Figure 9.21 Optical pathway of epi-fluorescence microscope.

The fluorescence microscope has wide applications in medical microbiology, environmental microbiology, immunology and other related fields. Many pathogenic bacteria such as tuberculosis-causing *Mycobacterium tuberculosis* can be identified after staining them with fluorochromes.

In environmental studies, fluorescence microscope is used to analyse samples for the presence of specific microorganisms. The sample is stained with fluorochrome-labelled probes or fluorochromes such as acridine-orange and diamino-2-phenylindol (DAPI, a DNA-specific stain). When the stained sample is observed under a fluorescence microscope, the specific microorganism is identified by the fluorescence and can be counted. It is also possible to distinguish between dead and live bacteria by using combinations of fluorochromes to produce difference in the fluorescing colours.

Immunofluorescent antibody technique or, simply, *fluorescent antibody technique* (*FAT*) is useful in localizing microorganisms, proteins and other molecules with the help of specific antibodies. *Indirect immunofluorescence (indirect fluorescent antibody)* technique is used to detect insoluble antigen (e.g. bacterial capsular polysaccharide) in histological or smear preparation on a microscopic glass slide (Figure 9.22). When the insoluble antigen is exposed to antiserum, the antibody combines with antigen. The slide preparation is washed to remove the antiserum leaving only the specific antibody (immunoglobulin)–antigen complex. An antibody to immunoglobulin is conjugated with a fluorescent dye (e.g. fluorescein isothiocyanate) and applied to the tissue section or smear preparation. This tagged antibody binds with the antigen–antibody combination in the material and fluoresces when illuminated with UV light in a fluorescence microscope.

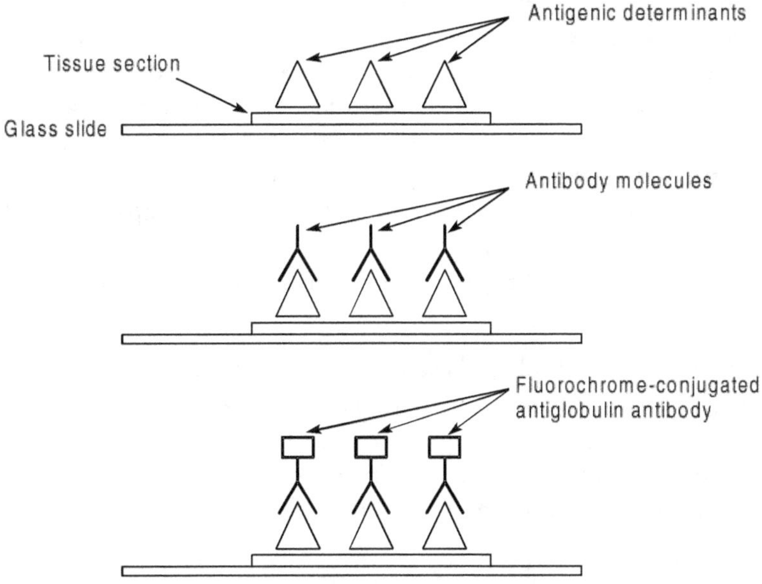

Figure 9.22 The indirect immunofluorescence technique.

In the *direct immunofluorescence technique,* the tissue or cell containing the antigens are first washed to remove free immunoglobulins and then treated directly with a fluorochrome-conjugated (e.g. fluorescein) antiglobulin reagent. This would demonstrate the binding of antibody or the deposition of antigen–antibody complexes *in vivo.*

Polarization Microscope

Polarization microscopes are common in geology laboratories where they are used to study minerals in thin sections of rocks. But this microscope has several applications in biology. For example, molecular organization of crystalline or highly organized biological molecules such as DNA, starch, wood, urea, etc. can be studied with the help of polarization microscope.

Since a polarization microscope uses *plane-polarized light*, it contains parts that convert ordinary white light into polarized light. *Polarized light* is one that vibrates in only one plane, in contrast to the ordinary white light, which may vibrate in any plane with equal probability. The human eye-brain system cannot perceive the vibration directions of light. Plane-polarized light can be detected by an intensity or colour effect.

A filter called *polarizer*, which is placed between the light source and the specimen stage, achieves the polarization of the light. There is another filter called *analyser*, which is also a polarizing element, located above the objective. Both the polarizer and the analyser can be rotated through 360 degrees. If the analyser filter is rotated, maximum light will traverse when its linear alignment is the same as that of the polarizer filter. Thus when a preparation is observed in a polarized microscope it will appear brightest when the polarizer and the analyser are aligned. It will appear totally dark when the two filters are unaligned by 90 degrees. In between these two extremes of maximal brightness and total darkness, there are infinite shades of brightness.

Polaroid films or doubly refracting crystals such as calcite are used to polarize the light. The common *Nicol prism*, made from transparent form of calcite, is used as both the polarizer and the analyser in most of the polarization microscopes. In addition to the polarizer and the analyser, there are other essential components in a polarization microscope. They are:

1. A rotating specimen stage

2. Strain free objectives

3. An eyepiece fitted with a cross wire graticule to mark the centre of the field of view

4. A Bertrand lens for easy examination of the objective rear focal plane, to adjust the illuminating aperture diaphragm and to view interference figures

5. A slot for the insertion of compensators/retardation plates between the polarizers. These plates enhance the optical path difference in the specimen. A polarization microscope can be used both with reflected and transmitted light.

As already indicated, the main application of polarization microscopes is the determination of molecular orientation in the specimen. It is the molecular orientation of a material that determines how the light traversing it will be transmitted. For example, in a cell, certain parts exhibit *birefringence*, i.e., *double refraction*. When polarized light passes through a birefringent part of a cell, it will be split into two rays, an ordinary ray and an extraordinary ray. These two rays have different velocities and are vibrating in two different planes. The polarization microscope determines the degree and the sign of the birefringence. If the extraordinary ray travels at a greater velocity than that of the ordinary ray, the birefringence is said to be positive. On the other hand, if its velocity is less than that of the ordinary ray, it is said to be negative.

Materials that exhibit birefringence are referred to as *anisotropic*, while those that do not are said to be *isotropic*. Using polarization microscope we can determine the orientation of molecules in anisotropic biological materials. Such information will be useful in studies as, for example, to determine the orientation of the protein and lipid molecules in the cell membrane.

The other applications of polarization microscope are:

1. Study of minerals in rocks.

2. Determination of the molecular structure of refined, extracted or manufactured natural and industrial minerals.

3. Analysis of composite materials such as cements, ceramics, mineral fibres and polymers.

Confocal Scanning Microscope

In an epifocal fluorescence microscope the specimen is completely illuminated by the excitation light, and this causes the entire area of the specimen to fluoresce at the same time. However, the maximum intensity of the excitation light is at the focal point of the objective lens. The fluorescence of areas other than the focal point contributes to a background haze in the image. This problem is overcome in *confocal microscope* by the addition of a pinhole aperture-screen combination at the point where the image is formed. That is, the focal point of the objective forms an image of the spot of the specimen at the minute aperture (pinhole/slit) of a screen at the back of the objective. Thus the two points, focal point and the pinhole, conjugate, and hence the name "confocal" microscope. The image is formed only by the light rays that pass through the pinhole because the rays from other areas of the specimen are blocked by the screen. Consequently, the image of the fluorescing spot is free from the "out-of focus" flare and has higher resolution.

Although unstained specimens can be viewed using light reflected back from the specimen, they are usually labelled with one or more fluorescent probes.

In a *laser scanning confocal microscope (LSCM)*, a laser is employed to provide excitation light. The laser light, blue in colour and of very high intensity, is reflected off a dichroic mirror onto two other mirrors, one after the other. These mirrors are mounted on motors and they scan the laser across the sample, through the objective. The emitted light from the fluorescing element of the specimen travels back to the scanning mirrors, gets reflected to the dichroic mirror which allows it to go through to focus onto the pinhole. A detector, which is a photomultiplier tube (PMT), measures the light that passes through the pinhole and forms an image of the element on a high-resolution monitor. A schematic representation of the light pathway of an LSCM is shown in Figure 9.23.

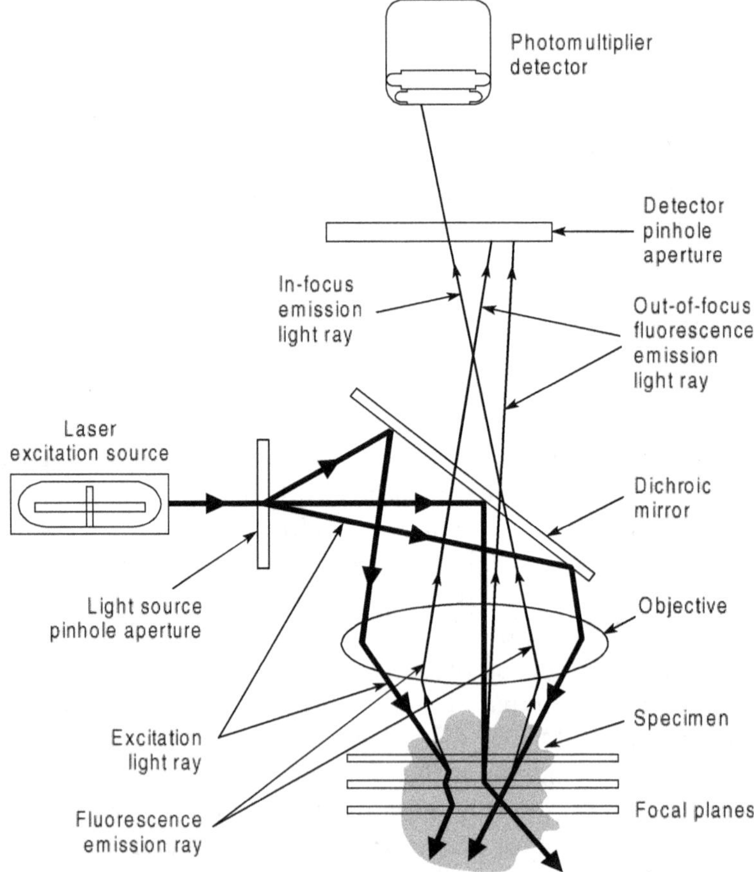

Figure 9.23   Schematic diagram of the light pathway of an LSCM.

There is never a complete picture of the specimen formed, at any given point of time. The PMT is connected to a computer that processes the information received from the photmultiplier and constructs the complete image of the specimen, one pixel at a time, at a rate of 3 frames per second. The scanning mirrors are the limiting factors, being rather slow in scanning the laser and in descanning the emitted radiation. Therefore, they are replaced, in more recent versions of the confocal scanning microscopes, with special Acoustic Optical Deflector (AOD), in order to increase the speed of the scanning. An AOD uses high-frequency sound wave in a special crystal to create a diffraction grating. By varying the frequency of the sound wave, the AOD modifies the angle of the diffracted light and thus speeds up the scanning process. With AOD it is possible to form images at a much faster rate.

There are several advantages of confocal scanning microscope.

1. A clear image, free from glare caused by out-of-focus fluorescence.

2. Higher resolution, both lateral (about 0.2 mm) and vertical (about 5 mm).

3. Confocal microscopy is noninvasive in the sense that the images formed by it are optical sections, that is, they are formed by the use of focused laser light rather than physical sectioning of the specimen.

4. By scanning several consecutive thin sections of a specimen it is possible to build up a very clear three-dimensional image of the specimen.

5. Relatively thick specimens (tissue sections or smears) can be scanned to get images of objects not only on the surface but also those at deeper planes; a computer correlation of a stacked series of images generates three-dimensional information about the specimen.

6. Imaging of living specimens can be achieved with ease because of the automated collection of three-dimensional data; this process can be improved if the live specimens are labelled with multiple probes.

7. Study of the interaction of bacteria with or within eukaryotic cells is a major use of this type of microscopy.

8. Since a computer is associated with this microscopic system, it is possible to store and if necessary transfer enormous amount of data with ease.

Stereomicroscope

We the human beings, are capable of focusing both eyes on a single object and perceive all its three dimensions. This three-dimensional perception is referred to as stereoscopic vision. The stereoscopic vision is made possible by our brain's interpretation of the two slightly different images made from two different angles. Our eyes are separated by a distance of about 6.5 cm. Each eye perceives an object from a slightly different angle. When the images formed in the retina of each eye is transmitted to the brain, the images are fused together, but still retain a high degree of depth perception. This principle is adapted in the design of stereomicroscopes. A stereomicroscope transmits twin images to our eyes from two slightly different angles, differing by about 10–12 degrees, to produce a true stereoscopic effect.

In a simple design (Greenough design), stereomicroscopes have two separate compound optical trains, each consisting of an eyepiece, an objective, and intermediate lens elements. Another design of stereomicroscope has a single objective but two optical channels for binocular vision. This design is referred to as common main objective (CMO) stereomicroscopic design. Both designs have a pair of prisms located before each eyepiece, to invert the image and give the correct orientation of the specimen.

Modern stereomicroscopes have zoom lens system or a rotating drum with telescopic optics that can be utilized to increase or decrease the overall magnification. The rotating drum system works as an integral intermediate tube containing paired sets of lenses that can be put in place into optical pathway by rotating the drum. Simplified optical pathways of the Greenough and CMO stereomicroscopes are shown in Figure 9.24.

The working distance in a stereomicroscope is large enough (2 to 14 cm, depending on the objective) so that the stage can accommodate relatively large containers (e.g. petri plates, cavity blocks, culture bottles, tubes). Thus it is useful for observation, dissection, etc. of small organisms and other materials. Most of the stereomicroscopes have objective magnifications of $0.5\times$, $1.0\times$, $1.5\times$, $2.0\times$, $3.0\times$, etc. Interchangeable eyepieces with different magnifications ranging between $1\times$ and $30\times$ are provided. The total magnification of a stereomicroscope may range between $2\times$ and $500\times$. Stereomicroscopes with trinocular head for attaching photographic equipment have also been designed.

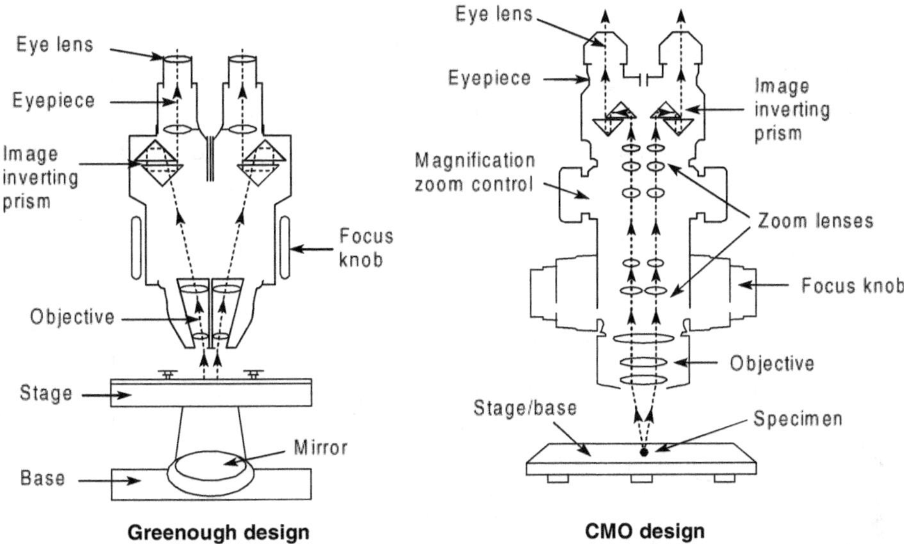

Figure 9.24   Optical pathways of stereomicroscopes.

Micrometry

The practice of measuring the dimensions of microscopic objects (e.g. diameter of RBC, thyroid follicles, ova, nuclei, etc., lengths of intestinal villi, thickness of connective tissue layers) is generally referred to as micrometry or morphometry. The most common method of making such measurements is the use of ocular micrometer and stage micrometer. We can make measurements in ordinary compound microscopes only in the range of 0.2 mm to 25 mm. We cannot measure dimensions smaller than 0.2 mm because it is less than the resolving power of a compound microscope. Likewise, measurement above 25 mm is also not practical because it will be above the average field diameter of a widefield eyepiece. Larger objects can however be measured with a stereomicroscope.

The ocular micrometer (OM) is a glass disk with a diameter of 1 cm. It is engraved with an arbitrary scale of 100 divisions or less (Figure 9.16). It is also referred to as reticle, reticule or graticule. Since it is fitted into the eyepiece of the compound microscope it is more appropriate to call it an ocular micrometer. This is the scale that is used for all measurements. Since the scale is arbitrary, we have to calibrate (standardize) it using a known standard scale, the stage micrometer (SM).

A stage micrometer is a standard microscope slide having a scale of defined length (Figure 9.25). Usually, the scale is 1 mm (1000 μm) divided into 100 divisions, so that one stage micrometer division = 10 μm. Such a micro-level scale is made by methods such as photographic process, physical engraving or electrodeposition of a metallic film directly onto the glass surface. A protective cover glass may be mounted on the scale. The scale may be encircled by a black line for easy location and focusing under the microscope.

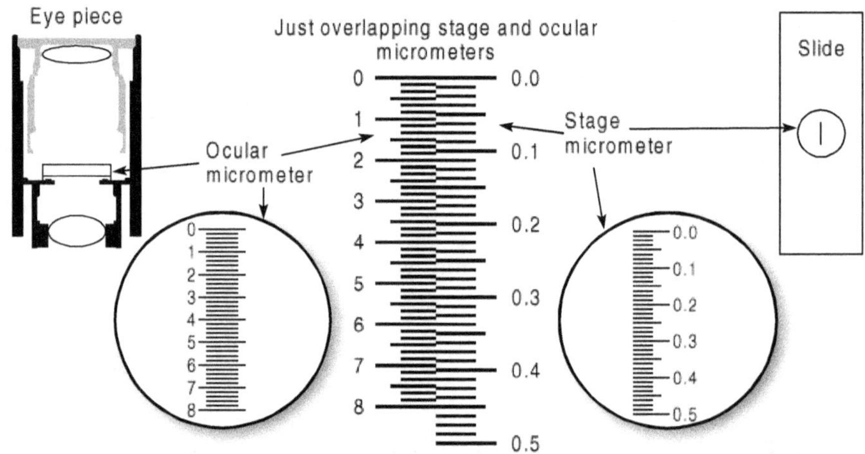

Figure 9.25 Stage and ocular micrometers.

The calibration of the ocular micrometer refers to determination of the distance of one division in terms of the absolute distance of a stage micrometer. A simple vernier principle is used for this purpose. How many of OM divisions are equal to how many of the SM divisions under a particular microscope–eyepiece–objective combination is found out. Suppose, it is found that 2 OM divisions are equal to 1 SM division which means that 2 OM divisions have a value equivalent to the absolute distance of 1 SM division, i.e., 10 μm. This is given by

$$1 \text{ OM division} = 10 \text{ }\mu\text{m}/2 = 5 \text{ }\mu\text{m}$$

$$\text{or}$$

$$= 10 \text{ }\mu\text{m} \div 2 = 5 \text{ }\mu\text{m}$$

This value is often known as micrometer value or calibration factor. Once this value has been determined, the dimension of any specimen can be calculated by multiplying the number of OM divisions spanned by the specimen with the calibration factor. It must be remembered that a calibration factor only applies to a specific microscope–eyepiece–objective combination. The procedure for the calibration of a microscope for micrometry is briefly described in Box 9.5.

BOX 9.5   CALIBRATION OF OCULAR MICROMETER

1. Note down the number of the microscope, which will be used for morphometric measurements. The same ocular micrometer and microscope with the specific combination of the eyepiece and objective should be used. For example, if we calibrate a specific ocular micrometer for use with specific 10× eyepiece and 10× objective, the micrometer value obtained is applicable only for this combination. That is, we cannot use this value for any other OM, eyepiece or objective, even if they have the same magnification powers. It is needless to say, we cannot use the obtained value for an eyepiece or objective of different powers such as 15× and 100×, respectively. In such cases, we will have to do the calibration separately for this combination.

2. Align the microscope for Köhler illumination.

3. Insert the proper and clean OM into the eyepiece. In a common microscope, the eye lens is threaded out and the OM is slipped into the tube to rest on a circular shelf in between the eye lens and the field lens. Ensure that the engraved side of the OM is facing up so that we can read the numbers correctly. We may have to insert the OM in the specific recommended position in different designs of the eyepiece. The eye lens is replaced and fitted into the microscope tube.

4. Place the stage micrometer on the stage and focus the SM scale through the specific power objective (e.g. 10×). A dark ring around it facilitates location of the SM scale. The SM scale should lie in the centre of the microscope optical path (stage aperture). Both OM and SM should be in sharp focus.

5. Using the mechanical stage and by rotating the eyepiece, the two scales are brought into parallel alignment. They should be close but not overlapping. It is further adjusted such that the

---

BOX 9.5   (Continued)

0 of the OM and 0 or any one line of the SM scale coincide. The point of initial coincidence should be as far as possible to the left end of the SM scale.

The SM is magnified under different powers of the objectives. Consequently, we can notice that under low-power objective, the entire SM scale is visible and in higher magnifications, only a part of it is seen. Under oil immersion, only a few lines of the SM scale can be observed. However, whatever be the objective magnification, the size of the image of the OM remains the same, magnified by the eye lens only.

6. Now, moving our observation from left to right from the initial point of coincidence, we identify the next point of coincidence, i.e., where lines of the two scales overlap. We count the number of divisions between the two points of coincidence, in both the scales and record the values.

7. Assuming that the second point of coincidence is the initial point of coincidence, we proceed further right to identify the next point of coincidence, and so on, every time counting the number of divisions between the two points of coincidence in both scales.

8. The counted divisions give us the information how many SM divisions are equal to how many OM divisions. Now we can calculate how many SM divisions are equal to one OM division as: 1 OM division = No. of SM divisions ÷ No. of OM divisions.

9. Since we know the value of the SM scale, we can easily convert the above value (i.e., number of SM divisions equivalent to one OM division) into microns by multiplying with the value of one SM division. For example, if we had counted that 3 SM divisions are equal to 5 OM divisions. Then

1 OM division = 5 ÷ 3 = 1.67 SM divisions.

Since 1 SM division = 10 μm (SM scale is 1 mm divided into 100 divisions, i.e., 1000 μm divided into 100 divisions, and therefore

1 SM division = 10 μm)

BOX 9.5 (Continued)

1 OM division = $1.67 \times 10 = 16.7 \, \mu$m

16.7 μm is the micrometer value or calibration factor or calibration constant.

This value is usually verified with the additional countings we had made. If they are not concurring, the mean of four or five readings may be taken as the final value.

10. The above procedure is repeated for every combination of eyepiece–objective of a microscope and only that specific factor should be used while making measurements under the respective combination.

11. To simplify the counting procedure, we can adjust the two scales so that 0 of the OM scale coincides with the 0 of the SM scale. Scanning towards the right, we identify the subsequent points of coincidences and note down the numbers of the coinciding lines in the respective scales. For example, we can note that 0 & 0, 3 & 5, 6 & 10, 9 & 15, and 12 & 20 are the points of coincidences. These values are usually recorded and the micrometer value calculated in the following table format.

Microscope make and number: **Ernst Leitz GmbH – 522114**

Eyepiece: 6×

Objective: 10×

OM	SM	SM ÷ OM	(SM ÷ OM) × 10	Micrometer value
0	0			
3	5	5/3 = 1.67	1.67 × 10 = 16.7	
6	10	10/6 = 1.67	1.67 × 10 = 16.7	16.7 μm
9	15	15/9 = 1.67	1.67 × 10 = 16.7	
12	20	20/12 = 1.67	1.67 × 10 = 16.7	

BOX 9.5   (Continued)

Eyepiece: 6×

Objective: 45×

OM	SM	SM ÷ OM	(SM ÷ OM) × 10	Micrometer value
0	0			
5	2	2/5 = 0.40	4.0	
14	5	5/14 = 0.36	3.6	
27	10	10/27 = 0.37	3.7	
51	19	19/51 = 0.37	3.7	3.7 μm
67	25	25/67 = 0.37	3.7	

Eyepiece: 6×

Objective: 100×

OM	SM	SM ÷ OM	(SM ÷ OM) × 10	Micrometer value
0	0			
6	1	1/6 = 0.17	1.7	
12	2	2/12 = 0.17	1.7	1.7 μm
18	3	3/18 = 0.17	1.7	
24	4	4/24 = 0.17	1.7	

Using the calibration factor obtained for a specific eyepiece–objective combination of a specific microscope, we can measure the dimensions of the specimen. After removing the SM from the stage, the specimen mounted on a glass slide is placed on the stage. The entire specimen or the specific object in the specimen, which we want to measure is focused under appropriate eyepiece–objective combination. For example, if we want to measure the diameters of mature ova in a histological section of frog's ovary, we can use a 10×-10× combination, or even lesser combination such as 6×-10× combination. On the other hand, to measure the diameter of the RBCs in human blood smear preparation we may require at least 10×-100× combination. The entire specimen or an object in the specimen should always lie within the OM scale range. Adjusting either the OM or the mechanical stage the 0 or any other marking of the OM scale is brought into alignment with the edge of the specimen of the object from where we want to begin the measurement (Figure 9.26). The other end of the specimen or the object up to which we need to take the measurement would lie somewhere within the OM scale. We count the number of OM divisions between the two limits of the object being measured. Suppose we count 15 OM divisions between the equatorial edges of a human RBC under 10×-100× combination with a calibration factor of 2.0 μm. The diameter of the RBC is obtained by multiplying the number of OM divisions counted with the calibration factor as $15 \times 2 = 30$ μm.

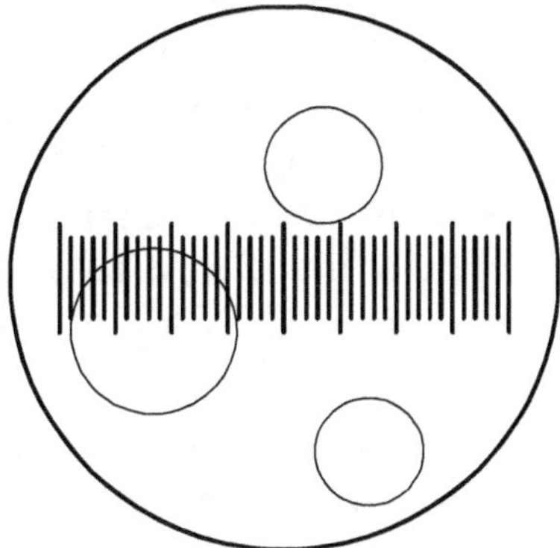

Figure 9.26 Micrometric measurement of an RBC.

One problem we commonly encounter during morphometric measurements is that while we align one of the edges of the object with the 0 or any one line of the OM scale, the other edge may not exactly lie at a specific line of the OM scale. That is, it may be observed, especially under high-power magnifications, that the edge lies in between two lines. This leads to inaccurate measurement. This problem is overcome by the use of drum micrometer or Filar eyepiece micrometer.

Drum micrometer A *drum micrometer* or, more specifically *filar micrometer*, is a specialized eyepiece micrometer functioning on the same principle as a standard eyepiece–reticle combination. But it includes a mechanism to move the OM scale by means of a precision screw mechanism that is operated by rotating an external drum. The drum micrometer should also be calibrated for the ocular micrometer using an SM. The drum of the micrometer screw is also divided into 100 intervals, so that one interval of the drum division corresponds to 0.1 interval of the eyepiece scale. Full rotation of the drum translates the measuring rule (line) across one interval of the eyepiece scale. The drum micrometer eliminates the necessity to estimate values of fractions of an OM division.

The initial steps in the procedure are same as for the standard micrometer. However, when we encounter the second edge of the specimen lying in between two OM lines, the drum is rotated so that the scale moves from left to right and the last OM line before the specimen edge reaches the edge. The number of divisions rotated in the drum scale to reach the end point is noted and added to the number of divisions of the main OM. For example, let the calibration factor of drum OM be 2.00 μm, i.e., 1 OM division = 2 μm. This would mean that one full rotation of the drum is equal to 2 μm and that 1 drum division = 2/100 = 0.02 μm. An RBC may occupy 15 OM divisions plus a fraction. The 15th line of the OM is moved using the drum till it reaches the edge of the RBC up to which the measurement has to be taken. If the number of drum divisions rotated to achieve this is 62, then the fractional value (62 × 0.02 = 1.24 μm) is added to the value of the 15 OM divisions (15 × 2 = 30 μm). Thus the actual dimension of the RBC is 31.24 μm. We can also calculate it as 15.62 × 2 = 31.24 μm.

Besides the above conventional morphometric techniques, we can also use digital imaging or traditional photomicrography techniques to determine the linear dimensions of a specimen. It is done by direct measurement of specific features and comparison to the image of a stage micrometer at the same magnification. For instance, a specimen is photographed with a 10× objective which can be measured by consecutively acquiring a second photograph of the stage micrometer at the same magnification. Using a ruler or similar measuring device, we can then directly measure the specimen feature and calculate the dimensions using the photograph of the stage micrometer. The technique is also useful for digital images, where computer software is used in place of a ruler to compare specimen images with a stage micrometer when the two images are captured at identical pixel resolution.

Camera lucida   A camera lucida is an equipment that helps to draw images of the objects that are viewed under a microscope. Originally it was an artist's accessory to sketch outlines of subjects in front of him or her in correct perspective. It is basically a four-sided reflecting prism with specific angles so that the light from the object is reflected through an eye lens.

The observer can clearly see the image of the object. A white paper is laid flat on the drawing board. The light from the paper also passes through the prism to the eye lens. The artist would look through peephole above the lens containing the prism, so that both the paper and the image of the object could be seen and the outline of the object could be sketched (Figure 9.27). Since no dark-room is required for this set-up (in contrast to an earlier set-up which required a dark-room and hence called camera obscura) the instrument is called camera lucida, meaning "bright room".

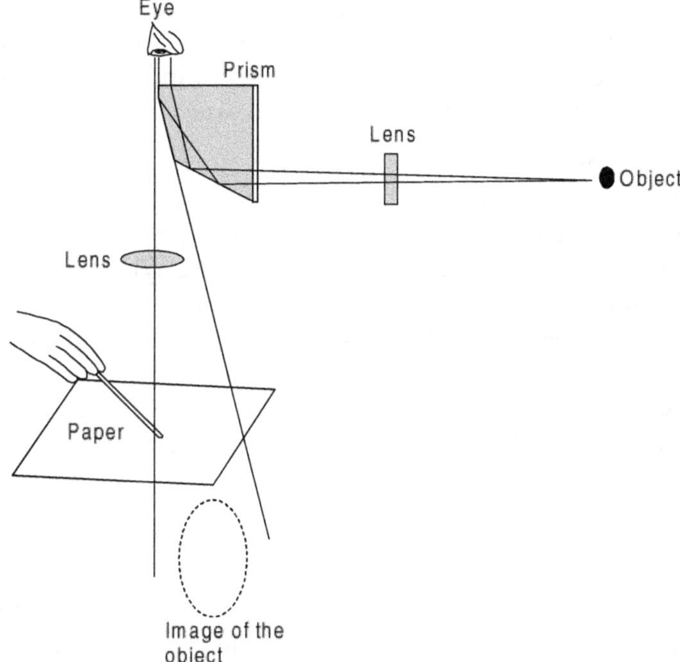

Figure 9.27 Working principle of camera lucida.

The microscope of the camera lucida is mounted above the eyepiece in such a way that the large mirror and small prism bring the object and the paper into the same view. It permits the observer to see the specimen along with the paper when he looks through the eyepiece. The observer can then trace the image while looking through the microscope.

ELECTRON MICROSCOPES

Electron microscope (EM) is one of the most powerful tools used in the study of the cell biology at the ultrastructural level of both prokaryotes and eukaryotes. During the last 50 years or so it has undergone several modifications but the basic principle remains the same. That is, it uses electrons instead of light to illuminate the specimen in order to obtain a resolution several times greater than that of any model of light microscope.

Electron is one of the three elementary particles that constitute atoms and molecules, the other particles being protons and neutrons. In a vacuum tube, a heated cathode emits a stream of electrons called cathode rays. Cathode rays can be deflected and directed to focus using magnetic fields or electrostatic fields. The wavelength of the electron beam can be manipulated by altering the accelerating voltage. These properties of an electron beam are utilized in electron microscopes and in other instruments like the cathode-ray oscilloscope and the television picture tube.

The main concept based on which the electrons are used in microscopy is that the electrons have wavelike properties. The wavelength of the electron radiation used in an EM is about 0.005 nm, which is approximately 100,000 times shorter than that of the visible light. When such a radiation is used to illuminate the specimen, the resolution achieved is about 1000 times superior to that realized by visible light. Consequent to the higher resolution, the magnification (useful magnification) capacity of the EM is enormously increased. Since electrons cannot pass through glass, electrical fields and magnetic fields are used as condensers and lenses.

Types of Electron Microscopes

There are two common types of EM, the transmission electron microscope (TEM) and scanning electron microscope (SEM). As the names indicate, a TEM transmits electrons through the specimen while a SEM scans the electrons on the specimen.

Transmission electron microscope   A modern TEM consists of (1) an electron source, (2) electromagnets to function as "lenses", and (3) a system that projects an image onto a fluorescent screen or photographic film. An electron gun consisting of a tungsten filament (cathode) is the source of electron. When the tungsten cathode is heated, it releases electrons. A high-voltage electrical field (in the range of 50,000 to 100,000 volts) applied between pair of metal plates accelerates the electrons. Electric or magnetic lenses condense into a narrow beam. When the narrow beam of electrons passes through the specimen, individual electrons are scattered in various directions depending on the density of the materials on which they strike. Materials having higher density scatter the electrons more, so that only a few of the scattered electrons reach the objective magnetic lens. Such dense areas would appear darker in the final image. In contrast, electron-transparent areas of the specimen would appear brighter. All the reflected and unreflected electrons that enter the objective magnetic field are focused to form an image of the specimen. The image is enlarged by projector lens magnet and projected on a fluorescent screen or photographic film. A simplified ray-diagram to illustrate the functioning of a TEM is shown in Figure 9.28.

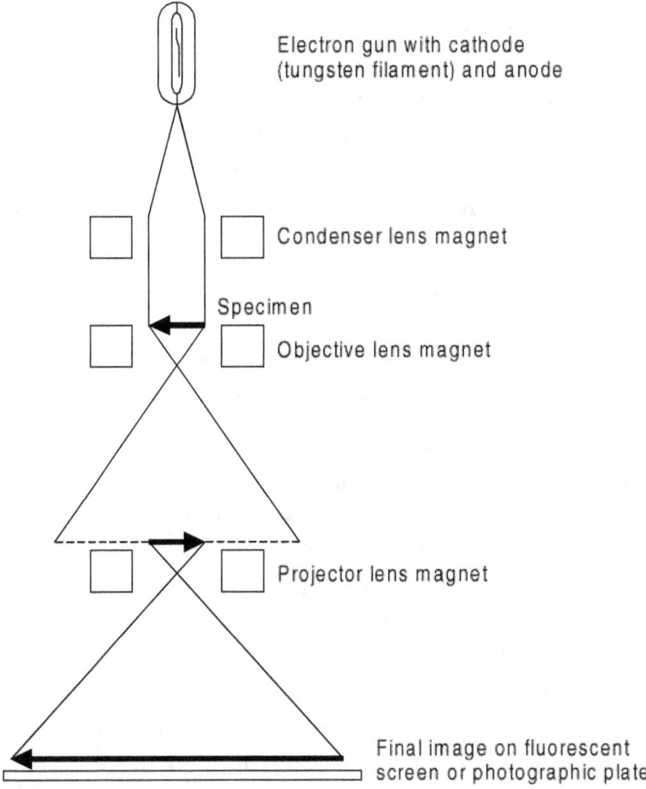

Figure 9.28 Working principle of a transmission electron microscope.

The entire column of a TEM consisting of electron gun and electromagnetic lenses is encased in a high vacuum. The specimen is also illuminated in vacuum. This is necessary to get a clear image of the specimen, because electrons are deflected when they collide with air molecules. Since electrons are not visible to human eyes, the image is focused onto a phosphorescent screen, which is excited by the incident electrons to emit visible light. The emitted light is of low intensity and therefore the images can be clearly viewed only in a darkened room. The photographic film required for recording the image is of special type that is sensitive to electrons.

The main limitations of TEM are:

1. The high vacuum condition in which the specimen is examined causes the specimen to dry up. Therefore, special cryospecimen holders are used to keep the specimen in frozen state at extremely low temperatures.

2. The very high voltage electric current (60,000 to 100,000 volts) used to accelerate electron rays causes the specimen to gain so much energy. As a result many chemical bonds break, molecular conformations alter, and substantial organic mass is lost. Therefore, the duration of the exposure of the specimen to electron rays has to be kept at minimum.

3. The increased acceleration of the illuminating electrons causes a decrease in the contrast of the image, and therefore the specimen has to be ultrathin and selectively stained by salts of heavy metals. Since heavy metals scatter electrons to a greater extent, the projected image has useful contrasts.

The uses of TEM are manifold especially in the field of biology and medicine. Our understanding of the shapes and ultrastructural details of microorganisms including virus and bacteria and the eukaryotic cells is largely due to the use of TEM. These microscopes are routinely used in clinical pathology laboratories of medical centres for diagnostic purposes. Application of various tissue preparation techniques also contributes to the better utilization of TEM. For example, negative staining method is a superior way to study the structure of viruses and bacterial gas vacuoles. Negative staining involves the spreading of a thin film of the specimen with heavy metals such as phosphotungstic acid or uranyl acetate. The heavy metals do not penetrate the specimen but provide a dark background. The specimen appears bright in photographs.

Another technique is *shadowing*, which is useful in understanding the virus morphology, bacterial flagella and plasmids. In this technique, the microorganism is coated with a thin film of platinum or other heavy metal. The coating is made at an angle of 45° from horizontal so that the coating metal strikes the microorganism only on one side. The regions coated with metal scatters electrons and appear much more brighter than the uncoated areas, which appear darker. The photographic image of the specimen appears as if light is shining on it from the side casting a shadow.

Freeze-etching is yet another technique applied to TEM to obtain a detailed, three-dimensional view of the intracellular structures. In this technique, cells are swiftly frozen using liquid nitrogen. Then the frozen cells are warmed to –100°C in a vacuum chamber. The cells are now very brittle. Using a precooled knife the cells are fractured. The cells break along lines of maximum weakness, usually down the middle of internal membranes. The specimen is allowed to sublimate for a minute in a high vacuum chamber, so that the ice can sublimate away and uncover more structural details. Shadow casting is applied to the exposed surfaces, which are then coated with a layer of platinum and carbon to form replica of the surfaces. The specimen is removed chemically, and the replica is studied in TEM. Since in the freeze-etching technique the specimens are frozen quickly and are not subjected to chemical fixation, dehydration and plastic embedding, the problem of artifacts is almost completely eliminated.

Scanning electron microscope   While the transmission electron microscope forms an image from radiation that is transmitted through the specimen, the scanning electron microscope (SEM) creates an image from electrons emitted from the surface of the specimen. The SEM is generally used to examine the surfaces of microorganisms and those of eukaryotes.

When compared to a TEM the resolving power of a SEM is lesser (Resolving power refers to the capacity of the microscope to resolve; resolution refers to the actual resolution achieved by a microscope; when we say resolving power is lesser, we mean that the resolution, the distance between two closely placed objects which it can distinguish, achieved by the microscope is high). It is in the range of 3 to 7 nm. More recent designs of SEM strive to attain a resolution closer to that of a TEM.

The specimen preparation for SEM is less complicated than that of TEM. It is even possible to examine an air-dried specimen under a SEM. However, microorganisms are fixed, dehydrated, and dried to preserve the surface structure and to prevent disintegration of cells when they are placed in high vacuum chamber of a SEM. The dried sample is mounted and coated with a thin layer of metal to prevent the accumulation of electrical charge on the surface and to obtain better image.

The working principle of a SEM is illustrated in Figure 9.29. A narrow, tapered beam of electrons, the primary electrons, produced by an electron gun, accelerated and condensed by magnetic lenses, scans back and forth over the specimen. A scanning coil located between two condenser lenses facilitates the scanning process. When the electron beam strikes a specific spot it induces emission of radiation, a tiny shower of low-energy electrons called secondary electrons or backscattered primary electrons. The secondary electrons are captured by a detector which is connected to a photomultiplier. Secondary electrons entering the detector strike a scintillator causing it to emit light flashes. The photomultiplier converts the light flashes into an electrical signal and amplifies. The electrical signals are sent to a cathode-ray tube and produce an image of the specimen. Each point on the specimen corresponds to a pixel (picture element) on the screen. The more electrons that strike a particular pixel on the screen, the brighter the pixel appears.

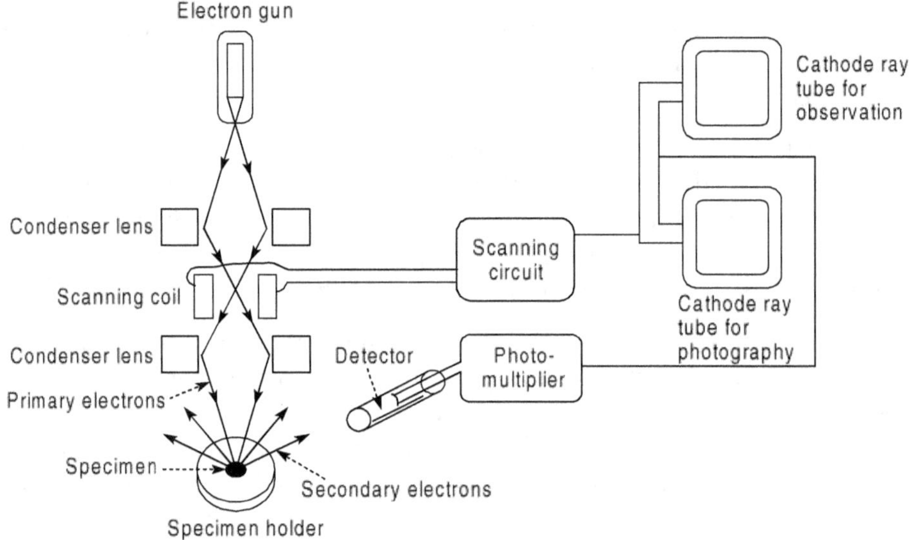

Figure 9.29   Working principle of a scanning electron microscope.

The number of secondary electrons entering the detector from each spot depends on the nature of the surface of specimen. If the electron hits a raised area, a large number of secondary electrons are backscattered, and therefore the detector traps more of them. On the other hand, when the primary electron strikes a depression, fewer secondary electrons are released and correspondingly fewer of them enter the detector. Thus, when the final image is constructed and displayed on the monitor, the raised areas of the surface of the specimen appear brighter (lighter) and the depressed areas appear darker. The image produced can be viewed and photographed. The magnification range of SEM is between 20× and 200,000×.

As we have already seen, the main application of a SEM is to configure the ultrastructural details of the surface of prokaryotic and eukaryotic cells. SEM facilitates the creation of a realistic three-dimensional image of the surface of a cell. It is also possible with SEM to examine the actual *in situ* location of microorganisms in their ecological niche such as the skin, gut, etc. of animals and plants. SEMs also have applications in fields such as metallurgy, where surface features of metals and alloys are studied.

SCANNING PROBE MICROSCOPES

A new family of microscopes, generally referred to as scanning probe microscopes, has evolved in the last twenty-five years (Figure 9.30). Such microscopes function by scanning the surface of the sample with a probe-tip that may be as small as a single atom. Because of the extremely fine nature of the scanning tip, these microscopes can achieve magnifications of 100 million times, which would enable us to view atoms on the surface of a solid sample. We shall consider the principles and applications of a few of these microscopes.

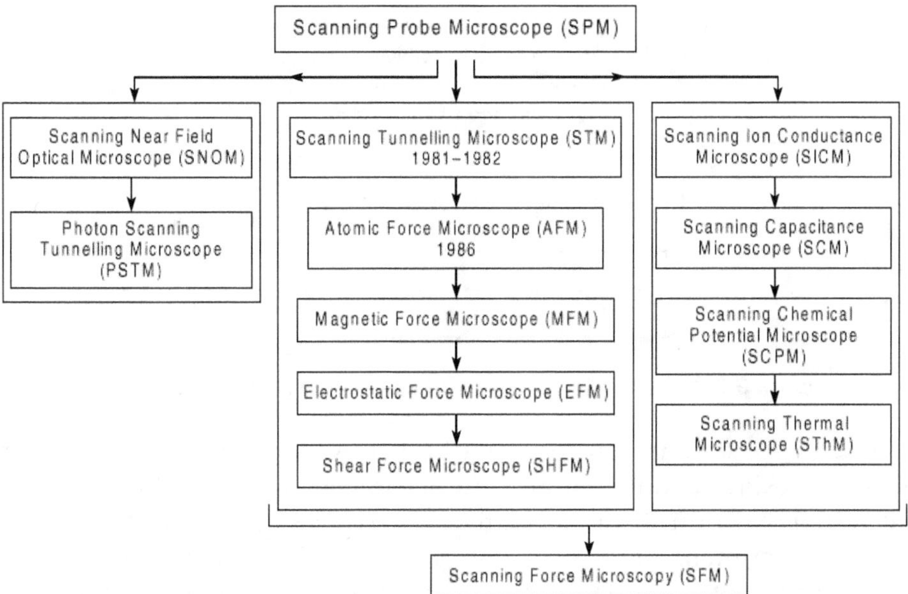

Figure 9.30 Family tree of scanning probe microscope.

The basic concept of all types of SPM is almost the same (Figure 9.31). A specific probe, which is extremely sharp (3 to 50 nm radius of cuvature) is used to scan the surface of material. The tip is mounted on a flexible cantilever, allowing it to follow the surface contour.

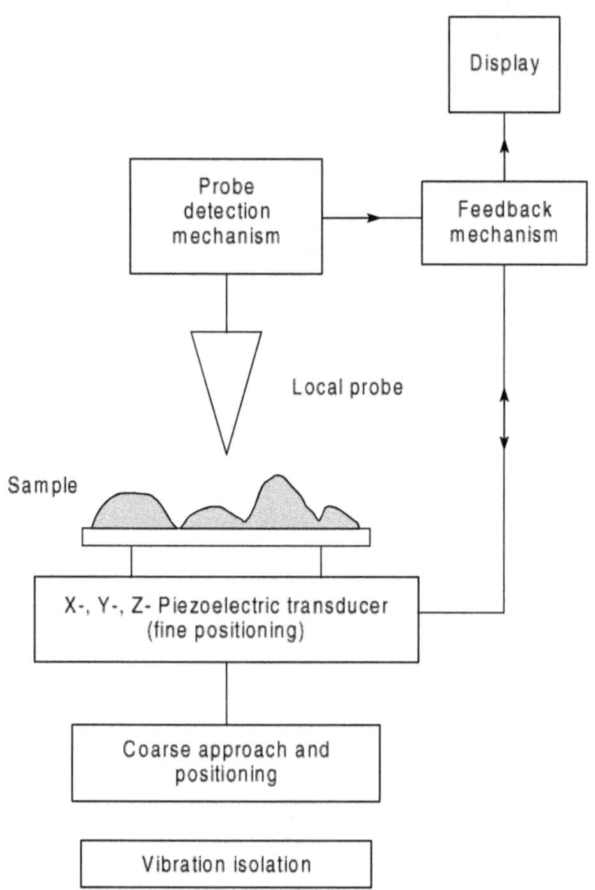

Figure 9.31 Generalized scheme of a scanning probe microscope.

When the scanning probe (tip) is brought very close, as close as about 1 nm, to a surface, different physical phenomena occur as a result of the interaction between the atoms at the probe tip and those on the surface. These phenomena depend on the nature of the probe used and the characteristics of the surface. As the surface is scanned very precisely (just like an electron beam scans a CRO screen or TV screen) and systematically, a lot of information at nanometric level about the surface topography is generated. The scanning is effected with a piezoelectric transducer attached either to the probe or to the vibration-free stage on which the sample is mounted. During scanning, variations in the physical phenomenon, which are proportional to the tip–surface distance, influence the movement of the cantilever. These movements are detected by selective sensors, amplified, converted to electronic signals, and transmitted to a computer, where they are processed. The computer-generated digital image of the surface contour is projected onto a monitor for viewing or is photographed. Various interactions can be studied depending on the mechanics of the probe.

The recorded images are digital, generally 512×512 point plots. Each point has a value between 0 and 255 so that greater the value, the more elevated is the point, and lesser the value, more plane it is. Accordingly, three-dimensional information is presented in two dimensions. Different shades of colour help to differentiate the height differences.

Scanning probe microscopy has adapted several related technologies for imaging and measuring surfaces of samples on an enormous three-dimensional scale, which ranges between fractions of a nm (e.g. dimensions of molecules and groups of atoms) at one end and about 100 μm at the other, in the X and Y directions. Further, the scanning is also made in the Z direction (the third dimension, i.e., the depth or height). The SPM technology is therefore having profound effects on many areas of science and engineering.

Nearly 50 years after the fabrication of the transmission electron microscope, by Ernst Ruska, in 1930, the first SPM, the scanning tunnelling microscope (STM), was invented in 1981, by G. Binning and H. Rohrer (IBM, Zurich, Switzerland). Interestingly, all the three shared the Nobel Prize in 1986 for their inventions. G. Binning, C. F. Quate and Ch. Gerber introduced another version of the SPM in 1986, viz., the atomic force microscope (AFM).

Scanning Tunnelling Microscope

The principle of the scanning tunnelling microscope (STM) is based on a concept of quantum mechanics, viz., the tunnel effect of electrons (Box 9.6). The electrons surrounding the surface atoms tunnel or project from the surface boundary at a very short distance.

BOX 9.6   SCANNING PROBE MICROSCOPE—EXPLANATION OF A FEW TERMS

Tunnelling effect   This is quantum mechanical effect. It is an effect of the wavelike property of electrons. A tunnelling current happens when an electron wave moves through a barrier. The wave does not end abruptly at the barrier, but tapers quickly. However, if the barrier is thin and enough electrons are available, a few electrons pass through the barrier and appear on the other side of the barrier. When an electron moves through the barrier in this way, it is called tunnelling. The number of electrons that will tunnel across a barrier depends upon the thickness of the barrier. The actual current through the barrier decreases with the increase in the thickness of the barrier.

Now, we can extend this electron-tunnelling concept to STM. The starting point of the electron is either the probe tip or the sample surface. The barrier is the gap between the probe tip and surface. It may be air, vacuum or liquid. The other side of the barrier is therefore either the tip or the surface. That is, the electrons from the surface may tunnel pass across the gap into the tip or the electrons around the atoms in the tip may tunnel across the gap into the surface. By monitoring and manipulating the current through the gap it is possible to control the tip–sample distance.

Piezoelectric materials   This effect was discovered by Pierre Curie (husband of Mary Curie) as early as 1880. It is created by pressing the sides of certain crystals (e.g. quartz and barium titanate). As a result of compression of a piezoelectric crystal, opposite charges are created on the sides. The effect can be reversed, that is, by applying a low voltage of current across a piezoelectric crystal it can be made to vibrate, i.e., elongate or compress. In an STM, piezoelectric material such as Lead Zirconium Titanate (PZT) is used to scan the tip of the probe.

Electronics and the Feedback Loop   Electronics is needed to measure the current, scan the probe-tip, and to translate the information obtained into digital format and create an image of the scanned surface. A feedback loop continuously monitors the tunnelling current and adjusts the tip to maintain a constant tunnelling current. The adjustments are recorded by a computer

> BOX 9.6   (Continued)
>
> that creates the image in the STM software. The entire set-up is called a "constant current" image. However, if the surface that is scanned is a flat one, the feedback loop is turned off and only the current is displayed. This is a "constant height" image.

A probe with a tip as thin as an atom is moved close to the sample surface until its electron cloud touches the electron cloud of the atoms of the surface. A small voltage of current is applied between the tip and the specimen. This causes a flow of electrons through a narrow channel in the electron clouds. This tunnelling current is very sensitive to distance. It will decrease about a thousand-fold if the probe is raised from the surface by a distance as small as the diameter of an atom (exponential decay). The tip is mechanically connected to the scanner, an XYZ positioning device realized by means of piezoelectric materials. Scanning is the systematic line-by-line movement of the probe tip back and forth over the sample. A feedback system keeps the tip at constant height by adjusting the probe distance to maintain a steady tunnelling current. As the tip moves up and down while following the surface contours, the variations in tunnelling current, which are proportional to the tip–surface distance, are recorded and analysed by a computer to produce an accurate three-dimensional image of surface atoms. The image is displayed on a computer monitor. It can also be plotted on a paper or photographed. A schematic representation of the working principle of STM is shown in Figures 9.32 and 9.33.

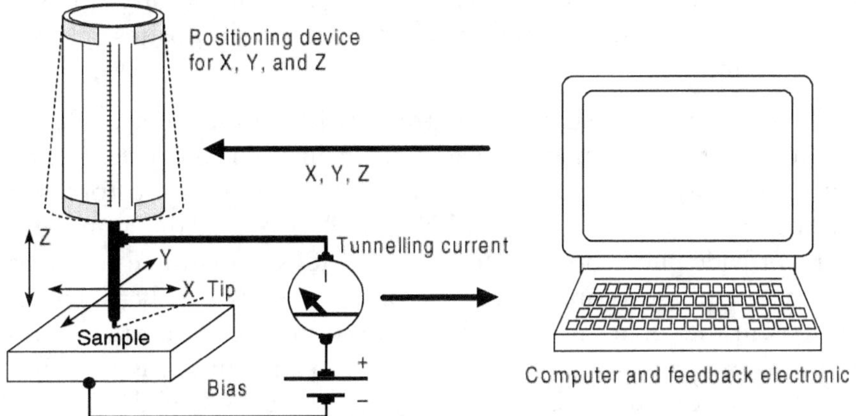

Figure 9.32   Schematic illustration of scanning tunnelling microscope.

The scanning tunnelling microscope has several applications in biology. It is used to directly view DNA. Since the microscope can examine objects when they are immersed in water, it is particularly useful in studying biological molecules.

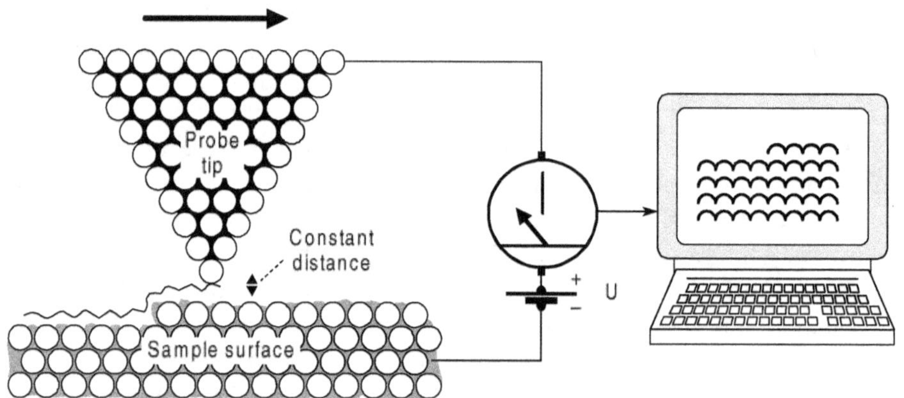

Figure 9.33   Tip–sample tunnelling contact in an STM.

Atomic Force Microscope

The main disadvantage of STM is that it can only image conducting or semiconducting surfaces, i.e., it can only scan those samples which have at least some surface features that are electrically conductive to some degree. The atomic force microscope (AFM) was developed to overcome this basic drawback. AFM does not use a tunnelling current, so the sample does not need to conduct electricity. It is capable of imaging almost any type of surface, including polymers, ceramics, composites, glass and biological samples.

The AFM moves a sharp metal probe (with a tip diameter of about a few nm) over the specimen surface while maintaining a constant distance between the probe tip and the surface. The interatomic forces (the electrons in the probe are repelled by the electrons of the atoms in the sample) provide interaction mechanism. The interaction force will depend on the nature of the sample, the probe tip and the distance between them. Variations in the force cause the tip to move up and down as it scans the surface. The interaction force between the tip and surface is not directly measured but calculated by measuring the deflection of the lever of a cantilever mechanism attached to the probe. The tip–cantilever assembly is microfabricated from Si or Si_3N_4. The calculation takes into account the stiffness of the cantilever. In modern AFMs the vertical motion of the tip usually is followed by measuring the deflection of a laser beam that strikes the lever holding the probe (Figure 9.34). A laser beam strikes the cantilever and is deflected to a mirror. The up-and-down movement of the cantilever causes variations in the deflection of the laser beam. The variations in the deflection of the laser beam striking the cantilever is passed onto a computer through a photodetector and an amplifier. The computer processes the data and creates a three-dimensional image of the surface of the sample.

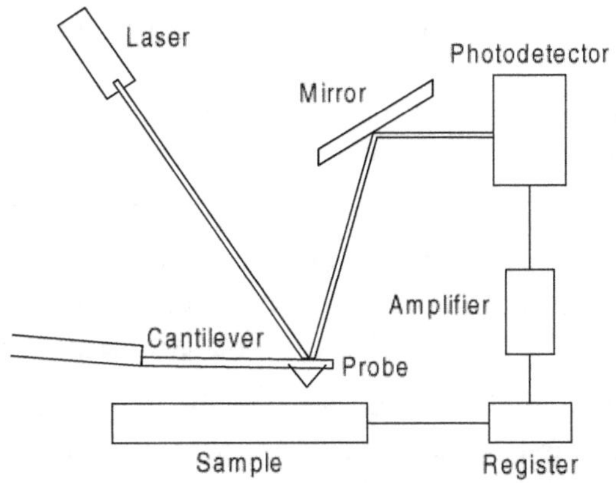

Figure 9.34 Working principle of an AFM.

Because of its versatility, AFM has been applied to a large number of research fields. Consequently, the AFM has also gone through many modifications for specific application requirements. To suit the different requirements, the AFM may operate in several modes, which differ according to the force between the tip and surface, as follows:

- Contact mode for strong (repulsive)—constant force or constant distance
- Non-contact mode for weak (attractive)—vibrating probe
- Intermittent contact mode for strong (repulsive)—vibrating probe
- Lateral force mode for frictional forces exert a torque on the scanning cantilever
- Magnetic force for the magnetic field of the surface is imaged
- Thermal scanning for the distribution of thermal conductivity is imaged

The range of applications of the AFM is much wider than that of STM. AFM is used for studies of non-conductors and is more commonly employed for studies of macromolecules and biological specimens. AFM has been used for measurements on a wide variety of biological sample types, including polymers and polymer matrix, natural resins and gums, muscle proteins, biological structures, DNA, plant cell walls, bacterial flagellae, chromosomes, cell surface and membrane surface. AFM has been applied for similar studies on inorganic and synthetic materials. We shall consider a few of the biological applications in some detail.

Analysis of biological systems   Since AFM can image the non-conducting surfaces, its application was easily extended to biological systems such as analysing crystals of amino acids and organic monolayers; DNA and RNA; protein–nucleic acid complexes; chromosomes; cellular membranes; proteins and peptides; molecular crystals; polymers and biomaterials; ligand–receptor binding.

Imaging of genetic materials   The ability of the AFM to generate nanometer-resolved images has contributed to obtaining images of unmodified nucleic acids and those involved in translation, and small molecule–DNA interactions such as intercalating mutagens.

Cell biology   AFM is applied in the study of the dynamic behaviour of living and fixed cells such as RBC and WBC, bacteria, platelets, cardiac myocytes, living renal epithelial cells and glial cells. For instance, images of plasma membrane in migrating epithelial cells have been obtained. The dynamic membrane invagination process was observed in the presence of calcium; it was noted that this invagination process was prevented and 30 nm lipidic pore formation occurred when the calcium levels were reduced. AFM is found suitable for imaging cell surface features such as rearrangement of plasma membrane or movement of submembrane filament bundles.

Molecular biology   It has been possible using AFM to obtain images of individual proteins and other molecules such as collagen. Employing selective affinity procedure, several smaller biomolecules such as PCR products and immunoglobulin G can be imaged.

Quantification of biological processes   Many biological processes such as DNA replication, protein synthesis and drug interaction are governed by intermolecular forces. AFM can be used to measure these forces in the nanonewton range. Thus it is possible to quantify the molecular interaction in biological systems as in, for instance, ligand–receptor interactions. It is also possible to image and measure electrical surface charge. The dynamics of several biological systems depends on the electrical properties of the sample surface.

Measurement of micromechanical properties In addition to measuring binding forces and electrostatic forces, the AFM can be used to probe the micromechanical properties of biological systems. For example, AFM can observe the elasticity and viscosity of samples such as live cells and membranes of bone and cartilage.

Microbiology The AFM has been effectively used to probe the structural and physical properties of living microbial cells. Cell surface nanostructures (e.g. rodlets, appendages) have been directly visualized using topographic imaging. Changes that occur during physiological processes (e.g. germination, division) can be quantified. Force–distance curves constructed with measurements obtained through AFM would provide useful information on surface forces, adhesion and nanomechanics. Such information might help to explain the mechanisms of biological events such as cell adhesion and aggregation. Further improvements and innovations in sample preparation techniques, instrumentation and recording conditions may enhance the AFM imaging quality to such an extent as to monitor molecular conformational changes in microbial systems.

Immunology Another evolving area in atomic force microscopy is the fabrication of tips to measure specific force interactions in cells. Antibody modified tips that could measure or localize antigens on the surface of a cell (by vertical/lateral force detection) or relocate them in the plasma membrane are being constructed. Such specific probe-tips will make it feasible to study the biophysics of molecular interactions and its role in important processes such as signal transduction, the process by which information flows throughout a biological system.

The AFM has been effectively combined with other microscopes such as inverted optical, bright-field, fluorescence microscopes to obtain very high-resolution images of structures on surfaces.

Near-Field Scanning Optical Microscope

Near-field scanning optical microscope (NSOM) uses a very small light source close to the sample for scanning. Detection and digital interpretation of this light energy forms an image. The resolution of NSOM is below that of any light microscope. NSOM uses quasipoint light source with a diameter less than the wavelength of light. The probe is very close to the surface, the gap between the tip and the surface being less than the wavelength of the light. This region is referred to as the "near-field' and hence the name near-field scanning optical microscope. Generally, laser light is used for illumination. It is fed to the aperture through an optical fibre. The aperture is a tapered fibre coated with aluminium, a microfabricated hollow AFM probe, or a tapered pipette.

Two types of feedback are used to maintain proper working distance of the probe and sample. One is cantilever mechanism used in AFM. The second is the shear-force feedback in which changes in the amplitude of the reflected light is monitored through a tuning fork, which oscillates at its resonance frequency. Depending upon the nature of the sample the mode of operation of NSOM may be transmission, reflection, collection or illumination. In the transmission mode, the light source travels through the probe aperture, and transmits through the sample. Therefore, the sample has to be transparent. Opaque sample require reflection mode in which a low intensity light source travels through the probe aperture, and reflects from the surface. The collection mode illuminates the sample from a large outside source, and the probe collects the reflected light. In the illumination/collection mode the probe both illuminates the sample and collects the reflected light.

The collected signal is handled in a number of ways such as spectrophotometer, Avalanche Photo Diode (APD), Photomultiplier tube, or Charge Coupled Device (CCD). Several contrast mechanisms like polarization, topography, birefringence, index of refraction, fluorescence, wavelength dependence, and reflectivity are used in NSOM.

Magnetic Force Microscope

In a Magnetic Force Microscope (MFM) a magnetic tip is used to probe the magnetic stray field above the sample surface. The magnetic tip is mounted on a small cantilever which translates the force into a deflection that can be measured. The MFM can sense the deflection of the cantilever which will result in a force image (static mode) or the resonance frequency change of the cantilever, which will result in a force gradient image. The sample is scanned under the tip that results in a mapping of the magnetic forces or force gradients above the surface.

CHAPTER TEN

Centrifugation

INTRODUCTION

Centrifugation is a separation method which uses the action of centrifugal force to accelerate sedimentation of particles in a solid–liquid mixture.

Sedimentation is the deposition of particles suspended in a liquid medium, to the bottom of the container, according to their density and size. For example, when water with suspended particles of varying sizes and densities (e.g. sand) is left to stand undisturbed in a container, say, a measuring cylinder, the process of sedimentation due to gravitational force would occur. As a result, particles of higher density would settle first to the bottom, then particles of lesser density, and so on, thus forming a gradient according to density. If a medium other than water is used, such as glycerol or castor oil, the rate of sedimentation (i.e., the speed with which the suspended particle descend to the bottom) will be reduced because of the higher viscosity of the medium. Thus, the sedimentation process is dependent on

i)   the density and size of the suspended particles,

ii) the viscosity of the medium in which the particles are suspended, and

iii) the gravitational pull. The gravitational force under normal conditions is about 980 cm s^{-2} (1 g unit).

This sedimentation process due to the force of gravity takes very long time and the process is generally incomplete because very minute particles of the order of macromolecules are insensitive to gravitational setting. However, if the liquid with the suspended particle is subjected to centrifugal force, which is several thousand times greater than the gravitational force, the sedimentation process can be completed in a short time. The instrument that is used to apply the centrifugal force to a solution with suspended material is a centrifuge.

CENTRIPETAL AND CENTRIFUGAL FORCES

The force that maintains an object in circular motion is called centripetal force (force seeking to move towards the centre). If no force is exerted on an object, it tends to move in a straight line at a constant speed. In order to make the object to deviate from that straight-line path to a circular path, a centripetal force must be applied at right angles to the object's velocity. The centripetal force causes a corresponding centripetal acceleration, which is also towards the centre. In general, the centripetal force that needs to be exerted to an object of density m to keep it moving in a circular path of radius r at a constant velocity v is

$$\frac{mv^2}{r}.$$

According to the Newton's third law, for every force of action there is an equal and opposite reaction force. Thus the force that is equal and opposite to the centripetal force applied to an object moving in a circular path around the centre is the centrifugal force, meaning a force tending to move away from the centre.

The existence of a centrifugal force is not real. The centrifugal force is imagined as a force acting on an object moving in a circular path, tending to pull the object out, away from the centre and that it is a force which equals and thus balances the centripetal force that pulls the object towards the centre. In reality, there is no centrifugal force acting on the object, there is no balancing of the centripetal force and the object does not tend to pull away from the centre. Suppose the centripetal force that makes an object to move in a circular path is removed, then the object will not accelerate but will move in a straight line tangential to the circular path. This shows that no other force is acting upon the object. For this reason, the centrifugal force is often described as a fictitious force. A fictitious force is experienced by the object only when it is in an accelerating frame of reference. However, from the viewpoint of the moving object, the centrifugal force exists.

The centrifugal force is directly proportional to the speed of rotation (angular velocity) and the radius of the rotation. That is, higher the angular velocity and the radius, greater will be the centrifugal force. The relationship between these parameters can be expressed as

$$\text{Centrifugal force} = (\text{angular velocity})^2 \times \text{radius}$$

or,

$$F = \omega^2 \times r$$

Angular velocity is related to revolutions (rotations) per minute by the following equation

$$\text{Angular velocity } (\omega) = \frac{2\pi \times rpm}{60} \text{ radians per second}$$

RELATIVE CENTRIFUGAL FORCE

The centrifugal force F divided by 980 (earth's gravitational force) is generally referred to as Relative Centrifugal Force, RCF, which means number of times g units ($\times$ g). Thus,

$$\text{RCF} = \frac{\omega^2 r}{980} g \text{ units}$$

$$= (1.119 \times 10^{-5}) \times (\text{rpm})^2 \times (r) g \text{ units}$$

It can be seen in the relationship (1.119×10^{-5}) $(\text{rpm})^2$ (r) g units, that all factors other than rpm and r are constant. Thus RCF is dependent on rpm and r. The radius of rotation r is also constant for a given centrifuge and, therefore, rpm is the only factor which determines the RCF.

The radius of rotation, r, is the distance between a particle in the sample contained in the centrifuge tube and the axis of rotation. The position of a given particle in the sample is not constant during centrifugation, because it would be progressively sedimenting towards the bottom of the centrifuge tube and, thus, progressively increasing the r, the distance from the axis of rotation. Because of this reason, r is calculated as the distance between the rotor axis and the middle of the liquid column in the centrifuge tube. This radius is referred to as the average radius, r_{av}, because it is an average of the minimum and maximum radii (Figure 10.1).

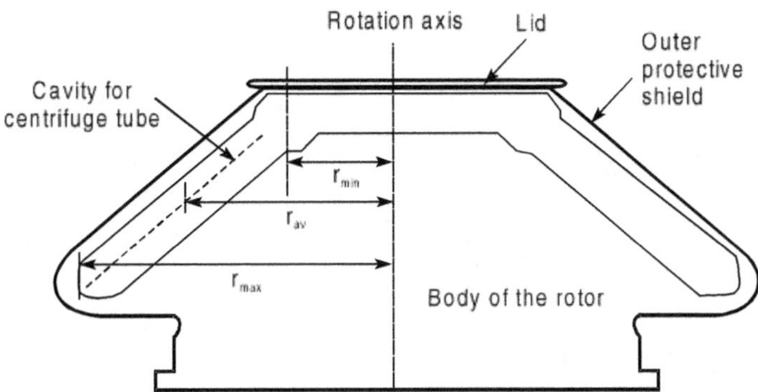

Figure 10.1 Cross section of the rotor of an angle-head centrifuge, showing the minimum (r_{min}), average (r_{av}) and maximum (r_{max}) radii. The rotor is built of heavy metal plates to withstand very high speeds of rotation.

The speed of a centrifuge can be expressed in rpm or RCF, but RCF is a better expression. Since different centrifuges might have different r's, and, for a given rpm, the different centrifuges will have different RCF's, therefore, in order to avoid any ambiguity, the centrifugal force acting upon a particle is expressed in terms of RCF.

FACTORS AFFECTING SEDIMENTATION RATE DURING CENTRIFUGATION

In addition to RCF, there are other factors that influence the sedimentation rate of particles during centrifugation (i.e., the rate of descent of particles suspended in a liquid). These are density and size (radius) of the particle and the density and viscosity of the medium in which the particles are suspended. The general relationships of these factors to the rate of sedimentation are as follows:

1. The larger sized particles sediment faster than smaller particles.
2. Particles having higher densities sediment faster than less dense particles.
3. Medium with higher density will slow down the sedimentation of a particle.
4. Medium of higher viscosity will slow down the sedimentation rate.

The relationship of the sedimentation rate of a particle to the various factors can be expressed as follows.

$$v = \frac{d^2(\rho_p - \rho_l) \times g}{18\eta}$$

where,

v = sedimentation rate
d = diameter of the particle
ρ_p = particle density
ρ_l = liquid density
g = centrifugal force
η = viscosity of the liquid
18 = shape factor, constant for a spherical particle

At constant centrifugal force and velocity, the sedimentation rate is proportional to the size of the particle and to the difference between the density of the particle and the density of the liquid. The above relationships are exploited in differential centrifugation and density gradient centrifugation techniques of cell biology.

SEDIMENTATION COEFFICIENT AND SEDIMENTATION CONSTANT

The sedimentation rate of a particle expressed per unit of centrifugal field is called sedimentation coefficient, i.e., the rate of descent per unit of centrifugal field per second. It is expressed in Svedberg unit (S) as follows.

$$S = \frac{v}{\omega^2 r}$$

where,

S = sedimentation coefficient

v = sedimentation rate

ω = angular velocity

r = radius of rotation

The sedimentation coefficient standardized for a medium of water at a temperature of 20°C is referred to as sedimentation constant or standard sedimentation coefficient ($S_{20.w}$).

$$S_{20.w} = S_{obs} \times \frac{\eta_T}{\eta_{20}} \times \frac{\eta_C}{\eta_0} \times \frac{(1 - \bar{v}\rho_{20.w})}{(1 - \bar{v}\rho_T)}$$

where,

$S_{20.w}$ = standard sedimentation coefficient or sedimentation constant

S_{obs} = experimentally observed sedimentation coefficient

η_T = viscosity of the water at temperature T

η_{20} = viscosity of water at 20°C

η_c = viscosity of the solvent at the given temperature

η_0 = viscosity of water at the given temperature

$\bar{v}$ = partial specific volume of the solute

$\rho_{20.w}$ = density of water at temperature 20°C

ρ_T = density of water at the given temperature

The value of 1S (Svedberg unit) is 1×10^{-13} sec. When we say that a molecule (e.g. rRNA) has a sedimentation coefficient of 5S, what we mean is that the molecule has a descent rate 5×10^{-13} sec. The value of S for a particle or molecule, its size and its sedimentation rate are generally directly proportional. That is, larger the S value, larger the molecular size and faster the sedimentation rate. It is a common practice to characterize ribosomes and their subunits by their sedimentation coefficients.

CENTRIFUGE

A centrifuge is a mechanical device used for separating substances of different densities using the principle of centrifugal force. A common centrifuge is a container that is revolved at high speeds. The principle of separation is similar to that of sedimentation by gravity. But in a centrifuge the driving force is much higher because the force results from the rotation of the liquid in the container. In the case of sedimentation, the driving force comes from the difference in density between the solid particles and the medium in which the particles are suspended. In a centrifuge, on the other hand, separation is effected by a force that is 1000 to 100,000 times that of gravity.

The principal components of a centrifuge are a rotor and motor. The rotor is connected to a central shaft and has, at the distal end, provision for holding the container. The shaft is connected to a motor, which can be operated at different speeds. There are four types of rotors

    i)   swinging (swing-out) bucket

    ii) fixed angle

    iii) vertical tube

    iv) zonal

In the **swinging bucket rotor**, the centrifuge tube holders, the buckets, are hinged to rods extending from the shaft such that they can swing freely. During centrifugal rotation, the buckets swing out to different angles proportional to the speed of the rotation, ultimately attaining a horizontal orientation. In the horizontal position, the medium in the centrifuge tube orients to a plane perpendicular to the axis of rotation, and returns back to the original plane as the bucket swings back to its vertical position. Because of the relatively long path length, the swinging bucket rotors are used for most preparative rate–zonal separations.

The **fixed angle rotors** have sockets for placing the centrifuge tubes at a fixed angle, which is about 30° (14°–40°) to the axis of rotation. During centrifugation, the particles in the solution travel radially outward, moving only a short distance before striking the wall of the centrifuge tube, and then slide down along the wall to settle as a pellet at the bottom. The sliding of the particles along the wall of the centrifuge tube is referred to as the wall–effect. The advantage of the wall–effect is faster sedimentation. The disadvantage is the inability to resolve particles that do not have significant difference in their sedimentation properties.

Vertical-tube rotors also have fixed sockets to accommodate the centrifuge tubes, but they are vertical, oriented parallel to the axis of rotation. During centrifugation, the medium in the centrifuge tube orients at an angle of 90°, which is perpendicular to the direction of centrifugal force, and regains its original orientation when the centrifugation stops. The advantage of the vertical-tube rotor is that it has quicker sedimentation because the particles have to move only a short distance along the diameter of the tube and the sedimentation occurs across the diameter of the tube. The disadvantage of this type of rotor is that the pellets of sedimentation are formed along the outer wall of the centrifuge tube, which makes the recovery of these pellets somewhat difficult.

Two major disadvantages of the rotor types discussed above are the convection current due to wall-effects and the small quantity of sample that can be contained in the centrifuge tubes. To overcome these problems, the zonal rotor is designed as a sector-shaped container (Figure 10.2), in which the sedimenting particles do not come into contact with the wall, and large quantities of sample can be loaded.

The zonal rotor is a flattened sphere, the interior of which is divided into four sectors by four septa. Unlike the conventional rotors, zonal rotor is started to rotate even while its sectors are empty. The density gradient is first loaded into the rotor while the rotor is spinning, through the gradient loading channels in the septa. The density gradient is established such that the lowest density is close to the centre of rotation and the highest density is at the rotor wall. The highest dense gradient, loaded last, acts as a fluid cushion.

Once all the sectors are full with gradient, the sample is loaded through the sample ports, located in between the bases of the septa. The sample is loaded such that they are introduced above the gradient and close to the axis of rotation, i.e., where there is lowest density gradient. After loading of both gradient and sample is completed, the speed of the centrifuge is increased to the desired level to apply the centrifugal force. Under the influence of the centrifugal force, the particles in the sample start moving toward the periphery and get sedimented at a density (in the gradient)

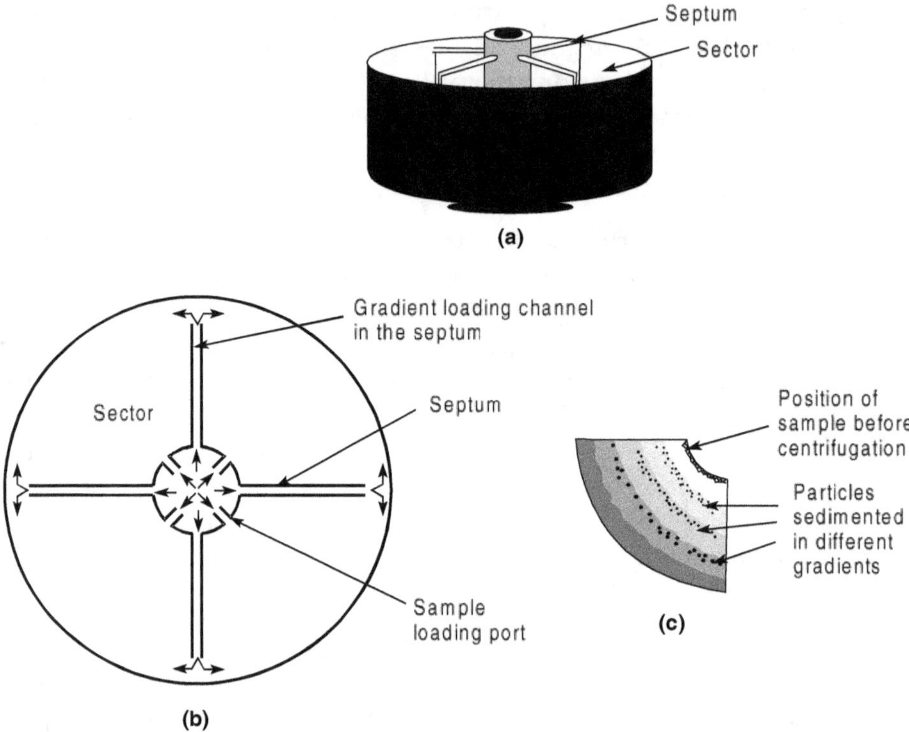

Figure 10.2   Simplified diagram of a zonal rotor. a) overview of zonal rotor; b) cross-section of the rotor; c) a sector with density gradient.

that is similar to their density. When the centrifugation is complete, a still higher dense liquid is pumped in through the gradient loading channel to displace centrifuged gradient (with sedimented particles). Each fraction of the effluent is collected through the sample ports in separate tubes.

GRADIENT MEDIA

There are several gradient materials used in centrifugation and include

    i)   sucrose and Ficoll

    ii) cesium chloride

    iii) potassium bromide

    iv) Percoll

    v) Metrizamide

vi) Nycodenz

vii) Renografin

The most common media is the 54% (wt/wt) sucrose solution, prepared by dissolving 54 g of sucrose in 46 mL of water (assuming water has a density of 1 g/mL. Ficoll (Ficoll 400) is a commercial preparation of a hydrophilic polymer of sucrose, having a molecular weight of 400 000. The advantage of Ficoll over sucrose is its lower osmotic pressure, which helps in preserving the morphology and activity of subcellular fractions.

Cesium chloride and potassium bromide are useful in isopycnic density gradient centrifugation technique. Percoll is again a commercial preparation of density gradient medium containing colloidal silica particles (30 nm) coated with polyvinylpyrrolidine. This medium, because of its low osmolarity, low viscosity, and large particle size, is suitable for separating cells, bacteria, viruses and subcellular organelles. Metrizamide and Nycodenz are useful for the isolation of membrane fractions by floatation. Renografin is also used for cell fractionation.

The choice of the gradient medium depends on the requirement, but a medium having certain properties is to be considered as an ideal one. It should not affect the nature and biological activity of the sample molecules and should not interfere with the assay technique. It should be readily removed from separated particles, and should be non-corrosive, easily sterilizable, inexpensive, readily available and easily recovered and recycled.

TYPES OF CENTRIFUGES

Several models of centrifuges are manufactured to suit different requirements. They range from the simple hand-centrifuge to the highly sophisticated analytical ultracentrifuge.

Hand-centrifuge with two out-swinging buckets is obsolete but still may find use in field studies and in emergency situations when there is power failure.

Desktop or **clinical centrifuges** are simple designs either with swinging-bucket or fixed angle rotor. They have a maximum speed of about 3000 rpm with a relative centrifugal force of 7000 g. Their applications include the preparation of serum from whole blood, separation of RBC, yeast cells, etc.

High-speed centrifuges, which have about 25000–30000 rpm, a centrifugal force of about 90000 g and refrigeration facility to function in low temperatures (0°C–4°C), are used for cell fractionation, i.e., to isolate cellular organelles like nuclei, mitochondria, etc., and to isolate and collect microorganisms and precipitates of chemical reactions, antigen–antibody reactions, etc.

Ultracentrifuges are capable of operating at very high speeds of about 75000 rpm with a centrifugal force of about 500,000 g. Obviously, the rotor chamber is fully refrigerated (0°C–4°C) and evacuated to counter and minimize the heat generated due to friction between the spinning rotor and the ambient air. Ultracentrifuges are useful for the separation and collection in pure form of cellular organelles and macromolecules.

Analytical ultracentrifuges are more powerful than the preparative ultracentrifuges, in the sense that they have a built-in optical system to measure the sedimentation characteristics of the macromolecules in the centrifuge. They are useful for measurement of sedimentation coefficients and molecular weights of macromolecules.

APPLICATIONS OF CENTRIFUGATION

The applications of centrifugation can be distinguished into two categories, preparative and analytical. Accordingly, the centrifuges used for these purposes are categorized as preparative centrifuges and analytical centrifuges. However, ultracentrifuges are useful for both preparative and analytical applications.

Preparative Centrifugation

The objective of preparative centrifugation is to prepare solutions or pellets of a single particle type from a mixture of different types. Generally there are three different approaches, viz., differential centrifugation, rate-zonal centrifugation and isopycnic banding. The rate-zonal and isopycnic techniques are together referred to as density gradient centrifugation.

Differential centrifugation In differential centrifugation, a sample solution in which particles of different densities are suspended is centrifuged for a specific time at a specific speed, resulting in a supernatant and a pellet fraction. This centrifugal technique is useful for separation of particles with very different sedimentation velocities. An example of the separation of the cellular fractions by successive centrifugation is illustrated in Box 10.1.

BOX 10.1   PREPARATION OF SUBCELLULAR FRACTIONS

Homogenization   The first step in the preparation of subcellular fractions is to disintegrate the cells. This is achieved in a homogenizer (Figure 10.3a), which consists of two concentric cylinders separated by a narrow gap. The cell suspension passes through the gap in which the cells are subject to a shearing force sufficient to rupture the cell membranes.

Another method for cell disintegration is to irradiate the cells with ultrasonic waves. The ultrasonic waves can be generated by piezo-electric crystals and transmitted into solution of cell suspension through a stainless steel rod. The vibrations created by the transmitted ultrasonic waves rupture the cells.

Centrifugation   The homogenate containing subcellular organelles is centrifuged successively with increasing centrifugal force and duration, as shown in Figure 10.3b.

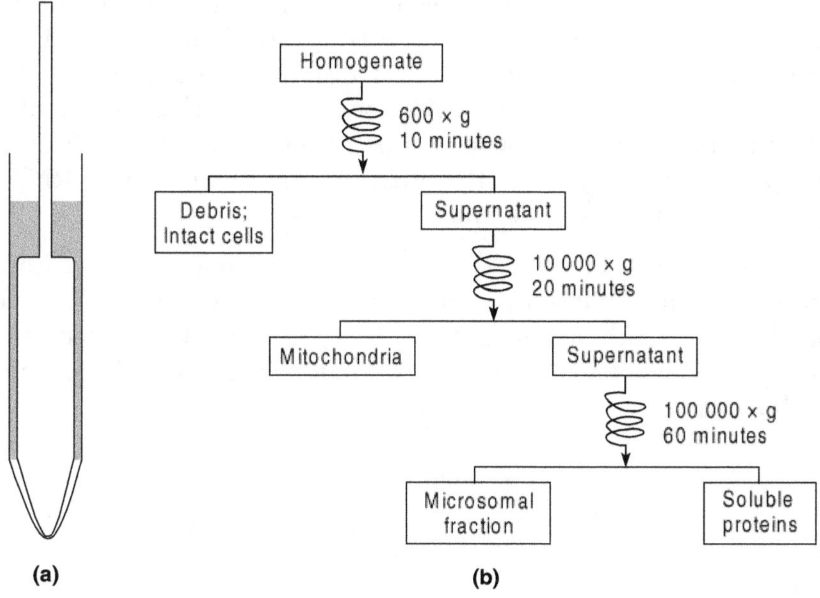

Figure 10.3   Cell fractionation by homogenization and centrifugation. a) homogenizer; b) scheme of successive centrifugation.

Rate-zonal centrifugation Rate-zonal centrifugation is effectively used for the separation of subcellular components that have similar buoyant density but different shapes or particle sizes (e.g. ribosomal subunits, different classes of polysomes, and various forms of DNA molecules). Zonal centrifugation can be performed in either swinging-bucket rotors or zonal rotors. A shallow linear gradient is used to prevent convection in the centrifuge tubes or the chambers of the zonal rotor during centrifugation.

The preparation of a density gradient (e.g. sucrose gradient) involves the flow of the solutions of different concentrations into the base of the container, beginning with the low-density solution. When a higher density solution flows, it displaces upward the previously flown concentration. The final gradient is such that the lowest density is at the top and the highest density is at the base of the container, the centrifuge tube or the zonal rotor chamber. The density range chosen is such that during the separation, the particle densities are always higher than those in the gradient. For example, a gradient range of 15 to 40% (wt/vol) sucrose is used for the separation of subcellular fractions of prokaryotic and eukaryotic cells.

The sample is applied to the gradient as a zone or a narrow band at the top of the gradient, and centrifuged. The speed and time of centrifugation is determined empirically. The centrifugation is terminated before the separating particles reach the bottom of the container. Generally, centrifugation at $100,000 \times g$ for about 1–4 hours is sufficient for the separation of subcellular fraction. Apart from the use in cell fractionation, rate-zonal centrifugation is applied for the separation of RNA–DNA hybrids and ribosomal subunits.

Isopycnic centrifugation Isopycnic centrifugation, also known as sedimentation equilibrium centrifugation, is used to separate particles only on the basis of their buoyant densities, instead of their sedimentation velocities. We had earlier seen that the sedimentation velocity (sedimentation rate) is expressed as

$$v = \frac{d^2(\rho_p - \rho_l) \times g}{18\eta}$$

where,

ρ_p = density of the particle and

ρ_l = density of the medium.

If the density of the particle and density of the medium become equal, then the sedimentation velocity will be zero. This is achieved, in the isopycnic (isodensity), by using a gradient, in a tube, such that at a certain distance from the top of the solution, the fluid density increases to that of the particle. At that point the sedimentation velocity will disappear. During centrifugation, a particle initially in a region where the fluid density is less than the particle density will migrate down the tube to the point of equal density, the isopycnic point. A particle in a region where the density is higher than its own density will migrate up (float up) to the isopycnic point. At equilibrium, the particles will be distributed along the length of the tube according to their density.

In this isopycnic method the sedimentation coefficient is not involved in the separation, and the separation depends only on the buoyant density differences of the particles in the sample. For this reason, this method is not suitable for separation of proteins, as many of them have similar buoyant densities but different molecular weights. However, it is highly useful for separation of cellular organelles, which are membrane-bound and vesicular in nature, such as mitochondria, lysosomes, etc., because, these organelles do not differ much in their size but have different buoyant densities. Isopycnic centrifugation method is also used in the separation of nucleic acid fractions.

Analytical Centrifugation

Analytical centrifugation, mostly carried out with analytical ultracentrifuges, is used to characterize the properties of the particles in pure sample. The design of the rotor of an analytical ultracentrifuge incorporates an optical system and special centrifuge tub holders with which the concentration of the particles can be measured as a function of the radius from the axis of rotation. Analytical centrifugation is used in the determination of

    i)   molecular weights of molecules such as nucleic acids and proteins,

    ii)   the purity of DNA preparations, proteins and microorganisms including viruses, and

    iii) conformational changes in DNA and protein.

pH and pH Meter

INTRODUCTION

Qualitative and quantitative analyses of physical and chemical constituents of biological samples have yielded a lot of data that were useful in unravelling to a large extent the functioning of living systems. The modern analytical methods are designed to find the presence of a particular molecule and to quantify it as accurately as possible, even in a minute amount of sample. The introduction of ready-made "test-kit" system has simplified the usage of many analytical methods. The advent of electronics into the analytical techniques has not only increased the accuracy but also cut down the complicated procedures involved in the preparation of the sample and in the operation of the instruments.

Research students in different branches of biology use different analytical methods according to their requirements. All may not use all the techniques and many may use only one or two. However, it is always advantageous if we have basic knowledge on the principles and applications of as many methods as possible. Though we may not use a technique in our investigation, an awareness of its principle will help us in appreciating the published results obtained using that method. In this and the following chapters we shall consider, as briefly as possible, the principles and applications of some of the basic analytical techniques.

pH

All living organisms are very vulnerable to changes in the acidity or alkalinity of not only the surrounding medium but also the internal milieu. This is important for the growth and survival of the organism. While many organisms such as bacteria show dynamic growth within a fairly wide range of acidity or alkalinity, there are others requiring this to be adjusted within narrow limits before multiplication takes place. Moreover, all physiological and biochemical activities (e.g. enzyme action) of an organism are pH-dependent.

The acidity and alkalinity of a medium can be conveniently expressed in terms of molar hydrogen-ion concentration $[H^+]$. The square brackets around the hydrogen ion, etc. signify concentrations expressed in molarity. For reasons of practical convenience, $[H^+]$ is usually expressed on a logarithmic or pH scale (from French *pouvoir hydrogène,* "hydrogen power").

In water there is always a small number of dissociated H^+ ions (or more correctly the hydrated ion H_3O^+, the hydronium ion) and OH^- ions (the hydroxide ions). This may be represented as the reversible dissociation

$$H_2O \rightleftharpoons H^+ + OH^-$$

The heavy arrow in the above equation indicates that the equilibrium is lying very much to the left. If we apply the law of mass action (i.e., the rate of chemical reaction is proportional to the masses of the reacting substances) to this equilibrium, we have, at constant temperature,

$$\frac{[H^+][OH^-]}{H_2O} = K$$

where K is a constant, and $[H^+]$ and $[OH^-]$ are the concentrations at equilibrium of the ions in moles per litre, and $[H_2O]$ that of the undissociated water in molarity. The water is only very slightly dissociated, and therefore even a large relative change in the concentration of ions makes very little percentage difference to the concentration of undissociated water molecules, which may therefore be regarded as constant. Thus,

$$[H^+][OH^-] = K_W$$

where K_W is the ionic product for water. Since equal numbers of H^+ and OH^- ions are formed in pure water, $[H^+] = [OH^-]$. This has been determined experimentally and found to be 1×10^{-7} moles/litre at 25°C.

$$K_W = [H^+][OH^-] = (1 \times 10^{-7})(1 \times 10^{-7}) = 1 \times 10^{-14} \text{ (at 25°C)}$$

The H^+ concentration is always equal to the OH^- concentration only at neutrality. The ionic product ($[H^+] \times [OH^-]$), however, is always constant at constant temperature. If the hydrogen ion concentration rises above 10^{-7} moles/litre, the solution is described as acidic, and in such a solution the hydroxyl ion concentration is less than 10^{-7} moles/litre. If the hydrogen ion concentration is less than 10^{-7} moles/litre the solution is referred to as alkaline or basic, and the hydroxyl ion concentration in this solution is greater than 10^{-7} moles/litre. To describe the acidity or alkalinity of a solution it is only necessary to state the hydrogen ion concentration, since the product of hydrogen and hydroxyl ion concentrations is always constant, K_W.

Since extensive variations in hydrogen ion concentration are possible in nature, the use of a logarithmic scale to express pH value is convenient. The accepted scale is the negative logarithm to the base 10 of the hydrogen ion concentration in moles/litre. This is termed as the pH of a solution. Accordingly, at neutrality, $[H^+] = 10^{-7}$ moles/litre, thus

$$pH = -\log_{10}[H^+] = -\log(10^{-7}) = 7$$

The minus sign is reversed to plus simply for convenience, to give a positive scale of readings. In an acidic solution the pH will be less than 7, for instance, where $[H^+]$ is as high as 10^{-3} moles/litre, the pH is $-\log_{10}(10^{-3}) = 3$. On the other hand, in an alkaline solution the pH will be greater than 7; thus, if $[H^+]$ is 10^{-9} moles/litre, the pH is $-\log_{10}(10^{-9}) = 9$.

We should remember the following two points about the pH scale.

1. Since the pH scale is a logarithmic scale, a change of one unit of pH is equivalent to a tenfold change in hydrogen-ion concentration, that is, a 10-fold change of acidity; thus, for example, a liquid of pH 5.0 is 10 times more acid than a solution at pH 6.0; likewise, a liquid of pH 9.0 is 10 times more alkaline than one of pH 8.0. Therefore, it is necessary to give values correct to the first decimal place, e.g. 6.0, 6.1 or 6.2.

2. The pH scale is a reciprocal scale and, therefore, the lower the pH, the greater will be the acidity. A pH value of less than 7.0 indicates an acid solution and greater than 7.0 indicates an alkaline solution. For example, 1 mol/litre HCl has an approximate pH value of 0; 0.1 mol/litre HCl has an approximate pH value of 1; 0.01 mol/litre has an approximate pH value of 2.

The pH of a solution can be measured by titration, which involves neutralization of the acid (or base) by a measured quantity of base (or acid) of known concentration, in the presence of an indicator (a compound the colour of which depends on the pH). The pH of a solution can also be determined somewhat approximately by the use of pH papers.

The most useful and accurate method of pH determination is the electrometric method by using a pH meter, i.e., by measuring the electric potential arising at special electrodes immersed in the solution.

pH METER

Introduction

Numerous types of pH meters and pH electrodes are now available. For example, we have miniature battery-operated pH meters which can measure the pH of even a few microlitres of sample; bench-top models with microprocessor control with facilities for measuring several other parameters such as temperature, electrode potential (Eh), and, when fitted with ion selective electrodes, different ion concentrations.

The basic principle of a typical pH meter is determination of the activity of the hydrogen ions by potentiometric measurement with the help of a glass electrode (standard hydrogen electrode) and a reference electrode (calomel electrode).

Glass Electrode

A glass electrode (Figures 11.1 and 11.2) is the most common electrochemical sensor. It consists of a glass bulb membrane, hence its name, and an electrically insulating tubular body that separates an internal solution and a silver/silver chloride electrode from the sample solution. The Ag/AgCl electrode is connected to a special voltmeter, the pH meter, through a cable. The internal solution (fill solution) is a fixed concentration of HCl or a buffered chloride solution (e.g. KCl buffered to 7.0 pH).

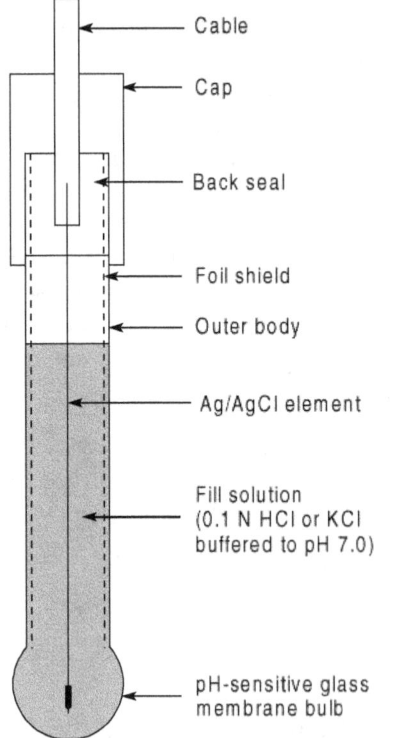

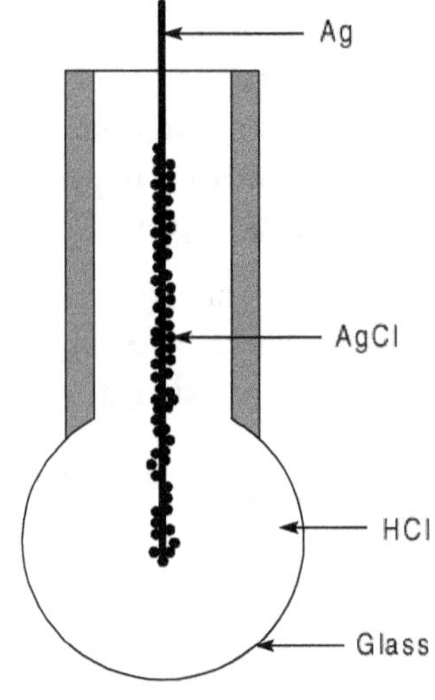

Figure 11.1   Glass electrode of a pH meter.

Figure 11.2   Schematics of a glass electrode.

Reference Electrode

A reference electrode provides a constant electrode potential. Either calomel (mercuric chloride, Hg_2Cl_2) or silver (silver chloride) is used as the electrode. The calomel electrode (Figure 11.3) has KCl wire and a column of mercuric chloride. The reference solution for the calomel electrode is KCl. The reference electrode has a liquid junction between the electrode and the sample solution. The liquid junction, which is made of some porous material, is the point at which the electrode forms a salt bridge with the sample. As a result of the formation of this bridge, a liquid junction potential is generated. This potential affects the potential produced by the reference electrode.

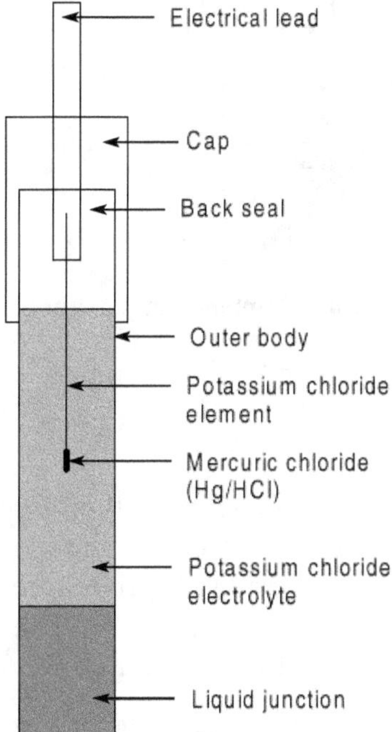

Figure 11.3   Calomel reference electrode of a pH meter.

Combination Electrode

This incorporates both the glass and the reference electrodes in a concentric double barrel body (Figure 11.4).

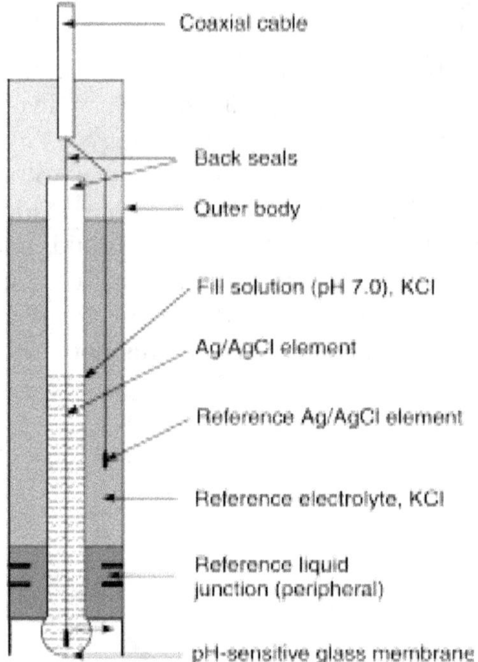

Figure 11.4   Combination electrode of a pH meter.

Working of pH Meter

The pH meter is a potentiometer and, therefore, two inputs are required. The first input is from the glass electrode and the second one is the reference electrode. Even in a combination electrode, the two inputs arise from the same device in the form of a coaxial cable. The complete circuit of a glass electrode with a reference electrode can be represented as

Ag/AgCl | HCl | glass | | test solution | reference electrode

The glass/test solution interface (marked | | in the above expression) is the site where the potential difference relating to the measurement of pH is built up. The glass membrane (the bulb) is thus the key functional part. It is about 0.1 mm thick and is sealed to a thicker glass or plastic tube. The filling solution (e.g. 0.1 M HCl) has constant Cl^- concentration, which keeps the Ag/AgCl inner electrode at fixed potential. The glass membrane is mostly amorphous silicon dioxide, with embedded oxides of alkali metals.

When the glass bulb is immersed in water, some Si^-O^- groups accept protons from the hydronium ions, i.e., the outer surface of the glass electrode gets hydrated, as follows:

$$Si^-O^- + H_3O^+ = Si^-O^-H^+ + H_2O$$

The pH sensing ability of the glass electrode is based on this exchange of H^+ between the solid glass membrane and surrounding medium, and the equilibrium nature of this exchange.

When the electrode is placed in a solution of unknown pH, the activity of the H^+ ions in the test solution is likely to be different from that of the H^+ ions in the outer hydrated layer. This "potential difference" is a linear function of pH and is measured with a pH meter, a special voltmeter with circuitry that enables direct readout of pH. Since the pH measurements using a pH meter are affected by temperature, a pH measurement is always reported with the temperature.

The existing designs of glass electrodes have many disadvantages such as fragility and calibration problems. To overcome such drawbacks, new generation of calibration-free, solid and strong electrodes based on quinhydrone are gaining usage. Such electrodes can be used to measure a wide range of pH, from concentrated acid to strong bases. These electrodes have very short reaction time, are temporally stable over several weeks, and can be easily regenerated.

CHAPTER TWELVE

Chromatography

INTRODUCTION

Chromatography is a wide range of physical methods used for the separation and analysis of complex mixtures. The components of a mixture may have different properties—physical, chemical, molecular, immunological, etc. Exploiting one or more of the differences in these properties, it is possible to physically separate the components. For example, differences in the size or molecular weights of the constituent proteins of a sample such as a tissue or body fluid would enable us to separate these proteins using appropriate chromatographic techniques.

The basic principle of all types of chromatography is the differential movement of a mixture over or through an adsorptive material. The differential movement is related to the differences in the physical, chemical, and other properties of the components of the mixture. The mixture itself does not move but is carried by another material which may be a liquid or a gas. The material carrying the mixture is referred to as the *mobile phase*. The material over or through which the mobile phase is flushed is called the *stationary phase*. The stationary phase may be solid, gel, liquid, or a mixture of liquid and solid. This phase is immobilized, hence called the stationary phase. The stationary and mobile phases are mutually immiscible.

The components of the mixture interact with both the stationary phase and the mobile phase differentially according to the differences in their properties. The differential interaction determines the rate of the migration of the components over or through the stationary phase. Thus, different components migrate at different rates. Those components that react more (i.e., have higher affinity) towards the mobile phase and less towards the stationary phase migrate faster. On the other hand, components reacting less with mobile phase and more with stationary phase migrate slowly. The faster or slower migration of the different components is reflected in the distances they have travelled over or through the stationary phase in a given time. The differential migration of the different components results in their separation.

The ratio of the concentration of the solute (a component of the sample mixture) in the stationary phase (C_s) to the concentration of the same in the mobile phase (C_m) of a chromatographic system is called *distribution coefficient* (*K*) or *partition coefficient*.

$$K = \frac{C_s}{C_m}$$

K describes the way in which a substance distributes or partitions itself between the two well-defined immiscible phases, the stationary phase (e.g. liquid) and the mobile phase (e.g. solid) of a chromatographic system. This coefficient is also referred to as *adsorption coefficient* or *Henry constant* with specific reference to HPLC (high performance liquid chromatography). Each substance has a specific partition coefficient for specific stationary and mobile phases. For example, if the partition coefficient of a substance between cellulose (stationary phase) and CCl_2 (mobile phase) is 0.1, it means that the concentration of the substance is more in the mobile phase than in the stationary phase (as much as 10 times the concentration in the stationary phase). It indicates that the substance has higher affinity to the mobile phase than to the stationary phase. Consequently, the substance would migrate faster and a longer distance in the stationary phase.

Some of the terms commonly used in chromatography have been defined in Box 12.1.

BOX 12.1   CHROMATOGRAPHY—TERMINOLOGY

Adsorption   As dictated by the cohesive forces acting between the molecules, surfaces of liquids and solids tend to attract and retain on them gases or dissolved substances with which they come into contact. This process of retention of molecules or ions on the surfaces of all kinds is called **adsorption**. The substance, which is thus attracted and retained adhered to the surface, is described to be the **adsorbed phase** or **adsorbate** while the substance to the surface of which it is attached is called the **adsorbent**. The release of the adsorbate from the surface of adsorbent is called **desorption**.

(*Note*   We should distinguish adsorption from absorption. **Absorption** is a process in which a substance is not only retained on the surface, but penetrates to the interior to become distributed throughout the phase. Both adsorption and absorption may take place side by side and it may be difficult to distinguish between the two processes. The general term **sorption** is implied in such cases. Absorption may refer to the taking up of fluids or dissolved substances by living tissue, especially, the takingup of digested contents of the intestine into the blood and lymphatic system; the swallowing up or

---

BOX 12.1 (Continued)

engulfing of bodies; the chemical or physical process of absorbing substances, energy, light, etc.)

Adsorbent The stationary phase, solid or liquid or a mixture of liquid and solid, possessing the property of adsorption.

Sorbent Any of the stationary phases, liquid or solid.

Support Solid scaffolding (physical framework) which holds a liquid stationary phase.

Solvent or developer Mobile phase which causes the **solute** or **sample** to migrate through the stationary phase.

Solute or sample The component in a mixture that is being subjected to chromatography.

Solvent front The front edge of the migrating solvent.

Migration Movement of the solvent or solute through the stationary phase.

Origin The point on the stationary phase at which the solute or sample is introduced into the solvent.

Rf value Retention factor; the position of a solute on a chromatogram. The ratio of the distance the solute migrated, to the distance the solvent migrated, $Rf = \dfrac{d(solute)}{d(solvent)}$.

Detection and visualization Visually identifying the separated solute on a chromatogram, if necessary, using a stain or a detector.

Elution The process of collection of solute that has been separated on a chromatogram, by allowing the solvent to overflow the stationary phase so that the desired solute is made to migrate further and ultimately runs out.

CLASSIFICATION OF CHROMATOGRAPHIC METHODS

We can classify the different chromatographic methods on the following bases:

1. Nature of the mobile and stationary phases

2. Principle of separation of the components
3. Geometry of the stationary phase
4. Mode of operation

Nature of the Phases

On the basis of the nature of the mobile phase used in a chromatographic system, we can classify it as gas chromatography or liquid chromatography.

Gas chromatography employs a gaseous fluid as the mobile phase, which is called the **carrier gas**. The carrier gas is generally an inert gas. The stationary phase of gas chromatography is generally a solid adsorbent or a liquid distributed over the surface of a porous, inert support. Depending on the nature of the stationary phase of a gas chromatographic system, we can further classify it as **gas–solid** chromatography and **gas–liquid** chromatography.

Liquid chromatography employs a liquid as mobile phase and a solid stationary phase, and therefore is described as **liquid–solid** chromatography. The liquid has a low viscosity and flows through the solid stationary phase bed. The stationary phase may be an immiscible liquid coated onto a porous support, a thin film of liquid phase bonded to the surface of a sorbent, or simply a sorbent of specified pore size.

Principle of Separation

When the mobile phase flows through the stationary phase, the different solutes are retained at different distances from the origin. On the basis of the specific process of retention, which leads to the separation of the different solutes, we can classify the chromatographic methods as adsorption, partition, ion-exchange, molecular exclusion, and affinity.

In **adsorption chromatography** (Figure 12.1) the components of a sample are separated on the basis of the differences in their adsorption-desorption behaviour between the mobile and the stationary phases.

Partition chromatography (Figure 12.2) is one in which the separation is based on differences between the solubility of the sample components in the stationary phase (gas chromatography), or on differences between the solubilities of the components in the mobile and stationary phases (liquid chromatography).

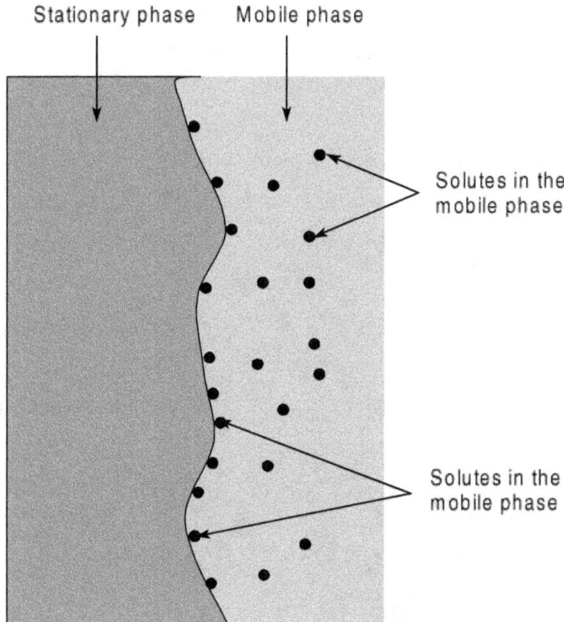

Figure 12.1 Adsorption chromatography.

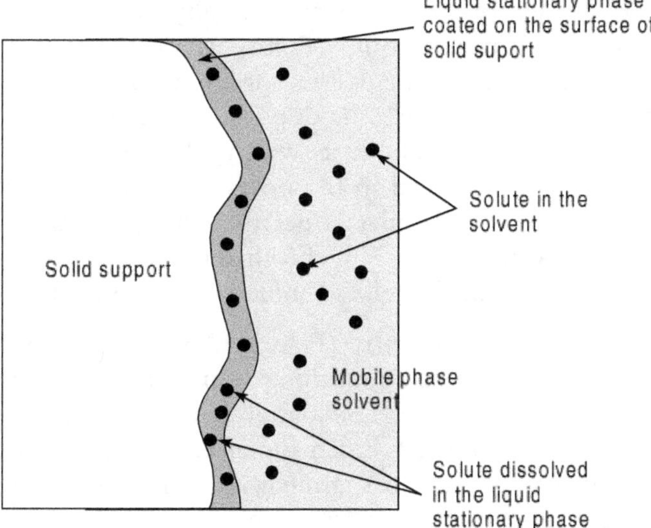

Figure 12.2 Partition chromatography.

Ion-exchange chromatography (Figure 12.3) separates the sample components on the basis of differences in their ion-exchange affinities.

For example, anions like SO_3^- or cations like $N\,(CH_3)_3^+$ are covalently attached to a stationary phase, such as resin

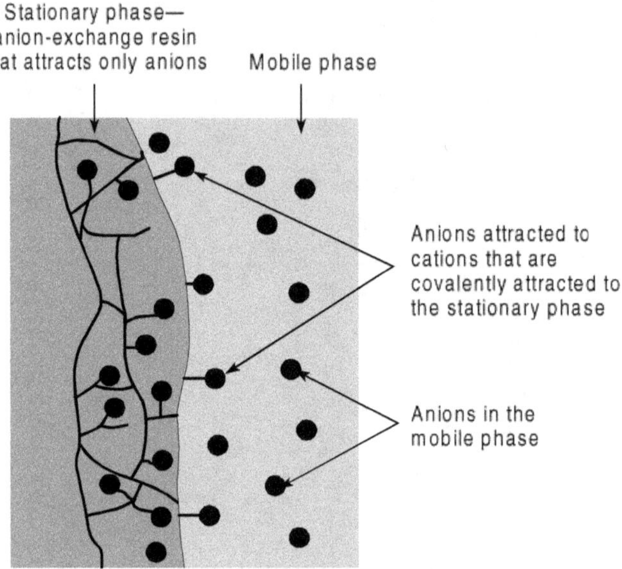

Figure 12.3   Ion-exchange chromatography.

Molecular exclusion chromatography or **gel-filtration** or **gel-permeation chromatography** (Figure 12.4) is a technique in which retention of sample components is according to their molecular size. The stationary phase is a gel consisting of porous beads with strictly controlled pore size. The sample components with molecular size smaller than that of the pore size, get trapped in the pores and as a result migrate slowly. On the other hand, molecules larger than the pore dimension migrate faster without any hindrance. This chromatographic technique is also useful in the determination of the relative molecular weight of a macromolecule.

Affinity chromatography (Figure 12.5) uses column packing that has been chemically altered by attaching a compound with a specific affinity for the desired molecules (e.g. biological compounds such as an enzyme). The packing substance used, called the affinity matrix, is inert and easily modified. Agarose is the most familiar substance used for this purpose. The **ligands (or "affinity tails"**, are substances capable of forming complexes with specific molecules) that have been genetically engineered to possess a specific affinity are inserted into the matrix. The desired molecules in the mobile phase adsorb to the ligands on the matrix because of the unique biological specificity of the analyte and ligand interaction

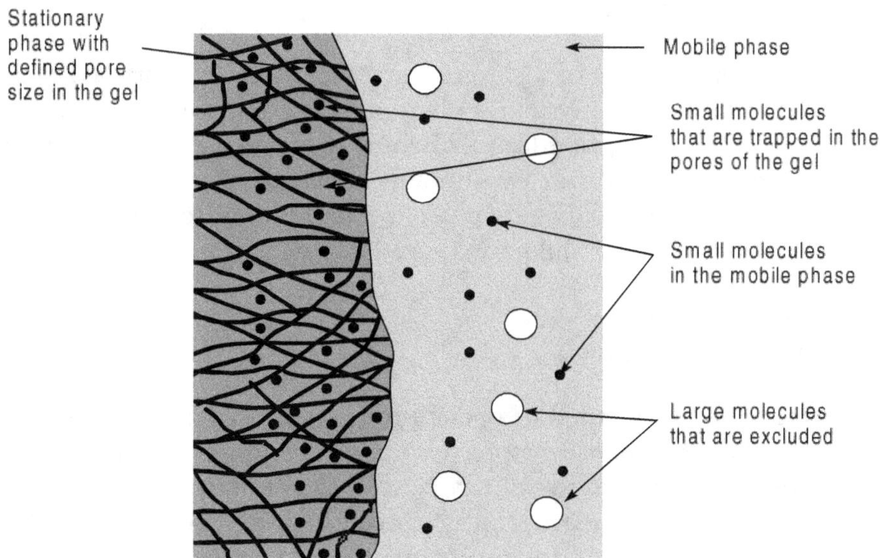

Figure 12.4   Molecular exclusion chromatography.

(e.g. antigen–antibody reaction). A solution of high salt concentration is then passed through the column to cause desorption of the molecules from the ligands, and to elute from the column.

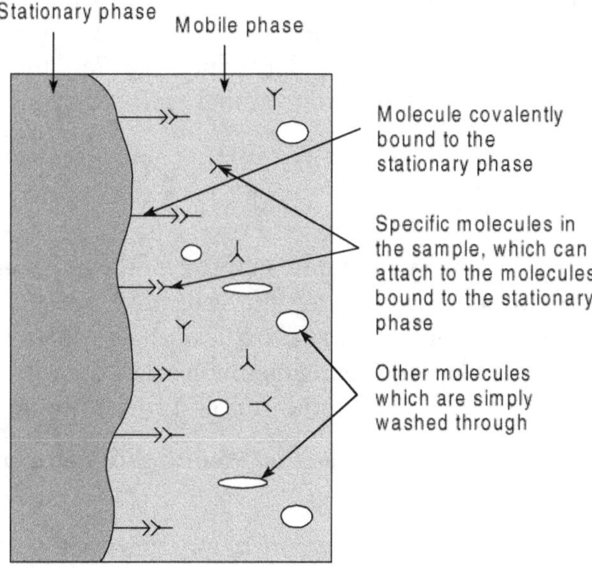

Figure 12.5   Affinity chromatography.

Geometry of Stationary Phase

If the stationary phase of a chromatographic system is contained in a tube called the column, the system is referred to as **column chromatography.** On the other hand, in **planar chromatography** the geometry of the stationary phase is a thin two-dimensional type. For example, in paper chromatography the stationary phase is a thin film of solid particles bound together with a binder and coated on a glass plate or plastic or metal sheet.

Mode of Operation

According to the method of operation, i.e., whether the migration of the mobile phase is stopped before reaching the end of the stationary phase or allowed to overrun it, we can classify the chromatographic system as development chromatography and elution chromatography. In **development chromatography** the flow of mobile phase, called the **developer**, is stopped before the solutes reach the end of the stationary phase bed. The migration of the developer liquid along the stationary phase bed is called **development**. In **elution chromatography**, which is generally used with columns, the sample components are allowed to migrate through the entire system. As the different components emerge from the column they are identified by a **detector** and quantified. The sample component that emerges from the column is called **eluate.** A detector monitors the quantity of each eluate and transduces the information to a strip chart recorder, which registers the different eluates as peaks of different heights proportional to their quantities.

Retention Mechanism

We can also classify chromatography in terms of the mechanism of retention of the solutes in a sample. Since the retention process is a mixture of more than one mechanism, this classification is somewhat approximate. Some of the interacting mechanisms involved in the retention process are hydrophobic (non-specific), dipole–dipole (polar), ionic, etc.

A general scheme of the classification of chromatography is shown in Figure 12.6.

We are not going to discuss in detail the protocols of the various chromatographic techniques. However, we shall review briefly the principle and applications of a few important chromatographic techniques.

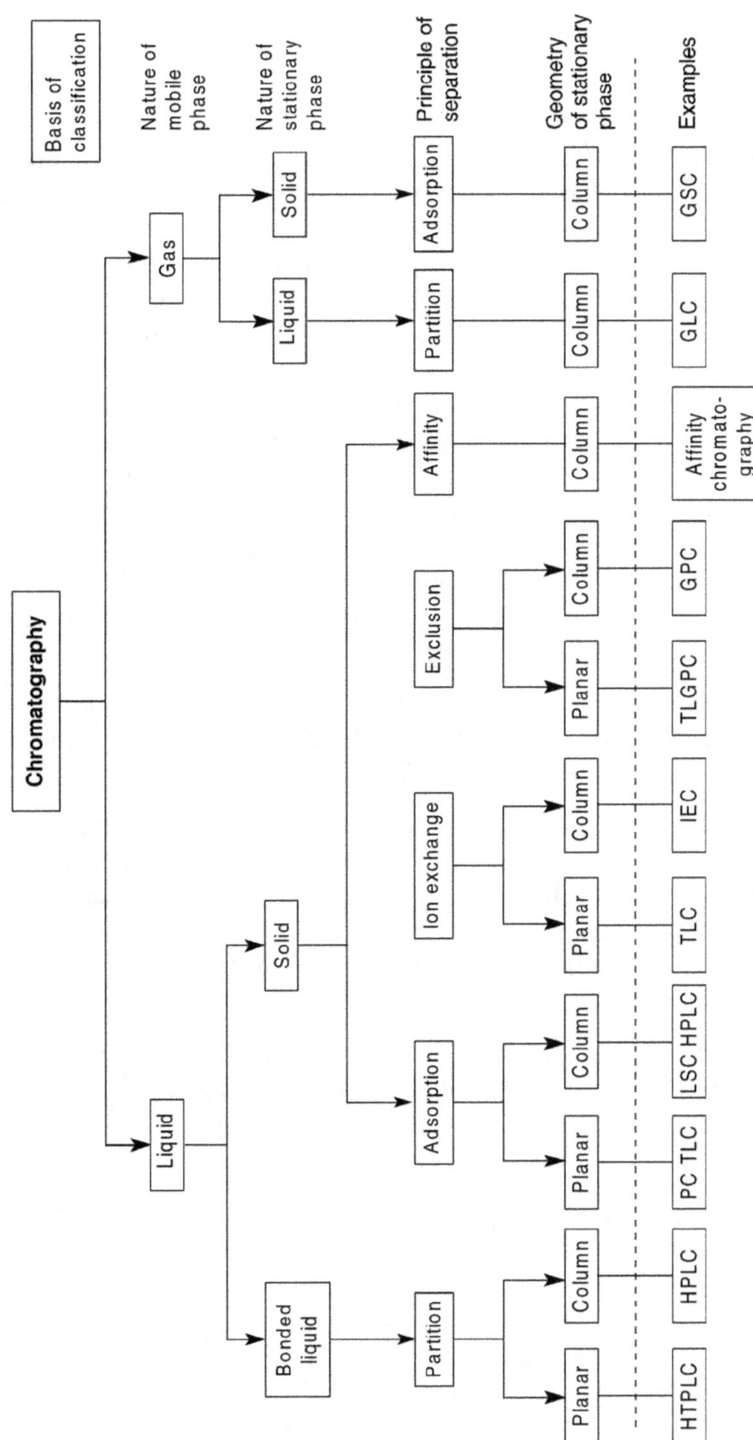

Figure 12.6 General scheme of classification of chromatography.

PAPER CHROMATOGRAPHY

Principle

It is a method suitable for separation of small amounts of low molecular weight compounds that are soluble in liquid phase. The paper used is a high quality filter paper (e.g. Whatman filter paper), which is made of purified cellulose. Cellulose is a polysaccharide composed of glucose molecules which have a large number of hydroxyl groups. The cellulose fibres in the filter paper hold moisture tightly by the formation of hydrogen bonds between their hydroxyl groups and water molecules. Even a sheet of 'dry' paper contains about 10% water by mass. Thus water acts as a liquid stationary phase. The mobile phase is any solvent consisting of an aqueous solution or an organic liquid such as ethanol that is partially miscible with water. Choice of the solvent depends on the nature of the solutes to be separated. A mixture of two or three solvents may also be required. The mobile phase moves along the paper by a mechanism of capillary action resulting from the forces between the solvent and the solid fibres of the paper.

The mode of separation of sample components is partition. Solutes are separated based on the interactions between two non-miscible liquid phases according to their relative solubility.

However, cellulose itself assumes a negative charge in the presence of water. Therefore, the paper exhibits ion exchange and adsorptive properties. Specialized chromatographic papers that are impregnated with alumina, silica gel, ion-exchange resin, etc. are also available. The separation method in these modified papers may vary according to the nature of the stationary phase.

The solutes that are highly water-soluble or have the greatest hydrogen-bonding capacity move slower along the paper, while less polar compounds will travel faster with the solvent.

Mobile Phase of Paper Chromatography

As already seen, unless modified, the stationary phase of a paper chromatographic system is water which is strongly adsorbed to the cellulose. The mobile phase of a paper chromatography is generally a mixture of solvents such as alcohols, acids, esters, ketones, phenols, amines, hydrocarbons, etc. The factors which determine the choice of the mobile phase for paper chromatography are:

i)   It must yield high degree of separation of the sample components.

ii) It should contain fewer number of constituents in the solvent mixture.

iii) The constituents of the solvent mixture should have more or less equal degree of evaporation.

iv) Solvent phase should be immiscible with the stationary phase.

v)   The sample components should possess distinctly different solubilities in the mobile and the stationary phase.

vi) Partition coefficient of the sample components should have a range between 1 and 100, with a higher affinity towards the stationary phase, i.e., water.

vii) It should have lower boiling point so that it can be easily removed from the paper, so that another solvent can be used for second development as in two-dimensional chromatography.

viii) It should be stable and should not get oxidized during migration.

Some of the commonly used solvent mixtures of the paper chromatographic system are:

i)   butyl alcohol + acetic acid + water

ii) butyl alcohol + pyridine + water

iii) methyl alcohol + pyridine + water

iv) propyl alcohol + petroleum ether

v)   chloroform + petroleum ether

Types of Paper Chromatography

The paper chromatographic technique may be an ascending type or a descending type. In **ascending chromatography** the paper is suspended from the top of the chromatographic chamber (developing chamber) and the base of the paper is in contact with the mobile phase (solvent) at the bottom of the chamber (Figure 12.7). The sample is spotted on the paper in a position just above the level of the solvent. The solvent ascends the paper by capillary action and carries the spotted sample along for the different solutes to be separated at different distances.

In **descending chromatography** the paper is suspended from a trough at the top of the developing chamber (Figure 12.7). The paper is held in position with glass rods. The lower end of the paper hangs vertically down but does not come into contact with the solvent at the bottom of the chamber, which acts as an equilibrating solvent. The sample is spotted at the top of the hanging part of the paper. The solvent in the top trough rises for a short distance in the paper by capillary action and then flows downward under gravitational force. The downward flow carries the sample for separation into its components.

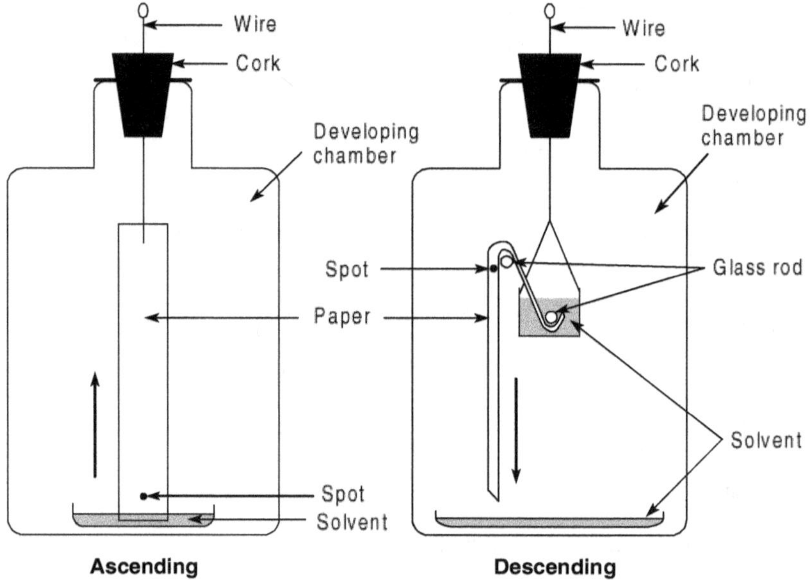

Figure 12.7   Ascending and descending types of paper chromatography.

The advantages of ascending chromatography are that

1.   assembly of the apparatus is simple, and

2.   it yields better resolution.

Two forces, capillary force for moving up and gravitational force opposing it, act on the sample in an ascending chromatography and therefore a better resolution is obtained in the chromatogram. The main disadvantage of ascending type is that

1.   it is very slow, and

2.   the solvent can rise 20 to 25 cm and separations of the solutes, which require longer migration distance, might be limited.

The descending chromatography has the advantage of separation over a longer distance and, therefore, a high degree of separation is achieved. Its main disadvantage is its complicated assembly.

Another variation of paper chromatography is **radial chromatography** in which the development spreads radially from the centre of a circular paper (Figure 12.8). The circular paper (e.g. Whatman filter paper) is mounted on a trough containing the solvent. The centre of the paper is pierced and inserted with a wick, which hangs down into the solvent. The whole assembly is covered by a jar. A pair of petri plates generally serves the purpose well. The samples are spotted radially around the centre of the paper. The solvent rises through the wick by capillary action and spreads over the paper carrying the samples to be separated.

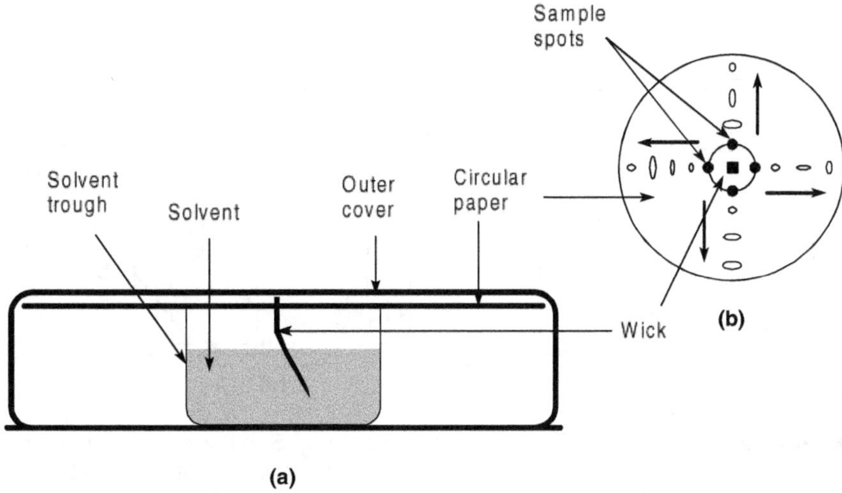

Figure 12.8 Radial chromatography a) assembly and b) chromatogram.

A useful adaptation of paper chromatography is to develop the paper in two dimensions, sequentially using two different solvents. This technique is referred to as **two-dimensional chromatography**. In **one-dimensional chromatography** the sample is developed only once in one direction and using one solvent. This may not yield desired resolution in the chromatogram. In order to increase the resolution, the same paper is developed after turning it 90° clockwise, and using a different solvent (Figure 12.9). After a second run at right angles to the first, the various sample components will be spread out at distinct spots across the sheet, forming a **chromatogram.**

Detection

The location and identification of the developed spots in a chromatogram is called detection.

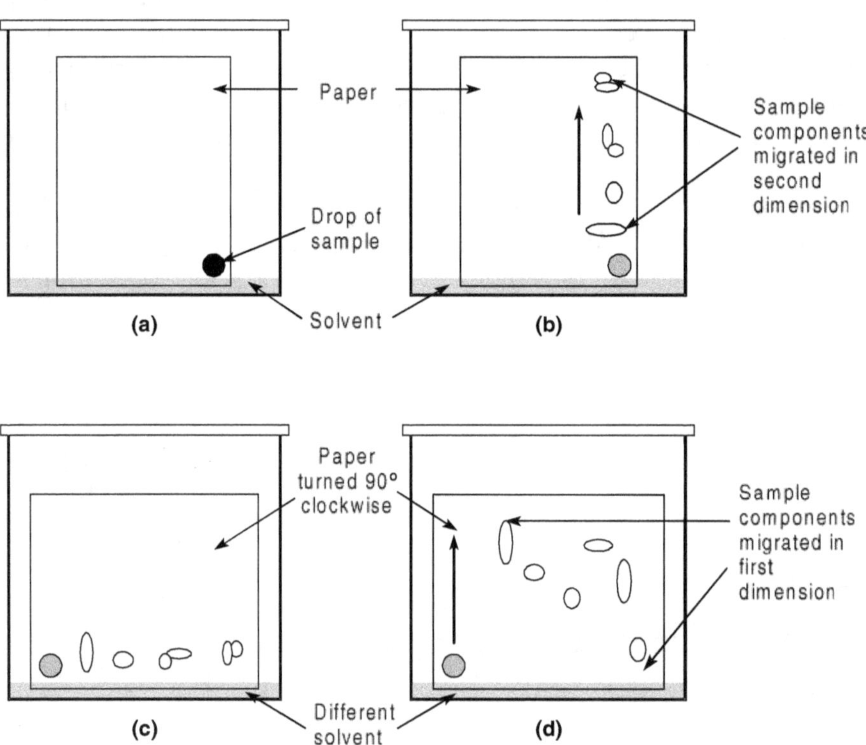

Figure 12.9 Two-dimensional chromatography. a) Spotting the sample in one corner of the paper, b) Migration of sample components in the first dimension, c) Paper turned 90° and developed with different solvent, d) Migration of sample components in the second dimension, yielding better resolution.

The location process is done by staining. If the sample components are themselves coloured the location of the spots does not pose any problem. However, if the components are colourless, spots are also colourless and therefore we will have to stain them specifically to make them visible. Usually, specific colour-producing chemicals (i.e., those which specifically react with the component on the chromatogram and produce a colour) are sprayed on the chromatogram. For example, ninhydrin is sprayed to impart a blue or purple colour to the spots of amino acids. Benzidine is used to locate sugars, and Sudan black to locate lipids.

The identity of each spot can be determined by

1. comparison of retardation factors
2. physical and chemical analysis
3. autoradiography

Retardation Factor

Retardation factor value (Rf) is a variable used to characterize the position of a substance in a chromatogram. It is an index of the degree of retention of a component. It is defined as the distance between the start line and the solvent front line divided by the distance between start line and solute front line. That is, the *Rf* of a solute (a component of sample mixture) is a ratio of distance migrated by solvent (eluent) to the distance migrated by the solute, in a chromatogram.

$$Rf = \frac{\text{Distance migrated by solvent}}{\text{Distance migrated by solute}}$$

The *Rf* values always lie between 0 and 1. They depend on the specific chromatographic conditions, nature of the stationary phase, matrix of substances in the sample, etc. They are only appropriate to chromatograms that have been developed once. Good resolution results when *Rf* is about 0.4 to 0.8.

For comparison, a known sample (e.g. sample containing known amino acids) is developed alongside the unknown sample (Figure 12.10). Developed spots of unknown sample located at identical distances (i.e., those having same *Rf* values) are identified as the same substance of known sample.

The identity of the developed spot in a paper chromatogram can also be achieved by carefully cutting out the paper containing the spot and subjecting it to physical and chemical analysis such as UV and IR absorbance, fluorescence, etc.

Autoradiography is also used in the identification of developed spots of chromatogram of paper chromatography. If the mixture contains molecules that have been labelled with a radioactive tracer, these can be located by placing the chromatogram next to a sheet of X-ray film. The location of dark spots on the developed film (because of radiation emitted by the tracer) can be correlated with the position of the substances on the chromatogram.

Applications of Paper Chromatography

Though the paper chromatographic technique is becoming obsolete, because of its simplicity and ability to separate very small quantities of substances from each other, it still has many applications in biochemistry, biotechnology, food industries, etc. In the biochemistry and biotechnology, paper chromatography is a routine technique for the separation and analysis of proteins, amino acids, sugars, lipids, etc. In the food industry, paper chromatography is used for separation of highly polar compounds such as sugars, amino acids and natural pigments. It is very often used as a quality control test especially to detect adulteration and contamination in food, drinks, gasoline, cosmetics, dyes, etc.

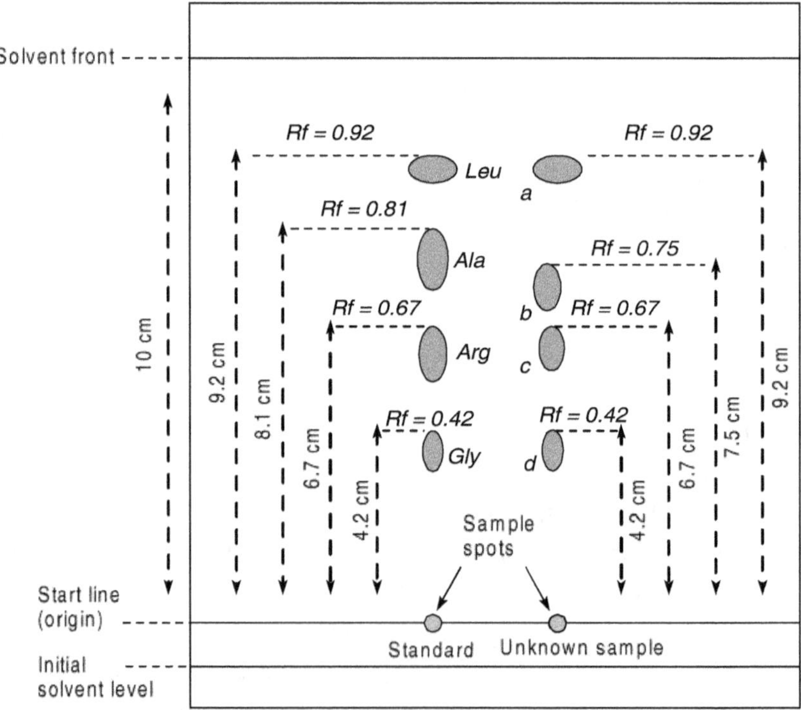

Figure 12.10   A paper chromatogram of separation of amino acids. Substance *a* is identified as leucine, *c* as arginine, and *d* as glycine. Substance *b* is identified as unknown because it has an *Rf* value that cannot be compared to any one in the standard substances.

THIN LAYER CHROMATOGRAPHY

Principle

Thin layer chromatography (TLC) is an adsorption or partition and planar type and, therefore, works on the same principle as that of paper chromatography.

Stationary Phase

The stationary phase is a thin layer of adsorbent (Table 12.1) attached on a piece of glass, aluminium or polyethylene plate. A slurry made of the powdered adsorbent in water is applied on the plate with the help of a *spreader* so as to form a thin layer. The thickness of the adsorbent layer varies according to the use, about 0.25 mm for analytical separation to about 5 mm for preparative separation. A *binder* such as calcium sulphate is mixed in the slurry to make the stationary phase firmly attach to the plate.

Table 12.1   Examples of stationary phases used in TLC and their specific applications.

Adsorbent	Application: Compounds separated
Alumina	Alkaloids, food dyes, phenols, steroids, vitamins, carotenes, amino acids
Celite	Steroids, inorganic cations
Cellulose powder	Amino acids, food dyes, alkaloids, nucleotides
Ion-exchange cellulose	Nucleotides, halide ions
Kieselguhr	Sugars, oligosaccharides, dibasic acids, fatty acids, triglycerides, amino acids, steroids
Polyamide powder	Anthocyanins, aromatic acids, antioxidants, flavonoids, proteins
Sephadex	Amino acids, proteins
Silica gel	Amino acids, alkaloids, sugars, fatty acids, lipids, essential oil, inorganic anions and cations, steroids, terpenoids
Starch	Amino acids

Mobile Phase

A few examples of the solvent mixtures used in the TLC are listed in Table 12.2. The choice of the mobile phase and the method of elution are similar to those of paper chromatography.

Table 12.2 Examples of solvents used in TLC and their applications.

Solvent mixture	Application: Compounds separated
Butanol/Acetic acid/Water	Amino acids
Butanol/Acetone/Phosphate buffer	Mono-and disaccharides
Carbon tetrachloride/Chloroform	Cholesterol esters
Chloroform/Methanol/Water	Phospholipids
Ethanol/Water	Amino acids
Ethyl acetate/Propanol	Mono-and disaccharides
Petroleum ether/Diethyl ether/Acetone	Triglycerides
Petroleum ether/Propanol	Plant pigments

Sample Application

Sample compounds are placed on a TLC plate as a spot that should be as small as possible at one end of the plate. Sample volumes spotted are generally 1–2 μL. However, in high performance TLC plates, which have a well-defined stationary phase particle diameter of 5–7 μm, the sample is loaded in much smaller quantity, about 100–200 nL. The sample area should be circular, not more than 1 mm in diameter. Concentration of sample can be accomplished by local heating causing evaporation of the solvent in which sample is dissolved. Application of such minute quantity of sample necessitates the increased concentration of the sample and special applicators such as micropipette and microsyringe. Automation of the sample application using computer-controlled syringes is also possible.

Plate Development

The plate is developed in a closed chamber in order to prevent evaporation of the solvent (Figure 12.11). The migration of the solvent is in ascending direction, the solvent rising through the adsorbent layer by capillary action. Two-dimensional TLC can also be performed to separate complex mixtures.

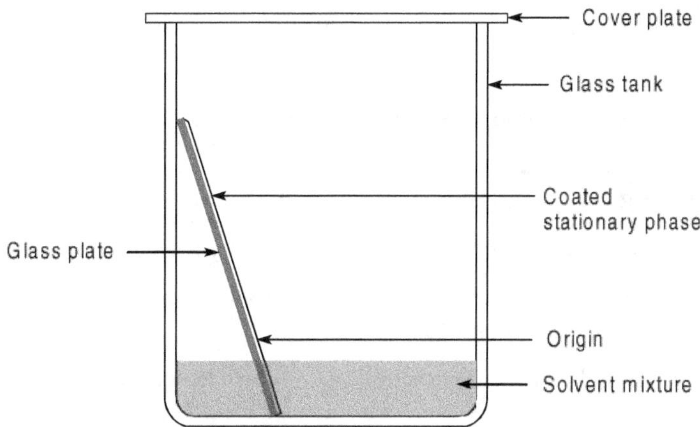

Figure 12.11   TLC assembly.

A modification of the normal TLC development is the **continuous development** of the plate. The apparatus (Figure 12.12) includes TLC plate held horizontal and inverted so that the stationary phase layer faces downwards and rests on a second glass cover plate. Thus, the stationary phase is sandwiched between two glass plates. A wick transfers the solvent from the reservoir to the stationary phase. The whole system is placed in a suitable chamber to prevent evaporation of the solvent. The solvent migrates along the stationary phase by surface tension. As the solvent reaches the end of the plate, it gets evaporated by the heat of an electrical heater. Thus, the development continues till a better resolution of the separated sample components is obtained.

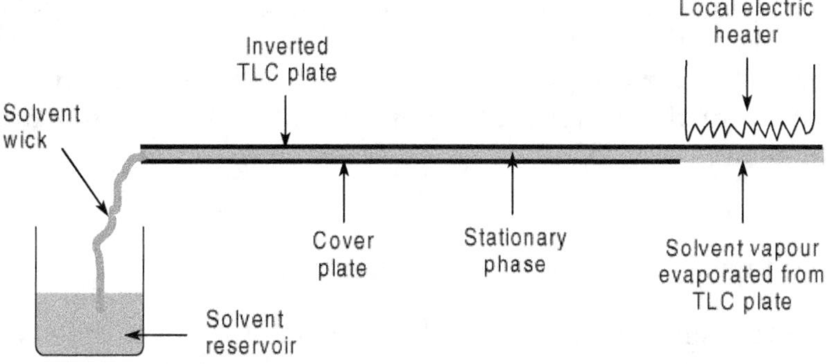

Figure 12.12   Continuous development of TLC.

Detection

The detection methods employed for paper chromatography such as UV-absorption, fluorescence and autoradiography can be applied to TLC as well. Commonly used TLC detecting agents and their specific uses are listed in Table 12.3.

Table 12.3   Detecting agents used in TLC.

Detecting (locating) agent	Used to detect
2,4-dinitrophenyl hydrazine	Aldehydes and ketones
Acridines (Acridine orange, proflavin, acriflavin) and fluorescence	Nucleic acids
Antimony trichloride in methanol	Steroids, terpenoids, glycosides and vitamin A
Bromine vapour	Olefins
Dragendorff's reagent	Alkaloids
Iodine vapour with benzidine spray	Organic compounds
Ninhydrin	Amino acids, amino sugars
Potassium permanganate in sulphuric acid	Hydrocarbons
Rhodamine B	Lipids
Sulphuric acid with heating	Any organic compound

The detected spots are then selectively scraped out, dissolved in suitable solvent and analysed qualitatively and quantitatively. Densitometers are used for the quantification of the separated sample components. The densitometers can also provide measurements of UV and visible absorptions of the separated components and the complete absorption spectrum of the sample. These are useful in the identification of the components.

Advantages Over Paper Chromatography

i)   TLC is simpler and requires lesser time to complete the development.

ii) The resolution of the components of a sample is much higher.

iii) Since a wide variety of materials can be used in the preparation of the stationary phase, it is more versatile.

Applications

It is used in:

i) Determination of the complexity of a mixture; separation of molecules like amino acids, nucleic acids, lipids, steroids, terpenoids, hydrocarbons, and carbohydrates.

ii) Tracing the course of reactions such as the incorporation of radio labelled carbon compounds into metabolites.

iii) Identification of drugs, contaminants and adulterants in food, drink, etc.

iv) Biochemical preparation of plant extracts, etc.

v) Analysis of body fluids in medical laboratories.

COLUMN CHROMATOGRAPHY

Introduction

In contrast to the planar chromatography that we discussed above (paper chromatography and TLC), which has a sheet of stationary phase, column chromatography has a cylindrical or tubular stationary phase. Most of the modern research and industrial applications of chromatography employ column type.

Column

The column, where the actual separation of the components of sample takes place, is usually a glass or metal tube of adequate strength to withstand the pressure and temperature that may be applied across it. The column encloses the stationary phase. The column may be either a packed bed or open tubular column.

A **packed bed column** contains a stationary phase that is in granular form and packed tightly into the column as a homogeneous bed. The column is thus a cylinder with a solid core.

An **open tubular column**, on the other hand, has a stationary phase applied as a thin film or layer onto its inner wall. Thus the centre of the column is hollow.

The mobile phase of column chromatography is a solvent into which the sample is injected. The solvent, referred to as **carrier fluid**, and the sample flow through the column. The stationary phase is made up of a material for which the components of the sample have varying affinities.

Depending on the nature of the materials which make up the mobile and stationary phases of column chromatography, we distinguish two types, viz., gas chromatography and liquid chromatography.

In **gas chromatography** the mobile phase is a gas, generally an inert gas. The stationary phase is a solid adsorbent or a liquid distributed over the surface of a porous, inert support.

In **liquid chromatography** the mobile phase is a liquid of low viscosity, which flows through the stationary phase bed. The stationary phase may be

i)   an immiscible liquid coated onto a porous support

ii) a thin layer of liquid bonded to the surface of a solid sorbent, or

iii) a sorbent of controlled pore size.

Plate Concept of Column

The efficiency of the column depends on a number of factors such as dimension (length and diameter) of the column, particle size of the column packing, quality of the mobile phase, temperature of the column, flow rate of the mobile phase, and proper packing of the column. Apart from these factors, the number of plates in the column plays a significant part in the effective separation of the solutes. The **concept of plates** is based on the assumption that the column can be separated into a number of segments that are identical and that one equilibrium distribution of a solute between the solid and mobile phase takes place in each segment. Each of these segments in a column is referred to as **theoretical plate**. The number of equilibrium distributions that take place in a column is thus proportional to the number of plates in the column. For example, two solutes in a sample which have negligible difference in their partition coefficients would require a large number of plates in the column for their complete satisfactory resolution.

Therefore, a column built with greater number of plates is more efficient than that with fewer plates. The number of plates can be increased either by increasing the length of the column or by better preparation of the column and optimization of the other factors that influence the solute equilibration between the two phases. The latter option is better because it yields better resolution, with narrow distinct peaks in the chromatogram.

Basic Operation of Column Chromatography

The basic components of a typical column chromatographic assembly are shown in Figure 12.13. The general working of a column chromatography is illustrated in Figure 12.14.

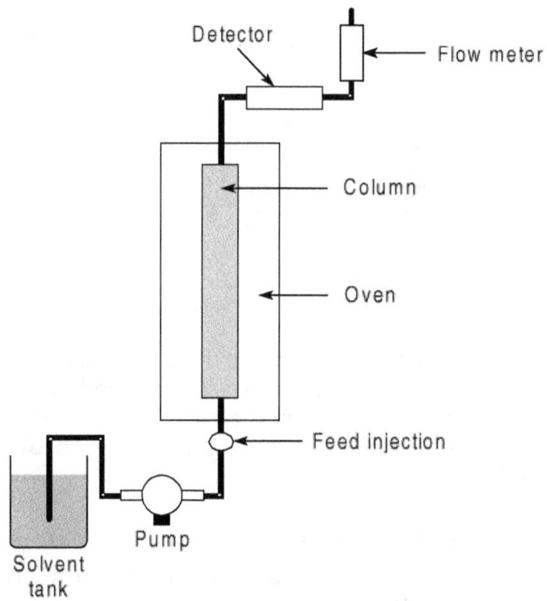

Figure 12.13   Column chromatography.

The following are the basic operations involved in column chromatography:

1. ***Feed injection***   The sample is injected into the mobile phase. The mobile phase is forced to flow through the column by the action of pump.

2. ***Separation in the column***   As the injected sample, now dissolved in the mobile phase, flows through the column, the different sample

components will adsorb to the stationary phase to varying degrees. Those solutes that have a strong attraction to the stationary phase move more slowly than those with weak attraction. In this way the components of the sample get separated.

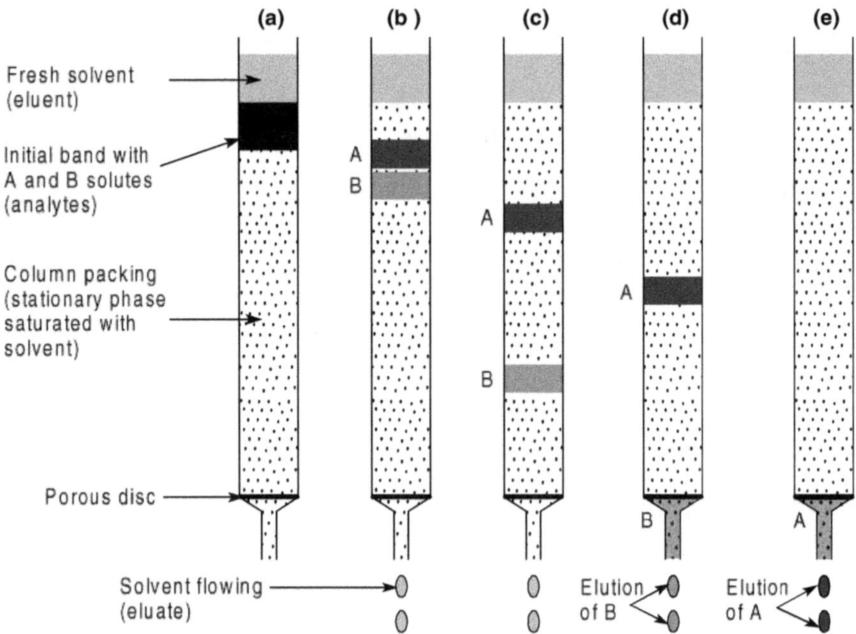

Figure 12.14 A simple separation in column chromatography. A continuous flow of solvent carries sample with solutes A and B down the column. a) Initial band with poor resolution of the solutes, b) & c) At a later point of time, solute B is seen migrating faster than solute A, d) Elution of B, e) Elution of A.

3. ***Elution from the column***   After the mobile phase carrying the sample reaches the other end of the column, the different sample components will elute (emerge out) from the column at different times. The components with minimum affinity for the stationary phase and therefore weakly adsorbed, will elute first. Those solutes with maximum affinity for the stationary phase and therefore strongly adsorbed, will elute last. By varying the pH of the solvent or temperature of the column, the elution pattern, i.e., the timing of the emergence can be significantly altered.

4. ***Detection***   The different sample components are collected as they emerge from the column. A detector analyses the emerging solutes by measuring a property which is related to concentration and

characteristic of their chemical composition. For example, the refractive index or ultraviolet absorbance is measured.

Detector

Once the component solutes of a sample are separated in a chromatographic system, it is necessary to locate and identify the separated substances. In simple chromatographic techniques such as paper chromatography and TLC, visual location and comparison of the separated spots with standards are useful in qualitative identification of the substances. Subsequent colorimetric or densitometric measurement may help in the quantitative estimations of the solute concentrations. However, in column chromatography, elaborate electronic devices called detectors are used for this purpose.

A column chromatographic **detector** is a device that locates in the dimensions of space and time, the positions of the components of a sample that has been subjected to a chromatographic process.

Detectors are generally classified into two types, viz., bulk property detectors and solute property detectors. The bulk property detector measures certain bulk physical property of the eluent (e.g. dielectric constant, refractive index). The solute property detector measures specific unique physical or chemical property of the solute (e.g. heat of combustion, fluorescence).

Detectors are also classified as **concentration sensitive devices** (e.g. Katharometer) or **mass sensitive devices** (e.g. flame ionization detector).

A third method of classification defines detectors as specific or non-specific.

Nitrogen-phosphorus detector is an example of **specific detectors**, as it detects only those substances that contain nitrogen and phosphorus. Katharometer detector is an example of **non-specific detectors**, and detects all vapours that have specific heats or thermal conductivities different from those of the carrier gas. A non-specific detector has lower sensitivities than a specific detector.

Chromatogram

The output from the detector of column chromatography, the chromatogram, shows the separation of the components of a sample in the form of a line graph with peaks of different heights and widths, each peak corresponding to a specific solute. A chromatogram and its characteristic nomenclature are shown in Figure 12.15.

The following information can be inferred from a chromatogram.

i)   The performance of a column is indicated by comparison with standards.

ii) The degree of complexity of the sample is shown by the number of peaks in a chromatogram.

iii) Qualitative information about the sample components is obtained by comparing peak positions with those of standards.

iv) Quantitative information about the relative concentrations of the components of the sample is gained from peak area comparisons.

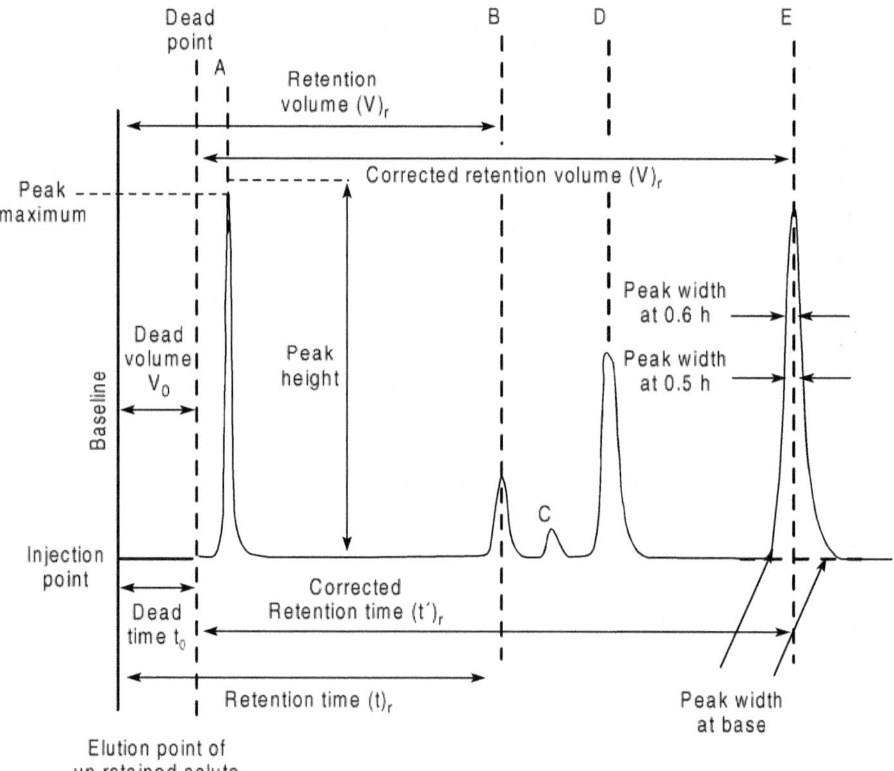

Figure 12.15   The nomenclature of a chromatogram that shows the separation of five solutes (A, B, C, D and E).

The *baseline* is any part of the chromatogram where only mobile phase is emerging from the column. The *peak maximum* is the highest point of the peak. The *injection point* is that point in time/position when/where the sample is placed on the column. The *dead point* is the position of the peak-maximum of an unretained solute. The *dead time* is the time elapsed between the *injection point* and the *dead point*. The *dead volume* is the volume of mobile phase passed through the column between the *injection point* and the *dead point*. The *retention time* is the time between injection and the appearance of a peak on the chromatogram. It is a characteristic for each solute. Identical solutes will have the same retention time under the same circumstances. The *retention volume* is the volume of the mobile phase required to elute a solute.

GAS CHROMATOGRAPHY

In gas chromatography (GC), the separation of solutes is carried out by the gaseous mobile phase passing through a column. The high degree of resolution of the different solutes in a GC is due to the efficiency of mass transfer in a mobile gas phase, and the very long columns used. The efficiency of a GC depends on the selection of appropriate column packing/coating and temperature programming.

Carrier Gas

The mobile phase, i.e., the carrier gas, is supplied constantly and at an appropriate flow rate. The carrier gases must be chemically inert and the most common carrier gases are helium, nitrogen and hydrogen. Oxygen cannot be used as a carrier gas because it may induce oxidation in the stationary phase and lead to a loss in column performance. The carrier gases must be highly pure and therefore a molecular sieve is used to remove moisture and other impurities. The choice of carrier gases is determined by the type of detector employed in the system.

Sample Injection System

Direct injection of sample using a microsyringe (1 μL–25 μL) is a common method of injecting sample into the column. However, this method is less precise and is therefore replaced by an automated sample injection system in sophisticated GC systems. The sample is usually injected into an injection port where it is immediately vaporized so that it can pass through the column along with the carrier gas, for separation. In order to vaporize the sample, the temperature in the injection port is maintained at about 50°C above the boiling point of the least volatile component of the sample.

Column

The column of a GC is either a packed column or capillary or open tubular column (OTC). The OTC is better than packed column in resolution, speed and column efficiency.

Packed column   A packed column is made up of glass, stainless steel, copper, aluminium, nickel or Teflon tubes. The length of the column varies from 0.5 to 5.0 metres. The outer diameter of the column ranges between 1.6 and 12.7 mm, and the inner diameter between 0.2 and 4.0 mm. The long column is accommodated in a chromatographic oven in the form of a coil of about 15 cm diameter.

The column tube is packed with a solid support material. An ideal support material should have the following properties:

i)   should be small, uniform, spherical particles with a surface area greater than 1 m^2/g

ii)   should remain inert at increased temperatures

iii)   should possess good mechanical strength

iv)   should be capable of being uniformly wetted by the liquid phase

In general, smaller the diameter of the packing material, higher is the efficiency of the column. However, with decrease in the particle size of column packing, greater pressure is required to maintain the flow rate of the carrier gas.

A few examples of materials used as solid support of packed column are:

i)   Diatomaceous earths

ii)   Crushed firebrick

iii)   Glass microspheres

iv)   Porous silica beads

v)   Polytetrafluoroethylene

vi) Powdered Teflon

vii) Graphitized carbon

viii) Carborundum

The supporting material is coated with a thin layer of liquid as the stationary phase. The following are the characteristics of a good stationary phase of a GC:

i)   should be non-volatile at the temperature maintained in the oven

ii)   should be thermally stable

iii) should have appropriate partition coefficient for the sample components

iv) should be inert towards the solutes in the sample

Common liquid stationary phases used in GC are listed in Table 12.3.

Table 12.3   Examples of liquid stationary phase of gas chromatography and their uses.

Stationary phase	Used to separate
Dinonyl phthalate	Any type of sample
Hexadecane	Hydrocarbons, fluorides
Paraffin (high molecular weight)	Esters, ethers, hydrocarbons
Polyethylene glycol (Carbowax)	Amines, ketones, alcohols
Silicon rubber gum	Steroids, alkaloids, pesticides
Silicone oil	Any type of sample
Silver nitrate	Olefins, hydrocarbons
Sqalane	Hydrocarbons

Capillary or open tubular column   When compared to packed column, open tubular column (OTC) is considered superior because it offers:

i)   higher resolution

ii)   shorter analysis time

iii) greater selectivity

iv) lower sample capacity

Tubular columns are generally made of fused silica (SiO_2) and coated with polyamide, which is a plastic that can withstand a temperature of 350°C. The length of OTC ranges from 15 to 100 metres. The inner diameter of the column varies from 0.10 to 0.53 mm.

There are two types of OTC, wall-coated open tubular and support-coated open tubular. In the **wall coated open tubular (WCOT) column**, a thin film (0.1–0.5 mm) of stationary phase is coated on the inner wall of the column. With the decrease in the thickness of the coated stationary phase layer, there is increase in the resolution, decrease in retention time and decrease in sample capacity.

In the **support coated open tubular (SCOT) column,** a thin layer of liquid stationary phase is coated on the solid support attached to the inner wall of the column. Such columns have increased surface area and consequently can handle larger samples. The SCOT columns are considered to be intermediate between WCOT and packed columns.

The nature of the stationary phases used in open tubular columns should be appropriate to solutes that are being separated. The stationary phase should have the same polarity as that of the solutes, for example, non-polar columns yield better results for non-polar solutes.

Oven

In GC, the elution time and resolution of the solutes are highly dependent on the temperature of the column. Therefore, the temperature of the column is maintained as accurately and precisely as possible by enclosing the column in an oven. The oven is heated electrically with forced air circulation by a fan. The usual range of temperature maintained in the oven is from room temperature to about 400–500°C. The operating temperature of the oven is generally at least 50°C higher than the maximum boiling point of the solutes in the sample. For samples having a broad boiling point range, temperature programming is employed. Such programming enables the increase of the column temperature either continuously or in steps as the separation proceeds.

Detection System

Once the solutes of the sample have been separated in the column, they are detected, identified and quantified as they come out of the columns with the help of a detector. An effective GC detector has the following characteristics:

i)   Sufficient sensitivity

ii) High degree of stability and reproducibility

iii) Linear response to solutes that extend over several orders of magnitude

iv) Temperature range between room temperature and about 400°C

v)   Quick response time, independent of flow rate

vi) Reliability and simple use

vii) Uniform response time towards all solutes or predictable selective response towards one or more classes of solutes

The following are the most common GC detectors:

i)   Flame ionization

ii) Thermal conductivity

iii) Electron capture

iv) Flame photometric

The **flame ionization detector** (FID) uses hydrogen as the combustion gas, which is mixed with carrier gas (helium, nitrogen, etc.) eluting from the column, and burnt at a small jet situated inside a cylindrical electrode. An electric potential of a few hundred volts is applied between the jet and the electrode. When a carbon-containing solute is burnt in the jet, the electron/ion pairs that are formed are collected at the jet and cylindrical electrode. The current is amplified and fed into a recorder or to an analog/digital converter of a computer data gathering system. The basic organization of an FID is shown in Figure 12.16. This detector is best suited for compounds containing C—C or C—H bonds. It is commonly used for the analysis of fatty acids, flavour components, carbohydrates and antioxidants.

The FID is capable of detecting almost all carbon containing solutes, with the exception of a few small molecular compounds such as carbon disulphide, carbon monoxide, etc. Because of its diverse and comprehensive response, FID is considered as a universal detector.

In **thermal conductivity detector** the eluted solutes cause changes in the temperature inside the detector, which generate electrical signals that are fed into a recorder or a computer system. The advantage of this detector is that it is non-destructive to the sample, and permits the use of sample for further analysis. The main disadvantage is that it is less sensitive.

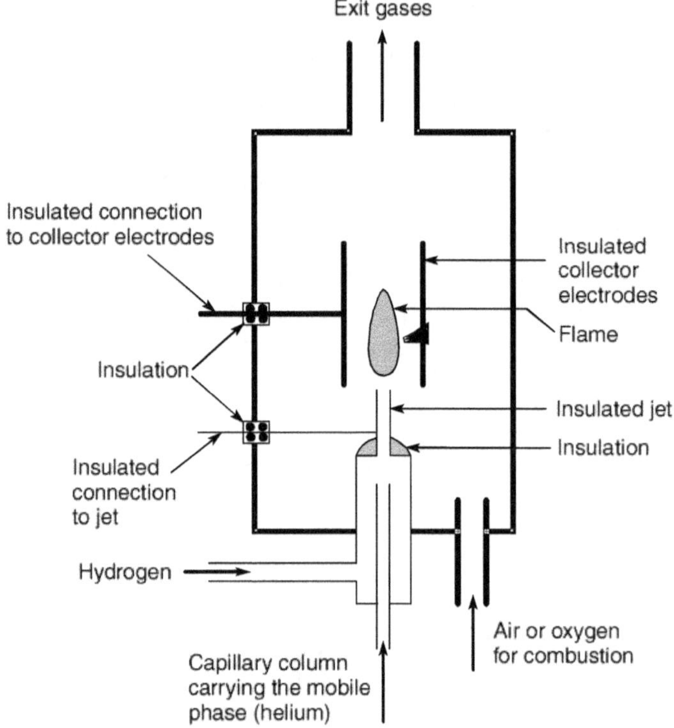

Figure 12.16   Flame ionization detector.

Electron-capture detector is highly sensitive to molecules containing electronegative functional groups such as halogens, peroxides, quinines and nitro groups. It is ideal for the analysis of pesticide residues. This detector is also non-destructive to the sample, but has a narrow detection range.

Sulphur-and phosphorus-containing compounds emit light at specific wavelengths, when heated. The **flame photometric detector** measures the intensity of the light emitted by such solutes in eluent when passed into a low-temperature flame. The intensity of the light is directly proportional to the concentration of the solutes.

Types of gas Chromatography

We can distinguish three types of GC on the basis of the stationary phase in the column, viz., gas-solid, gas-liquid and capillary-gas chromatography.

Gas-solid chromatography (GSC) or **gas-adsorption chromatography** employs a packed bed of adsorbent as the stationary phase. Zeolite, silica gel and activated alumina are the common adsorbents. This method is generally used for the separation of a mixture of gases.

Gas-liquid chromatography (GLC), the most common analytical GC, uses a column which contains an inert porous solid coated with a viscous liquid that acts as the stationary phase. Diatomaceous earth is often used as the solid. Sample injected into the column dissolves into the liquid phase and eventually vaporizes, and the separation of the sample components is based on the relative volatilities.

Capillary-gas chromatography is yet another common analytical method. The capillary wall is made up of glass or fused silica. The wall is coated with an adsorbent. Because of the small amount of the stationary phase, the column has only a limited capacity but yields quick separation of mixtures.

Applications of Gas Chromatography

The chief application of gas chromatography is the separation of volatile materials. Even non-volatile substances such as amino acids, steroids and high molecular weight fatty acids can be separated using GC after converting them into volatile materials. The capillary columns in GC have much higher efficiency and therefore can handle multicomponent samples such as essential oils. The following are a few specific applications of GC.

1. *Analysis of hydrocarbon* Aromatic hydrocarbons are highly toxic and carcinogenic. Therefore the presence of these compounds in all spheres from where they might enter into human food chain, has to be continuously screened. For example, drinking water and irrigation water are such potential areas. Water samples from areas where contamination might occur, need to be constantly analysed for the presence of aromatic hydrocarbons. Very often, the level of these contaminants is very low, in ppb (parts per billion). Only gas chromatography employing a high-sensitive detector can measure such low levels of contaminants.

2. *Analysis of essential oils* Essential oils contain several components that are commercially important. For example, there are components of essential oils that have flavours similar to that of peach, melon and other fruits. These flavoured components have

uses in the food industry. Only gas chromatography is capable of separation and analysis of such components of the essential oils.

3. ***Identification of bacteria***   The bacterial species belonging to *Clostridium* are found in the soil. Three species, *C. tetani*, *C. botulium*, and *C. gangrene* are highly dangerous to humans as they cause tetanus, botulism and gas gangrene, respectively. Gas chromatography is useful in the identification of these bacteria. Samples containing these bacteria when subjected to GC yield three distinct volatile fatty acid profiles. The GC for this purpose uses a glass column with packing consisting of phosphoric acid supported on Chromosorb, a form of diatomite. The distinct fatty acid profile is used as a means of identification.

LIQUID CHROMATOGRAPHY

Liquid chromatography (LC) is another column chromatography in which the mobile phase is liquid (e.g. methanol). More specifically, an LC may be a **liquid-adsorption chromatography** in which adsorbent stationary phase is used, or a **liquid-liquid chromatography** in which a liquid stationary phase is coated onto a solid support.

There are many types of liquid chromatographic methods such as reverse phase chromatography, high performance liquid chromatography (HPLC), size exclusion chromatography, supercritical fluid chromatography, ion-exchange chromatography, affinity chromatography, etc.

Reverse Phase Chromatography

Reverse phase chromatography involves a hydrophobic, low polarity stationary phase that is chemically bonded to an inert solid (e.g. silica). The separation of the sample components is by extraction process. This type of chromatography is used for the separation of non-volatile components of a sample.

High Performance Liquid Chromatography

High performance liquid chromatography (HPLC) is similar to the reverse phase chromatography except that the extraction process is conducted at a high velocity and pressure drop. The column of an HPLC is shorter and has a smaller diameter. However, the column is equivalent to having large number of equilibrium stages, i.e., plates.

HPLC is the most common type of liquid chromatography. It is called "high performance" because it has a very effective separation and high resolution. The high performance is due to the improvement in the technology of columns and instrumental components. In this technique, the stationary phase is held in a column and the mobile phase is forced through under pressure. Separation of sample solutes is much faster. A commonly used stationary phase is silica particles with long-chain alkanes adsorbed onto their surfaces. A common mobile phase of HPLC is methanol. The general scheme of an HPLC assembly is shown in Figure 12.17.

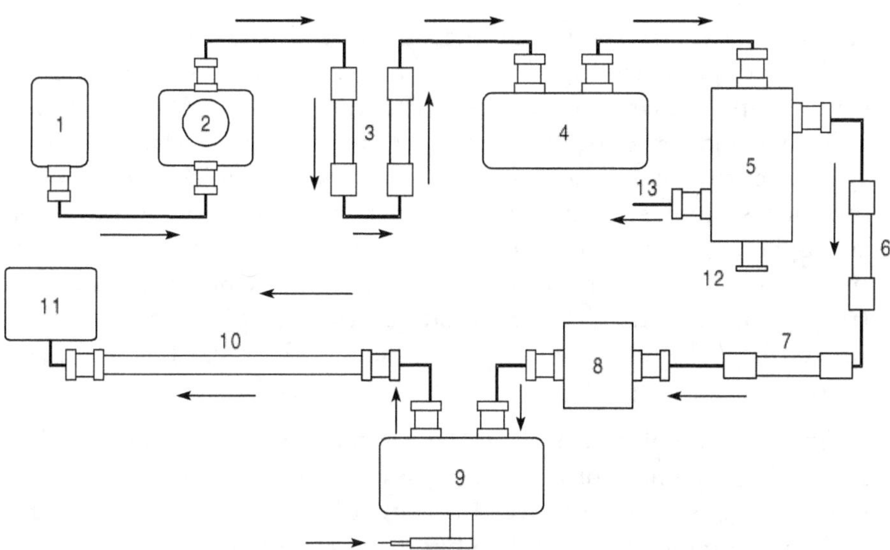

Figure 12.17 Main parts of an HPLC. 1. Mobile phase solvent reservoir; 2. Solvent proportionating valve; 3. Pumping system; 4. Pulse damper; 5. Drain valve; 6. Filter; 7. Backpressure regulator; 8. Pressure transducer; 9. Sample injection system; 10 Column; 11. Detector system; 12. Drain valve control; 13. Outlet for waste.

Solvent Reservoir An HPLC apparatus has one or more glass or stainless steel mobile phase solvent reservoirs. Each reservoir has a capacity of about 200 to 1000 mL. The mobile phase solvent is purified by removing gases and filtered to remove particulate matter. The elution process in a HPLC system may be isocratic or gradient. In **isocratic elution** only a single solvent of constant composition is used. On the other hand, in gradient elution a mixture of up to four solvents with different polarities is used as mobile phase. During the gradient elution process, the ratio of the solvent is varied in a programmed way. Thus gradient elution increases the performance of the system.

Pumping system   The mobile phase solvent is delivered into the HPLC system in a controlled, accurate and precise manner. The pumping system of an HPLC is made of corrosion-resistant components such as stainless steel or Teflon and it can

i)   generate pressure up to 6000 psi

ii)   provide pulse-free output, and

iii)   maintain flow rates ranging from 0.1 to 10 mL per minute.

Sample injection system   Samples are introduced into HPLC system, without depressurizing the system. This is achieved by the use of microsyringes that can withstand pressures up to 1500 psi. Since the efficiency and reproducibility of a microsyringe are not very high, **loop-type injection valves** are widely used for sample injection. In loop-type injection system, the sample is injected from the syringe into a loop at atmospheric pressure. At high pressure, the valve of the system is switched from "load" to "inject" position as a result of which the mobile phase travels through the loop and carries the sample onto the column. Interchangeable loops with varying capacities, from 10 μL to 10 μL, can be used.

Column   The column of an HPLC is made up of stainless steel, glass, fused silica, titanium, or polyether resin. However, smooth-bored stainless steel tubing is preferred over other materials because it can withstand high pressure (6000 psi) while other material will not be able to withstand pressures more than 600 psi. Generally, the column of an HPLC has a length range of about 10–30 cm with an internal diameter of 4.6 or 5 mm. The particle size of the column packing is 5 or 10 μm. More recently, columns with an internal diameter range of 1–4.6 mm are being used, and such columns yield higher resolution in shorter time with minimal solvent consumption.

Column packing materials   The packing material of an HPLC column has a well-defined particle size with narrow particle size distribution, sufficient mechanical strength to withstand high pressure and high chemical stability. Its main function is to provide the column support, as it may or may not be involved in the actual separation process. There are four basic types of HPLC column packing materials:

i)   Bonded phase silica

ii) Pellicular packing

iii) Microporous

iv) Macroporous

Bonded phase silica (Figure 12.18a) contains silica beads made up of silanol (Si-OH). The beads are bonded with hydrocarbon groups. The nature of the bonded phase determines the chromatographic behaviour.

Pellicular packing (Figure 12.18b) has an inert core that provides the physical support. A thin layer of functional groups that is coated on the inert core is involved in the chromatographic separation process.

Microporous packing (Figure 12.18c) is a gel-type resin consisting of cross-linked polymers.

Macroporous packing (Figure 12.18d) is a resin that has much higher cross-linking than microporous resin. It is stable in a wide range of pH, 1 to 14. It is available in a variety of particle and pore sizes.

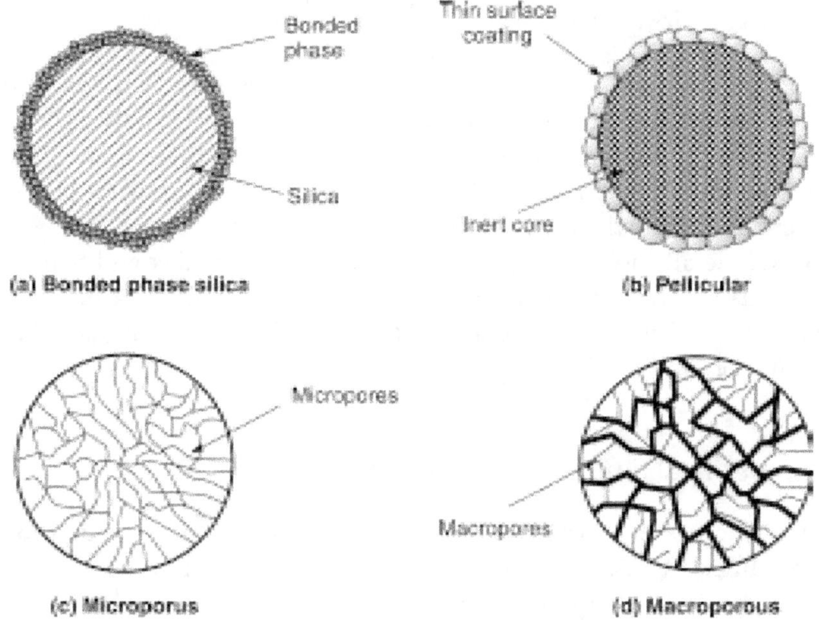

Figure 12.18   Types of HPLC column packing materials.

Detector system   An HPLC system may use a bulk property detector or a solute property detector. The choice depends on the solute type and its concentration. A **bulk property detector** responds to changes in properties of mobile phase due to the presence of solutes. For example, properties such as refractive index, dielectric constant and density of the solvent may be changed due to the presence of different solutes in different concentrations.

A **solute property detector** responds to the property of the solutes in the mobile phase. For example, properties such as UV absorbance, fluorescence and diffusion current of the solutes may vary and be proportional to their concentrations. The separated components of the sample are usually detected by UV spectroscopy as they pass through a flow cell between a UV source and detector. The output of the detector is recorded as a chromatogram. In addition to these two basic types of detectors, other types such as **light scattering detector, radioactive detector** and **viscosity detector**, have been designed to respond to solutes more sensitively. Increase in specificity and sensitivity of multiple type solutes in a sample is also often achieved by the use of more than one type of detector.

Size Exclusion Chromatography

Size exclusion chromatography (gel permeation or filtration chromatography) has no adsorption phase. It is an extremely fast technique. The column packing is a porous gel that is capable of separating molecules on the basis of their sizes. The larger molecules elute first because they cannot penetrate the pores of the gel. The porosity of the gel can be adjusted to exclude all molecules above a certain size. Sephadex and Sepharose are trade names for gels that are available commercially in a broad range of porosities. This chromatographic technique is commonly used in protein separation and purification.

Supercritical Fluid Chromatography (SFC)

This type uses a supercritical carrier fluid (e.g. carbon dioxide mixed with a modifier). Compared to the usual mobile phase liquids, a supercritical fluid has higher solubility, density, and diffusivity.

Ion-exchange Chromatography

Ion-exchange chromatography is normally used in the separation and purification of biological molecules. Two types of ion exchange, viz., cation exchange and anion exchange, are used. In cation exchange, the stationary phase carries a negative charge. In anion exchange, the stationary phase has positive charge. Charged molecules in the carrier liquid phase pass through the column and bind to cation or anion in the stationary phase. The bound molecules in the column are eluted by flushing the column with a solution of varying pH or ionic strength. Separation by ion-exchange process is highly selective.

Affinity Chromatography

Affinity chromatography has column packing that has been chemically modified by attaching a material having specific affinity for the molecule to be separated. This method is useful in the separation of biological molecules. For example, specific antibodies in a serum sample can be isolated using corresponding antigens in the column packing. The packing material of affinity chromatography is called the affinity matrix. It is inert and easily modified. The ligands ("affinity tails") are substances that are inserted into the matrix, and that have or are genetically modified to have specific affinity for a desired molecule. The desired molecules adsorb to the ligands on the matrix. A solution of high salt concentration is flushed through the column matrix to cause desorption of the adsorbed molecules and to elute them from the column.

Preparative Liquid Chromatography

LC is often used as a method for the isolation of sample components in sufficient quantities so that they can be further purified and studied or put to specific use. There are two major approaches for preparative LC, viz., (i) elution, and (ii) displacement chromatography.

Elution chromatography is carried out under one of the following conditions:

i)   isocratic, i.e., with constant mobile phase composition,

ii)   gradient, i.e., continuous change of mobile phase composition, and

iii) step elution. Under these conditions, the sample is injected into the column as a finite volume pulse. The solutes in the sample feed migrate through the column at different speeds, as dictated by the mobile phase velocity and the distribution of the solutes between the mobile and stationary phases.

In order to increase the output of a separated solute, the preparative elution chromatography uses mass and/or volume overload conditions. In volume overloading the concentration of the sample is maintained but its volume is increased until an optimum input and, therefore, output is attained. In mass overloading, the sample concentration is increased to gain increased output. A combination of volume and mass overloading can also be employed to maximize the output.

Another preparative chromatography is **displacement chromatography** which is more useful in the separation of protein components of a sample. The main feature of the displacement chromatography is the use of a displacer compound, which has a higher affinity for the stationary phase than the solutes in the sample. First, a large volume of the sample is loaded onto the column. Next, a displacer solution is pumped constantly into the column. The displacement process is based on the competition of solutes for the adsorption sites on the stationary phase according to their relative binding affinities and the displacer solution concentration. The action of the displacer solution, which has a higher affinity to the stationary phase than the solutes, causes the sample components to migrate faster through the column. The sample components elute from the column as adjacent zones of highly concentrated pure solute, in the order of increasing affinity of adsorption.

Applications of Liquid Chromatography

LC is useful in the separation of highly polar and large molecular weight molecules that are nonvolatile or have low volatility. The following are a few specific applications of liquid chromatography.

1. *Trace analysis*   Trace analysis is analysis of a sample in which the material to be identified and quantified is present in minute quantity. The trace analysis involves a concentration procedure. First, the sample is subjected to LC with short adsorbent-packed column to selectively remove the trace material. The adsorbed material is extracted from the column, dissolved in suitable solvent and injected onto an LC column containing higher number of theoretical plates (as many as 25,000) for further finer separation and analysis.

An example of trace analysis is the detection of presence of drugs and drug metabolites in blood samples. Another example is the determination of tetrahydrocannabinol carboxylic acid in the urine samples of persons suspected of smoking marijuana.

2. ***Ionic interaction chromatography***   Liquid chromatography in which ion exchange resins are used as the stationary phase is referred to as ionic interaction chromatography or, simply, ion-exchange chromatography or ion-chromatography. The polyesterene divinyl benzene cross-linked polymers, which are highly stable in a wide range of salt concentrations and can function effectively in a pH range of 2.0 to 12.0 is:

3. ***Food industry***   Liquid chromatography is used in the analysis of aroma composition of white wine, colour and pigment stability in juices, fatty acid content in fish, lactosyl beta-lactoglobulin conjugates in heat-treated milk and whey, and sugars and free amino acids in stored potatoes.

4. ***Biotechnology***   Liquid chromatography can be used to separate delicate products since the conditions under which it is performed are not typically severe. For these reasons, chromatography is quite well suited to a variety of uses in the field of biotechnology, such as separating mixtures of proteins.

CHAPTER THIRTEEN

ELECTROPHORESIS

INTRODUCTION

Electrophoresis is an analytical method commonly used in molecular biology. This method is applied for the separation and characterization of biological macromolecules such as proteins and nucleic acids. It is also used for the isolation and analysis of minute particles like viruses and small organelles. Electrophoresis works on the principle that charged particles of a sample migrate in an applied electrical field. If electrophoresis is conducted in a solution, the molecules (particles) are separated according to their net charge. However, electrophoresis uses paper or, most frequently, gel as a support medium. The use of gel (agarose, polyacrylamide) matrix provides a sieving effect so that molecules are characterized by both charge and size. Addition of a charged detergent (e.g. SDS, sodium dodecyl sulphate) to the gel equalizes the surface charge and, therefore, permits differential migration of protein molecules only according to their sizes. Such electrophoresis helps in the estimation of molecular weight of proteins.

PRINCIPLE

Electrophoresis (*electro* means the energy of electricity and *phoresis*, from the Greek word *phoros*, means "to carry across.") is based on the ionic property of the molecules, which enables them to migrate in an electrical field. Biomolecules such as proteins and nucleic acids carry certain electrical charges due to their component groups that are capable of dissociating electrolytically (see Box 13.1 for the explanation of electrolysis and related terms). For example, the net charge of a protein molecule depends on its amino acid composition. If the protein has more positively charged amino acids so that the sum of the positive charges is greater than the sum of the negative charges, it will have an overall positive charge. Such a protein with a net positive charge is a cation and therefore will migrate towards a cathode (negative electrode) in an electric field. Variations in the composition of amino acids, even if it is with reference to one amino acid, will result in different net charges of a protein. Such variations are used to distinguish proteins electrophoretically.

The charged molecules migrate in an electrical field with a velocity and in a direction depending on their **electrophoretic mobility**.

BOX 13.1   ELECTROLYSIS—TERMINOLOGY

An **electrolyte** is a compound which, when in solution or melted, conducts electric current and is decomposed by it. Examples: sulphuric acid, hydrochloric acid, nitric acid, sodium hydroxide and most salts are strong electrolytes. Water, carbonic acid, ammonium hydroxide and acetic acid are weak electrolytes.

Ionic theory   Electrolytes consist of ions which are positively and negatively charged particles. An **anion** is a **negatively charged ion** that moves to the **anode**, the **positive electrode**, during electrolysis. A **cation** is **positively charged** and moves to the **cathode**, the **negative electrode** during electrolysis. Substances with similar charges repel one another and those with opposite charges attract one another.

A **non-electrolyte** is a compound which, when in solution or melted, does not conduct electricity. Examples: sugar, alcohol, petrol, benzene and most organic compounds are non-electrolytes.

Electrolysis is decomposition of an electrolyte by passing an electric current through it. Example: Sodium chloride (NaCl) decomposes into Na^+ (cation) and Cl^- (anion).

The **electrodes** are two pieces of metal or graphite by which the electrical current (i.e., electrons) enters and leaves an electrolyte. Examples: platinum, carbon, nickel, and copper. The **cathode** is the **negative electrode** by which electrons enter the electrolyte; the **anode** is the **positive electrode** by which electrons leave the electrolyte.

Molecules that bear both positive and negative charges are called **zwitterions** (German: *Zwitter* = hermaphrodite) or **dipolar ions**. For example, some amino acids are dipolar ions. Proteins, which are made up of amino acids, also have zwitterion characteristics.

Electrophoresis is the movement of particles by an electric force.

An electrophoresis unit is a simple electrical circuit that works on Ohm's law: $V = IR$, where V is the electrical field measured in volts, I, the current measured in milliamps, and R, the resistance in ohms. Thus, electrical field is proportional to the product of current and resistance. In a given amount of V, a constant amount of current flows through the elements of the circuit. However, the total applied voltage decreases across any element as its resistance increases. Resistance (the ability of a substance to resist the flow of electricity through it) is dependent on the thickness of the material (e.g. thickness of the gel layer) through which the current flows. A good conductor has low resistance while a poor conductor has high resistance. In an electrophoretic set-up, other factors such as ionic strength of buffer also influence the resistance.

Molecules in a sample migrate according to voltage density (V cm^{-1}) of the gel. The velocity of migration (cm sec^{-1}) of a molecule at 1 V cm^{-1} is referred to as the **electrophoretic mobility** that is expressed as cm^{-1}V^{-1}sec^{-1}. The other factors that influence the electrophoretic mobility of a molecule are its net electrical charge and its frictional coefficient. Frictional coefficient is determined by the shape and molecular weight of the molecule.

COMPONENTS OF AN ELECTROPHORESIS UNIT

An electrophoresis unit consists of the following basic components:

1. Power supply with voltmeter and voltage regulator.
2. Electrophoresis tank that holds electrophoresis buffer, an anode and a cathode connected through the power supply.
3. A glass plate that holds the gel and is submerged into the buffer.
4. A comb that is used to make the sample wells, into which the samples are introduced.

The voltage delivered is controlled by the power supply. The current flowing between the anode and the cathode sets up an electric field in the electrophoresis tank and in the gel.

FACTORS AFFECTING ELECTROPHORETIC MOBILITY

i) *Charge, size and shape of sample molecules*   The relationship between the net charge of a molecule to its size (molecular weight) and shape is expressed as charge/mass ratio. This ratio affects the electrophoretic mobility of the molecule. Molecules having higher net charge have greater electrophoretic mobility than those with lesser net charge. Bigger molecules face higher frictional and electrostatic forces in an electrophoretic field and, therefore, exhibit lesser electrophoretic mobility when compared to smaller molecules. Molecules having smooth contours in their shapes (e.g. globular proteins) encounter least frictional and electrostatic force in an electrophoretic field and, therefore, have higher electrophoretic mobility when compared to molecules having rough contours (e.g. fibrous proteins).

ii) ***Ionic strength of buffer***   In order to maintain physiological pH and control the net charge of the sample molecules, the supporting medium of the electrophoresis is perfused with a buffer solution. The ionic strength of this buffer has an inverse influence on the migration of the sample molecules. If the buffer ionic strength is higher, it reduces the effective charge of the molecules and results in their slower migration. On the other hand, a buffer with low ionic strength causes high rates of migration of the sample molecules. Electrophoresis in higher ionic buffer yields sharper bands of sample molecules. However, higher ionic strength of buffer may lead to excess power generation, which in turn may cause increase in the temperature of the supporting matrix (gel).

iii) ***pH of the buffer***   Ionization of the organic molecules is related to the pH of the buffer system of electrophoresis. With increase in the pH of the buffer the ionization of organic acids increases while that of organic bases decreases. Consequently, the rate and direction of migration are affected.

iv) ***Temperature***   Electrophoresis is generally run in a surrounding with lower temperature because increase in temperature adversely affects the gel system. During electrophoresis, the temperature may rise for several reasons such as increased voltage of the current supply and increase in the ionic strength of the buffer and the resulting higher power generation. The temperature increase leads to decrease in the viscosity of the medium, which in turn would produce more current and further increase in the temperature. Higher temperature may cause buffer evaporation, sample denaturation, convective mixing and other similar problems.

v) ***Power supply***   An increased voltage produces a greater proportional current through a given resistor medium such as the gel of an electrophoresis. Increase in current flow leads to excessive heat generation, which, as already seen, might cause convection current, buffer evaporation and even melting of the matrix as in the case of agarose electrophoresis. Special power supply units are needed to provide a constant and regulated voltage of the current. Modern commercially available power packs have safety features like (i) automatic shutoff facility to ensure that proper voltage and current are not exceeded, (ii) more than one output terminals so that one unit can be connected to more than one electrophoresis unit simultaneously, (iii) LCD display of voltage, current turnover and time, (iv) voltage range of 10 to 3000 V and current range of 5 to 200 mA, and (v) continuous power supply for about 16 hours.

In addition to the above factors, others such as the nature of the supportive medium, composition of the buffer, etc. also play an important role in the effective electrophoretic separation of the molecules in a sample.

SUPPORT MEDIUM

Electrophoresis can be performed in a liquid medium (moving boundary electrophoresis) or a solid medium (zone electrophoresis) such as filter paper, cellulose acetate paper and gel.

Filter Paper

Any good quality filter paper (e.g. Whatman) that contains about 95 per cent cellulose and has minimum adsorption property is suitable for electrophoresis. Special chromatographic papers can also be cut in desired dimensions (squares, rectangular strips, etc.) and used for electrophoresis purpose. Paper as a support medium of electrophoresis is useful for routine, quick diagnostic analysis. Paper electrophoresis is cheaper and easy to use. Its main disadvantage is that the paper has substantial adsorptive property and, it may cause background staining while locating the bands. That is, obtaining clear-cut bands with high degree of resolution is difficult with paper.

Cellulose Acetate Strip

Cellulose acetate as a support medium for electrophoresis was developed from bacteriological cellulose acetate membrane filters, which are used for sterilization by filtration of culture media and other solutions. These filters are available as highly pure cellulose acetate strips, which are thin, about 130 μm, and have uniform microporous structure. The hydroxyl groups in the cellulose acetate contain derivative groups, which make the strips least adsorptive. The disadvantage of background staining encountered in paper electrophoresis is eliminated in the cellulose acetate strips because of their minimum adsorptive nature.

The other advantages of this type of support medium are as follows:

i) It is chemically pure as it is free from lignins, hemicelluloses, or nitrogen.

ii) Because of its translucent nature, direct photoelectric scanning of the bands is possible.

iii) Since the glucose content of cellulose acetate strips is very low, they can be used for the electrophoresis of polysaccharides in combination with PAS staining reaction.

iv) Because of its least hydrophilic property, cellulose acetate strips hold only a minimum buffer, which leads to better resolution within a short duration.

The disadvantages of cellulose acetate strips are as follows:

i) The lower buffer-holding property of the strips causes higher heat generation which may lead to drying and denaturation of molecules. Therefore, temperature regulation is a critical factor in cellulose acetate electrophoresis.

ii) It is unsuitable for preparative electrophoresis.

Gels

Most of the modern electrophoresis techniques make use of gels as the supporting medium. Though there are several types of gels such as starch gel, agar gel, agarose gel and polyacrylamide gel, only agarose and polyacrylamide are most commonly used.

Starch Starch obtained from potato is hydrolysed in acidified acetone at 37°C. The suspension is neutralized with sodium acetate, washed with plenty of distilled water, and dried with acetone. The hydrolysed starch is dissolved in a buffer, heated and cooled to set as a gel by intertwining of amylopectin chains. Gels prepared using 2% (w/v) starch have high porosity, while those prepared with higher concentration (10–15%) of starch have low porosity. However, the pore size of the starch gel cannot be controlled by varying the concentration of the starch.

The main advantage of the starch gel is its high resolving power. However it has several disadvantages.

i)   As already indicated above, the pore size of the starch gel cannot be controlled.

ii) It is unsuitable for the electrophoresis of basic proteins.

iii) It gets easily contaminated with microorganisms.

iv) While staining it to locate the separated proteins, it turns opaque and therefore unsuitable for measurements of the bands.

The important application of starch gel is the analysis of isoenzyme patterns of biological samples.

Agar   Agar is a compound of two galactose-based polymers, agarose and agaropectin. Agar dissolved in water at 80°C is mixed with an equal volume of 40 per cent polyethylene glycol to form a precipitate. The precipitate is washed several times with distilled water and dried with acetone. Agar dissolves in aqueous buffer at 40°C and solidifies to form a gel at 38°C. Low concentration of agar yields gels with larger pore size, which virtually has no molecular sieving effect. These gels, therefore, have low diffusion resistance and are useful in immunoelectrophoresis. Agar gels are also useful in the electrophoresis of high molecular weight biomolecules such as proteins and nucleic acids. The main disadvantage of agar gel is the severe electroosmosis produced as a result of the sulphated nature of the agaropectin.

Agarose   Agarose is a natural colloid that is extracted from seaweed (red algae). It is very brittle and easily damaged during handling. The pore size of the agarose gel matrix is relatively large and therefore this gel is mainly used to separate large molecules which have a molecular mass of about 200 kdal or more. Though the agarose gels have the advantage of being processed faster than the polyacrylamide gel, the resolution is poor. The bands formed in the agarose gels are blurry and spread far apart. The disadvantages of agarose gels are due to their pore size which it cannot be controlled.

Another significant disadvantage of agarose electrophoresis is electroendosmosis (EEO) which means flow of water under the influence of an electric field. This phenomenon of EEO is caused by immobilized charge groups (e.g. sulphate and carboxyl groups) on the matrix. These ions do not migrate but their counter ions migrate towards the cathode, resulting in the flow of water through osmosis in the same direction. EEO has the effect of badly smeared bands. This problem can be obviated by the use of high quality electrophoresis grade agarose.

Chemically, agarose is a linear polysaccharide with an average molecular mass of about 12000. It is made up by polymerization of the basic units of agarobiose, which contains alternating units of galactose and 3,6-anhydrogalactose.

For electrophoretic purpose, agarose is generally used at concentrations ranging between 1 and 3 per cent. It is prepared by suspending dry agarose in aqueous buffer and then boiling the mixture till a clear solution is obtained. It is poured into a suitable mold and allowed to cool to form a rigid gel. Agarose gel is easier to handle and safer to prepare than the polyacrylamide gel. Agarose gel is used for the electrophoretic separation of large nucleic acid molecules, which have more than 400 base pairs.

Polyacrylamide The polyacrylamide gel is produced by the polymerization of acrylamide in water, in the presence of a crosslinker such as methylenebisacrylamide (bis, or MBA). It is the same material used in the soft contact lenses. The copolymerization of acrylamide and MBA creates a mesh-like network. This three-dimensional network consists of acrylamide chains which are interconnected by MBA. In addition to MBA, there are other crosslinkers such as PDA (piperazine diacrylate), BAC (N, N´-bisacrylylcystamine) and DATD (N, N'-diallyltartardiamide). Each of these crosslinkers has specific advantage over the others. For example, PDA is useful to reduce silver stain backgrounds in SDS-PAGE gels. BAC and DATD produce a gel matrix that can be easily solubilized.

A standard nomenclature is used to describe the composition of polyacrylamide gels, the most important of which are T and C. The T of polyacrylamide gel represents the total percentage concentration (weight/volume) of the monomer (acrylamide + crosslinker) in the gel. The letter C represents the percentage of the crosslinker alone in the total monomer. For instance, polyacrylamide gel may be described to have a T value of 8% and a C value of 5%. The T value of 8% indicates that the total acrylamide/bisacrylamide content in the gel is 8 g in 100 mL of distilled water. The C value of 5% indicates that the acrylamide–bisacrylamide ratio is 19:1. That is 19 parts of acrylamide and 1 part of bisacrylamide are mixed, and 8 g of this mixture is dissolved in 100 mL of distilled water.

The polymerization process of acrylamide/bisacrylamide is catalysed by either ammonium persulphate and TEMED (a tertiary aliphatic amine, N,N,N´,N´- tetramethylethylenediamine) or riboflavin and TEMED.

The main advantage of polyacrylamide gel is that the pore size of the matrix can be altered in a reproducible manner by changing the total concentration of monomer (T) and the ratio of the acrylamide to bis (C), that is by varying the T and C values of the gel.

There is an inverse linear relationship between the T value and the pore size: increase in T results in decrease in pore size. A small pore size in polyacrylamide gel would yield a superior resolution to the extent of detecting molecules with a difference of 1 bp. Thus such gels with small pores are useful in the separation of smaller molecules.

The relationship between C value and the pore size of a gel matrix is somewhat complicated. When the C value is about 5% (a 19 parts acrylamide : 1 part bis), minimum pore size is obtained. Further decrease in C value will not yield further decrease in pore size and, in fact, may produce larger pore size, because of insufficient crosslinkers. Increase in the C value beyond 5% also increases the pore size due to nonhomogeneous bundling of strands in the gel. A C value of 5% has been found to be ideal for most forms of denaturing DNA and RNA electrophoresis, 3.3% (29 parts acrylamide : 1 part bis) for native DNA and RNA gels, and 2.6% (37.5 parts acrylamide : 1 part bis) for SDS-PAGE electrophoresis of proteins.

BUFFERS

The buffer system (Box 13.2) of an electrophoresis regulates the pH of the gel, prevents any damage to the sample molecules, and controls the ionization state of the molecules. Another function of the buffer system is its role in flow of current. Majority of the current flow through the electrophoresis gel is carried by the buffer ions. In those electrophoresis systems in which the type and concentration of buffers in the tank and the gel are the same (e.g. denaturing PAGE electrophoresis of DNA), the buffer prevents any marked change in the pH and controls the conductivity of the gel. In native electrophoresis of proteins, the buffer system also controls the state of ionization of the sample by maintaining a constant pH. In general, the ionic strength of the buffer in the gel should be maintained at a level sufficient to keep the sample in solution and to provide enough buffering capability. Though higher concentration of the gel buffer would produce sharper bands by slowing down the migration of the sample molecules, it has certain disadvantages. At any given voltage, higher buffer concentration produces higher electrical conductivity and consequent higher heat generation. For this reason, in order to avoid the problems associated with high temperature, a low voltage gradient is maintained when high buffer concentration is used.

BOX 13.2   BUFFER SYSTEM

In order to appreciate the role of buffers in electrophoresis let us first understand what a buffer solution is and how it functions. A buffer is a mixture of a weak acid and its conjugate base.

An **acid** is a molecule or an ion that can function as a proton donor. Proton is actually a hydrogen ion. Therefore, an acid is a molecule or ion that can supply a hydrogen ion to a solution. For example, hydrochloric acid, HCl, is an acid because it ionizes in water to form hydrogen and chloride ions. Similarly, carbonic acid, H_2CO_3, also ionizes into H^+ and HCO_3^- and contributes hydrogen ion to the solution.

A **base** is a molecule or ion that can function as a proton acceptor. It combines with hydrogen ions to remove them from a solution. For example, the bicarbonate ion, HCO_3^- can combine with hydrogen ions to form H_2CO_3, and therefore is a base. In the same way, HPO_4^- is also a base as it can combine with H^+ to form H_2PO_4. Proteins also can function as bases. Some of the amino acids in a protein molecule have negative charge and therefore behave as a negative ion and bind with hydrogen ions. Haemoglobin of RBCs and other proteins in the body fluids and cells are capable of functioning as bases. However, it should be remembered that the negative ions in molecules such as sodium bicarbonate and sodium phosphate the real bases according to the definition of the word "base".

The term "alkali" is often used as a synonym of the word "base". An **alkali** is a combination of one alkaline metal (e.g. sodium, potassium, etc.) with a highly basic ion such as OH^-. The basic ions of these molecules can combine with H^+ and therefore remove them from a solution. For this reason the term "alkali" is used synonymously with the term "base".

BOX 13.2   (Continued)

A **strong acid** is one that has a strong tendency to dissociate into ions and consequently release its H^+ into the solution. Hydrochloric acid is a strong acid. A **weak acid** is one that has lesser tendency to dissociate and release H^+ into the solution. Carbonic acid, sodium acid phosphate, etc. are weak acids. A **strong base** is one that reacts strongly with H^+ and removes them from the solution. Hydroxyl ion (OH^-) is a strong base. A **weak base** (e.g. bicarbonate ion, HCO_3^-) combines more weakly with hydrogen ions.

Functioning of acid-base buffers

Generally, an acid-base buffer is made up of two or more chemical compounds that can prevent marked alterations in pH when either an acid or a base is added to a solution. For instance, if a few drops of concentrated HCl are added to 100 mL of distilled water, the pH of the water would immediately fall to about 1.0. On the other hand, if the same amount of acid is added to a buffer solution, the drop in pH would be insignificant.

We would better understand the functioning of a buffer system with an example. Bicarbonate buffer system is a mixture of carbonic acid, H_2CO_3, and sodium bicarbonate, $NaHCO_3$. Carbonic acid is a very weak acid because (i) it has a poor degree of dissociation into H^+ and HCO_3^-, and (ii) most of it (as much as 399/400 parts) dissociates into CO_2 and H_2O.

When strong hydrochloric acid is added to bicarbonate buffer it is converted into a very weak carbonic acid, as in the following equation.

$$HCl + NaHCO_3 \rightarrow H_2CO_3 + NaCl$$

Consequently, there is only a slight change in the pH.

On the other hand, when sodium hydroxide, a strong base, is added to a buffer solution containing carbonic acid, its hydroxyl ion (OH^-) combines with the hydrogen ion (H^+) of the acid to form water, as shown below.

$$NaOH + H_2CO_3 \rightarrow NaHCO_3 + H_2O$$

Further the sodium combines with bicarbonate ion to produce a weak base, sodium bicarbonate.

BOX 13.2   (Continued)

Dynamics of buffer system

All acids are ionized to a certain extent, and the percentage of ionization is called the degree of dissociation. A weak acid ionizes into H^+ and its conjugate base. In general, if HA is a weak acid, it dissociates into H^+ and A^-, in a solution. This reversible dissociation can be represented as follows:

$$HA \rightleftharpoons H^+ + A^-$$

The ionization of weak acid follows the law of mass action, which can be expressed in terms of Henderson–Hasselbalch equation. The status of acid/base equilibrium is represented by the **acid dissociation constant, K_a**.

$$K_a = \frac{[H^+][A^-]}{[HA]}$$

The above equation states that in any given solution of a weak acid, the product of the concentration of the hydrogen ion and the concentration of the conjugate base, divided by the concentration of the undissociated weak acid is a constant.

The value of K_a is large if the acid is stronger and equilibrium is more towards dissociation. The K_a value will be small, if the equilibrium is toward association, i.e., proton capture. Buffers that are commonly used in biological research have K_a value in the range of 10^{-4} to 10^{-10}.

The K_a value is usually expressed as its negative logarithm ($-\log K_a$) and is represented as pK_a.

$$pK_a = -\log K_a$$

A buffer having a K_a value of 10^{-2} has a pK_a of 2, which indicates that the equilibrium is more in favour of dissociation. A buffer having a K_a of 10^{-12} has a pK_a of 12, favouring proton capture.

Without going into the details of its derivation, the Henderson–Hasselbalch equation can be stated in its useful form as follows.

BOX 13.2   (Continued)

$$pH = pK_a + \log\frac{[\text{basic form}]}{[\text{acidic form}]}$$

The above equation is often described to be the centre of buffer design. The equation states that the pH of buffered solution will differ from the buffer's pK_a by an amount, which is determined by the ratio of the base to the acid form in solution. If the concentrations of the base and acid are equal then pH = pK_a. If the concentration of the concentration of the base is greater than that of the acid, then pH will be greater than pK_a. If the concentration of the acid is greater than that of the base then pH will be less than pK_a.

Choice of Buffer

We have to consider several factors while choosing a buffer system for electrophoresis. They are as follows:

i) **pK_a value**   An electrophoresis buffer is selected with a pK_a that is as close to the desired pH as possible. Such a buffer will have maximum capacity to absorb and to release protons. In native protein electrophoresis, the basic proteins are efficiently separated at acid pH. Most of the other proteins have isoelectric points below 7.5 and are therefore effectively separated in mild alkaline conditions, i.e., at a pH of 8 to 9. This slightly alkaline pH range has been found to be effective for DNA electrophoresis also. Thus, buffers having a pK_a value in the range of 7 to 9 are ideal for most electrophoretic applications.

ii) **Charge**   Generally, buffers having high ionic charges are not preferred for electrophoresis, because such buffers do not possess high buffering capacity in spite of high ionic strength. Therefore, buffers that remain uncharged part of the time at the desired pH are selected. For example, Tris base and borate buffers are found to be good for electrophoresis because they are represented by uncharged species in their acid-base equilibrium and as a result exhibit relatively low electrophoretic mobility.

iii) Molecular size   Buffers with ions having relatively larger molecular size, in addition to being lowly charged, are good for electrophoresis. Such ions having low charge-to-mass ratio move slowly in electrophoresis, due to their large molecular size. For example, Tris, which has large molecular size, moves more slowly than small ions such as chloride or phosphate. An added advantage of buffers with such large molecules is that these molecules are not sieved by the gel matrix, and their migration rates are determined solely by their charge–mass ratio.

In addition to the above three factors, there are other factors to be considered while choosing a buffer. They are toxicity, solubility, UV absorption and possible interaction with other molecules in the solution. A few commonly used electrophoresis buffers and pK_a values and molecular weights are given Table 13.1.

Table 13.1   Commonly used electrophoresis buffer salts and their characteristics.

Buffers	Molecular weight	PK_a
Acetic acid	60.05	4.80
Boric acid	61.83	9.23
Citric acid	192.10	6.40
Glycine	75.05	9.80
Phosphoric acid	98.00	7.20
Taurine	125.10	9.10
Tricine	179.18	8.15
Tris	121.10	8.06
MOPS	209.26	7.20

Homogeneous Buffer System

In a homogeneous electrophoresis buffer system, only one type of buffer is used in the gel and the tanks. Thus the concentrations of the components of the buffer are same in all the regions of the electrophoretic field. Electrophoresis of DNA and RNA use homogeneous buffer systems. A homogeneous buffer system performs the dual function of protecting the sample and carrying the current. The disadvantage of homogeneous buffer system is the possible differences in the ionic strength of buffer in the tanks and the gel. The consequence of such a concentration gradient between the tank and the gel is a "salt wave" that migrates through the gel during electrophoresis. Concentration gradient may also produce changes in the electrical parameters, which in turn may result in localized distortions in the band pattern.

Multiphasic Buffer System

Multiphasic electrophoresis design employs discontinuities in the gel and buffer, in order to improve resolution and produce sharp separation of sample components. This design is generally used in electrophoresis of proteins. A multiphasic system employs two distinct phases in a single gel slab, a stacking gel and a separating gel. The buffer compositions are different in the stacking gel, separation gel and the tank. The sample components are first stacked into very thin, sharp zones in the stacking gel. The sample components are fractionated in separating gel. An example of electrophoresis that uses multiphasic buffer system is "Multiphase SDS-PAGE protein electrophoresis".

Buffer Additives

In most of the electrophoresis techniques the buffer solution perfusing the gel matrix includes one or more additives, in addition to the buffer salts. The main function of these additives is to modify the properties of sample molecules. There are three categories of additives viz., hydrogen bonding agents, surfactants and reducing agents.

Hydrogen bonding agents   Hydrogen bonds are dipole–dipole attractions occurring between hydrogen containing polar groups (e.g. amine and hydroxyl groups). Hydrogen bonding has significant influence on the conformation and solubility of biomolecules. In order to standardize the conformation of the sample molecules or to solubilize the sample molecules, hydrogen bonding agents (e.g. urea and formamide) are added to the electrophoresis buffer. These agents are capable of cleaving hydrogen bonds, and they disrupt hydrogen bonds in the sample molecules by occupying the bonding sites themselves.

One or more of the hydrogen bonding agents are introduced to electrophoresis samples before loading or they can be incorporated into the gel buffer. For instance, in denaturing DNA and RNA electrophoresis, the agent formamide is introduced during sample preparation stage, and the urea is incorporated into the gel buffer. These additives disrupt the hydrogen bonding between base pairs of the nucleic acid molecules, and occupy the disrupted places. As a result, these agents cause the separation of the complementary strands in double stranded DNA and RNA molecules. In addition, the additives disrupt the kinks and loops in single stranded nucleic acid molecules. The nucleic acid molecules that were treated with the additives are long and straight and are free from the effect differences in conformation during electrophoresis. The electrophoresis gel that employs the hydrogen bonding agents is often referred to as "urea gel" or "**denaturing gel**".

In protein electrophoresis, if the sample protein molecules are insoluble or aggregated, urea is added to segregate them and to make them soluble. A disadvantage of using urea is the formation of cyanate ions, which will react with some proteins. Tris buffers are used to overcome this disadvantage and thus protect the protein molecules in the sample. An alternative to the use of Tris buffer is the pre-running of the gel for 30–40 minutes before loading the samples. Yet another solution is treatment of the solution with ion exchange resin prior to mixing the gel solution.

Surfactants (Solubilizers)   A surfactant is a surface-active agent, the one that affects, especially reduces, the surface tension of liquids containing it. Many detergents can function as surfactants. The surfactants are used in electrophoresis as agents to promote solubility of the sample molecules and for separation of the subunits of a molecule. Many biomolecules have more than one unit and have extensive nonpolar interactions between the units. For example, an oligomeric protein molecule contains more than one polypeptide chain. The tertiary and quaternary structures of such proteins are maintained by hydrogen bonding, disulphide bridges and hydrophobic interactions. RNA polymerase, for instance, is an enzyme that has three nonidentical subunits, a, b and b_1. During electrophoresis, an untreated oligomeric protein would migrate as a single band. On the other hand, if treated with a solubilizer, an oligomeric protein would migrate into a number of bands corresponding to the number of subunits.

Non-ionic, anionic and cationic detergents are commonly used as surfactants in electrophoresis. Nonionic detergents (e.g. Tween-20 and Triton X-100) are less strongly denaturing than anionic and cationic detergents. Such non-ionic detergents are used when there is a need to preserve enzyme activity or some immunological properties of the protein molecules that are being subjected to electrophoresis. Use of anionic and cationic detergents might destroy these properties. Even the non-ionic detergents such as Tween-20 and Triton X-100 are used in low concentrations (about 1%) in the gel buffer. However, higher concentration (1%) may be used for the pretreatment of the sample.

A major disadvantage of nonionic detergents is that they do not produce consistent charge to mass ratio among sample molecules for electrophoresis. Therefore, direct determination of molecular weight of proteins separated using non-ionic detergents. This disadvantage is overcome by the use of charged surfactants.

The most commonly used detergent additive in protein electrophoresis is the anionic surfactant SDS (sodium dodecyl sulphate). When proteins in a sample are treated with SDS, they become completely covered by negatively charged SDS ions and unwind to assume an extended conformation (Figure 13.1). The number of SDS molecules covering the protein molecules is very large, as many as half the number of amino acid residues. As a result, the intrinsic charge of the SDS treated proteins becomes overpowered by the charge of the detergent. Even proteins that have widely different tertiary and quaternary structures tend to have consistent and uniform charge to mass ratio. Electrophoretic separation of such SDS treated proteins is strictly on the basis of their molecular weights.

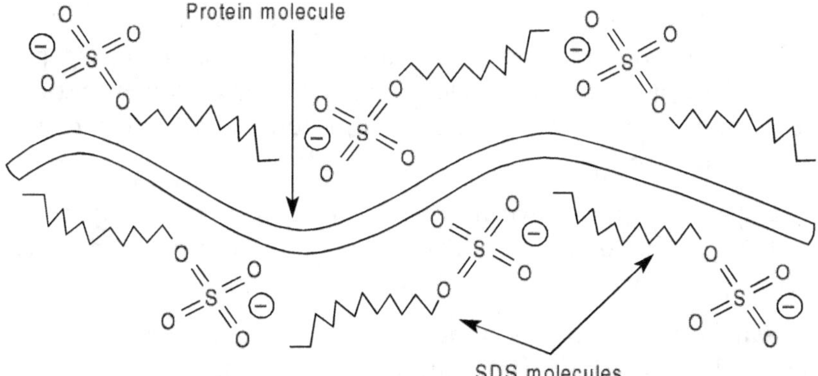

Figure 13.1   SDS treated protein molecule. SDS molecules blankets the protein molecule, makes it unwind, and gives it a uniform charge-to-mass ratio.

Cationic surfactants are rarely used. A common example of cationic surfactant is CTAB, cetyltrimethylammonium bromide. It is used for the electrophoresis of samples that pose difficulties for SDS-PAGE. For example, CTAB is used in the electrophoresis of extremely acidic or extremely basic samples. Because of their very high negative charge, the extremely acidic samples may not bind with SDS. The extremely basic samples may precipitate when SDS is added. The use of the cationic CTAB overcomes these problems and produces a uniform charge-to-mass ratio among the sample molecules. However, the electrophoresis apparatus must be adjusted to allow sample molecules to migrate toward the negative pole.

Reducing agents   Disulphide bonds are formed between or within a protein molecule and they are responsible for the formation of aggregates, and they play a role in binding the subunits. It is necessary that these disulphide bonds are cleaved before electrophoresis of the protein molecule. Disulphide bond reducing agents (e.g. 2-mercaptoethanol and dithiothreitol) are added to sample buffers. These agents can also be added to the cathode tank. However, they are never added to the gel casting solution because they tend to inhibit gel polymerization.

DETECTION AND ASSAY

Once the electrophoresis has been run, it is necessary to locate, identify and quantify the molecules that have been separated. The separated molecules are in the form of bands on the support medium. A recording of the separated components of a mixture produced by electrophoresis is called electropherogram or electrophoretogram. However, the support medium carrying the separated molecules is also referred to as electrophoretogram.

Usually, the individual molecules in paper or gel electrophoretogram are detected and identified *in situ*, exploiting their physical properties such as staining, fluorescence, UV absorption, and radioactivity.

Staining   The bands in an electrophoretogram can be located and identified using appropriate staining (Table 13.2).

Table 13.2   A few common stains used to identify molecules in electrophoretogram.

Stain	To identify
Alcian blue	Glycoproteins
Bromophenol blue	Proteins
Coomassie brilliant blue	Proteins
Iodine	Polysaccharides
Lanthanum acetate–acridine orange	RNA and DNA
Lissamine green	Proteins
Methyl green-pyronine	RNA and DNA
Methylene blue	RNA
Nigrosine	Proteins
Periodic Acid Schiff reaction	Glycoproteins
Sudan black	Lipoproteins
Toluidine blue	RNA

When enzymes are electrophoresed on a strip of paper or gel, they can be identified *in situ* by special staining techniques. A substrate of one of the enzymes that have been separated on an electrophoretogram is bound to an indicator dye and is impregnated onto a strip of paper. The enzyme-impregnated paper and electrophoresis strip are of similar dimensions. The two are placed together in juxtaposition so that the enzyme in the electrophoretogram can act upon the substrate. That part of the substrate paper corresponding to the enzyme band in the electrophoretogram turns to a specific colour, enabling the location of the enzyme.

Fluorescence   Staining of the bands of the separated molecules in an electrophoretogram with specific fluorescent dyes and subsequent observation under UV light also enables us to detect and identify proteins and nucleic acids. Dyes such as dansyl chloride, fluorescamine, and anilinonaphthalene sulphonate (ANS) are useful for the identification of protein molecules. Ethidium bromide is a very sensitive fluorescent dye for the identification of nucleic acids. **Fluorescent transilluminator** built in a PVC housing with glass window and equipped with 6 W fluorescent bulbs are useful for viewing gels marked with fluorescent dyes and proteins.

UV absorption   The property of the protein and nucleic acid molecules to absorb light in the wavelength range of 260–280 nm is taken advantage for the identification of these molecules in electrophoretogram.

Autoradiography   Sample components specifically labelled with appropriate radioactive isotopes can be electrophoresed and the electropherogram can be applied to X-ray films to identify the specific band containing the desired molecule.

Quantitative assay   "Eye balling" is a crude method of quantification of the separated compounds of an electrophoretogram. This method is a simple visual comparison of the stained bands with a standard colour chart.

A more accurate quantitative assay is to elute the separated bands with suitable solvents and read the colour density of the resulting solution in a colorimeter. Comparison with colour density of known quantities (e.g. a standard graph) of the specific compound would yield the quantity of the substance in the band.

The band portions of paper or gel can also be cut, separated and dissolved in an appropriate solvent for colorimetric reading.

Use of radioisotopes for labelling the component molecules and subsequent counting of the eluted bands with scintillation counter is also a method for the quantification.

Electrophoresis **densitometers** are used to record results from almost all types of electrophoresis support media, including cellulose acetate, polyacrylamide, agar, and starch cells. Densitometers are spectro-photometers fitted with gel scanning accessories. A monochromatic beam of light passing through a stained gel scans the different bands, recording the percentage of transmission of the light, which is inversely proportional to the density of the bands. The reading and printing of the results is very quick, about 10 seconds, in densitometers.

RECORDING AND STORAGE

The papers and gels used in an electrophoresis can be photographed and stored. Taking photographs of electrophoresis papers with stained bands is relatively easier. It can be done using any good quality 35-mm camera fitted with close-up lens system. Diffuse bright light and white background are necessary to produce colour pictures of sufficient contrast. For black-and-white pictures, in addition, appropriate filters will have to be used to produce sufficient contrast. For example, yellow filters are used to photograph Coomassie-blue-stained bands to enhance the blue colour of the bands. The same filter is used to photograph the fluorescence of DNA bands stained with ethidium bromide. Green filters are used while photographing bands stained with silver salts.

Polaroid MP-4 camera fitted with 120 or 135-mm lens is useful in photographing UV fluorescence of DNA bands stained with ethidium bromide. This system, used in combination with a UV transilluminator, offers easy viewing and focusing of the bands and simple operation. Precautions must be taken against the harmful effects of the UV radiation and the mutagenic, toxic ethidium bromide.

Gel documentation systems (GDS) are computer-based systems designed to perform several functions such as photographing stained gels, printing out photographic data and saving photographic data. Samples such as fluorochrome-stained samples requiring UV excitation, those samples having bands stained and viewed on a white transilluminator and those having membrane or TLC colouring and viewed with an in-cabinet lamp, can be photographed and documented using GDS. Image data can be displayed in real-time.

While electrophoresis papers can be stored in dry condition, gels are usually stored in dilute acetic acid in watertight containers. Drying of the gels is also possible, and the dried gels can be rehydrated when required.

SAFETY

Many of the chemicals used in electrophoresis including buffers, staining solutions, gels, etc. are hazardous to human health. Such materials must be decontaminated before disposal. Gloves and goggles are mandatory while performing electrophoresis.

UV radiation used for viewing stained DNA is hazardous to eyes and skin. Wearing of protective UV-blocking goggles and safety mask is strongly recommended. The source of UV should be properly and adequately shielded.

Acrylamide and bis-acrylamide are easily absorbed by the skin and they are highly powerful neurotoxins that have cumulative effects. When weighing acrylamide or methylene-bis acrylamide or handling solutions of these chemicals, gloves and mask should be used. Similar precautions must be taken while handling polyacrylamide because it contains minute quantities of toxic unpolymerized acrylamide.

Ethidium bromide is commonly used for fluorescence staining electrophoresed DNA. It is highly mutagenic, capable of inducing mutation in human DNA. It is also moderately toxic. Gloves must be used while handling this chemical. Ethidium bromide should be disposed in one of the following ways: (i) It can be decomposed by incineration at 262°C. (ii) Concentrated solutions of ethidium bromide should be diluted with water and then treated with Amberlite XAD-16, filtered with Whatman No.1 filter paper and disposed. (iii) Dilute ethidium bromide solution can be treated with powdered activated charcoal, stored at room temperature with intermittent shaking, filtered with Whatman No.1 filter paper and disposed. (iv) Contaminated surfaces are usually cleaned with slurries of Amberlite XAD-16 or activated charcoal.

TYPES OF ELECTROPHORESIS AND THEIR APPLICATIONS

We can classify electrophoresis into three main types, viz., free electrophoresis, zone electrophoresis and gel electrophoresis. **Free electrophoresis** is performed in a liquid medium, i.e., it does not have a solid support medium. It includes **microelectrophoresis** and **moving boundary electrophoresis**. **Zone electrophoresis** uses a solid support, which is inert and homogeneous. Paper strips, cellulose acetate strips, starch blocks and agar blocks are the main supporting solid media used in zone electrophoresis. Individual bands of the zone medium can be fractionated into components with different electrical mobilities and thus individual components can be separated. **Gel electrophoresis** was originally grouped under zone electrophoresis, but now treated as a separate category. The solid support medium in gel electrophoresis is starch gel, agarose gel, or polyacrylamide gel. In contrast to the support media of the zone electrophoresis, the gel has pores, which can act as sieve and selectively separate molecules by size and shape. The gel electrophoresis is most commonly used for the separation of proteins and nucleic acids.

MICROELECTROPHORESIS

As the name suggests, this type of free electrophoresis involves microscopic observation of movement of minute particles (e.g. blood cells, bacteria, etc.) in an electrical field. The apparatus includes a flat transparent microelectrophoretic cell (e.g. Abramson's microelectrophoretic cell), electrode compartments, ocular micrometer, and stop clock. This technique is useful in measuring the zeta potential (electrokinetic potential, measurement of the velocity of the particle under an electrical field) of blood cells and microorganisms (e.g. bacteria).

MOVING BOUNDARY ELECTROPHORESIS

The basic principle of moving boundary electrophoresis is that when an interface is formed between two homogeneous electrolyte solutions, and an electric current is passed through the system, moving boundaries are formed. When the current has been passed for a long enough time, there are homogeneous regions between the boundaries.

The main features of moving boundary electrophoresis are the absence of a solid support and, therefore, migration of the molecules dissolved or suspended in a liquid buffer in electrical field. Originally this method, as described by Lodge in 1886, used dyes as markers. Tiselius refined this method in 1925 with the involvement of optical system to follow the movement of boundaries and to measure refractive indices of colourless solutions.

A moving boundary electrophoresis consists of U-tube (Tiselius tube, observation cell) fitted with electrodes at the two ends (Figure 13.2). A buffered solution containing macromolecules is introduced into the tube. The same buffer but without the macromolecules is introduced into the two arms of the U-cell, such that the sample solution is under a layer of pure buffer. The pH of the buffer is so adjusted that the net charge of all the molecules in the sample becomes negative. This would ensure the migration of all the molecules towards the anode. A constant electric current is passed through the system under constant temperature and vibration free condition. In the electrical field created, the molecules in the sample begin to migrate towards the anode forming a moving boundary or front in the pure buffer. The rates of movement of the different molecules differ according to their respective electrophoretic mobility property (charge, size, shape, etc.). After a time in the electric field, the observation cell would contain homogeneous boundaries containing specific macromolecules. The refractive indices of the solutions within each boundary will vary according to the physical properties of the molecules they contain. The change in refractive index along the tube can be measured with appropriate optical system (e.g. Schleiren optics; Rayleigh interference optics). The resulting electrophoretic pattern would exhibit the relative mobility and concentration of the sample molecules.

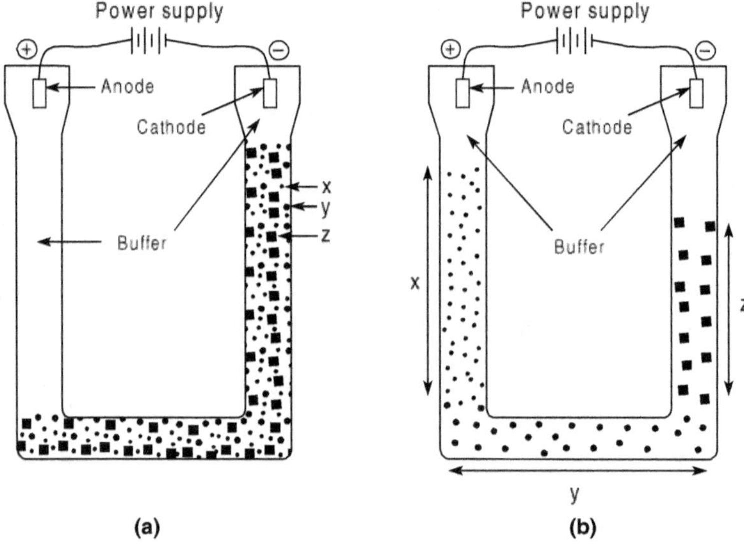

Figure 13.2   Diagram of a moving boundary electrophoresis. In the initial stage (a), the molecules (x, y and z) of the sample are mixed, forming a single boundary between the pure buffer in the two arms. After electrophoresis (b), separate boundaries of the buffer containing the molecules x, y, and z are established. The boundaries may not be visible to naked eye, and require special optical system to measure their refractive indices. The separation of molecules on the basis of size and shape may not be so distinct as depicted.

Moving boundary electrophoresis is primarily used in separating components of biological fluids (e.g. serum proteins). It is also used in water purification. The advantages of this electrophoresis system are its high degree of purity and application to wide variety of high molecular weight substances. The disadvantages of this system are that it cannot separate individual components, it does not separate size and shape as well and it requires large quantities of sample. Convection current occurs due to density stabilities when components become separated. Moving boundary electrophoresis is limited to partial resolution in a vertical U tube to minimize density instabilities.

PAPER ELECTROPHORESIS

Paper electrophoresis is a member of the group of procedures called zone electrophoresis, which are distinguished from moving boundary electrophoresis by the use of a supporting medium. The supporting medium of paper electrophoresis is obviously the paper, more specifically high quality filter paper (e.g. Whatman) or special electrophoresis paper. The paper is cut into desired dimension (e.g. 5 mm strips).

The paper electrophoresis can be run either horizontally or vertically. The assemblies of the two models are shown diagrammatically in Figure 13.3 and Figure 13.4, respectively. The apparatus is generally a tank (glass, plastic or Perspex) consisting of two buffer reservoirs separated by a platform. The platform provides support for placing the paper. Electrodes, fixed to the reservoirs, are connected to a power pack.

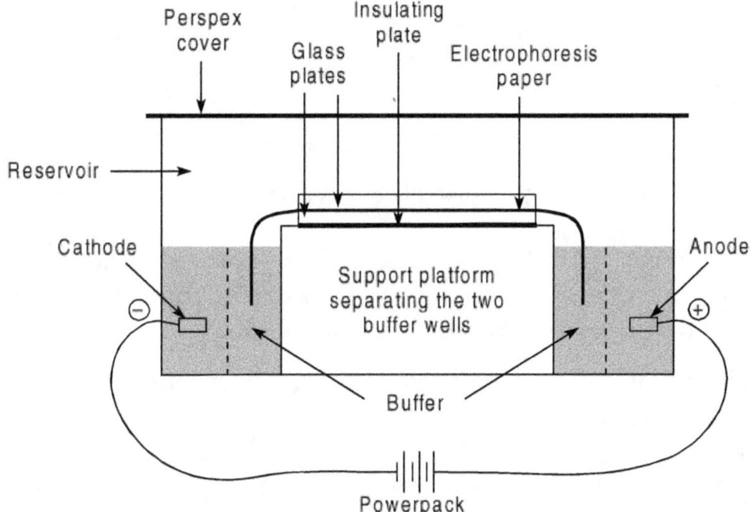

Figure 13.3   Assembly of horizontal paper electrophoresis.

In horizontal electrophoresis, the paper is mounted on the platform, held in between a pair of glass plates. The ends of the paper strip may directly dip into the buffer reservoirs on either side or may be connected to them through a wick made up of filter paper or gauze. In the vertical electrophoresis, the paper is hung in the electrophoresis tank in such a way that the two ends of the paper are dipping into the upper and lower reservoirs, respectively. In both the designs, the paper is saturated with the buffer before the electric field is set up.

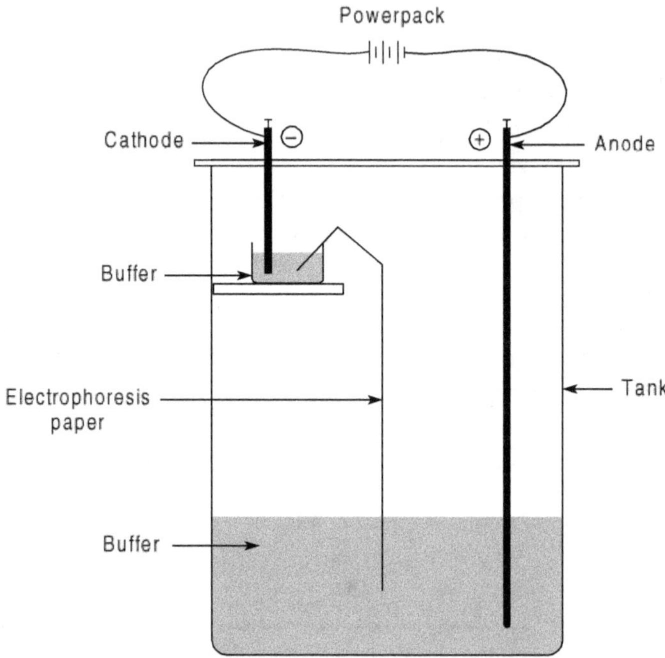

Figure 13.4   Assembly of vertical paper electrophoresis.

The most critical part of the electrophoresis process is the sample loading onto the paper. The sample is loaded as a spot or as a streak before or after the paper is saturated with the buffer.

The whole set-up is covered with Perspex sheet to prevent evaporation of the buffer. A constant, regulated supply of electric current is provided for a specified period, with the help of a power pack. In order to avoid possible overheating, the electrophoresis run may be conducted in a cold ambient temperature. When the electrophoresis is completed, the appropriate detection and assay (e.g. staining, fluorescence, UV absorption, etc.) is applied. The stained bands in the paper can also be cut out, dissolved in suitable solvent for colorimetric reading. Direct densitometric reading of the paper strips is also possible if they are made transparent by immersing them in suitable organic mixtures (e.g. paraffin oilbromonaphthalene, anisol, etc.). Electrophoresis bands of radiolabelled compounds can be detected and assayed by autoradiography.

Paper electrophoresis has wide applications such as the separation of proteins; amino acids and nucleic acids.

High Voltage Paper Electrophoresis

The voltage of the current in the conventional paper electrophoresis is low, in the range of 2 to 50 V, in order to minimize heat generation. The disadvantage of such low voltage is the marked diffusion effects and the consequent poor resolution of sample compounds, especially low molecular weight substances having high diffusion rates (e.g. amino acids, small peptides, etc.). This problem is overcome by high voltage electrophoresis, which involves the application of high voltage in the range of 2500 to 5000 V and a current of about 500 mA.

The main feature of the high voltage electrophoresis assembly is the cooling system to counter the enormous heat generated by the high voltage. The cooling system covers the electrophoresis paper and keeps it in cool, constant temperature. It is made up of cooling plates (aluminium) through which a cooling agent (water or water + glycol) is circulated.

High voltage electrophoresis can be performed in two directions using buffers of different pH values. The first run is in one direction at a specific pH, and after drying the paper, the second run is in a direction at right angles to the first using a buffer of different pH. It is also possible to combine the first high voltage electrophoresis with a second chromatographic run.

CELLULOSE ACETATE ELECTROPHORESIS

Cellulose acetate which is used as the supporting medium of electrophoresis has several advantages over paper:

i)   It has no adsorption property.

ii) It is chemically pure, as it does not contain impurities such as lignins, hemicellulose, or nitrogen.

iii) Since it is translucent, it can be subjected to direct photoelectric scanning of the bands.

iv) Very low glucose content in it makes it suitable for electrophoresis of polysaccharides and subsequent PAS (periodic acid Schiff) staining.

Cellulose acetate electrophoresis has applications in research and clinical investigations such as the separation of glycoproteins, lipoproteins, and haemoglobin of blood samples. It is also useful in immunoelectrophoresis and immunodiffusion.

GEL ELECTROPHORESIS

The main difference between the gel and the supporting media of the zone electrophoresis is the presence of pores of specific dimension in the gel, which has sieving effect on the migrating molecules. Thus, while the main factor of separation of molecules in zone electrophoresis is the net charge of the migrating molecules, in gel electrophoresis, additional factors, viz., size and shape of the molecules, and the pore size of the gel play significant roles.

Principle   Gel electrophoresis is a technique in which molecules in a sample are forced across a span of gel, by an electrical current. Separation of macromolecules (nucleic acids and proteins) depends on two forces, viz., charge and mass. When a sample (e.g. proteins or DNA) is mixed in a buffer solution and applied to an electrophoresis gel, these two forces act together. The electrical current from cathode repels the molecules that have net negative charge while the anode simultaneously attracts the molecules. The frictional force of the gel functions as a "molecular sieve", separating the macromolecules by size. In addition, the macromolecules are forced to move through the pores of the gel in an electrical field, and this adds to the molecular sieving effect. Thus, the following factors determine the rate of migration of the molecules in gel electrophoresis:

i)   Strength of the electric field

ii)   Size and shape of the molecules

iii)   Relative hydrophobicity of the sample

iv)   Ionic strength and temperature of the buffer in which the molecules are migrating

The theory behind the electrophoretic mobility of molecules in a gel is explained in Box 13.3. After electrophoresis is completed, the gel can be stained to visualize the separated macromolecules in each lane as a series of bands from one end of the gel to the other end.

BOX 13.3   GEL ELECTROPHORESIS THEORY

Electric current   All substances are made up of atoms. Each atom consists of a nucleus containing positively charged protons and neutral neutrons, and negatively charged electrons circling around the nucleus in orbits. While most of the electrons are tightly bound to the atom, a few of them in the outer orbit are loosely bound. These loosely bound electrons can move from one atom to another. Such free atoms are many in "conductors" (e.g. copper wire), and nil in "non-conductors" or "insulators" (e.g. glass)

The direction of movement of the free electrons in conductor is random, right to left or left to right, such that the net flow of electrons across any point is zero (Figure 13.5a). However, when an electric field is applied across a conductor (i.e., when the conductor is connected to the positive and negative terminals of a power supply such as a battery) a net flow of electrons is created because all the electrons (negatively charged) are attracted to the positive side of the conductor, i.e., the end that is connected to the positive terminal (Figure13.5b). The rate of flow of electrons through a point is called

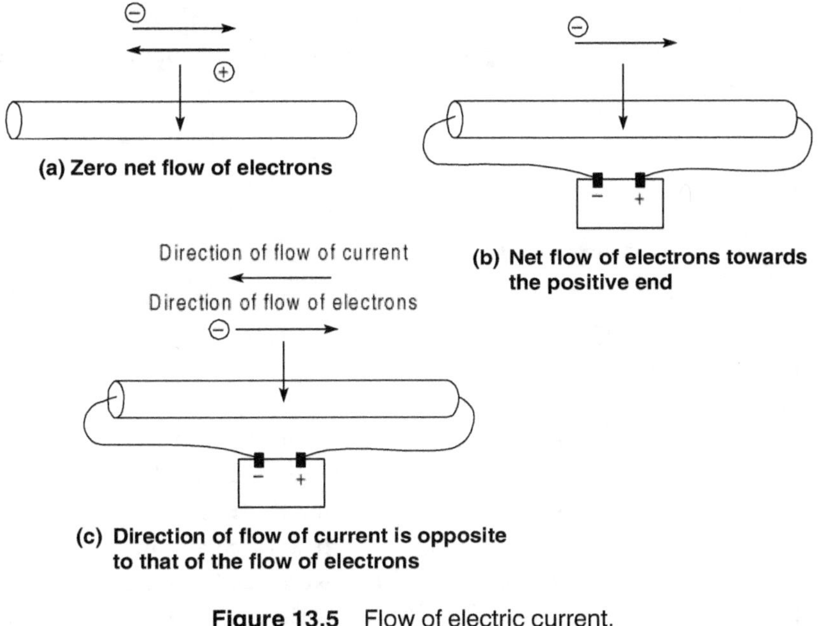

Figure 13.5   Flow of electric current.

BOX 13.3   (Continued)

current. Only the negatively charged electrons move in one direction, toward the positive end. No positively charged particles move toward the negative end of the conductor. However, the effect of the movement of the negatively charged electrons toward the positive end is equal to that of the movement of positive charged particles (protons) to the negative end. Conventionally, the direction of electric current is defined as the direction of positive charge movement, i.e., opposite direction to that of the flow of electrons (Figure 13.5c).

Coulomb's law   All substances have an electrical charge, either positive or negative. Those substances with opposite charges attract one another and those with similar charges repel each other. The force of this attraction or repulsion is a function of distance; the force decreases with square of the distance. The force of attraction or repulsion also depends on the medium separating the charged particles. For example, the force of attraction between two oppositely charged particles would be much greater in air than in water. The relationship between charged particles in terms of their distance and the medium is expressed by Coulomb's law, which states that the force exerted between two electric charges is directly proportional to the product of the charges and inversely proportional to the square of the distance between them. The equation is

$$F = \frac{q_1 \times q_2}{x^2 \times K}$$

where F = force in dynes

q = magnitude of the charge of the particle in statcoulombs

x = distance between the particles in cm

K = the dielectric constant

A dyne is defined as the force that gives a mass of one gram an acceleration of one centimetre per second. A statcoulomb is defined as that charge which, placed one centimetre from an equal and like charge in vacuum, will repel it with a force of one dyne. The dielectric constant is a value which expresses the characteristic of the medium with reference to attraction or repulsion between

BOX 13.3   (Continued)

specific charged particles. The dielectric constant of vacuum is 1.00, air, 1.00059 and water, 80.

Potential difference   According to the Coulomb's law, there is a force exerted by two charged particles. If the particles have opposite charges they attract each other. In order to keep such attracted particles separated, work must be done. Likewise, work must be done to bring together two similarly charged particles repulsing each other. The measure of the work done to move the charged particles is the **electrical potential**. If the two particles are similarly charged they repel each other and the work done is positive and the potential is positive. On the other hand, if the particles are oppositely charged, they attract each other, the work done is negative and the potential is negative. To determine the potential, we need to know the charge of both the particles and the distance between them, and the equation giving the potential is

$$V = \frac{q}{r}$$

where V = the potential in statvolts, that is, ergs per statcoulomb charge of the moved particle

r = the distance in centimetre between the particles

q = the charge in statcoulombs of the stationary particles.

While defining the potential, we mean the movement of one of the particles either from infinite distance to a specific distance close to the other particle. Or, the movement of one of the particles may be from a specific distance from the other particle to an infinite distance. If, instead of moving one of the particles from infinity to a specific point within an electrical field, the particle is moved from a point already in the field to another point in the field, then the work done is called the potential difference. The equation for the potential difference is as follows:

$$\text{Potential difference} = \frac{q}{r_1} - \frac{q}{r_2}$$

BOX 13.3 (Continued)

where q = the charge in statcoulombs of the field r_1 r_2 = initial and final distances in centimetre between the charged particles.

A potential difference between two points in an electric field gives rise to a "force" called an **electromotive force** or **emf** that tends to push electrons or other negatively charged particles or molecules from one point to the other. In the SI system of units, potential difference, electrical potential and electromotive force are measured in volts, leading to the commonly used term **voltage** and the symbol V. Named after Alessandro Volta, one volt is defined to be one joule of energy per coulomb of charge.

Electrophoretic separation of particles in a gel

When a potential difference (voltage) is created across the electrodes, it produces a potential gradient (E), which is the applied voltage divided by the distance between the electrodes.

$$E = V/d$$

When the potential gradient E is applied, the force on a molecule bearing a charge q coulombs is Eq newtons.

$$F = Eq$$

This is the force that pushes a charged molecule towards an electrode.

The movement of the molecule towards an electrode is opposed by frictional force, which slows down the movement. The frictional force is a measure of the hydrodynamic size of the molecule, the shape of the molecule, the pore size of the medium (gel) in which the electrophoresis is taking place and the viscosity of the buffer. The velocity (v) of a charged molecule in an electric field is expressed as

$$v = Eq/f$$

where f is the frictional coefficient.

In an electrophoretic field, the force moving a macromolecule (nucleic acid or protein) is the electrical potential, E. The electrophoretic mobility (m) of an ion is the ratio of the velocity of the particle, v, to the electrical potential.

BOX 13.3   (Continued)

$$\mu = v/E$$

Electrophoretic mobility is also equal to the net charge of the molecule, Z, divided by the frictional coefficient, f.

$$\mu = Z/f$$

In an electrophoresis field of potential difference, molecules with different overall charges will begin to separate due to their different electrophoretic mobilities. Even if molecules have similar charges, they will tend to separate if they have different molecular sizes, since they will encounter different frictional forces. While some forms of electrophoresis depend almost totally on the different charges on molecules to produce separation, other methods make use of the differences in size and shape of the molecules and therefore the frictional effects to cause separation.

The current in the solution between the electrodes of the electrophoresis field is conducted mostly by the buffer ions with a small proportion being conducted by the sample ions. Ohm's law expresses the relationship between current (I), voltage (V), and resistance (R):

$$R = V/I$$

The above relationship reveals that it is possible to accelerate an electrophoretic separation by increasing the applied voltage, which would effect a corresponding increase in the current flow. The distance migrated by molecules will be proportional to both current and time.

However, increasing the voltage would result in greater generation of heat. During electrophoresis the power (W, watts) generated in the supporting medium is given by

$$W = I^2 R$$

The power generated is dissipated as heat. Heating of the support medium of electrophoresis can have the following effects:

i)   Higher rate of diffusion of sample and buffer ions leading to broadening of the band of the separated samples.

> BOX 13.3   (Continued)
>
> ii) The development of convection currents which might cause mixing of separated samples.
>
> iii) If the sample components are temperature-sensitive they may be denatured. For example, proteins may be denatured or they may lose their enzyme activity.
>
> iv) The viscosity of the buffer may decrease and lead to a reduction in the resistance of the medium.

A gel electrophoresis apparatus is designed to maintain uniform electric field across the gel, providing sufficient cooling to prevent any distortion of the bands due to heat, and permitting easy access to the gel for sample loading and monitoring the electrophoresis run. There are two types of gel electrophoresis, vertical and horizontal. The vertical gel may be a slab gel or a tube gel. In general, agarose gel is run in a horizontal design, while acrylamide gels are run in vertical design.

Horizontal Gel Electrophoresis

In a horizontal gel electrophoresis apparatus (Figure 13.6), the gel rests on a platform which divides the electrophoresis tank into two chambers that act as buffer reservoirs. However, the buffer level is maintained higher than the surface of the gel so that the two buffer chambers are connected.

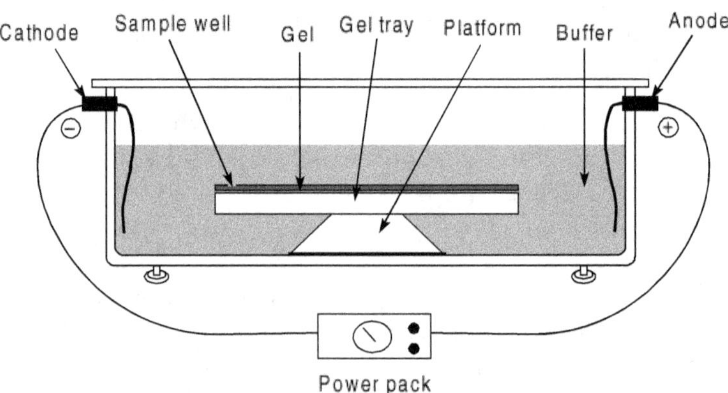

Figure 13.6   Horizontal gel electrophoresis apparatus.

Samples are loaded into the wells of the gel and they migrate towards the positive electrode.

Casting of horizontal gel   Trays with removable ends are used for casting agarose gel slabs. The ends of the tray are set in position. Very often adhesive tape of suitable width is used to seal the ends of the tray. Molten agarose solution is poured into the tray. The thickness of the gel is generally 0.5 to 1.0 cm. A well comb is inserted into the gel at one end so that its teeth penetrate into the gel deep enough but not piercing it. The wells are at least 1–2 mm above the lower surface of the gel. The gel is allowed to cool, the comb is removed and the polymerized gel is mounted in the electrophoresis apparatus.

Advantage of horizontal electrophoresis   Horizontal system of electrophoresis can be run only with agarose gel, which is not affected by atmospheric oxygen. The main advantage of this system is its simplicity and the ease with which it can be handled.

Limitations of horizontal electrophoresis   Since the buffer level in the tank is maintained above the gel surface, it is not possible to use discontinuous buffer systems. Another disadvantage of this system is its inability to use polyacrylamide gel. Polyacrylamide gel cannot cast in an open tray because it will not polymerize when exposed to atmospheric oxygen.

Vertical Gel Electrophoresis

As already mentioned, two types of gel, slab gel and tube gel can be used in the vertical gel electrophoresis system.

Slab gel electrophoresis   The assembly of a slab gel electrophoresis apparatus is shown in Figure 13.7. The components of this system are in fact common to most of the vertical slab systems.

Casting of vertical slab gel   Vertical gel slabs are cast in a cassette made up of two glass plates separated by spacers. The spacers, which are less than 2 mm thick, run along the sides of the plates. The bottom of the cassette is temporarily sealed with the help of adhesive tape, agarose or a gasket. The gel monomer solution is treated to initiate polymerization, and poured into the cassette. A well-comb is inserted into the top of the cassette to create the sample wells. The gel is allowed to polymerize completely, and then the temporary seal at the bottom of the cassette is removed.

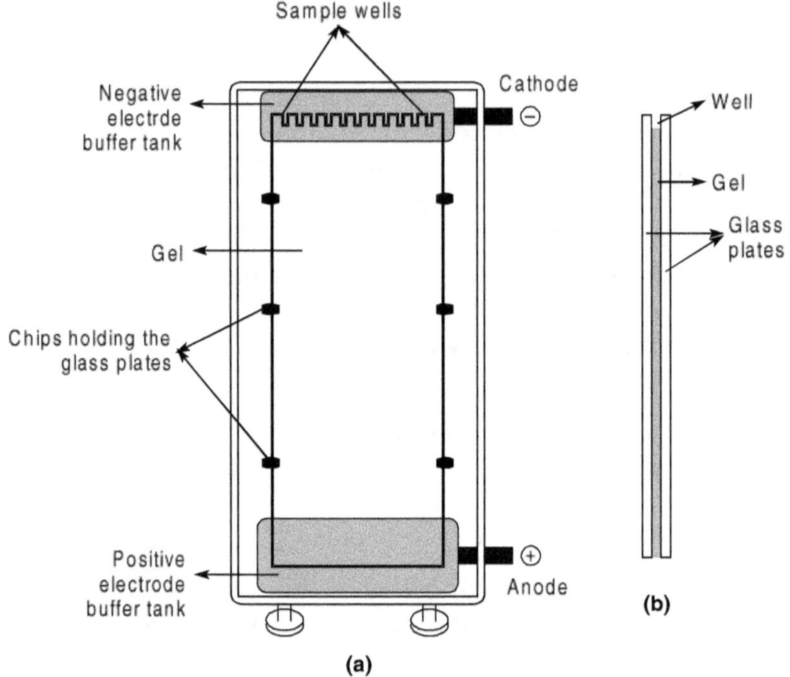

Figure 13.7 Vertical gel electrophoresis apparatus (a) and side view of the gel held between glass plates (b).

The slab gel cassette is firmly clamped into the apparatus so that its lower end is immersed in the lower buffer chamber, and the upper end forms one wall of the upper chamber. Thus the upper end of the slab gel is in contact with negative electrode chamber and the lower end is in contact with the positive electrode chamber. It may be seen that unlike the horizontal system, the only connection between the buffer chambers is through the gel. This has the advantage of producing precise and reproducible voltage gradient across the gel. Because of the high resistance of the thin gel, excess heat is generated. To counter this heat, a cooling mechanism is incorporated into the apparatus. While the front of the gel cassette is open, its back is held against a metal plate (metal heat sink), which dissipates the heat rapidly. In some designs, the upper buffer chamber extends almost to the bottom of the gel and the buffer in it acts as a cooling agent.

Mini vertical gel apparatus A 'mini' vertical gel electrophoresis apparatus has also been designed. In this system, the gel slabs are 10 cm × 10 cm or even smaller. The advantages of such smaller slabs are the ease of preparation and handling, and shorter duration of electrophoresis run. Mini slabs are also cast between glass plates. The cassettes are mounted onto the upper negative electrode chamber in such a way that the cassettes themselves form the sidewalls of this chamber. This smaller chamber is placed in a larger tank that contains buffer with positive electrode. This ensures that the gels are effectively submerged in buffer during the electrophoresis run, and the buffer itself provides the cooling. As in the larger vertical system, the gels are the only connection between the two chambers.

Vertical slab gels are loaded through the top, under a layer of buffer. The gels can be monitored during the electrophoresis run through the front glass plate.

Disadvantage of vertical gel electrophoresis   Since the body of the gel in these systems cannot be accessed until the end of the run, sample recovery in the middle of the run is not possible.

Advantage of the vertical gel electrophoresis   The resolution and reproducibility of vertical polyacrylamide gel electrophoresis is far more superior to those of the horizontal system.

Tube gels   Tube gels are cast in glass tubes of 1–3 mm. The gels are run in a cylindrical box, which is separated into two chambers horizontally. The horizontal partition has gasketed mounting holes in which the tube gels are mounted. The two chambers, upper and the lower, are filled with buffer and fitted with electrodes. The main disadvantage of this system is that only one sample can be loaded per gel, whereas a slab gel can accommodate as many as 96 samples on a single slab. The tube gels have limited applications such as isoelectric focusing.

Applications of Gel Electrophoresis

Gel electrophoresis in general has many applications in biochemistry, cell and molecular biology, genetic engineering, biotechnology, microbiology, medical diagnostics, forensics, etc. Basically, all these applications relate either to nucleic acid electrophoresis or to protein electrophoresis.

Electrophoresis of nucleic acids   DNA fragments, produced by cutting large strands of DNA molecules with the help of restriction enzymes can be sorted according to their size by electrophoresis. Restriction enzymes, are produced by bacteria as a defence mechanism against phages. These enzymes inactivate the viral DNA by cutting it in specific places, i.e., they restrict the multiplication of the virus. Taking advantage of this property, the restriction enzymes are used in recombinant DNA technology as molecular scissors to cut large strands of DNA into small fragments.

Different genes have different nucleotide sequences and, therefore, the restriction enzymes can be used to cut a chain of genes in a large DNA strand into small fragments, each representing a gene. Using gel electrophoresis, it is possible to identify the genes according to their sizes. In an agarose or polyacrylamide gel electrophoresis, the loaded DNA fragments migrate toward the anode, the rates of migration being determined by their size, larger fragments migrating slowly and smaller fragments moving faster.

The confirmation of the identity of the sorted fragments of DNA and the comparison of these fragments with known molecular probes are done by blotting technique. A blotting technique involves transfer of the sorted bands of nucleic acids or proteins to nitrocellulose membrane, and subsequent hybridization with the probe. When DNA bands are blotted to the nitrocellulose membrane, it is called **Southern blotting**, named after E. M. Southern, the inventor of this technique. When the RNA bands are blotted onto nitrocellulose membrane, it is **Northern blotting** (not after any scientist, simply a jargon of "Southern") and it is **Western blotting** for protein bands.

The Southern blotting technique involves the following sequence.

i)   Digestion of extracted DNA with appropriate restriction enzyme.

ii) Electrophoresis of the digested DNA.

iii) Denaturation of the DNA bands in the gel into single strands using an alkali solution.

iv) Placing the gel on a buffer-saturated filter paper laid on a solid support, the edges of the filter paper dipping into the buffer in the container.

v)   Placing nitrocellulose membrane on top of the gel.

vi) Placing stacks of paper towels over the nitrocellulose membrane.

vii) Placing a weight of about ½ kg over the absorbent paper towels.

In this set-up (Figure 13.8) the buffer solution in the tank is drawn up by the filter paper and passes through the gel, nitrocellulose membrane and finally into the paper towels. While passing through the gel, the buffer carries with it the bands containing single-stranded DNA fragments, which get filtered in and attached to the nitrocellulose membrane. The blotted bands are fixed permanently to the membrane by heat treatment. The bands are hybridized with labelled DNA or RNA probes. The nitrocellulose membrane is then washed to remove the unbound DNA and exposed to X-ray film to visualize those bands, which had hybridized with the probe.

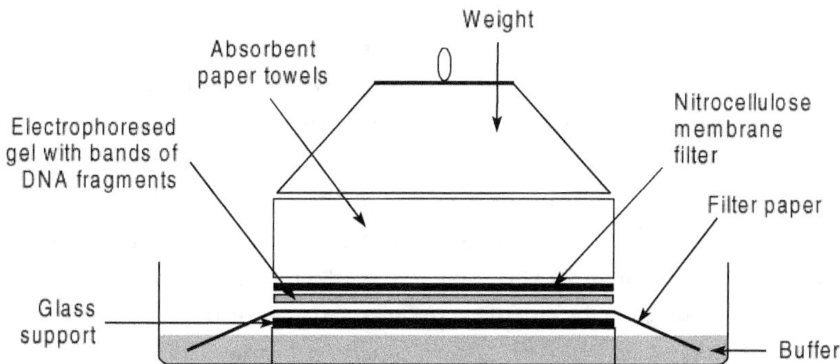

Figure 13.8 Southern blotting technique of transferring bands from gel to nitrocellulose membrane.

The electrophoresis of nucleic acids has been applied to several techniques such DNA fingerprinting, DNA footprinting, DNA sequencing, DNA shift analysis and RNase shift analysis.

Electrophoresis of proteins Gel electrophoresis is a useful method for the analysis of proteins from plant, animal and microbial sources. The advantage of gel electrophoresis is that it permits visualization and separation of proteins, estimation of the number of proteins in a mixture, and assessment of the degree of purity of a protein preparation. It also enables determination of certain physical properties of a protein such as isoelectric point and molecular weight.

Each amino acid differs from the other with reference to its R-group, ionic charge and molecular weight. The net charge of a protein molecule depends on its amino acid composition. If in a protein molecule, there are more positively charged amino acids than negatively charged amino acids, then the net charge of the molecule will be positive. Such a positively charged protein molecule will migrate towards the cathode (negatively charged electrode) in an electrophoresis. Proteins even with a difference of one amino acid will exhibit a different overall charge, and thus are electrophoretically distinguishable.

If a protein molecule is electrophoresed in its compact form (intact tertiary structure), i.e., without breaking the disulphide bonds between the component polypeptide chains, it will migrate as one entity and appear as a single band of, for example, 125 000 Daltons molecular weight on the gel. However, if the protein molecule is treated with appropriate reducing agent (e.g. 2-mercaptoethanol) before or during electrophoresis, it will separate into, for example, four distinct bands with a combined molecular weight of 125 000 Daltons.

In the different lanes of a single gel we can elctrophorese samples treated with reducing agent and samples not treated with reducing agent. The band pattern in such electrophoresed gel would show that some of the high molecular weight bands seen in the lane of samples not treated with disulphide reducing agent are missing in the lane of treated sample. Instead, the treated lane would show two or more bands indicating that the high molecular weight protein is in fact made up of that many polypeptide chains. Thus the break-up of complex proteins into their respective polypeptide chain permits us to analyse the structure of proteins that are synthesized by the interaction of several genes. A single gene codes for only one polypeptide chain. Therefore, a protein consisting of four polypeptide chains is probably synthesized by the interaction of four genes.

Determination of the approximate molecular weight of an unknown protein molecule is possible with gel electrophoresis. As we have already seen, treatment of a sample with appropriate detergent (e.g. SDS) results in the complete and excessive enveloping of the protein molecules in the sample by the anionic surfactant and in the unwinding of the molecule to assume an extended conformation. The result is the electrophoretic separation of all the protein molecules only on the basis of their molecular weights. A standard separation curve is prepared using protein markers of known molecular weights for the estimation of the molecular weight of unknown molecule.

Analogous to the Southern blotting technique followed for the transfer of nucleic acid bands of the gel to nitrocellulose membrane, the **Western blotting** technique transfers protein bands of the gel to nitrocellulose membrane, for further analysis. The procedure of blotting protein or polypeptide bands to nitrocellulose membrane is the same as that of Southern blotting. In addition, a transfer technique called **electroblotting** is used in Western blotting. In this blotting technique a sandwich of gel and nitrocellulose is encased in a cassette, which is immersed in a buffer between two parallel electrodes. When electric current passes through the cassette submerged in buffer, the protein bands in the gel get detached and deposited onto the nitrocellulose membrane, in identical positions. The nitrocellulose paper with blotted protein bands is referred to as a blot. The blot is used for further analysis.

Carbohydrate-binding proteins can be detected and characterized using a PAGE design called carbohydrate affinity-PAGE. Immobilized neoglycoconjugates are covalently cross-linked into polyacrylamide gel. The neoglycoconjugates have two active groups, viz., saccharide and allyl. The allyl group cross-links with the polyacrylamide gel matrix, while the saccharide group is free for specific protein interaction.

This neoglycoconjugate-gel is prepared as a thin layer within the stacking region of a polyacrylamide gel and electrophoresed according to native, non-denaturing conditions. Carbohydrate-binding proteins, specific for the immobilized neoglycoconjugates, are retarded during the electrophoresis migration, while the non-binding proteins move according to size and charge.

Electrophoresis of amino acids Yet another application of gel electrophoresis is separation of amino acids of a polypeptide. Each amino acid has a specific isoelectric pH (pHi), the pH at which the amino acids are electrically neutral. At any pH other than the pHi, an amino acid bears an overall electric charge and therefore migrates under the effect of an electric field. This property of the amino acids is exploited for their separation electrophoretically, i.e., by using a buffer with appropriate pH, the different amino acids can be electrophoretically separated.

When the pH of the buffer is greater than the pHi, the amino acids bear a negative total charge and therefore migrate towards the positively charged electrode, i.e., anode. On the other hand, when the pH of the buffer is less than the pHi, the amino acids bear a net positive charge and move towards the cathode. When the pH of the buffer is equal to the pHi, the amino acids are in their zwitterionic state, i.e., neutral form, and therefore, do not migrate and remain in the starting point. Thus by adjusting the pH of the buffer solution of an electrophoresis system, we can separate the amino acids.

Electrophoresis of proteins and amino acids has specific applications in peptide mapping.

SPECIALIZED ELECTROPHORETIC TECHNIQUES

The electrophoresis techniques that have been developed to meet specific applications are many, and continue to increase. We shall briefly consider the principle and applications of a few such recent developments.

POLYACRYLAMIDE GEL ELECTROPHORESIS

Commonly referred to as the PAGE, the polyacrylamide gel electrophoresis is one of the most versatile electrophoresis techniques. It is generally used to separate proteins and nucleic acids, specifically in processes like protein characterization and purification, DNA sequencing, RNase protection assays and DNA shift assays.

As we have already discussed, PAGE uses a synthetic gel made by polymerization of the monomers of acrylamide (forming long chains) and bisacrylamide (N, N´-methylene bisacrylamide, forming cross-links). The polymerization process is initiated by ammonium persulphate (AP) or by riboflavin, which induces polymerization reaction stimulated by light. The polymerization process is often accelerated by the addition of TEMED (N, N, N, N´-tetramethylethylenediamine). Factors like low pH and oxygen inhibit the process of gel polymerization. Increased concentrations of TEMED or AP increase the rate of polymerization.

One of the main advantages of polyacrylamide gel is that we are able to control its matrix pore size. The pore size increases with the increase in the percent composition of the acrylamide. Commercial gel mixtures are available as 30:1 or 15:1 acrylamide: bis ratio. The PAGE can be performed as a rod (tube) or slab, which may be vertical or horizontal.

The PAGE can be run with dissociating or nondissociating buffer system. **Dissociating buffer system** breaks the molecules down into subunits. SDS is the most common dissociation agent, and the PAGE using SDS is specifically referred to as the **SDS-PAGE**. BME (β-mercaptoethanol) or DTT (dithiothreitol) may be included in the buffer to break the sulphide bonds in the sample molecules. Urea is used to break the hydrogen bonds. Heating may also help in denaturation of the proteins in the sample.

However, because of the denaturing effect of the SDS-PAGE system, it is not suitable to separate enzymes with their biological activity intact. Therefore, for enzyme electrophoresis nondissociating buffer system is used and such an electrophoresis is often referred to as **native gel electrophoresis**. **Nondissociating buffer** system keeps intact the secondary and tertiary structures of the proteins in the sample. The gel of a native gel electrophoresis contains only the native polyacrylamide gel without addition of any surfactants. The proteins in the sample carry their native charge at the pH of the electrophoresis buffer, and separate according to their respective electrophoretic mobility, and the sieving effect of the gel.

The buffer system in PAGE may be continuous or discontinuous (multiphasic). Continuous system has constant buffer composition, pH and pore size throughout gel. It is used more commonly for the separation of nucleic acids and less commonly for the protein separation.

PAGE with discontinuous buffer system is the common method for protein electrophoresis. It can be used to separate larger volumes of sample. The discontinuous system involves discontinuities in buffer composition and pH for manipulating pKs of the sample molecules to get better separation. It consists of stacking gel with large pore size (top phase) and resolving gel with smaller pore size (bottom phase).

The electrophoresed gel is processed either by staining or by autoradiography. Stains such as Coomassie blue and silver stain are used for staining protein bands in the gel. Ethidium bromide is the common stain for nucleic acids. If radioactive markers are used, the gel can be exposed to X-ray film for autoradiography.

The electrophoresed gel can also be processed by removing the bands (specific molecules) for further purification and analysis. Even whole lanes can be removed for two dimensional gel electrophoresis.

The separated bands can be transferred to nitrocellulose or nylon by blotting and analysed. Southern blotting is used for transfer of DNA bands, Northern blotting for RNA and Western blotting for proteins. The blotted bands are probed with DNA, RNA or protein markers or antibodies.

AGAROSE GEL ELECTROPHORESIS

This electrophoresis method is commonly used for the separation of nucleic acids. It is used as preparative gel and has applications in Northern blots, Southern blots, apoptosis "ladder" assays and DNA mapping.

Agarose gel is prepared by dissolving agarose powder in appropriate buffer, by heating. The higher the agarose concentration the better the resolution of the DNA fragments. One per cent agarose gel is the most commonly used one. The common buffers used with agarose gel electrophoresis of DNA are Tris-acetate, Tris-borate, Tris-phosphate, and alkaline (NaOH) buffers. The common buffers for RNA electrophoresis are formaldehyde/MOPS and glyoxal/DMSO. The advantages of agarose gel electrophoresis are its easy and quick set-up and adequate resolutions. However, the resolution may not be as superior as that of the PAGE.

ISOELECTRIC FOCUSING

Isoelectric focusing (IEF) separates proteins in the sample on the basis of surface charge alone as a function of pH. Usually, the electrophoresis separation is carried out in a non-sieving medium (e.g. sucrose density gradient, agarose, or polyacrylamide gel) in the presence of carrier ampholytes. The ampholytes are special buffer substances that are used to establish a pH gradient increasing from the anode to the cathode. They are low molecular weight amphoteric molecules which, when present in a mixture, will migrate in an electric field to their pI (isoelectric point), thus producing a pH gradient. As we know, protein contains both positive (amines) and negative (carboxyl) charge bearing groups, and the net charge of the protein varies as a function of pH.

A protein applied to such a gel with pH gradient will be either positively charged or negatively charged depending on the local pH at that point in the gel. Upon application of an electric current, the proteins will move towards either the anode or cathode, depending on its charge, until it encounters that part of gel which corresponds to its pI. At this point the protein will have no net charge and will, therefore, stop migration. Proteins, which diffuse away from the isoelectric pH, become charged again and are electrophoresed back to their isoelectric pH, hence the name isoelectric focusing. An IEF gel is calibrated with proteins of known pI and a plot of position on the gel against pI allows an estimation of the pI of unknown proteins. This electrophoresis is used for the identification of the pI of a protein, as well as for separation and further purification of molecules.

TWO-DIMENSIONAL PAGE

Two-dimensional PAGE (2D-PAGE) is again a technique for separating complex mixtures of proteins. It is possible to resolve thousands of proteins in a single experiment of 2D-PAGE.

The key principles involved in 2D-PAGE are as follows:

i)   Proteins in a sample differ from each other in terms of mass and charge. These properties are used to separate proteins by gel electrophoresis.

ii) Successive application of electrophoresis separation by charge and then by mass in perpendicular directions (two-dimensions) leads to maximum separation and results in thousands of proteins being resolved.

iii) Staining of the gel reveals the position of the individual proteins as spots or smudges. These spots can be picked and analysed by mass spectrometry.

The 2D-PAGE is a combination of IEF and SDS-PAGE. The polyacrylamide gel is the supporting matrix, which has a pH gradient from top to bottom, the top being more acidic than the bottom. A sample of complex protein mixture is loaded in the middle of the left side of the gel, where the pH is neutral. A voltage is applied across (top to bottom) the gel. The proteins in the sample migrate through the gel until they reach their isoelectric point. This is isoelectric focusing.

The gel is then soaked in a denaturing solution containing SDS, to make the protein molecules unfold and have the same charge. When a voltage is applied across the gel from side to side with the anode at the right, the proteins already separated by IEF move towards the anode, according to their mass. Smaller proteins move faster through the gel than the larger ones.

2D-PAGE is generally used in combination with mass spectrometry to identify all the proteins present in a sample. It is also used for comparing related samples, such as healthy tissue against diseased tissue.

Comparative 2D-PAGE is also useful in identifying proteins from phylogenetically different origins but may have similar functions. It can also be used to identify proteins produced in response to drug therapy and relate them to drug-related side effects.

IMMUNOELECTROPHORESIS AND IMMUNOFIXATION ELECTROPHORESIS

The principle of immunoelectrophoresis is the antigen-antibody specificity (immunoprecipitation) in combination with molecular sieving of the electrophoresis gel and gel diffusion. The agar gel is usually prepared on a glass plate with a circular sample well in the middle (Figure13.9). In **immunoelectrophoresis**, the well is loaded with known antibodies, and the gel is subjected to electrophoresis. The antibodies migrate according to their respective charge and mass. At the end of the electrophoresis, a narrow trough is cut in the gel, parallel to the path of migration of the sample components. Now, the unknown sample is loaded into the trough, and incubated allowing the diffusion of the proteins, the antigens, in the sample and the electrophoresed antibodies to diffuse in the gel. If the unknown sample contains a specific antigen against any of the known electrophoresed antibodies, a precipitation reaction occurs and a characteristic arc (precipitation line or bow) appears in the gel between the trough and the specific antibody.

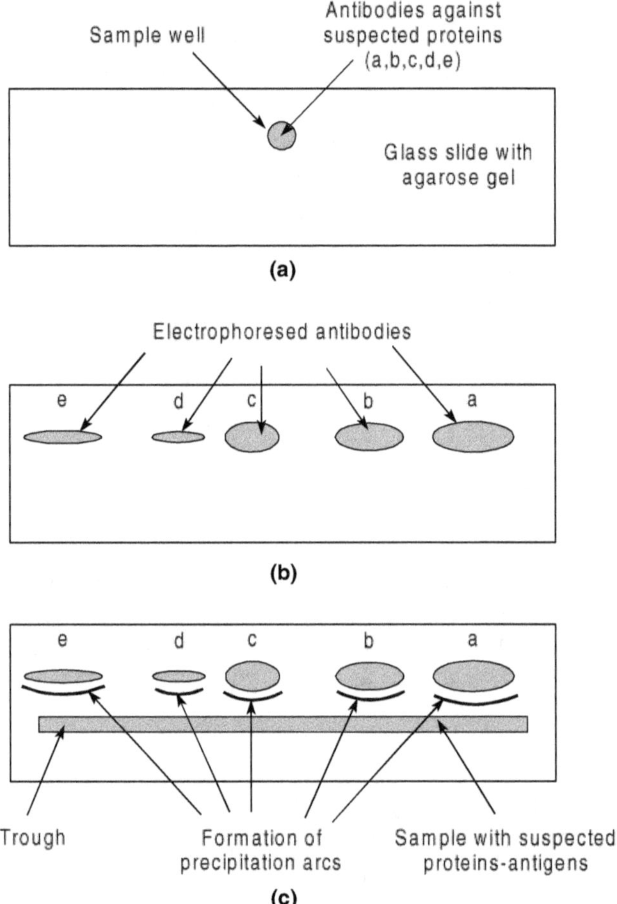

Figure 13.9   Immunoelectrophoresis. (a) Loading of specific antibodies against suspected proteins in the unknown sample. (b) Electrophoresis of the antibodies on the basis of their charge. (c) Loading of the unknown sample in the trough and the formation of precipitation arcs as a result of antigen–antibody reaction.

In immunofixation electrophoresis (IFE), the proteins in the unknown sample are first separated by electrophoresis according to their charge and after that the gel is overlaid with monospecific antisera reactive with specific protein. If the antigen (the protein in the unknown sample) is present, a characteristic immunoprecipitin band is formed. This technique is used to identify and characterize serum proteins. Immunofixation is more sensitive than immunoelectrophoresis in detecting a monoclonal protein in the urine, especially in recognizing a monoclonal heavy chain fragment in the urine with myeloma or amyloidosis.

DENATURING GRADIENT ELECTROPHORESIS

In denaturing gradient gel electrophoresis (DGGE), the gel is incorporated with a denaturing agent. When a small sample of DNA or RNA is applied to such a gel and electrophoresed, the nucleic acid melts at various stages. As a result, the nucleic acid spreads through the gel. The separated fragments which may be as small as 200–700 base pairs can be subjected to further analysis.

The unique feature of DGGE is that as the DNA is subjected to increasingly extreme denaturing conditions, the melted strands fragment completely into single strands. Using this technique, it is posssible to discern differences in DNA sequences or mutations in various genes. Small differences of few base pairs in the sequence of otherwise identical fragments cause them to partially melt at different positions in the gel. By comparing the melting behaviour of the polymorphic DNA fragaments side by side on the denaturing gradient gel, it is possible to detect fragments that have mutations. By parallel running of two samples that were allowed to denature together, it is possible to identify even the minutest differences in the two fragments.

TEMPERATURE GRADIENT GEL ELECTROPHORESIS

Though the DGGE, discussed above, is a more accurate and reliable method than many of the other gel electrophoresis methods, it has several inherent disadvantages. For instance, chemical gradients used in DGGE are not reproducible, are difficult to establish and do not completely resolve certain complex molecules. To overcome such deficiencies of the DGGE, the temperature gradient gel electrophoresis (TGGE) was developed. The TGGE is a form of electrophoresis that studies the behaviour of molecules under different temperatures. The main difference between the two is that the TGGE provides a temperature gradient instead of a chemical gradient. It is a new and powerful electrophoresis technique for the separation of nucleic acids like DNA or RNA and for the analysis of proteins.

The TGGE method exploits the temperature dependent change of conformation for separating molecules. It starts with double-stranded molecules. At a specific temperature, the DNAs begin to melt, resulting in fork-like structures. In this conformation the migration is retarded compared to double-stranded fragments. The melting temperature is highly dependent on the base sequence. Therefore, DNA fragments of the same size but having different sequences can be resolved. Thus, TGGE not only resolves molecules but also provides additional information about melting behaviour and stability.

The critical aspect in the TGGE technique is the preparation of the gel. The buffer system used should remain constant, withstanding increasing temperature. Urea is used in the preparation of the gel, in different concentrations and proportional to the overall temperature required to separate the DNA frgament.

The amount of sample required depends on the type of TGGE, vertical or horizontal. Vertical system requires large samples while horizontal system requires smaller amount of sample. The gel is loaded, sample injected, voltage adjusted and the electrophoresis is run. The electrophoresed gel is stained for detection and analysis of the separated molecules.

Silver staining is useful for staining the gel. The separated DNA fragments can be eluted from the stained gel for further analysis through PCR amplification.

The TGGE method has applications in the identification of mutations in DNA (e.g. mitochondrial DNA), proteins (e.g. pancreatic enzymes) and in studying mycobacterial diversity.

CAPILLARY ELECTROPHORESIS

Since its advent during 1980s, capillary electrophoresis (CE) has grown into a family of new powerful techniques which have wide applications in the fields of biology, analytical chemistry, biotechnology, biomedicine, pharmacology, toxicology and forensics. The fabrication of inexpensive narrow-bore capillaries for gas chromatography and development of highly sensitive detectors of the HPLC are the main contributing factors of the transformation of conventional electrophoresis to the modern capillary electrophoresis.

In CE, the sample molecules are electrophoretically separated in tiny capillaries at high voltages (10–30 kV), thus attaining very high efficiencies (handling more than 10^5 sample molecules at a time) and very high order mass sensitivities (up to 10^{-18} to 10^{-20} moles). The following are the major features of CE:

i) Versatility of application, ranging from inorganic ions to large DNA fragments

ii) Different separation modes with different selectivity

iii) Extremely low volume of sample

iv) Negligible operating cost

v) Use of different types of highly sensitive detection systems

vi) Easy automation

vii) Quantification

viii) Ruggedness

ix) Simplicity of the instrumentation

The basic instrumentation set-up of capillary electrophoresis is shown in Figure 13.10.

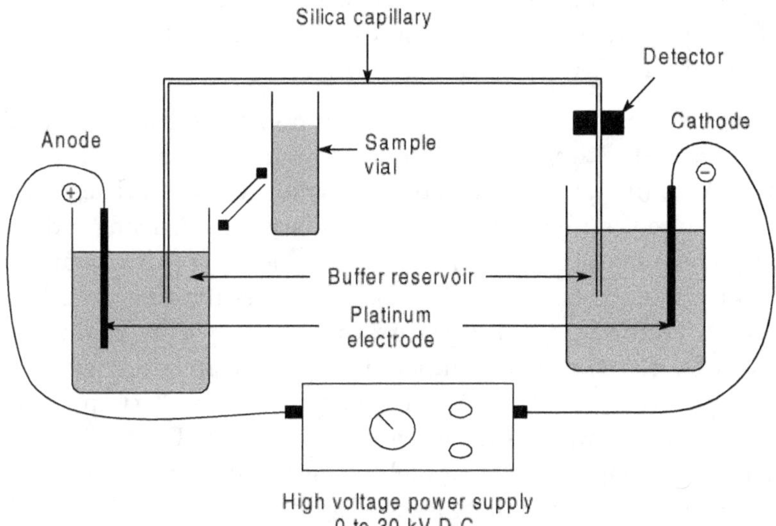

Figure 13.10 Capillary electrophoresis.

The capillary   It is the key component of the CE. It is made up of fused silica, which is chemically inert, UV-transparent, flexible and economical.

Since fused silica is brittle it is coated with polyimide to make it strong and easy to handle. The polyimide, a polymer, is not UV-transparent, and therefore it is removed in a part of the capillary to make a detection window. Materials such as silica and teflon are also used to fabricate the capillaries.

The inner diameter of the capillary ranges from 10 to 100 μm, though an inner diameter range of 25–75 μm is more common. The outer diameter of the capillary is about 350–400 μm. The length of the capillary is generally maintained short, 25–75 cm, in order to minimize the analysis time. It is about 10 cm for capillaries using gel and about 100 cm for those using complex samples. Capillary columns are conditioned to have consistent electrophoresis runs. A common procedure is base-conditioning with NaOH to remove the adsorbates and to deprotonate all the silanols from the capillary surface.

Thermostatting   Even slight variations in temperature may cause significant effect on buffer viscosity leading to inconsistency in the electrophoresis run. Therefore, effective control of capillary temperature is necessary. Thermostatting of capillary, i.e., maintenance of an optimum temperature, is achieved by a high velocity stream of air, or by liquid.

Power supply   A stable, high voltage of DC current (10–30 kV) must be supplied to obtain a reproducible capillary electrophoretic separation. The power supply should be capable of switching polarities to reverse the direction of EOF (electroosmotic flow).

Sample injection   The sample vial is temporarily substituted for the anode buffer tank at the time of introducing the sample. Extremely small volumes of sample in the range of picolitres to nanolitres are sufficient. The sample to be analysed is injected into the capillary system using either pressure (hydrodynamic) or electric field (electrokinetic). Hydrodynamic injection of sample is done by the application of a pressure difference between the two ends of the capillary. Pressure can be applied by raising the sample vial relative to the capillary outlet. This causes siphoning and injection of sample into the capillary. Alternatively an external gas pressure can be applied to the sample or a vacuum to the outlet. Electrokinetic injection of sample is done by applying high voltage of current for a specific period of time so that the sample is forced into the capillary.

Detection   An on-column detector is used to analyse the separated and eluting molecules, qualitatively and quantitatively. Several analysing techniques to measure the physical properties of the molecules are employed. The most common mechanisms are UV-visible fluorescence, laser-induced fluorescence, amperometry, conductivity, and mass spectrophotometry. Apart from the above methods, radioactivity, thermal lens, refractive index, circular dichroism and Raman effect are also used for detection.

Some of the commonly employed CE techniques are

i)   Capillary zone electrophoresis (CZE)

ii) Capillary gel electrophoresis (CGE)

iii) Capillary isoelectric focusing (CIEF)

iv) Micellar electrokinetic capillary chromatography (MECC)

v)   Capillary isotachophoresis (CITP) and

vi) Capillary electrochromatography (CEC)

Capillary Zone Electrophoresis

The CZE is the most widely-used and simple method of CE. It is used for the separation of both anionic and cationic molecules in a single electrophoresis run. In this technique, the sample is applied as a narrow zone (band) which is surrounded by the separation buffer. When a strong electric field is applied, each component in the sample zone migrates according to its electrophoretic mobility. Obviously, the cations and anions migrate in different directions, but they are all swept towards the cathode by the strong electroosmotic flow (EOF), which is higher than the velocity of the migrating molecules. Eventually, all sample component molecules will separate from each other to form individual zones of pure material. Cations will elute first because the direction of their migration is the same as that of the EOF. Neutral solutes which would remain unresolved, will elute next because they lack electrophoretic mobility and are carried by the EOF. Anions are the last to elute because they migrate in the opposite direction of the EOF.

Capillary Gel Electrophoresis

The CEG is an adaptation of the conventional slab gel electrophoresis to the capillary format. In this system, the capillary is filled with a non-crosslinked polymer such as linear polyacrylamide, polyethylene glycol or cellulose derivatives. Crosslinked polymers, polyacrylamide gel and agarose gel can also be used. Thus, the sample molecules are separated according to their size in a nonconvective medium. CGE is useful for the electrophoretic separation of large biological molecules like proteins and DNAs, which have similar electrophoretic mobility in free buffer solution because of their similar charge-to-mass ratio. The advantages of CEG over slab-gel electrophoresis are a wider range of gel matrix and composition, online detection, improved quantification and automation.

Capillary Isoelectric Focusing

This technique is used for the separation of amphoteric molecules (those which contain both acidic and basic groups) such as proteins, peptides, amino acids, and pharmaceuticals in a solution of pH gradient. The basis of the separation of sample components is the differences in the isoelectric point (pI) of sample components rather than differences in apparent velocity. CIEF is performed by filling the capillary with a mixture of ampholytes and the sample. The ampholytes (zwitterions) are used to generate a pH gradient inside the focusing capillary. Ampholytes that are positively charged migrate towards the cathode while those that are negatively charged migrate towards the anode. Consequently, the pH increases at the cathode side of the capillary and decreases at the anode side. When an ampholyte reaches its own pI and is no longer charged, its migration ceases. As a result, a stable pH gradient is formed. If a molecule in the sample has a net positive charge, it migrates towards the cathode. During this migration, it eventually encounters a pH at which it has a zero net charge and stops its migration. This process is known as "focusing". The pH gradient is smoother when a larger number of ampholytes are used. In order to prevent the migrations of buffers from the buffer reservoirs into the capillary, the pH of the electrolyte in the cathode must be higher than the pIs of all the basic ampholytes and the pH of the buffer in the anode must be lower than the pIs of all the acidic ampholytes.

Micellar electrokinetic capillary chromatography   MECC is an adaptation of CE for the separation of not only the anionic and cationic molecules of the sample, but also the neutral molecules. In MECC, micelles are used in the separation buffer. Micelles are amphiphilic monomeric surfactants, which have a hydrophilic head and a hydrophobic tail. The hydrophobic tail may be a straight or branched chain of hydrocarbon, or steroidal skeleton. The hydrophilic head may be cationic, anionic, zwitterionic or non-ionic. The micelles have a three-dimensional structure with the hydrophobic moieties of the surfactant in the interior and the charged hydrophilic moieties at the exterior. The separation of neutral molecules is based on the hydrophobic interaction of solutes with the micelles. The stronger the interaction, the longer the molecules migrate with the micelle. The selectivity of MECC can be controlled by the choice of surfactant and also by the addition of modifiers to the buffer.

Capillary isotachophoresis   Isotachophoresis is an electrophoresis technique in which the sample molecules are separated by displacement process in a multizonal medium. Isotachophoresis performed in a capillary system is called capillary isotachophoresis. In CITP, a sample is introduced between a leading electrolyte and a trailing electrolyte without EOF. The leading electrolyte possesses higher mobility while the trailing electrolyte has lesser mobility when compared to those ions in the sample zone. The separation of the CITP is on the basis of the differences in the velocities of the ions in the sample zone.

Capillary electrochromatography   The CEC is a combination of liquid chromatography (LC) and capillary electrophoresis. In this technique, the capillary is packed with stationary phase through which an electric field is applied. The electric field induces EOF, which moves the mobile phase through the packed column. The separation of the sample molecules depends on the partition between the stationary phase and the mobile phase. CEC yields high degree of resolution of neutral compounds in a very short duration. The main advantages of CEC are separation of closely related compounds, shorter analysis time, and increased sample throughput.

Chapter Fourteen

COLORIMETRY AND SPECTROPHOTOMETRY

PHOTOMETRY, COLORIMETRY AND SPECTROPHOTOMETRY

Photometry is the study of the measurement of intensity of light, colorimetry, the study of colour in a relatively broad range of wavelength of visible light, and spectrophotometry, the study of colour of a very narrow range of or specific wavelength in the UV, visible and IR regions of the spectrum. The term photometry is used to mean the use of photometers, colorimeters and spectrophotometers. The term colorimetry may also refer to the use of colorimeters and spectrophotometers. In the following discussion, we shall use the term colorimetry to the use of photoelectric colorimeters, spectrophotometry to that of UV and visible spectrophotometers.

Colorimetry is a common and useful method to determine the concentration of a chemical substance in a solution in terms of colour intensity that the substance has produced. It relates the intensity of the colour to the concentration of the substance in the solution. The instruments that are used for colorimetry are either colorimeters or the spectrophotometers. Though the instruments are different the basic principles of both are however the same.

PRINCIPLE

The basis of colorimetry is the relationship between the concentration of a substance in a solution and the colour intensity of that solution, the colour intensity being proportional to the concentration. There are many other factors which affect this relationship.

The intensity of absorbance of a colour depends on the chemical being used and the arrangement of the electrons and energy levels within the chemical. The colour we see in a solution is complementary to the colour absorbed by the chemical. When light passes through a solution, light of certain wavelength is absorbed by the coloured chemical substance (absorbance) and the remaining light is transmitted (transmittance). Let us consider the terms absorbance and transmittance in some detail with reference to colorimetry.

Transmittance and absorbance   When monochromatic light (light of specific wavelength) passes through a homogeneous solution in a cell (cuvette), the intensity of the emergent radiation depends on the path length (l) and the concentration (c) of the solution. Consider the Figure 14.1, where

I_0 is the intensity of the incident light passing through a solution in a cuvette with a diameter l (light path length) and I, the intensity of the transmitted light. The ratio I_0/I is the transmittance (T).

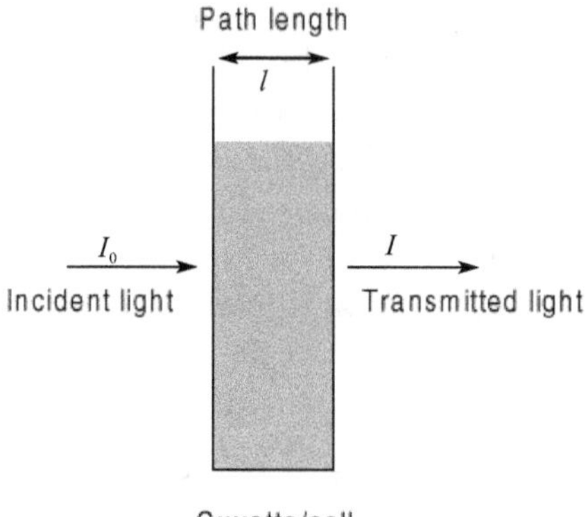

Figure 14.1   Relationship between the light transmittance and path length.

$$\text{Transmittance } (T) = \frac{I}{I_0} = \frac{\text{Intensity of transmitted light}}{\text{Intensity of incident light}}$$

Transmittance is often expressed in percentage, the percentage of transmittance (%T).

The difference between the intensity of incident light and that of the transmitted light is due to absorbance of some light by the solution in the cuvette. Absorbance (A) is given by the log (I_0/I).

$$\text{Absorbance } (A) = \log (I_0/I) = \log (1/T) = \log (100 \, \%T)$$

If we plot T or %T against the concentration of the solution in the cuvette we will get an exponential curve. On the other hand, a plot of absorbance against concentration would give us a linear graph (Figure 14.2).

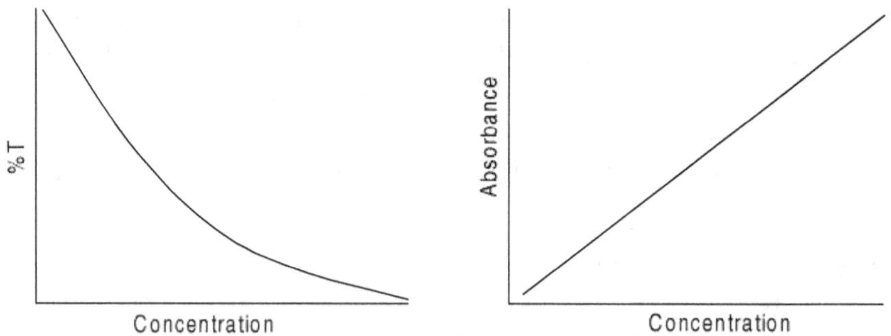

Figure 14.2   The relationships between concentration and %T and between concentration and absorbance.

The relationship between the percent transmission scale and absorbance scale is shown in Figure 14.3. It can be understood from these diagrams that if all the light passes through a solution without any absorption, then absorbance is zero, and percent transmission is 100. On the other hand, if all the light is absorbed, then the percent transmission is zero, and absorption is infinite.

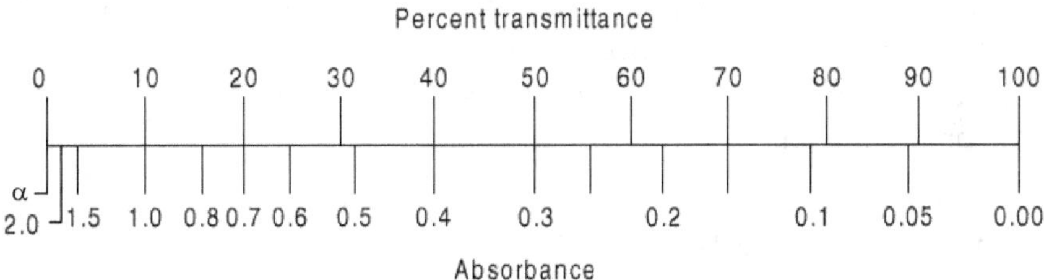

Figure 14.3   The relationship between percent transmittance scale and absorbance scale.

The relationship between absorbance of a solution, the concentration of the solute being measured and the path length is given by the **Beer–Lambert law** which states that for a given path length, the concentration of a substance in solution is directly proportional to the absorbance of that solution, provided that no other solute is absorbing at the specific wavelength. The Beer–Lambert law is valid only for monochromatic radiation (radiation of a single wavelength), and only when there is no change in the physical or chemical state of the solute with changes in concentration. The Beer–Lambert law is written as

$$A = kcl$$

where,

A *is* the absorbance (no units, since $A = log\,(I_0/I)$)

k *is* the molar absorptivity with units of $mol^{-1}\,L\,cm^{-1}$

c *is* the concentration of the compound in solution, expressed in mol L^{-1}

l is the path length of the sample, that is, the diameter of the cuvette in which the sample is contained, in centimetres.

The Beer–Lambert law tells us that the absorbance depends on the total quantity of the absorbing compound (concentration) in the light path through the cuvette.

The reason for expressing Beer–Lambert law using absorbance, instead of percent transmission, as a measure of absorption is the linear relationship between concentration and absorbance, and this is simple and straightforward.

Molar absorptivity is a significant constant factor in the Beer–Lambert law. The value of this constant depends on the nature of the molecule and the wavelength of the radiation. **Molar absorptivity** is defined as the absorbance of a 1 M solution (concentration of molecule in the solution expressed in moles per litre) measured over a path length of 1 cm, and it is expressed as cm^{-1} M^{-1}.

It will be useful if we understand the significance of molar absorptivity. Let us rearrange the Beer–Lambert equation $A = klc$ as

$$k = A/lc$$

This relationship means that k (molar absorptivity) is a measure of the amount of light per unit concentration. Molar absorptivity is a constant for a particular substance such that if the concentration of the solution is decreased by one-half, the absorbance will also decrease exactly to one-half.

For example, suppose we have a compound with a very high value of molar absorptivity (k), 100,000 mol^{-1} L cm^{-1}, which is in a solution in a cuvette of 1-cm diameter $(l,$ path length) and gives an absorbance of 1, it can be written as

$$k = 1/1 \times c$$

Therefore,

$$c = 1 / 100,000 = 1 \times 10^{-5} \text{ mol L}^{-1}$$

On the other hand, if we have a compound with a very low value of molar absorptivity, 20 mol^{-1} L cm^{-1} which is in a solution in a cuvette 1-cm diameter and gives an absorbance of 1, then

$$c = 1 / 20 = 0.05 \text{ mol L}^{-1}$$

Thus, a compound with a high molar absorptivity is very effective at absorbing light of a particular wavelength. Therefore a compound in a solution, even if it is present in minute quantity (low concentration), can be detected provided it has high molar absorptivity.

The molar absorptivity of Cu^{2+} ions in an aqueous solution of copper sulphate $(CuSO_4)$ is 20 mol^{-1} L cm^{-1}. The bright blue colour of the copper sulphate solution is not due to the molar absorptivity of the Cu^{2+} ions, but due to its concentration. β-carotene is an organic compound found in vegetables and imparts a reddish orange colour. Normally it is found in very low concentrations. However, its molar absorptivity is very high, with a unit of 100,000 mol^{-1} L cm^{-1}.

Variations in names and units of colorimetry/spectrophotometry   The names and units for the quantities used in colorimetry and spectrophotometry have undergone changes several times.

i)   **Absorbance** has previously been called as **Optical Density (OD)** or Extinction.

ii) **Molar absorptivity** was previously called the **Absorption Coefficient** or **Extinction Coefficient.**

iii) The units for expressing absorptivity vary with units in which the concentration of the molecule in the sample is expressed. The different units of absorptivity are given in Table 14.1.

Table 14.1   Units of absorptivity in relation to different units of concentration.

Units of concentration	Units of path length	Units of absorptivity
mol L^{-1}	cm	mol^{-1} L cm^{-1}
mol dm^{-3}	cm	mol^{-1} dm^{-3} cm^{-1}
M	cm	M^{-1} cm^{-1}

SI units of absorptivity   In the SI system, the unit for expressing the concentration of molecule in a solution is mole per cubic metre (mol m^{-3}) and for length, metres. If the concentration of the solute is expressed in moles per m^3 and the path length in m, the molar absorptivity is assigned a symbol τ (Greek alphabet, tau) and its unit is $mole^{-1}$ m^2. It is the absorbance of a solution having solute concentration of 1 mole per cubic metre over a path length of 1 metre.

COLORIMETER AND SPECTROPHOTOMETER

Both the instruments measure the intensity of the colour of a sample solution. However, spectrophotometers are more sophisticated than the colorimeter on several counts. The main differences between the two are as follows:

Mode of selecting the specific bandwidth of the incident light   While glass filters are being used in a colorimeter, a prism or a grating is used in a spectrophotometer for selecting the monochromatic light.

Range of spectrum   Colorimeter can use light only in the visible range. The spectrophotometer can use both visible and UV ranges.

Cuvette   Cylindrical glass tube is used in a colorimeter as cuvette to contain the sample solution. Rectangular, quartz or fused silica cuvette is used in a spectrophotometer.

Light source   Tungsten filament lamp is the source of white light in a colorimeter. In addition, hydrogen or deuterium lamp is used to provide light in the UV range, in a spectrophotometer.

Detector   A colorimeter is used in a photovoltaic cell as a detector to measure the intensity of the transmitted light. On the other hand, a spectrophotometer uses photomultiplier as the detector.

CONSTRUCTION OF COLORIMETER

A colorimeter has the following basic components:

    i)   Illumination source

    ii) Filter

    iii) Sample holder, cuvette

    iv) Detector

    v) Display

The organization of a simple colorimeter is shown in Figure 14.4.

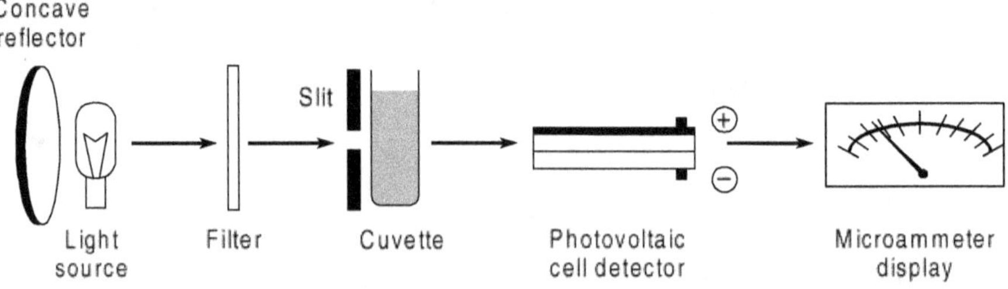

Figure 14.4   The main components of a colorimeter.

Light source   Since colorimeters are invariably employed for measuring the intensity of light in the visible range, a tungsten filament lamp is used as the source radiation. Tungsten lamps emit continuous radiation in the range of 350 nm and 2500 nm. Most of the common colorimeters have a reflector (concave mirror) to focus parallel beams of light through the filter.

Filter   A monochromator is required to select a specific wavelength of light. Filters are used for this purpose in a colorimeter. A filter absorbs light of all the wavelengths except that of a specific band of wavelength. A filter may be a thin gelatin layer stained with an organic dye, held between two thin layers of glass or simply a tinted plane glass. The bandwidth transmitted by filters is generally large, about 100 to 150 nm. Increasing the concentration of the dye in the gelatin or the tint of the glass may reduce the transmitted bandwidth.

Special filters called interference filters are capable of enhancing the specific needed wavelength and eliminating the unnecessary wavelengths. The transmitted wavelength band is much narrower than that of an ordinary filter, about 10 to 30 nm. Further, the interference filters are capable of transmitting radiations close to UV (150 nm) and to IR (5 mm). An interference filter consists of two glass plates, each mirror-coated on one side. The mirror coating ensures reflection of half of the incident light and transmittance of the rest. The glass plates are separated by a layer of transparent material such as calcium fluoride or magnesium fluoride, the thickness of which determines the specific bandwidth to be transmitted.

Cuvette   The sample solution, which is to be analysed for its absorbance (optical density), is contained in a tube, the cuvette (Figure 14.7) of specific diameter (1 cm) and made up of high quality glass.

Detector   A detector of a colorimeter is a device that converts electromagnetic energy (light energy) into electrical energy, which can be measured using an appropriate ammeter. Hence, the colorimeters are often referred to as photoelectric colorimeters. The common detector used in colorimeters is a photovoltaic cell (barrier layer cell). A photocell consists of thin film of silver coated onto the surface of a semiconductor material (e.g. cadmium sulphide, silicon and selenium), fitted onto a steel base (Figure 14.5). Electrodes are connected to the silver layer (anode) and the steel base (cathode).

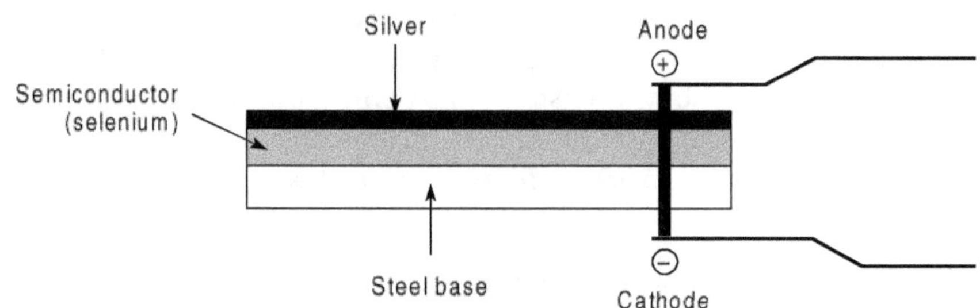

Figure 14.5   Photovoltaic cell.

When the transmitted light strikes the semiconductor layer, the bonding electrons in the crystal are dislodged out of their positions. These electrons move from the semiconductor layer to the silver film. Since the electrons cannot move back from the silver layer to the semiconductor layer, the silver functions as the collecting electrode (anode) for the displaced electrons. The steel plate acts as the cathode. The current flowing between the silver layer and the steel is measured with the help of a microammeter. Since the amount of electrons displaced is proportional to the intensity of the light transmitted by the sample solution, and the current flowing between the electrodes is proportional to the amount of electrons collected by the anode, the intensity of the transmitted radiation is displayed in an absorbance (optical density)/percent transmittance scale.

CONSTRUCTION OF SPECTROPHOTOMETER

A spectrophotometer consists of the following main components:

i)   Radiation source

ii) Monochromator

iii) Sample holder, cuvette

iv) Detector

v)   Amplifier, display and recorder

A simplified block diagram of a spectrophotometer is shown in Figure 14.6.

Radiation source   Since spectrophotometers are meant to measure absorbance of radiations in the visible, UV and IR regions of electromagnetic spectrum, different sources are provided for each.

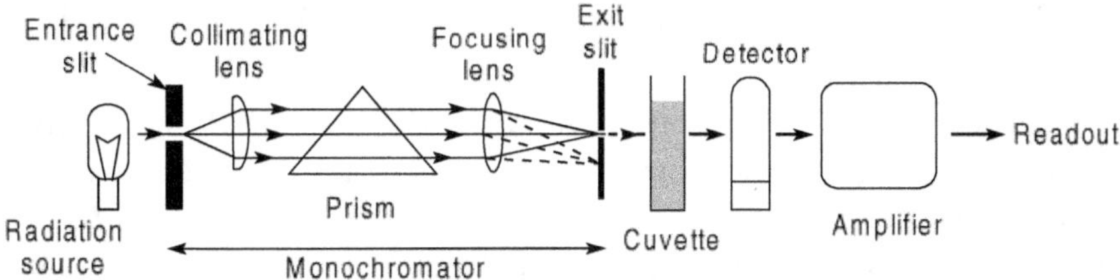

Figure 14.6   Block diagram of a spectrophotometer.

An incandescent lamp with tungsten filament which produces white light is used as the source of visible radiation in the wavelength range of 350–2500 nm. A gas-discharge tube which emits only one or more discrete wavelengths, such as hydrogen lamp and deuterium lamp are the common sources of UV radiation, in the wavelength range of 180–350 nm. Globars of carborundum and Nernst Glowers of rare earth oxides are the commonly used sources of IR radiation. Globar provides IR radiation in the bandwidth range of 1–40 mm and Nernst Glower, in the range of 0.4–20 mm.

Monochromator   In a spectrophotometer, a monochromator is used to resolve the polychromatic radiation into its component wavelengths, to isolate them into narrow bands and focus a specific band into the sample solution contained in the cuvette. A monochromator (Figure 14.6) consists of

i)   an entrance slit through which a narrow beam of the polychromatic light from the radiation source is admitted

ii)   a collimating lens or mirror that directs the radiation as parallel beams on to the scattering device

iii)   a prism or a grating for scattering the radiation into its component wavelengths

iv)   a focusing lens to focus the required specific narrow band of wavelength and

v)   an exit slit, through which the focused band is directed to pass through the sample solution.

A prism made up of simple glass is used for dispersion of light in the visible range. Fused silica or quartz or fluorite prism is used for dispersing radiation in the UV region. Ionic crystalline materials such as NaCl, KBr and CsBr are used for the dispersion of IR region. The advantage of prism is that it is simple and inexpensive to make. Its disadvantages are that the dispersion of the light is angularly non-linear, and that it is temperature-sensitive.

A diffraction grating may be used for the dispersion of polychromatic radiation in the monochromator of a spectrophotometer. A grating is a highly polished reflecting surface on which many thousands (6000–50000 per cm) of narrow parallel grooves have been etched. The dimensions of the grooves are of the same order as the wavelength of the light that is to be dispersed. A beam of radiation directed onto the grating is scattered, or diffracted, in all directions. Radiation waves reinforce each other in certain directions and cancel out in other directions. The direction of diffraction is different for each wavelength. The resolving power of a grating is superior to that of a prism. The main advantages of gratings are that they give a linear angular dispersion with wavelength, and that they are temperature-insensitive. A concave grating is used to perform the dual function of dispersion and focusing of radiation.

The monochromator may include both prism and grating. A prism placed in front of the grating (foreprism) permits the selection of a specific portion of the spectral band to be directed onto the grating for further resolution of the portion. Ideally the output from a monochromator is monochromatic light, i.e., radiation of a single wavelength. But in practice the output is always a band with symmetrical shape. The width of the band at half its height is called the Instrumental Bandwidth (IBW).

Cuvettes   Containers, referred to as cuvettes, made up of glass, quartz or fused silica are used to hold the sample and reference solutions in a spectrophotometer. Glass cuvettes can be used in the visible region but not in the UV region because glass absorbs UV radiation. Quartz or fused silica cuvettes are appropriate for measurements in the UV region. The cuvettes used in spectrophotometers are four-sided containers (Figure 14.7). Two opposite surfaces are precision-ground and polished to be optically flat for passage of the radiation. The other two opposite surfaces are rough-ground and meant for handling. The radiation path length of the cuvette is 1 cm.

Detector   The intensity of the radiation transmitted by the sample solution is measured with the help of a detector. The detectors used in spectrophotometers generally function on the principle of photoelectric effect. The photoelectric effect is the emission of electrons when light strikes the surface. The amount of electrons emitted is proportional to the intensity of the radiation.

A common detector found in many spectrophotometers is a **phototube** (Figure 14.8). Two conducting electrodes, anode and cathode, are enclosed in an evacuated glass tube. A source of potential difference (power supply) creates an electric field in the direction from anode to cathode. Light falling on the surface of the cathode causes a current (flow of emitted photoelectrons as a result of photoelectric effect) in the external circuit. The current is being measured by a galvanometer. Since the

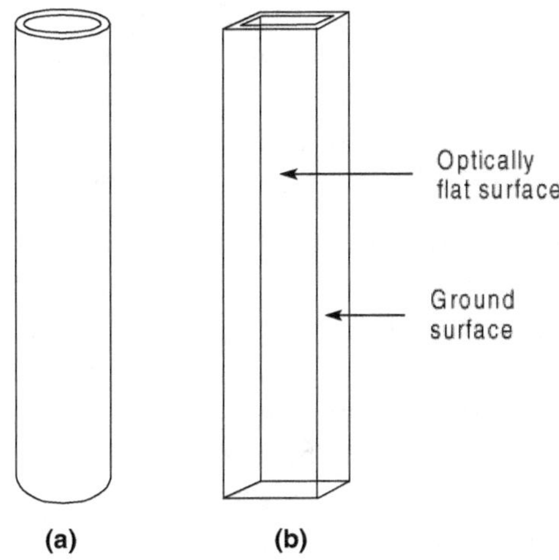

Figure 14.7   Cuvette of a colorimeter (a) and a spectrophotometer (b).

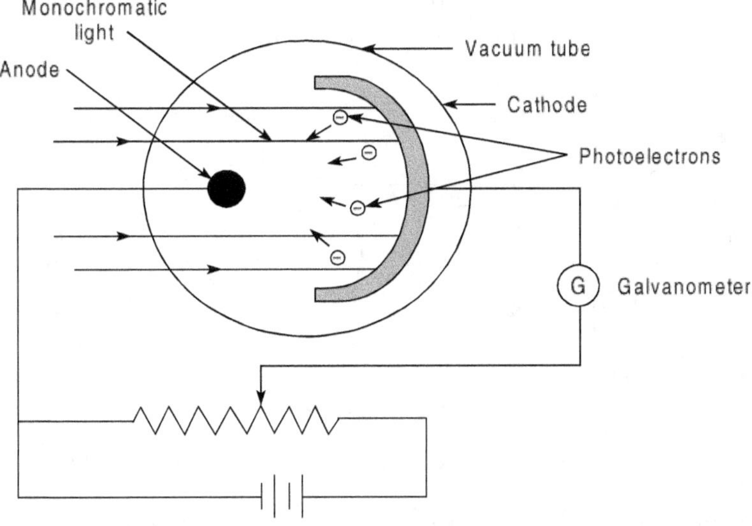

Figure 14.8   Schematic diagram of overview of a phototube circuit.

photocurrent is negligible, it is amplified by introducing high resistance in the circuit. The photocurrent varies with voltage and with the frequency and intensity of the light.

Another common type of detector used in spectrophotometers is the photomultiplier tube, which is the adaptation of the scintillation counter used for detecting gamma radiation. A diagrammatic representation of the photomultiplier device is shown in Figure 14.9. The initial portion of the device consists of a material called phosphor (sodium iodide crystal), which is transparent to photons, allowing the photons to pass through a light-pipe of Perspex onto a photoelectric surface. The phosphor and the light-pipe are sealed in an aluminum casing, which has a reflecting or diffusing inside wall so that no photons are lost by absorption.

The photoelectric surface, which is a photoelectric cathode, a series of inverting electrodes called dynodes, and a collector anode are enclosed in a vacuum tube called the multiplier tube. In this tube, the electrons emitted from the photoelectric cathode are accelerated by a potential difference to the first of the dynodes. At the first dynode, the impact of each of the arriving electron ejects about 2–5 further electrons in a process known as secondary emission. These secondarily emitted electrons are accelerated to the next dynode, where further secondary electrons are produced, and so on. In a ten-stage photomultiplier tube, the number of electrons reaching the final collector plate for every single electron emitted from the photo-cathode is of the order of 10^6. The device thus produces an appreciable pulse current and the intensity of the current is proportional to the initial ionization produced in the phosphor.

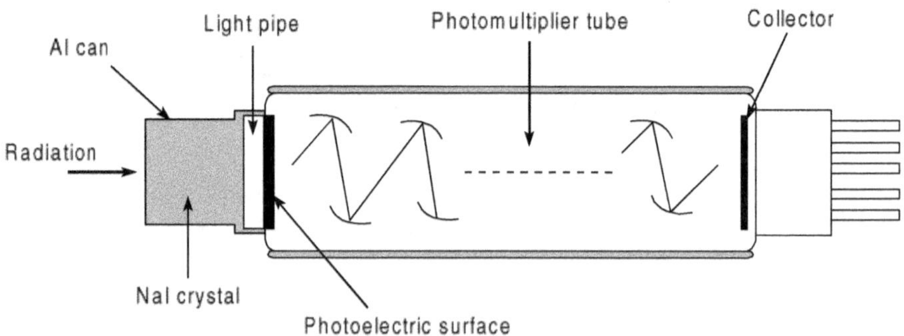

Figure 14.9   The design of the photomultiplier tube.

In addition to the above detectors, there are others such as photodiodes and IR detectors. Photodiodes are semiconductors which change their charged voltage when struck by the radiation. In a photodiode, light falling on the semiconductor material allows electrons to flow through it, thereby depleting the charge in a capacitor connected across it. The amount of charge needed to recharge the capacitor at regular intervals is proportional to the intensity of the light. Photoconductive cells are used to detect near IR radiation in the range of 0.8–3.0 mm. The semiconductor used in these devices is germanium, lead sulphide or lead telluride. The advantage of photodiodes is their greater dynamic range and, being solid-state devices, their robust nature.

Thermometers such as thermocouples, bolometers (resistance thermometers), pneumatic and Golay cells (gas thermometers) are used as detectors for middle and far IR radiation. The energy of the absorbed photons of the middle and far IR radiations is converted to thermal energy and measured by the thermometer.

Display and recording   Manual spectrophotometers apply an analog voltage output from the detector to a comparator circuit such as Wheatstone bridge or potentiometric recorder. In modern spectrophotometers, the analog voltage output is converted to digital data and fed into a computer and printer for display, recording and printout. The instrument itself may have an inbuilt computer for the operation, and for giving printout of the results.

SPECTROPHOTOMETRIC ANALYSIS

Apart from the operation of the instruments, spectrophotometric analysis involves calculations to obtain the concentration of a sample solution from the absorbance measurements made by using the spectrophotometer (or a colorimeter). There are three methods, viz., proportionality, graphing and Beer's law.

Proportionality

The proportionality approach to the calculation of the concentration of a substance in the solution from its absorbance is based on the principle that the absorbance of solution is directly proportional to its concentration. In this method, we must make sure that the values given are for different concentrations of the same chemical measured under the same conditions of wavelength and path length. Let us consider an example.

Example   A known sample solution with a concentration of 0.14 M is measured to have an absorbance of 0.43. Another solution of the same chemical is measured under the same conditions and has an absorbance of 0.37. What is its concentration?

The answer to this problem can be calculated by using the following equation, which states that the ratio of the concentration is proportional to the ratio of absorbance.

$$\frac{C_1}{C_2} = \frac{A_1}{A_2}, \frac{C_1}{0.14} = \frac{0.37}{0.43}$$

$$C_1 = \frac{A_1}{A_2} \times C_2, \ C_1 = \frac{0.37}{0.43} \times 0.14$$

$$C_1 = 0.12 \text{ M}$$

where,

C_1 = the concentration of the unknown sample,

C_2 = the concentration of the known sample,

A_1 = the absorbance of the unknown sample,

A_2 = the absorbance of the known sample.

Graphing Method

The graphing method of spectrophotometric analysis includes

   i)   plotting of the absorbance spectrum

   ii) choice of the appropriate wavelength, and

   iii) plotting of the calibration curve.

Absorption spectrum   Absorbance spectrum of a substance is often described as its "fingerprint", and it is necessary to have it for its quantitative determination. The molar absorptivity (extinction coefficient) for any specified substance is constant, provided the wavelength of the radiation is constant. The absorbance changes with the wavelength. For example, we can prepare two solutions of a compound (e.g. $KMnO_4$), one in higher concentration and the other in a lower concentration. We can take readings of absorbance (optical density) for each of these sample solutions at different wavelengths. If we plot the sample's absorbance of radiation at various wavelengths we will get a curve which is called the absorbance spectrum of the sample of a specific concentration. The absorbance spectra of the two concentrations of potassium permanganate solution are shown in Figure 14.10. It can be seen in the Figure that the two curves (Curve 1 and Curve 2), represent the absorption spectra measured under the same conditions except for the difference in the concentration. Both the curves have similar shapes. It can further be noted that at a specific wavelength the absorbance is more than in other regions. This specific wavelength where there is maximum absorption is referred to as *absorption maxima* (A_{max} or λ_{max}). In the case of the $KMnO_4$ the A_{max} is 525 nm.

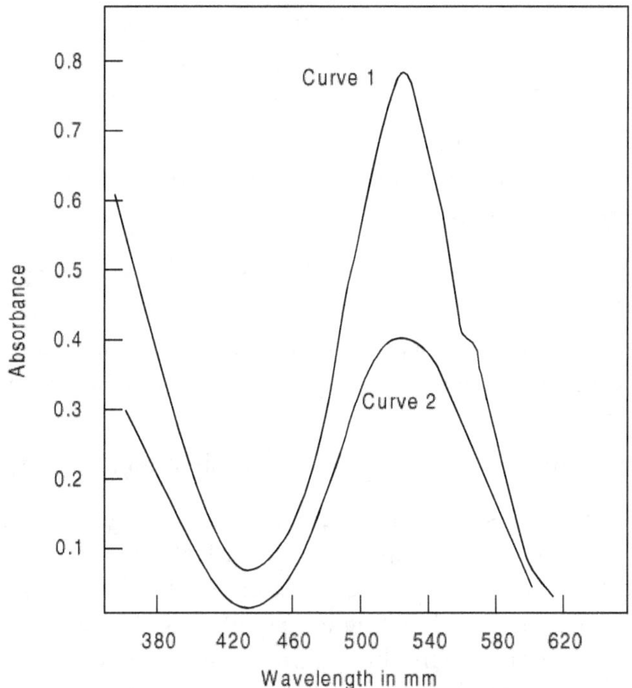

Figure 14.10 Absorption spectra of solutions of potassium permanganate at two different concentrations—Curve 1, higher concentration and Curve 2 lower concentration.

Choice of wavelength As we have already seen, absorbance is proportional to concentration at each wavelength (Beer–Lambert law). Theoretically we can choose any one wavelength for the quantitative estimation of the concentration of a compound. But the magnitude of the absorbancy is important, especially when we are detecting very minute amounts of the compound. Therefore, we have to choose the specific wavelength with maximum absorption for each concentration.

In the absorption spectra of the two concentrations of $KMnO_4$ solution shown in the Figure 14.10, it can be seen that the distance between curves 1 and 2 is at a maximum at 525 nm and that at this wavelength the change in absorbance is greatest for a given change in concentration. In other words, the measurement of concentration as a function is most sensitive at 525 nm. If we choose any wavelength other than 525 nm, say 500 or 580 nm, these wavelengths represent steep portions of the curve. Even a small fluctuation in the wavelength may cause significant error in the absorbance. Since most spectrophotometers tend to have slight fluctuation in the wavelength, errors in absorbance will be amplified if we use wavelengths other than that of maximum absorption. Therefore, we choose the wavelength of maximum absorbance for a given sample and use it in all measurements of absorbance of that sample.

Construction of calibration curve   Construction of the calibration curve (calibration plot) involves

i)   Preparation of a series of standard solutions of accurately known concentrations,

ii)   Careful reading of absorbance values of these standard solutions using the correct wavelength (wavelength of maximum absorption) and

iii) Plotting of absorbance (Y-axis) against concentration (X-axis).

Each standard solution is prepared in identical fashion. The only difference between the standard solutions should be the concentration. While plotting the calibration curves, care must be taken not to lose the accuracy of the data by choosing too small axes. The axes should represent the accuracy obtained in the readings of the spectrophotometer. For example, if we can read the absorbance correct to the second decimal place, say 0.34, then we should have an absorbance axis so that 0.34 can be located accurately on it.

Theoretically, the trend line of a calibration curve should pass through all the plots of concentration vs. absorbance. However, in spite of all the care and precautions taken in the preparation of standard solutions and in taking spectrophotometric readings, the plots may not fall in a perfect straight line. In such an event, it would be useful to work out a regression line of absorbance (Y) on concentration (X), using the regression equation

$$(X - \overline{X}) = \frac{rS_y}{S_x}(Y - \overline{Y})$$

where,

Y = variable absorbance

$\overline{Y}$ = mean of the variable Y

r = coefficient of correlation between X and Y

S_y = standard deviation of variable Y

S_x = standard deviation of variable X

X = variable concentration

$\overline{X}$ = mean of the variable X

The final equation giving the best estimation of absorbance (Y) for a given concentration (X) will be as follows.

$$Y = a(X) + b$$

where, a is the slope (regression coefficient) and b, the intercept. Using this equation, we can construct a straight-line calibration curve, or calculate the concentration (X) of an unknown from its absorbance (Y) as

$$X = \frac{Y - b}{A}$$

The values of the slope and intercept can be easily calculated with the help of any scientific calculator. Use of a computer with appropriate software (e.g. MS Excel) would make things much more easier. Examples are shown in Box 14.1. Once we have the calibration curve constructed, we can measure the absorbance of any unknown solution at the same wavelength and read off its concentration from the graph or calculate from the slope.

While plotting the calibration curve, we should bear in mind that according to the Beer–Lambert Law, when concentration is equal to zero, absorbance must also be zero. In other words, the calibration line must pass through the origin. An important source of error in spectrophotometric analysis is applying the Beer–Lambert Law at inappropriate concentrations. The Beer–Lambert Law is strictly applicable only for dilute solutions. It becomes less and less accurate as the concentration of the solution increases.

BOX 14.1   CONSTRUCTION OF CALIBRATION CURVE

Example 1

Concentration of standard solutions (M)	Absorbance (Optical density)
0.0	0.00
0.2	0.69
0.3	0.55
0.4	0.41
0.5	0.27

A simple graph of concentration vs. absorbance yields a straight line.

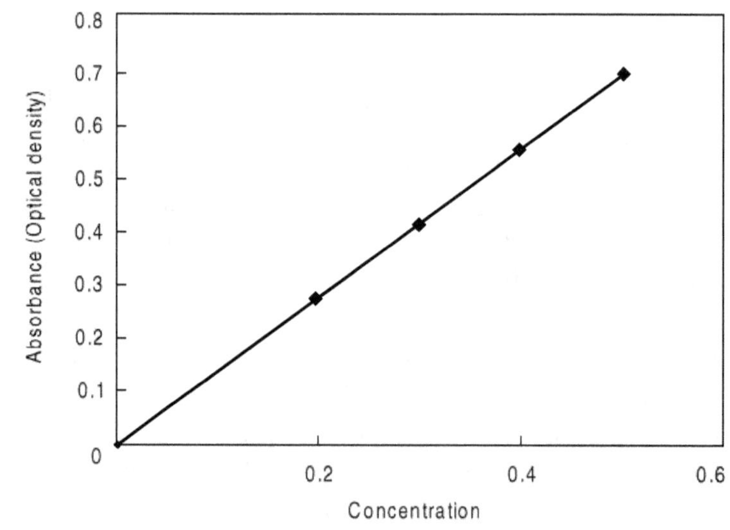

BOX 14.1   (Continued)

Example 2

Concentration of standard solutions (M)	Absorbance (Optical density)
0.0	0.00
0.1	0.15
0.2	0.28
0.3	0.30
0.4	0.51
0.5	0.62

A simple graph shows that plots of concentration against absorbance do not fall in a straight line.

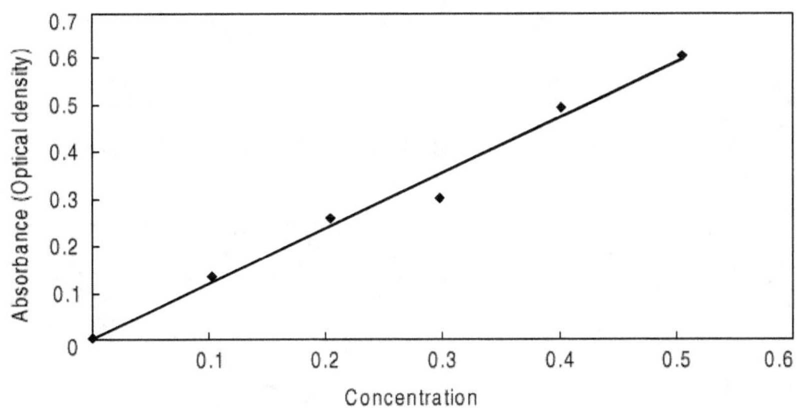

Therefore, a regression line of Absorbance on Concentration is constructed using the regression equation

$$(X - \bar{X}) = \frac{rS_y}{S_x}(Y - \bar{Y})$$

BOX 14.1   (Continued)

The equation obtained is

Absorbance (Y) = [Slope (A) × Concentration (X)] + Intercept (B)

$$Y = 1.2\,(X) + 0.01$$

Using the above equation best estimates of Y for various X is obtained as

$$X = (Y - 0.01) \div 1.2$$

And the estimated values are

Absorbance	Estimated concentration
0.0	0.000
0.1	0.075
0.2	0.158
0.3	0.242
0.4	0.325
0.5	0.408
0.6	0.492
0.7	0.575

Now, we can obtain a straight line for the plots of absorbance vs. concentration, as shown in the following graph.

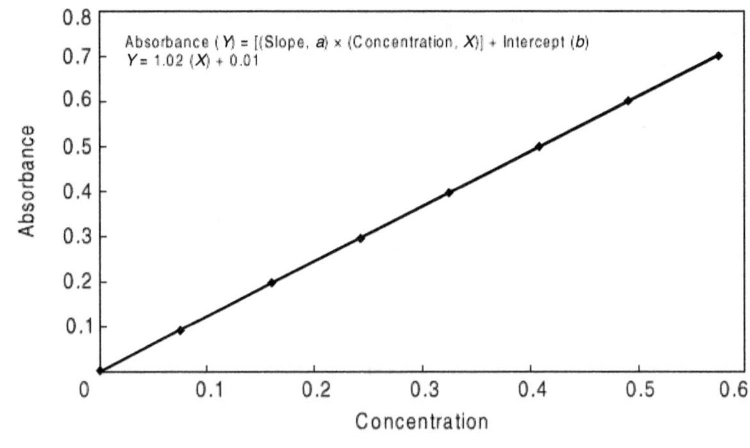

Beer's Law Equation Method

This method is used when data on the molar absorptivity (extinction coefficient) and radiation path length are available. The concentration is calculated using the Beer's law

$$A = klc$$

where A = absorbance

k = molar absorptivity

l = path length

c = concentration

Let us consider an example.

Example The absorptivity (k) of a compound is 1.5 M^{-1} cm^{-1}. What is the concentration of solution of this compound if a 2 cm sample has an absorbance of 1.20?

Rearranging the equation $A = klc$, as $c = A \div kl$.

$$c = 1.20 \div (1.5 \times 2) = 0.40 \, M$$

DESIGN OF SPECTROPHOTOMETERS

The conventional design of a spectrophotometer uses a single detector and a forward optics configuration, i.e., the dispersion device (prism or grating) comes before the sample. Such an optical system requires the sample area (cuvette) to be completely covered to prevent ambient light reaching the detector. We have to rotate the dispersion device to take reading at different wavelengths or to measure a spectrum. Thus, data acquisition is sequential, i.e., there is lapse of time between one reading and the next.

The basic conventional design has undergone several modifications to overcome problems, to improve the sensitivity and performance and to reduce errors. We shall consider only three variations of the basic design, viz., single beam, double beam, and split beam.

Single Beam

The single-beam spectrometer is the simplest design (Figure 14.6). In this design, we have to first measure the absorbance of the blank and then the absorbance of the sample, using the same cuvette. In practice, we set the absorbance of the blank at zero (100 percent transmittance) and then read the absorbance of the sample. Modern single-beam spectrophotometers have inbuilt electronic facilities to store the absorbance of the blank and to calculate the absorbance of the sample.

The main advantages of the single-beam spectrophotometer are

i)   low cost

ii) high throughput, and

iii) high sensitivity.

The disadvantage of the single beam spectrophotometer is that a significant amount of time elapses between the first reading (blank or reference) and the second reading (sample) so that there can be problems with drift (variations in the intensity of the radiation). However, modern single-beam spectrophotometers have better electronics and more stable lamps to overcome the problems of the earlier versions.

Double Beam

The design of the double-beam spectrophotometer (Figure 14.11) eliminates the problem of drift by measuring the absorbance of the blank and the sample virtually at the same time. A "chopper" alternately transmits

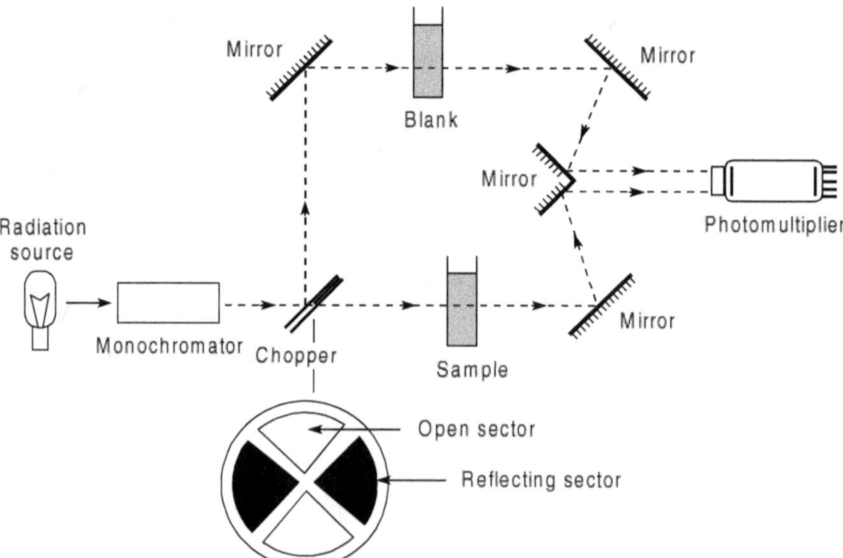

Figure 14.11   The schematic design of a double-beam spectrophotometer.

and reflects the radiation so that it travels down the blank and the sample optical paths to a single detector. The chopper causes the radiation to switch paths at about 50 Hz enabling the detector to receive signals from blank and the sample light paths, alternately, in short successions. These signals are processed electronically to calculate the ratio of the reference and sample signals and to give either transmittance or absorbance as output.

The advantage of the double-beam spectrophotometer is its high stability because of the almost simultaneous measurement of the reference and the sample. The disadvantage of this design is its high cost, lower sensitivity due to the complex optics which causes poor throughput of light, and lower reliability due to the complexity in the design.

Split Beam

The design of the split-beam spectrophotometer is almost similar to that of the double-beam spectrophotometer except that a beam splitter is used in the place of the chopper to direct the radiation along the reference (blank) and the sample paths. Thus, in this design, the measurements of both the blank and the sample take place at the same moment of time. The intensities of the transmitted radiation through the reference solution and the sample solution are measured by two different phototubes. The split-beam design, though stable, suffers from the disadvantage of using two detectors, one for the blank and one for the sample, which may drift independently.

ERRORS IN SPECTROPHOTOMETRIC ANALYSIS

There are several potential sources of error in the spectrophotometric quantitative analysis. These errors may be sample-related or instrument-related. Sample-related errors are generally due to

i)   errors in the preparation of the standard or sample solutions,

ii) presence of suspended particulates that cause scattering and

ii) unclean cuvettes.

These errors can be easily eliminated or, at least, minimized by following good laboratory practices.

Instrument-related errors are due to noise, drift, photometric accuracy, stray light or inadequate resolution, wavelength accuracy and wavelength reproducibility. The instrumental errors may be random or bias. A random error is one that is not reproducible, whereas bias is an error that is reproducible. For example, noise, wavelength reproducibility, and drift are random errors. Insufficient resolution, wavelength accuracy and stray light are bias. All the sources of errors are significant for absolute measurement. However, when a calibration is carried out before sample measurement, as in UV-Visible quantification, bias errors cancel out.

BIOLOGICAL APPLICATIONS OF UV-VISIBLE SPECTROPHOTOMETERS

In most of the life science laboratories the UV-Visible spectrophotometers are used for the quantification of nucleic acids and proteins, and to study the kinetics of enzymes. A spectrophotometer can also be used for turbidimetry and nephelometry. These uses are in addition to the numerous colorimetric assays that are carried out in an ordinary colorimeter.

Nucleic Acid Quantification

Checking the purity and quantification of the nucleic acid at various stages of the gene cloning process is routinely done using spectrophotometer. The absorbance ratio at 260/280 nm (UV region) for pure nucleic acids is a constant. The ratio is 1.8 for DNA and 2.0 for RNA. Any deviation from these values would indicate the presence of protein or phenol impurity because they absorb radiation at 280 nm, thereby reducing the ratio.

The nucleic acids extracted from cell is quantified using standard conversion factors. For example, at 260 nm, an absorbance value of exactly 1.00 is equivalent to 50 mg/mL of dsDNA and 40 mg/mL RNA, respectively, in a 10-mm path length cuvette.

Protein Determination

Quantitative estimation of protein in a sample solution can be easily performed using UV-Visible spectrophotometers. There are at least four standard methods that are commonly followed in many biology laboratories.

The Bradford method uses Coomassie brilliant blue to bind with the protein in the unknown sample and the absorbance of this sample (595 nm) is compared with that of different amounts of standard proteins (e.g. bovine serum albumin, BSA). This method is suited to quantify minute amounts (1–10 μg) of protein in the sample.

The Lowry method uses Folin–phenol reagent to react with tyrosyl residues of unknown protein in a sample to produce a blue colour, the absorbance of which is compared with those derived from a standard curve of a standard protein (BSA), at 750 nm. This method is useful for quantifying proteins in the range of 1–20 μg.

The basis of the Biuret method is reaction between cupric ions and peptide bond in an alkali solution, resulting in the formation of a complex absorbing at 546 nm. This method is routinely applied for measurement of proteins in serum and urine, up to a level of 130 mg/ml.

The BCA (bicinchonic acid) method involves the reaction between cupric ions and peptide bond in combination with the detection of cuprous ions using bicinchonic acid, giving an absorbance maximum at 562 nm. This method has a working range of 1–2000 μg/mL of protein.

Enzyme Kinetics

The spectrophotometric quantitative assay of an enzyme activity involves the following of the changes in the absorbance of the substrate as a function of time. If the substrate of an enzyme is coloured or capable of absorbing radiation (UV region), then it is possible to track the rate of its appearance or disappearance in a solution, using a spectrophotometer. If the substrate is not coloured and does not absorb radiation at the particular wavelength, a suitable light-absorbing derivative of the substrate is prepared, or alternatively, chromogen can be employed. The absorption of the coloured product formed is proportional to the concentration of the substance under investigation. For example, the activity of the enzyme LDH, which is involved in the transfer of electrons from lactate to NAD to form pyruvate, NADH and H^+, can be assayed.

$$\text{Lactate} + NAD^+ \rightleftharpoons \text{Pyruvate} + NADH + H^+$$

The NADH absorbs radiation in the UV range at 340 nm, but its oxidized form, NAD^+ does not absorb. The other components of the reaction, the substrate or the product do not absorb radiation at 340 nm.

Thus the increment in absorbance is proportional to the progress of the reaction in the forward direction.

Turbidimetry and Nephelometry

Spectrophotometers are also useful in the quantitative estimation of the number of bacteria or any other microorganism or any other particulate material in a liquid medium. When viable bacteria or other cellular or particulate matter is suspended in a liquid, the liquid becomes turbid. The turbidity of the liquid is proportional to the number of bacteria or other material suspended. When monochromatic light passes through such a turbid medium, certain amount of light is absorbed by the liquid medium, some light is scattered by the suspended materials, and the remaining light is transmitted. The intensity of the transmitted light is proportional to the degree of turbidity, which in turn is proportional to the number of bacteria or other materials suspended. Since bacteria do not absorb radiation but only scatter it, the term optical density (OD) is more appropriate to describe the measurement of turbidity due to bacterial suspension. The observed OD is interpreted with reference to the OD of a known bacterial concentration. When the turbidity of a solution is measured in terms of the intensity of the transmitted light, it is called turbidimetry. If the measurement is in terms of the intensity of the scattered light, it is referred to as the nephelometry.

Chapter Fifteen

Photography

INTRODUCTION

Photography is one of the powerful tools of research in Biology. It is widely used to record evidences for a variety of investigations, using various methods ranging from macrophotography to microphotography. A few of the applications of the photography in biological research are listed in Box 15.1. Because of its versatility, it is necessary that all research students who intend to use this method must have a knowledge of the basic principles of photography.

BOX 15.1   APPLICATIONS OF PHOTOGRAPHY

Macrophotography   The process of making photographs using a variety of cameras and lens attachments (Box 15.5), except the microscope, is referred to as macrophotography. It may be still or movie photography, and can be underwater or aerial.

The most common application of still photography in biology has been to record the biodiversity of the earth. Further, it is commonly used in biology laboratories to record the morphology and anatomy of animals and plants, complementing freehand drawing. Use of motion picture cameras has enabled the recording of the behaviour, life history, etc. of animals and plants, even those living in deep waters and in high altitude mountains. With the advent of fibre optics and digital photography, it is now possible to view, record and, even manipulate (endoscopy, laser surgery, etc.) the organization and functioning of organs.

The results of a number of laboratory investigations in the fields of microbiology, biochemistry, molecular biology, biotechnology, etc. can be recorded using appropriate macrophotographic techniques. Some such results are density gradients in centrifuge tubes, electrophoresis gels, immunodiffusion, and bacterial colonies.

Another common application of photography in research laboratories is copy work. Charts, graphs, photographs, radiographs, etc. can be photographed for making multiple copies to be included in theses, project reports, etc.

BOX 15.1   (Continued)

Libraries employ photography to make photocopies (microfilms) of articles in journals for the use of individuals. However, this use of photography has been largely replaced by the Xerox facilities. Computer generation of graphs, charts and, even, scanned photographs has made copy work by photography almost obsolete.

In *underwater photography*, underwater cameras use a watertight casing with a glass or plastic windowpane in front of the lens. If the water is clear and, therefore, sunlight penetration is good, photographs can be taken up to a depth of 10 metres. If water is not clear due to the presence of particulate materials and in deeper waters where there would not be sufficient light penetration, special artificial lighting such as electronic flash or floodlight are used. Underwater photography might become more difficult if the artificial light is reflected by the planktonic and nektonic organisms and other suspended particles. Wide-angle lenses are more suited for underwater photography especially to compensate the effect of higher refractive index (The refractive index of a medium is the ratio of the speed of light in empty space to the speed of light inside the medium) causing the underwater organisms and other materials to appear 25% closer than they are in reality. Underwater photography is used to record fauna and flora in the different strata of any water body. It has been of immense use in understanding the behaviour and life history of underwater organisms, especially those living in deep seas.

Making photographic images from aircraft or satellites is *aerial photography*. Specially designed cameras, equipped with several types of lenses and sufficient film stock, is mounted vibration-proof on aircrafts for aerial photography. These cameras are operated either from the aircraft or remotely from a ground station. Use of cameras having long focal lengths and large-format films enables the production images of very high resolution of a few metres width from altitudes of more than 15 kilometres. A common application of aerial photography is aerial survey.

Aerial survey is the study of the earth's surface using photographs taken from aircraft. It is commonly used in map-making, agriculture, environmental studies, and military operations. The science of

---

BOX 15.1   (Continued)

making precise measurements of height, area, distance, and volume from aerial photographs and using these measurements to create detailed physical maps is called *photogrammetry.* Using this method, it is possible to estimate

1. the size and distribution of wild fauna and flora on different parts of the earth;

2. the extent of afforestation and deforestation;

3. the agricultural activities such as extent of crop production, extent of pest infestation, etc.

4. the extent of natural calamity such as those due to cyclone, flood, fire, earthquake, avalanche, etc. Special infrared sensitive films are useful in gathering information about plant life on earth. Soil conservation and forest management programs also make use of aerial photography.

Weather forecasting is an important application of aerial photography from satellites. It is possible to view and record the formation and movement of weather phenomena such as cyclonic storm.

Microphotography   Photographic recording of images through a microscope is called microphotography or photomicrography. Wide-ranging subjects, from a whole-mount of a small insect to an organelle in a cell, can be photographed using an appropriate microscope, lens and lighting. The fields in which microphotography is used are cytology, histology, electron microscopy (TEM and SEM), autoradiography, etc.

LIGHT

Light is the basis of photography and it is, as we have already seen, an electromagnetic energy having both wave and particle properties. It is considered to be in the form of photons (the smallest units of light energy) which are discrete particles with zero mass and no electrical charge. Photons travel in waves and are able to strike objects as if they were particles and alter the physical and chemical nature of the molecules on the surface of the objects. Specifically, when they strike silver molecules they turn them black. It is this property of light that is made use of in photography.

FILM

In photography, the image of an object is captured on the surface of a film. The basic chemistry of the preparation, exposure, developing, fixing and printing of black-and-white photographic film is given in Box 15.2. The nature and principle of the colour film is given in the box 15.3. The various types of films, film format and other characteristics are listed in the box 15.4.

BOX 15.2   CHEMISTRY OF PHOTOGRAPHY

1. *Preparation of light-sensitive photographic plate or film*   A photographic plate is a glass plate and it is obsolete. A photographic film is a strip of colourless transparent material made of triacetate plastic base. Triacetate plastic base has replaced the less stable, flammable material made from nitrocellulose and camphor, which is commonly referred to as celluloid, originally a trademark name. The film is coated on one side with a thin layer of emulsion of silver bromide, AgBr, dispersed in gelatin (a transparent, brittle protein prepared from animal materials such as skin, bone, connective tissue, etc.). The silver bromide emulsion is prepared as a solution of silver nitrate, $AgNO_3$, mixed with a solution of ammonium bromide, NH_4Br.

$$AgNO_3 + NH_4Br \xrightarrow{\text{Gelatin}} AgBr \text{ (Emulsion)} + NH_4NO_3$$

The mixture of silver bromide and ammonium nitrate is maintained at a warm temperature of about 45°C, for *ripening*, during which period the silver bromide crystals grow bigger in size and, thereby, become more sensitive to light. After the required ripening, the emulsion is cooled on ice to solidify it into a jelly. The emulsion jelly is washed in water to remove the ammonium nitrate. It is then melted and spread over one surface of the glass plate or celluloid film. The entire operation of preparing photographic film is done in a dark and dust-proof room.

BOX 15.2   (Continued)

The size of the silver grains and, therefore the number of grains per unit area, determine the speed of the film. Larger silver grains are more sensitive to light and the films with such larger grains are described as *high-speed* films. On the other hand, smaller silver grains are relatively less sensitive to light and are used to make *slow-speed* films. Since, in slow-speed films, the number of smaller silver grains per unit area would be higher than in the same unit area of a high-speed film, the slow-speed film would be having more resolving power than a high-speed film. That is, slow-speed film would give a better picture of the finer details of the object. High-speed films are useful for capturing images of moving objects and those with poor lighting.

2. *Exposure*   The image of the object to be photographed is captured on the film by exposing it to the light reflected from that object. For this, the film is loaded into a camera, which has mechanisms to focus the reflected light from the object on the emulsion surface of the film and to admit specific amount of the reflected light to fall on the film for a specific duration.

Depending on the colour and intensity of the light falling on the photosensitive emulsion of the film, decomposition of the silver bromide to bromine and colloidal particles of metallic silver particle occurs. The gelatin of the emulsion acts as an adsorbent for the bromine produced, and prevents its recombination with the silver particles as long as the film is kept in the dark.

$$2AgBr \xrightarrow{\text{Exposure of light}} 2Ag + Br_2$$

Different parts of the object reflect different quantities of light (e.g. if we are photographing the face of a young girl, her black hair would reflect no or least light, whereas her white teeth would reflect maximum light) and, therefore, affect the film differentially, proportional to the intensity of the light. The silver bromide in the regions of the film where no light fell, that is, regions corresponding to the darker areas of the object which did not reflect any light, remains undecomposed. As a result of the exposure to light reflected from the object, an inverted image of the object is formed on the film. This image which is not yet visible is referred to as *latent image*.

BOX 15.2   (Continued)

3. *Developing exposed film*   The latent image formed on the film is made visible by developing the exposed film. Developing is a process in which the exposed film is immersed into a solution of developer, which is a weak reducing agent (e.g. Potassium ferrous oxalate; alkaline solution organic reducing agents like Quinol, Pyrogallol, Metol and Amidol). The process of developing is carried out in darkroom.

The developer reduces the exposed silver bromide to finely divided, black particles of metallic silver, the reduction being catalysed by the colloidal particles of silver produced during the exposure. The rate of reduction of the silver bromide is proportional to the amount of adsorbed silver. Thus, the effect of the developer is complementary to that of the light, the deposit of the silver being the thickest in regions where the light effect was the most intense during exposure.

The general reaction of a developer is as follows:

$$AgBr + Developer \rightarrow Ag(Black) + Oxidized\ developer$$

Specifically, if the developer is hydroquinone, the reaction is,

$$2AgBr + \underset{\text{Hydroquinone}}{C_6H_4(OH)_2} \rightarrow \underset{\substack{\text{Black silver} \\ \text{particles}}}{2Ag} + 2HBr + \underset{\text{Quinone}}{C_6H_4O_2}$$

As a result of developing, the latent image formed during exposure becomes visible but its shade is in negative relationship with the shade of the object that was photographed. The developed film is therefore called a *negative film* (in the negative of the photograph of the young girl, the black hair would appear white and the white teeth, black).

4. *Fixing*   Undecomposed silver bromide still remains in the developed negative film especially in regions of the film not exposed to light. This light sensitive silver bromide must be removed in order to make the image permanent and to view the negative film in light. The removal of the unaffected silver bromide from the negative film is called *fixing*. The process of fixing is done with the help of hypo, sodium thiosulphate ($Na_2S_2O_3$). When the negative film is immersed in a solution of hypo the unaltered silver bromide is

---

BOX 15.2 (Continued)

removed as a water-soluble the complex compound. The fixing is also carried out in a darkroom. The reaction is as follows.

$$2Na_2S_2O_3 + AgBr \rightarrow Na_3[Ag(S_2O_3)_2] + NaBr$$

Hypo $\qquad$ Sodium argento thiosulphate

The fixed negative film is thoroughly washed in water and dried, after which it can be taken outside the darkroom. The negative image is now permanent and the film is not affected by light.

As already indicated, the processes of preparation of the light-sensitive film, developing of exposed film and fixing of the developed film are carried out in the dark. However, a faint red light, which practically has no effect on silver bromide, can be used in the darkroom.

5. *Printing* In order to get the original picture of the object the negative image in the negative film is printed onto a light-sensitive printing-out-paper (POP) or a "Velox" brand bromide paper. A POP is coated on one surface with a mixture of silver chloride and silver nitrate. The printing can be done as a contact print or using an enlarger. In contact printing, the POP is placed beneath and in close contact with the negative film. Suitable frames, depending on the size of the film, to hold the negative and the POP in close contact are used. The negative and the POP together are exposed to light for a short duration. The negative image on the film is reproduced on the POP with the dark and bright parts reversed. Thus, the image in the print has the same shade as the object photographed. The exposed paper is developed, fixed, washed and dried as was done for the negative film. Since the image in the print has positive relation (the hair of the girl would appear black and her teeth, white), it is said to be a positive or actual photograph of the object.

When an enlarger is used, the film, usually 35 mm film, is mounted onto an enlarger, and the negative image is focused on the emulsion surface of POP, using red light or using a dummy POP. The image can be enlarged to the desired size and the intensity of the light can be varied to suit the quality of the negative. Further processing of the exposed POP is same as for the contact printing.

BOX 15.2　(Continued)

6. *Toning* Changing the tone (i.e., the colour and its shades) of the positive print from black to either purple or to steel gray is called *toning*. It is achieved by dipping the positive print in a dilute solution of gold chloride ($AuCl_3$) or potassium chloroplatinate (K_2PtCl_6). Silver being more electropositive than gold or platinum goes into solution and the gold or platinum is deposited as fine purplish or steel gray particles, respectively, to impart to the print a purple or steel gray colour. The reactions involved are as follows.

$$AuCl_3 + 3Ag \rightarrow 3AgCl + Au\,(purple) \downarrow$$

$$K_2PtCl_6 \rightarrow PtCl_4 + 2KCl$$

$$PtCl_4 + 4Ag \rightarrow 4AgCl + Pt\,(steel\ gray) \downarrow$$

7. *After treatment of negatives* The image density in the negative can be altered in order to obtain a better positive print by *intensification* or *reduction*. Intensification is done by using an intensifier such as mercuric chloride. When a negative is treated with 5 percent mercuric chloride solution containing a little hydrochloric acid, it is bleached. Blackening of the bleached negative is effected by treatment with a dilute solution of ammonia. A combination of uranyl acetate and potassium ferricyanide can also be used to make the image in the negative more opaque.

Reduction of the negative involves removal of some of the silver in the image by oxidation with ammonium persulphate or a mixture of potassium persulphate and potassium permanganate.

BOX 15.3　COLOUR PHOTOGRAPHY

Colour negative films are also transparent strips of celluloid coated with three layers of black-and-white emulsions with built-in dyes. These dyes chemically react with the silver salts that form the image. Each layer is sensitive to one of the three primary colours (Figure 15.1). The innermost layer is sensitive to the red colour and produces a cyan (greenish blue) tint in the negative. The next layer is

BOX 15.3   (Continued)

sensitive to green light and produces magenta (purplish red) tinge. And the outermost emulsion is sensitive to blue and produces yellow tint. Each of these layers also acts as a filter to screen the light it receives and to prevent scattering of light within the film.

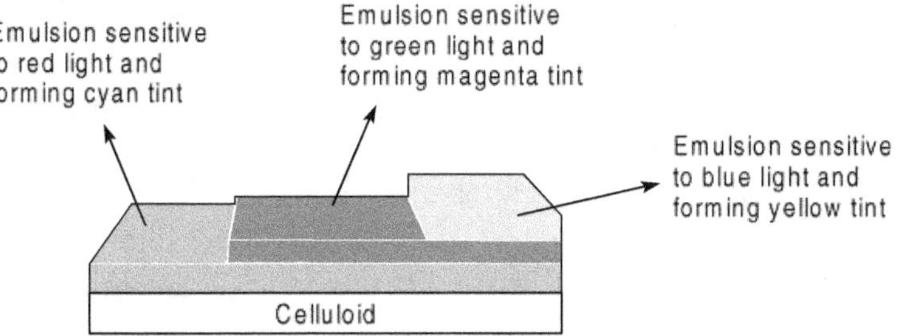

Figure 15.1   Construction of colour negative film.

In colour film processing the developer stimulates chemicals called *dye couplers,* which react to a specific colour of light and cause corresponding dyes to be released to form cyan, magenta, and yellow dye images in the negative. The silver is then removed, leaving a negative image in the three colours. Since the three colours are complementary to the primary colours and since these emulsions act as filters, different combinations of these colours similar to those reflected from the object appear on the final print. For example, if we are photographing a flag having stripes of green, yellow and red, the negative of the photo would appear as magenta (the complement of green), blue (the complement of yellow), and cyan (the complement of red) (Figure 15.2).

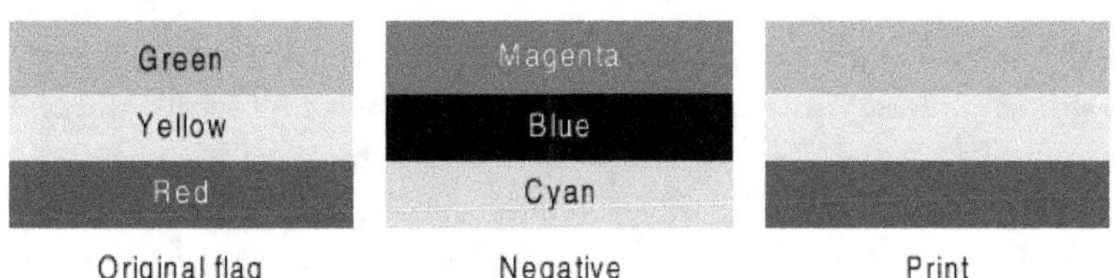

Figure 15.2   Colours appearing in the object, negative colour film and the print.

BOX 15.3   (Continued)

Colour transparency films when exposed and processed produce positive image that can be viewed with the help of a slide projector or a light table. The dyes may be built into the film itself or they may be added during processing of the exposed film.

BOX 15.4   FILM TYPES AND CHARACTERISTICS

Film types   A variety of black-and-white and colour films is available to be used for different purposes. It includes, print films, transparency films, polaroid films and speciality films such as X-ray and IR films.

Print films (*negative* films) commonly used for a variety of purposes in macro-and microphotography are designed to produce negative images of the subject when exposed. The negative images are then printed with required enlargement on to a print paper to get the positive image of the subject. The negative films bear brand names such as Kodacolor, Fujicolor and Agfacolor. For use in biological investigations, especially microphotography, slow speed films with fine grains are more suitable.

Transparency films (*Slide films* or *positive films* or *reversal films*) produce positive images of the subject to which the film is exposed and these images can be viewed with the help of slide projector. These films are useful in biological research especially for presentation of the results of an investigation in conferences and seminars. The brand names of transparency films end with a suffix chrome, such as Kodachrome, Ektachrome, Fujichrome and Agfachrome.

Polaroid films are designed to produce instant prints of images of the subject. These films are available in both black-and-white and colour, for both special polaroid cameras and for standard format cameras. In polaroid films, the silver halide emulsions and the processing chemicals are combined in a self-contained paper envelope or within the print paper itself.

Special purpose films such as UV, IR and X-ray films are sensitive to wavelengths beyond the visible spectrum of light. These films are

BOX 15.4 (Continued)

useful in medicine and biology for recording images that are not normally captured by the conventional black-and-white and colour films. Special films that can respond to other forms of electromagnetic radiation are also available.

Film characteristics There are several characteristics of photographic films that guide us to choose the appropriate one for a specific purpose. The most common are the format, colour sensitivity and balance, sharpness, tonality, grain and the speed.

The *format* of the film, determined by the format of the camera, refers to size (width or width × length) of the film. The common formats are 35 mm and 120 roll films, and 4 × 5, 5 × 7, and 8 × 10 inch sheet films. Each format might be having the other characteristics such as black-and-white, colour print, colour transparency, film speed, etc.

The *number of exposures*, i.e., the number of frames that can be exposed, is generally 36 for 35 mm film cassette and it might be 12, 10 or 8 for 120-film roll depending on the format of the camera.

Colour sensitivity of a film refers to the ability of the emulsion to react differentially to specific colours of the light spectrum. Films that are sensitive to all colours are called *panchromatic* and those sensitive to limited number of colours, generally blue and green, are called *orthochromatic*. While colour films are generally panchromatic, black-and-white films have greater sensitivity to green and red.

Colour films are also designed to be sensitive to the specific light source illuminating the object. The source may be sunlight, tungsten lamps, or electronic flash.

The light from each of these sources has a distinct characteristic referred to as *colour temperature*. The light colour changes corresponding to the temperature. For instance, if we heat a metal wire in a Bunsen burner, first it glows dull red, then bright red and finally blue. That is, with the increase in the temperature the colour also changes, from red to blue. This colour temperature is measured as an absolute temperature in Kelvin. In contrast to our eyes, the colour films are sensitive to slight variations in colour temperature.

BOX 15.4   (Continued)

Therefore, the colour films are *colour balanced* to yield best results in specific colour temperature. The commonly used daylight films are colour balanced for both outdoor photography and for use of electronic flash. The tungsten films are colour balanced to indoor photography without flash but with very bright tungsten photo lamps called *photofloods*.

Colour films are balanced to perform best in specific lighting conditions. The daylight films, the most widely used, are designed for both outdoor photography and for pictures taken indoors with electronic flash. Tungsten films are designed to be used indoors without flash, specifically with certain types of bulbs manufactured for such situations called *photofloods*.

Fluorescent and metal halide lamps are not suitable sources of light for photography because they are discontinuous (flickering). Modern colour films are designed to be less affected by lighting conditions with varying colour temperatures. For example, incorporation of a fourth emulsion layer has been found to overcome to a great extent the problems of varying colour temperatures.

Though the black-and-white films are less sensitive to variations in colour temperature, giving longer exposures while photographing in tungsten light or other low colour temperature light sources (e.g. candles) would yield better results, because these films are relatively insensitive to yellow, orange and red colours.

Exposure latitude is another characteristic of colour films. For any given lighting condition there is an optimal exposure that would record a perfect image on a film. If a film is exposed to light for a longer time than the optimum, i.e., overexposed, it would produce a bleached out and blurred print. On the other hand, if a film is underexposed, the negative will be very light and the print would fail to bring out the contrast of image.

The range of setting (exposure time and the aperture size) within which a film accurately records the colour and *tonal values* (contrast of light and dark) of the object that is photographed is called its *exposure latitude*.

BOX 15.4 (Continued)

Negative films have much greater exposure latitude than transparency films. High-speed films have greater latitude than slow-speed films. To achieve an image with good colour and tonal values in the final print, it is always better to expose the film within the exposure latitude specified for the film and to set the exposure time and aperture size appropriate to the lighting conditions.

Film speed is an important characteristic of colour as well as black-and-white films. It is a numerical measure of the film's sensitivity to light, and it determines the amount of exposure required to photograph an object under specific lighting conditions. The speed of the film is determined during the manufacture of the film, measured by exposure and developing under test conditions and is expressed as defined by the International Standard Organization (ISO). The ISO number for film speed consists of two numbers such as 50-18, 100-21, 200-24, 400-27 and so on. The first part of these combinations (50, 100, 200, 400) represents the arithmetic speed of the film as defined by ASA/BS (American Standard Association/British System). The second part of the ISO number (18, 21, 24, 27) represents the DIN (Deutsche Industrie Norm—German Industry Standard), German logarithmic standard of film speed. The corresponding values of ASA and DIN film speeds are as follows.

50ASA	100ASA	200ASA	400ASA	800ASA	1600ASA
18DIN	21DIN	24DIN	27DIN	30DIN	33DIN

It can be seen that the ASA speed index number increases by doubling whereas the DIN index increases by 3. A doubling of the ASA number suggests that the exposure time needed for a given object to be photographed is reduced by half. For example, if we are setting an exposure time of 1/250 for a 200 ASA film to photograph a subject under specific lighting conditions, then that exposure time would be 1/500 (i.e., half of 1/250 or one stop less) for a 400 ASA film (i.e., doubling of 200 ASA) to photograph the same object under same lighting conditions. In the case of the DIN system the stop (exposure time setting) decreases by one for every 3 DIN increase in the film speed. On the basis of the ISO film speed index, a film may be

BOX 15.4   (Continued)

classified as slow (ISO 25 and ISO 100), medium (ISO 125 and ISO 200) and fast (ISO 400 and above) films.

All the films, whether fast, medium or slow, have a pattern called *grain*, which is the visible unit of metallic silver in emulsion that forms the image. In faster films, these individual grains of silver are generally larger and more prominent than in slower films. As a result, photographs taken in fast film appear grainier, especially in higher enlargements, than those taken in slow speed films. Slow speed films, because of their finer silver halide grains, are suitable for conditions where high resolution is needed.

For example, in microphotography, to photograph the finer details of tissues and cells with greater sharpness, use of slow films (ISO 100 or less) would be ideal. In addition, slow-speed films produce a smoother range of tones and more intense colours than fast film. On the other hand, fast films (e.g. ISO 400) are more suitable to capture images of rapidly moving objects and when lighting conditions are very poor.

Exposure index (EI) of a film provides the information of speed setting to be followed while exposing and developing. It is expressed as, for example, E.I. 100/21°, where the first number (100) represents ASA film standard and the second part the DIN standard. Though the EI appears similar to ISO index, they are different. While film speed is based on standard exposure and development, the EI is meant for a variety of practical situations. For example, a film of ISO 125 may be exposed at EI 200. Many of the commercially available films recommend only EI rather than speed.

DX coding is a modern advancement in film and camera technology that removes the need to set the film speed by hand. It is printed on the of 35-mm film cassettes as a checkerboard pattern that corresponds to an electronic code. This code indicates to the camera's computer chip the ISO rating of the film as well as the number of frames on the roll. Cameras with electronic controls are fitted with DX sensors that can read this information and automatically adjust exposures accordingly. The DX code is also inbuilt within the film to tell the developing laboratory of this information.

CAMERA

The instrument needed for the capture of an image of an object on the film is the camera. Though there are a variety of cameras, most of them have become obsolete with advent of fully or partly automated point-and-shoot cameras. A brief survey of the different types of cameras and accessories is given in the Box 15.5. The common type of camera used in biological investigations, for both macrophotography and microphotography, is the 35-mm SLR camera. This type offers maximum adaptability because it is provided with facilities for changing lenses ranging from telephoto lenses, to close-up lenses, to those used in microphotography, to suit a variety of purposes. Further, the framing of object in a 35-mm SLR camera is more accurate because the image of the object that is seen through the viewfinder will be exactly reproduced on the film.

BOX 15.5 CAMERAS AND ACCESSORIES

Types of camera

1. *Box camera*   It consists of a rigid-light proof box fitted with a fixed, simple lens, a viewfinder window, and a shutter with limited speed options such as bright sun light, shadow and indoor. The use-and-throw cameras now available in the market are equivalents of the box cameras. These disposable cameras are light-proof plastic boxes preloaded with 35 mm film and may even provide additional features such as a flashlight. The use of box camera is limited because of lack of adjustable facilities.

2. *View camera*   It is designed to use large format films that have far greater resolving power than smaller format films such as 35 mm films. The body of the view camera has two independently movable parts, one front and one back, connected by an expandable leather bellows. The front part carries the lens and shutter. The back portion of the view camera is a ground-glass panel that allows focusing and framing of the subject to be photographed. The film, which comes in a lightproof frame, is inserted into the film holder in front of the glass pane. This type of camera is commonly used in studios and outdoor for photographing still objects, landscape, architecture, etc.

BOX 15.5   (Continued)

3. *Rangefinder camera*   A separate lens, the optical viewfinder, through which the subject can be focused and framed, is characteristic of rangefinder camera. The conventional viewfinder is provided with an adjacent rangefinder. When adjusted using "ring or collar" the two views, one through the viewfinder and the other through rangefinder, get superimposed to appear as one, indicating the precise focusing, matching with the distance of the object. Since the view seen through the viewfinder and that falls on the film through the camera lens are from slightly different angles, parallax error is common in rangefinder camera, and therefore less suitable for close-up photography. The lens of 35 mm format rangefinder camera is detachable and can be substituted with more suitable ones.

4. *Point-and-shoot camera*   This type of camera, which is generally of 35 mm format, has many automatic features and is highly user-friendly and capable of yielding high quality pictures. It has battery-operated electronic system with facilities for automatic controls for exposure, focusing, flash, film winding and film rewinding. The camera lens is not removable and it may be of fixed focal length or of zoom type. It is good for amateur photography.

5. *Single-lens-reflex camera*   Commonly known as SLR camera, this type derives its name from the fact that there is only one lens for both viewing and framing the subject and for exposure setting. Light rays from the subject come through the lens and falls onto a mirror, which acts also as a shutter. The mirror reflects the light rays into the viewfinder through a five-sided prism. When the camera is clicked for exposure, the mirror is lifted out of the path between the lens and film, thus opening the way for the light rays from the subject to strike the film. During the exposure period, the viewfinder is blocked. Because of the use of single lens for viewing as well as exposing, parallax error is eliminated.

The SLR cameras, because of their versatility, are most suitable for biological investigations including close-up photography and microphotography. Availability of a vast array of interchangeable lenses, macro lenses, adapters, etc. makes the SLR camera one of the essential equipment in any biology research laboratory.

BOX 15.5 (Continued)

Accessories of camera

The important accessories of a camera, especially the 35 mm SLR type, are the lenses, flashlights, filters, shutter release cable and tripod stand.

Lens The function of the lens in a camera is to refract the light from the object, without any optical distortion and to focus it on the film. Therefore, the lens of a camera is generally a ground and polished glass with specific focal length and diameter. The exterior and interior surfaces of the lens are coated with thin layers of reflection-absorbing material to prevent flare due to the unwanted light falling on the lens. These coatings increase the contrast of the film image. The characteristic green and purple hues seen on the lens surface are due to such coatings. In the professional 35 mm SLR cameras the lens is in fact a combination of several lenses, ranging from 6 to 14 in number, cemented together in groups and encased in a cylindrical *lens barrel*. The lens barrel has an *aperture ring* which helps in setting the size of the opening in the diaphragm, in front of the lens. It also has a *focus ring* for focusing the image of the object on the film, by altering the distance between groups of lenses in the barrel.

The lenses of a camera are classified according to their focal lengths and maximum apertures. With the increase in the focal length (i.e., the distance between the optical centre of a lens and the focused image formed) of the lens used, the image size formed on the film also increases. Lens having a short focal length is referred to as a *wide-angle lens* because it can cover an angle of view wider than that of the human eye. Lens having a long focal length is called a *telephoto lens*, which would make a distant object appear closer and magnified. A *normal lens* is one that has an angle of view almost equal to that of the human eye.

A popular type of lens is the *zoom lens* that can be adjusted to have different focal lengths. It can maintain the focus of the object even if the focal length is changed to cover different angles of view. However, if the zoom lens is of a *varifocal* type, then it would be necessary to focus the object every time the focal length is changed. Many of the modern cameras have facilities for autofocusing.

BOX 15.5   (Continued)

Highly magnified pictures of small objects such as a small insect, flower, anatomical parts of organisms, etc. can be made using *extension rings* and *macro lenses*. These accessories extend the focusing range (the distance between the camera lens and the object) of the lens of a 35 mm SLR camera. Extension rings, a group of detachable rings, are fitted between the base camera and the camera lens. The number of rings or combination of rings depends on the extent of close-up desired. Macro lenses are a group of lenses that can be attached to the front of the lens, individually and in different combinations, depending on the extent of close-up required.

Very high magnification photographs, microphotographs or photomicrographs, of small subjects such as those in histology, cytology and microbiology are made using a microscope. The camera used in modern microphotography is a 35 mm SLR type. The microscope is generally a trinocular microscope, the camera being fitted to the third, vertical ocular tube. If the microscope used is monocular type, the camera is fitted to the vertical tube through an adapter assembly that has a lateral viewfinder.

The lens in microphotography is the eyepiece lens of the microscope, and therefore can be changed according to the required magnification of the image. Fully automated microphotographic equipment are available. If manually operated, the ideal way to learn to use it is by trial-and-error. The important things to remember are the vibration-free set-up of the microphotographic assembly, the proper illumination, focusing and exposure.

Filters   A filter is a coloured/white, transparent/ground gelatin or glass and is fitted to the front of the camera lens. In microphotography, it is inserted in between the condenser of the microscope and the light source. It is also used in enlargers while making prints of negatives. The main function of a filter is to alter the quality and quantity of the light. Thus, they modify the colour balance of the light, contrast, brightness, and haze. Colour filters (red, blue, green, yellow, etc.) transmit light of one colour only. They are commonly used in black-and-white photography to enhance and reduce the tone of specific colour.

BOX 15.5 (Continued)

Conversion filters and *light-balancing filters* are used to change the colour balance of the light to suit the colour balance of the film. For example, if a film having a colour balance for tungsten light is exposed to sunlight would yield pictures with a bluish tint. This anomaly can be corrected by the use of an appropriate conversion filter. Similarly, a film with a colour balance for sunlight can be exposed indoors with suitable conversion filters.

Colour compensating filters are used to balance fluorescent light for films meant for daylight or tungsten light.

Skylight or *UV filter* helps to filter ultraviolet radiation, which is not visible to human eye but causes a blue tint in the film. It also serves as a transparent protective cover for the camera lens.

Polarizing filters are used to minimize reflections from glossy surfaces and to intensify the colours, in colour photography.

The reduction in the amount of light reaching the camera lens due to the use of a filter is referred to as *filter factor*. If the camera used is a manually operated one, then the filter factor must be taken into account while setting exposure. Invariably, exposure with slower shutter speed or larger aperture is required when a filter is used.

Artificial illuminations

While photographing indoors, the intensity of the light may be insufficient. Under such conditions artificial illumination such as flash, photofloods and quartz lamp are used.

Electronic flash is the common artificial lighting used for indoor photography. The flash unit is generally an accessory that can be fitted to the camera and the flash and the exposure can be synchronized. Many modern cameras have built-in flash. The power source of the electronic flash is either battery or, if the flash unit is large, electric mains. An electronic flash unit is a glass quartz tube filled with an inert gas, usually xenon. When electricity is applied to the electrodes that are sealed to the ends of the tube, the inert gas inside produces bright light for a very short span of time ranging from 1/1000 to 1/5000 of a second. The burst of intense light and the exposure are

---

BOX 15.5   (Continued)

synchronized at a shutter speed of 1/125 second or less or as indicated in the camera itself. Flash is also used in outdoor photography as *fill-flash*, i.e., to fill in a shadowy foreground in front of the object. Since the colour temperatures of a flash and sunlight are same, there will not be conspicuous colour difference in the pictures. However, when animals are photographed in dim light using a direct flash, the centre of the eye of the subject may appear red in the final print, a condition called *red eye*. Use of indirect flash, i.e., flash directed away from the face of the subject, towards a surface (wall, white screen, etc.) that can reflect the light on to the subject, might help to avoid the red eye, and other harsh, flat lighting effects of direct flash.

Photoflood or floodlight, used in studios and other professional places, produces an intensely bright and broad beam of light.

The illumination required for photomicrography is generally the tungsten lamp, which may be built-in. Special illumination sources such as high-pressure mercury vapour arc lamps and quartz-halogen-tungsten filament lamps may be used for specific purposes. The use of condensers of varying types (bright-field, dark-field, phase-contrast, etc.) in between the light source and the object may also be required for specific conditions.

In addition to the above-mentioned accessories there are others which are useful in recording images of biological interest. *Shutter release cord* or *cable* is attached to the click button of the camera and is used to operate it without causing appreciable vibration. It is very handy in photomicrography as well as in still photography when the camera is mounted on a tripod stand. *Tripod* is a three-legged portable stand, the legs being telescopically collapsible. It can be set with desired height on any uneven terrain. Tripod and shutter release cable together are ideal for photography requiring longer exposures with least vibration.

Operation of a Camera

The components of a standard, manually operated camera are shown in Figure 15.3. The important operations a photographer has to perform while photographing an object are:

- framing of the object
- focusing the object, and
- exposing the film.

Framing of the object refers to locating the object through the viewfinder in proper position in the field. Suppose, we are photographing a wild bird in its natural habitat, we would like to show not only the bird, but also its nest with eggs or chicks, and the location of the nest whether it is on a branch of a tree, in a bush or in the ground. In this case, the framing of the object could be such that the bird perching close to its nest with eggs, which is located in a bush, occupies the centre of the field in the viewfinder. We may have to zoom in or zoom out as necessary so as to capture all the needed details such as the image of the bird, the nature of the nest and its location, the colour and the number of eggs, etc. Here, a close-up of the bird and the nest alone will not be able to show the details of the habitat. On the other hand, a photograph from a long distance showing the entire area with the bird and its nest appearing as a speck in the field will also not help. In both close-up photography and microphotography, framing of the object is important. It is possible to edit the photograph at the time of printing using an enlarger. We can enlarge and centre the object of our interest in the negative. In photography, one has to learn the art of framing the object, by experience.

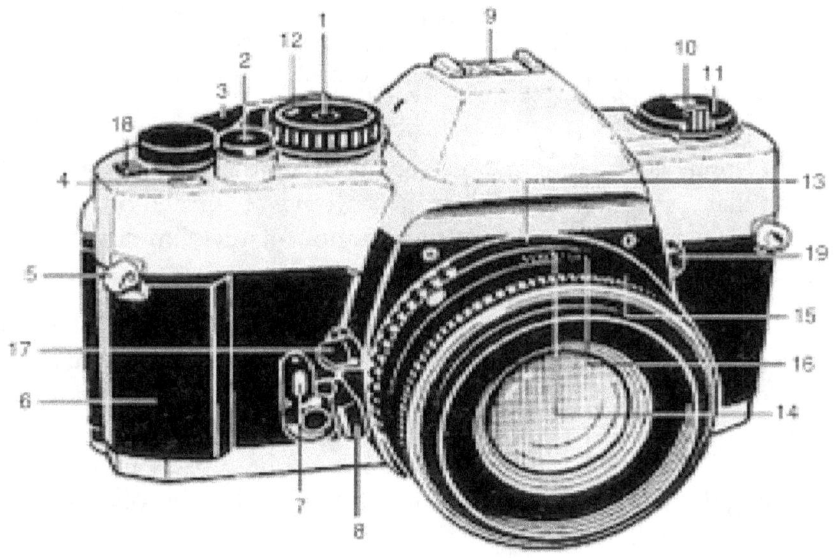

(a)

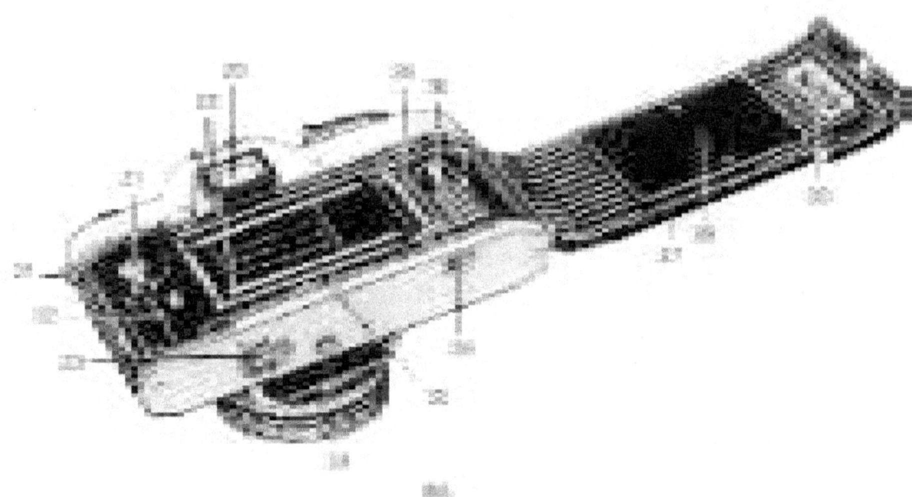

Figure 15.3   Single Lens Reflex Camera. a) front; b) back side open; 1. Shutter speed dial/Film speed set ring; 2. Shutter release button; 3. Film advance lever; 4. Frame counter; 5. Strap hook; 6. Hand grip; 7. Self-timer lever; 8. Lens release button; 9. Hot shoe (for flash unit); 10. Film rewind crank; 11. Film rewind knob/Film compartment opening knob; 12. Film speed window; 13. Aperture ring; 14. Distance scale; 15. Focusing ring; 16. Depth-of-field scale; 17. Preview lever; 18. Multiexposure lever; 19. Exposure measurement button; 20. Viewfinder eyepiece; 21. Rewind shaft; 22. Film chamber; 23. Battery compartment cover. 24. Tripod socket; 25. Film guide rails; 26. Sprocket; 27. Back cover; 28. Film pressure plate; 29. Film rewind button; 30. Film take-up spool; 31. Synchro contact; 32. Databack contact; 33. Film loaded window.

Focusing is adjusting the distance between the camera lens and the film, such that a sharp image of the object would be formed on the film. When the object is in proper focus the position of the film coincides with the position of the real image formed by the lens.

Since the camera lens is a converging lens, the image distance (i.e., the distance between the lens and the film) increases as the distance between the object and the lens decreases. Therefore, in focusing the camera, the lens is moved closer to the film for distant objects and away from the film for close-by objects. This movement of the lens is manually done by turning the lens in a thread mount, with the help of a collar or ring around the lens barrel.

In all the 35-mm SLR cameras, the viewfinders have built-in rangefinders that help in more accurate focusing. A circular area in the viewfinder shows a split image of the object when it is unfocused. The focusing ring (collar) around the lens barrel is adjusted until the split parts of the image become one. For example, when we focus a vertical lamppost, it would appear split into an upper half and a lower half, in the rangefinder area of the viewfinder. When the focusing collar around the lens barrel is adjusted, the two halves would merge and the post would appear as a single vertical post. Many of the modern cameras have automatic focusing facilities, which may have practical difficulties and limited use in biological research. For example, if we were photographing a colony of microbes on a culture medium in a petri plate covered by another petri plate, an auto-focusing camera would only focus the outer surface of the cover plate and not the colony underneath it.

Focusing in microphotography is much more important and has to be most accurate. Even a slight error, in terms of microns (μm), might result in a blurred picture of the object being photographed. It is needless to say that if the preparation of the material (e.g. histological or cytological preparation of a tissue) were not proper (e.g. unevenly spread sections, presence of air-bubble or dust particles under or above the sections, use of thick or bad quality cover glasses, etc.), focusing of the object under the microscope would be difficult. Choice of microscopic lenses, both objective and eyepiece, is also critical in microphotography. They must be spotlessly clean. The objective lens must be, as far as possible, flat surfaced in order to have perfect focusing of the entire area including the most peripheral area of the field. Uniform focusing of the whole area becomes more difficult while using high power objective lenses such as the oil-immersion 100x lens. Therefore, proper framing of the object, i.e., positioning of the object (e.g. a specific area in a tissue) in the centre of the microscopic field, is essential.

Exposure is allowing specific quantity of light rays from the focused object to fall on the film for a specific duration. Control of the quantity and duration of light is effected with the help of the aperture of the diaphragm and time setter, both adjusted in conjunction with each other. The proper exposure of the film to the focused object depends on several factors, the most important being the intensity of light reflected from the object. In bright sunlight, the light reflected from the object would naturally be of very high intensity as compared to that in a shady area or in the interior of a building. Accordingly, exposure with smaller aperture and for a shorter duration in bright light and with larger aperture with longer duration in dull light would be required. The meaning of the aperture size (*f*-stop) and its relationship with exposure time is explained in the Box 15.6.

While focusing an object we may be interested in knowing the other things in front and at the back of the object, which will also appear reasonably focused in the photograph. For example, when we photograph a wild animal, say a tiger drinking water in a waterhole, we would like to show the foreground (area between the camera and the tiger) and the background (area behind the tiger) in good focus. The range in which all objects will appear in adequate focus is the *zone of focus* or, in photographic parlance, the *depth-of-field*. A photograph of the tiger drinking in a waterhole, showing the entire pool of water in the foreground and the grassland with long blades of grass and distant mountain in the background in reasonable focus is said to have *deeper zone of focus* (Figure 15.4). On the other hand, a close-up photograph of the tiger, showing it lapping the water and the foreground and the background appearing blurred, is described to have a *shallow zone of focus*.

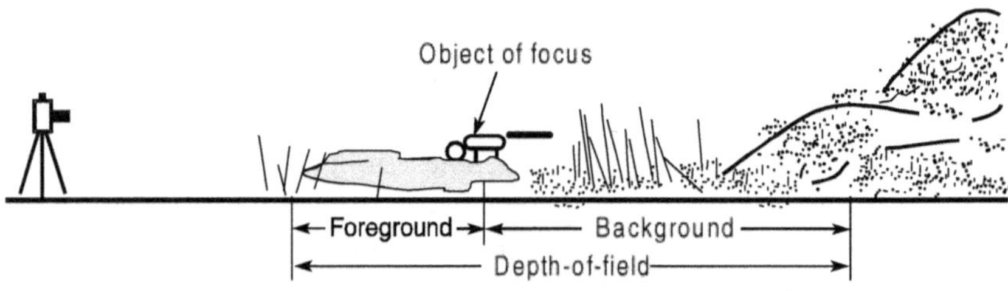

Figure 15.4   Depth-of-field.

BOX 15.6   EXPOSURE—*f*-STOP

The proper recording of the image of the object on the film depends on the correct exposure. *Exposure* is defined as the total light energy that falls on a unit area of the film. In a camera, the exposure is controlled by the *shutter* and *lens aperture*.

The *shutter* of a camera is a spring-activated mechanical device that keeps the light from falling on the film except during the exposure period. It regulates the time interval during which the reflected light from the object enters the lens. It may be of the *focal-plane type* or *leaf type*. A focal-plane type shutter has a black shade with a slit across its width. At the time of exposure, the shade moves across the film, exposing it progressively as the slit moves. In the case of the leaf shutter, when released, a group of interlocked blades move apart to expose the film to the incoming light rays. The *shutter speed*, i.e., the duration for which the shutter is kept open in order

BOX 15.6   (Continued)

to let the light pass through, is adjusted in steps corresponding to factors of about two, generally from 1 second to 1/1000 second. The shutter speed dial includes a "B" setting (B, originally, for bulb, a rubber bulb, which when squeezed keeps the shutter open, by air pressure, until the pressure is released; B now stands for " Brief time"), which enables to keep the shutter open for a duration as long as the shutter release button remains pressed. It is useful in photography in a very dull light and in photomicrography. Use of a tripod stand and a shutter release cable or remote control (infrared or radio) might ensure vibration-free exposure under B setting.

The size of the aperture, i.e., the diameter of the lens opening, determines the amount of light that should reach the film and is controlled by the lens diaphragm. Diaphragm is a mechanical device consisting of several thin overlapping metallic blades, operated by a narrow ring (aperture ring) around the lens barrel. The aperture diameter can be increased in steps from a minimum to a maximum that is equal to that of the lens. The aperture size is measured in numerical settings called *f-stops* or *f-number*. The usual *f*-stops are $f/2$, $f/2.8$, $f/4$, $f/5.6$, $f/8$, $f/11$, and $f/16$. The values of the *f*-number and the diameters of the aperture are inversely proportional. That is, the minimum *f*-number ($f/2$) corresponds to maximum aperture diameter and the maximum *f*-number ($f/16$) to the minimum aperture diameter.

The *f*-stop represents the light-gathering capacity of the lens and is the ratio of the focal length of the lens (f) to the aperture diameter (D).

$$f\text{-stop} = (\text{Focal length})/(\text{Aperture diameter}) = f/D$$

For instance, a camera lens having a focal length of $f = 50$ mm and an aperture diameter $D = 25$ mm, is described to have an *f*-stop of 2 ($= 50$ mm/25 mm), or an aperture of $f/2$.

The larger *f*-numbers represent smaller apertures and exposures, and each step corresponds to a factor of two in intensity. The actual exposure (total amount of light reaching the film) is proportional both to aperture area and the time of exposure. Thus $f/4$ and 1/500 s, $f/5.6$ and 1/250 s, and $f/8$ and 1/125 s all correspond to the same exposure.

The depth-of-field is determined by three factors, viz., lens aperture, focusing distance and focal length. Wide-angle lens with short focal length when used with a small aperture yields photographs of greater depth-of-field. Conversely, telephoto lens with long focal length used along with a larger aperture would produce photographs of lesser depth-of-field. In SLR cameras, the depth-of-field range at different aperture settings is obtained using the "depth-of-field scale" located between the focus ring and the aperture ring. It is engraved with *f*-stop values in pairs on either sides of central marking. For an object focused at a particular distance with a specific *f*-stop, the depth-of-field is read in the distance scale in the focusing ring. It corresponds to the range that lies between pairs of *f*-stops marked in the depth-of-field scale. For example, if an object is focused at a distance of 3 metres with an aperture set of *f*/8, the depth-of-field range is from 2.4 metres to 4.5 metres. These values (2.4 and 4.5) are read in the distance scale, corresponding to *f*/8 and *f*/8 in the depth-of-field scale. The meaning of these values is that in a photograph that is taken with an aperture of *f*/8, all objects within the range between 2.4 metres and 4.5 metres will appear acceptably focused.

DIGITAL PHOTOGRAPHY

All the basic information discussed above generally relate to cameras using celluloid film. The arrival of digital technology has revolutionalized the entire concept of photography. Digital cameras have entered the field of biological research and are making both macro-and microphotography much easier. A brief description and discussion about the various aspects relating to digital photography is presented in Box 15.7.

BOX 15.7   DIGITAL PHOTOGRAPHY

Digital photography is capturing images of objects in a digital format. This can be achieved either by scanning film-based photographs or by capturing images by a digital camera. A digital photograph is computer-based and is made up of minute squares called picture elements or *pixels,* numbering from several thousands to millions. A computer screen is divided into a grid of pixels and the values stored in a digital photograph are translated to specify the brightness and colour of each pixel.

BOX 15.7   (Continued)

The process of controlling individual pixels in terms of digital values is called bit mapping and the images thus formed are called *bit-maps*. The same principle applies to printing of digital images.

The *resolution* of the digital image, both as seen on a computer monitor and as printed on a paper, depends on the number of pixels used to make the image. Larger number of pixels give better quality to the image. If a digital image is enlarged past a certain point, the individual pixels of the image will become visible, an effect called *pixelization*. This effect is very similar to the appearance of grains in very outsized enlargements of film-based photographs.

The size of a digital photograph is specified in terms of its dimensions in pixels or by the total number of pixels it contains. For example, an image can be described to have 1800×1600 pixels (number of pixels in a row by number of pixels in a column) or to be made of 2.88 million pixels (number of pixels in a row multiplied by that in a column).

There are many advantages of digital photography over the conventional film-based photography. (1) Digital photography is an instant, economical and timesaving photography. As soon as the image is captured, a digital photograph is in a ready format that can be viewed, printed, copied any number of times, or distributed through e-mails and websites. (2) It can be edited with special software such as "Photoshop" or "Paint". (3) Since no film is involved, pollution due to toxic chemicals used in "darkroom" is avoided.

Digital cameras are similar to 35 mm film cameras with features such as lens, aperture, shutter, etc. However, they do not use films. Instead, they use a fingernail-sized silicon chip called *image sensor*. These light-sensitive sensors are technically referred to as charge-coupled devices (CCD). The surface of a CCD is a grid containing several photosensitive diodes called *photosites* or *photoelements*. Each photosite is designed to record a single pixel in the photograph that is being taken, that is, each photosite of a CCD is analogous to a pixel of a digital photograph.

BOX 15.7   (Continued)

During the brief exposure, each photosite on the CCD records the brightness of the light that falls on it by accumulating an electrical charge. If the intensity of the light that hits a photosite is more, the charge it records will be higher. Therefore, photosites capturing light from brighter regions of the object will have high charges. On the other hand, photosites that record light from shadowy and darker areas of the object will have proportionately low charges.

Immediately after the exposure, the charge from each photosite is measured, converted into a digital number and stored. The series of numbers thus stored is then used to reconstruct the image by transcribing the colour and brightness of photosite to the pixels of the digital image on the screen or printed page.

The photosites of an image sensor capture only the brightness of light and not the colour. They record as many as 256 increasingly darker shades ranging from pure white to pure black. These shades are converted to original colours by the use of red, green and blue filters over the photosites. Thus, each photosite records the brightness of the light that matches the specific filter placed over it. For example, a photosite with a green filter records only the brightness of green light. A process called *interpolation*, involving millions of calculations and performed by microprocessors of the camera, is used to recreate the real colour in each photosite. In this process, colours of the neighbouring photosites, which may be as many as eight and have different filters, are used to calculate the two colours that the given photosite did not record directly. Putting together the information about the two interpolated colours and that about the colour registered directly, the true colour of a photosite is calculated. The process of interpolation is completed within a few seconds after exposure and makes it possible for the camera to capture, preview, compress, filter, store, transfer and display the image.

Generally, three types of digital cameras are recognized according to the resolution capacity of their image sensors and other maneuverable facilities. Automatic point and shoot digital cameras with resolutions of 3 to 4 million pixels are economical and suited for general purposes

BOX 15.7   (Continued)

requiring smaller prints. The next level of digital cameras, referred to as *prosumer* cameras, have image sensor resolution of about 5 to 9 million pixels and facilities such as through-the-lens focusing and creative controls. They can yield prints up to about 8 × 10 in size. The third category of digital cameras is the professional cameras having an image sensor resolution of 6 to 12 million pixels, and therefore can generate larger prints. These cameras are designed similar to 35-mm film SLR cameras and possess most of their features such as exposure controls and accessories. The interchangeable lenses of 35-mm camera can be fitted to these cameras.

However, it must be noted that the resolution of digital camera is much lesser than that of a 35-mm film (estimated to be about 20 million pixels) and that of our eye (estimated to be about 120 million pixels).

Chapter Sixteen

Intellectual Property Rights

INTRODUCTION

We are familiar with "property rights", which are the rights we legally claim over our own, parental, grandparental and ancestral properties such as land, house, jewels, money, etc. These properties are tangible properties, meaning that they are materials that can be seen, touched and easily perceived. Intellectual property (IP) refers to creations of the mind such as inventions (e.g. a software program) and artistic (e.g. a composition of music) and literary (e.g. a poem, novel, essay, etc.) creations. The legal right of the ownership of such intangible intellectual properties is the Intellectual Property Right (IPR).

There is lot of argument over the definition of the terms "intellectual" and "property", but there is general agreement for the necessity to protect the rights of intellectual creations. International organizations such as UN's World Intellectual Property Organization (WIPO) and World Trade Organization (WTO) are striving to pressurize their member countries to enact stringent laws to protect IPR.

The statutory IPR once granted to a person (or persons, company, institution, etc.), he (or she) becomes the owner of the intellectual property and has the right to exclude others from exploiting the same commercially for a given period of time. The IPR permits the owner to enjoy the benefits when the property is exploited commercially. IPR is granted to an inventor, creator or designer and not to the discloser of the property.

All the arguments for and against IP and IPR are due to the monetary benefits it might bring to its owner. If there is no such benefit many owners may not care about their IPs. The value of an IP is determined by the financial income it might produce and not by its cost, because the two, value and cost, are independent. For example, a musical composition might have taken only a short duration and very little expenditure for its creator, i.e., its cost is less. But when this composition is commercially exploited (movies, cassettes, CDs, etc.) the return is enormous, i.e., its value is very high. Likewise, a software program, which might have costed very less in terms of money and time, may carry a very high value.

A problem associated with high value of an IP is piracy, such as the piracy of software, music, cinema, books, etc. For example, according to an investigation, about 80 billion dollar worth software is installed on computers worldwide, and only 51 billion dollar worth of it is licensed and the rest is pirated. The main reason for this high incidence of piracy is the prohibitively high cost of the software. The IP piracy is more prevalent in countries like India, China, Thailand and Russia.

Another issue associated with IP is the unethical exploitation of the traditional knowledge and biodiversity of a country. This issue still continues to be debated in various forums and fought in legal institutions.

Our discussion hereunder will confine to the definition and brief explanations of the important terms associated with IPR, the status of IPR in India, and aspects relating to biological inventions, biodiversity and biopiracy.

PROTECTION OF IPR IN INDIA

India has a well-established statutory, administrative and judiciary framework to protect intellectual property rights relating to patents, trademarks, copyrights, and industrial designs. This framework is continuously strengthened by enactment of additional laws and by amendment of the existing laws not only to protect India's own interests but also to comply with obligations to TRIPS (Trade Related Intellectual Property Rights) agreement of WTO. Some of the governing laws and their amendments, for IPR, are as follows:

1. The Patents Act of 1970, the Patents (Amendment) Act of 1999, the Patents (Second Amendments) Bill, 1999, and the Indian Patent Act, 2002.

2. The Trademarks Bill of 1999, repealing and replacing the Trade and Merchandise Marks Act of 1958.

3. The Copyright Act of 1957 and the Copyright (Amendment) Act of 1999.

4. Geographical Indications Goods (Registration & Protection) Bill, 1999, and a *sui generis* legislation for the protection of geographical indications.

5. The Industrial Designs Bill of 1999 replacing the Designs Act, 1911, and Plant Variety and Farmers Right Protection Act, 2001.

6. Bill for the Protection of Plant Varieties and Farmers' Rights, 1999.

7. The Biological Diversity Bill, 2000.

8. The Design Act, 2000.

Apart from the amendments to the legislations, the Government of India is taking several steps to streamline and strengthen the IP administration system. Modernization of the patent information services and trademark registry have already been carried out. Another project for the modernization of the patent office includes facilities for human resource development, recruitment of additional examiners, infrastructure facilities, computerization and elimination of backlog of patent applications.

The Trademark Registry is also further strengthened and modernized with emphasis on strengthening the infrastructure of Registry, removal of backlog of pending applications, transfer of records to CD-ROMs, and appointment of additional examiners.

In addition to upgrading and strengthening the existing laws, the Government of India is paying serious attention to the effective enforcement of these laws. It has intensified raids against infringers of patent and copyright laws. Special cells to enforce copyright laws have been set up in 23 States and Union Territories. Copyright societies have also been established for collective administration of copyright.

At the international level, India is taking active participation in conventions and conferences in order to protect the interests of the inventions, traditional knowledge, biodiversity and other IPR related matters of not only its own but also those of developing countries, in general. India is signatory of the following conventions.

1.   Bern Convention for Copyright—1886
2.   Paris Convention for the Protection of Industrial Property—1967
3.   Strasbourgh Convention on International Patent Classification—1971
4.   Convention on Biological Diversity (CBD)—1992
5.   General Agreement on Trade and Tariff (GATT)—1995
6.   Patent Cooperation Treaty (PCT)—1998

There are diverse views on the usefulness of the IPR and its legal protection. According to some, such protection will hinder a community's ability to discover new products. On the other hand, others feel that because of the monetary gains involved, IPR protection would attract more and more enterprising scientists to take up research and development activities. The Government of India is making all efforts to protect IPR, especially in the fields of computer software, cinematography, and designing.

TERMINOLOGY ASSOCIATED WITH IPR

Patent

A patent is granted by the Government to the creator of an invention, conferring the exclusive right to make, use and sell that invention for a limited period of time, usually 20 years provided a maintenance fee is paid annually. The invention for which the patent is given must be a product or process, and must have novelty, creativity and utility. A patent is registered in the office of the Controller General of Patents, Designs and Trademarks.

Territoriality of patent   A patent is protected by law. Patent laws are territorial, i.e., a separate patent should be obtained in each country. Thus Indian patent law protects inventions only in India.

Patentability search   To obtain a patent for an invention, patentability search is necessary. A patentability search is a search (e.g. through computer database) in order to ascertain that the product for which the patent is to be claimed was not already patented elsewhere and that it is not a traditional knowledge. Patentability search requires the following detailed information about the invention:

i)   Illustrations or photographs of the invention, showing how it is made, operated and used

ii) Any known earlier similar design

iii) A summary of the shortcomings of the earlier designs with an explanation of how these shortcomings are overcome in the present design

iv) List of other advantages, possible applications, etc.

v)   Details of any possible variants or modifications that could be made without deviating from the general concept of the invention.

Applicant of patent   The person who is applying for a patent for an invention should be a true and first inventor who holds the rightful ownership of the invention. Any person who is an assignee or legal representative of the first and true inventor can also apply for the patent. In case of the death of the first and true inventor, the person's legal heir can apply for patent.

What is not patentable   The following are not eligible for patenting:

i)   an invention that is frivolous or that claims anything contrary to established natural law

ii) the mere discovery of a scientific principle or the formulation of an abstract theory

iii) an invention, the use or commercial exploitation of which could be against the public

iv)   the mere discovery of any new property or new use for known substance or of the mere use of known process, machine or apparatus unless such known process produces a new product or employs a new reactant

v)   a substance obtained from mere admixture resulting into aggregation of properties

(vi)   mere arrangement or rearrangement or duplication of known devices each functioning independently

vii)   a traditional method of agriculture or horticulture

viii)   any process for the medicinal, surgical, curative, prophylactic or other treatment of human beings or animals

ix)   plants and animals (microorganisms exempted) in whole or any part thereof including seeds, varieties and species

x)   any biological processes for production or propagation of plants and animals

xi)   a computer program per se other than that of its technical application to industry or combination with hardware

xii)   a mathematical method or business method or algorithm

xiii)   a literary, dramatic, musical or artistic work or any aesthetic creation whatsoever including cinematographic works and television production

xiv)   a mere scheme or rule or method of performing mental act or method of playing game

xv)   a presentation of information

xvi)   topography of integrated circuit

xvii)   an invention which in effect is traditional knowledge or which is an aggregation or duplication of known properties or traditionally known component or components

xviii)   invention relating to atomic energy

Patent specification   A patent specification gives details of the inventions for which the patent protection is sought. The legal rights in a patent are based on the disclosures made in the specification. Patent specifications are of two types: provisional and complete. A provisional patent specification makes known incomplete invention requiring time to develop further. The purpose of provisional patent specification is to claim priority date of an invention.

The complete patent specification is a document that provides detailed description of the invention along with the drawings and claims. The description regarding prior design is also included in the complete specification.

Date of priority   The date on which the patent application either with the provisional specification or complete specification is filed at the patent office is called the date of priority.

Granting of patent   First, a patent examiner peruses the patent application and then communicates the objections, if any, to the applicant through the first examination report. The applicant must provide necessary explanation for the objections and comply with requirements of the patent office within the specified period. If the applicant fails to do so, the patent application will be abandoned. If all the conditions are complied with, the application will be published in the patent gazettes issued by the patent office. The published application is open for public scrutiny, and if there is no opposition from any quarter, the patent shall be granted.

Prior Inform Consent   Prior inform consent (PIC) is a consent obtained from an inventor or innovator or knowledge holder to develop, protect, explore, and commercialize an invention or innovation.

Invention and innovation   An invention is a new product or process involving a creative step and capable of industrial application. An innovation is a successful exploitation of new ideas and concepts, not necessarily one's own. All inventions are patentable, while all innovations are not patentable.

Patent cooperation treaty   The patent cooperation treaty (PCT) is an international treaty that facilitates a patent applicant to file a single patent application, specifying the countries in which he/she wants patent for his/her IP. Thus a single application is filed for the purpose of an international patentability search report and claim the priority date in all the specified countries. Following the receipt of the international examination report, the applicant can apply to each of the specified countries to consider his/her patent claim. This process is called national phase of a patent application. India is a member country to PCT.

Copyright

Definition   Copyright is an intellectual property protection granted under Indian law to the creators of original works such as

    i)   literary

ii) dramatic

iii) musical

iv) artistic

v) cinematographic films

vi) sound recordings

With respect to copyright, the word 'original' does not mean that the work must be the expression of the original or inventive thought of the creator. The Act governing the copyright does not require that the expression must be in an original or novel form, but the work must not be copied from another work and it should originate from the creator. The expenditure of original skill or labour is required in executing a work and not in the originality of thought. The term 'literary work' includes computer programs, tables and compilations including computer databases.

Copyright legislation   The provisions relating to Copyrights and their protection is governed in India by the Copyright Act, 1957. Copyright is granted to an expression of an idea and it is not a right in the novelty of an idea. It aims to protect skill, labour, and capital employed by the creator. It protects the creator from unlawful reproduction, plagiarism, piracy, copying and imitation. Only violations with regard to the form, manner, arrangement and expression of the idea by the creator are considered as infringement of the copyright.

Rights of the copyright holder   The copyright provides the creator the right for reproduction, public performance, recording, broadcasting, translation or adaptation. The copyright holder can sell the copyright in return for royalties, but always retains the "moral right". Moral right includes the right to claim authorship of a work, and the right to oppose changes to it that could damage the creator's reputation.

The rights granted under copyright may not hold under "fair use" conditions. For example, a copyrighted material can be used for research or non-commercial purposes, with due acknowledgement to the source.

Copyright of computer programs   At present, in India, computer programs are protected under the Copyright Act, 1957. Again, this is confined to its form and expression. The Act gives protection against piracy of software. However, Patent is a better option to protect a software as an invention or new idea. Accordingly, many IT companies and software developers file for patent in addition to copyright, for the same work.

Registration of copyright   A copyright need not be registered; it exists from the moment the work is created. However, registration is an evidence to support a person that he or she is the copyright holder of a particular creation. The following procedure has to be followed for the registration of copyright in India.

i)   Application should be made in triplicate, in specified format, to the Registrar of Copyright.

ii)   If an artistic work is to be registered and it can be used in relation to some other product (as a logo, emblem, mascot, trademark, etc.), then the application should include a statement to that effect. Further, a certificate from the Registrar of Trademarks to the effect that no Trademarks identical with or deceptively similar to the artistic work for which copyright application is made exist, must be attached.

iii)   Only one application should be filed for one work. The fees for registration is prescribed as per Copyright Rules of 1958, and it varies according to the nature of the artistic work. The prescribed fees should be paid along with the application.

iv)   It is the responsibility of the applicant to simultaneously provide copy of the copyright application to all the persons (e.g. co-authors, publishers) interested in the copyright of the work.

v)   A period of 30 days from the date of filing copyright application is allowed to receive objection from anybody regarding grant of copyright to a particular work. If the Registrar of Copyrights does not receive any objection during this period, he shall verify the correctness of the particulars given in the application and then enter these particulars in the Register of Copyrights.

vi)   If the Registrar receives objections against the grant of copyright to a particular work and/or he is not satisfied with the correctness of the particulars given in the application, then he may conduct an enquiry by way of calling for explanation from the applicant or by any other means which he may deem fit. Only if the Registrar is fully satisfied, entries will be made in the Register of Copyrights. Copies of the entries will be sent to all the parties concerned.

Duration of copyright   A copyright is granted for published literary, dramatic, musical or artistic work for a period of the life time of the creator plus 60 years from the beginning of the next calender year after the death of the copyright holder. If the copyright is held by more than one, the period of 60 years shall commence after the death of the last copyright holder.

Display of copyright A copyright notice is displayed with

i) the symbol © (letter c in a circle)

ii) the year of first publication, and

iii) the name of the copyright owner, as in, for example, © 1998 WHO.

Licensing of copyright A copyright holder can grant license to a third party to use his work. The license must be in writing and should have been signed by the copyright holder or by his authorized agent. A license should specify the following particulars:

i) Identification of the work

ii) Duration of the license

iii) The specific right (e.g. translation, copying, printing, etc.) licensed

iv) Territorial extent of license

v) The quantum of royalty amount payable

vi) The specific terms regarding revision, extension and termination of the license.

Protection against infringement of copyright Any of the following acts is construed as infringement of copyright:

i) Doing anything to the work without license from the copyright holder

ii) Providing any space (e.g. lecture hall, theatre, etc.) to be used for the infringement of a copyright.

iii) Selling, hiring, distributing, exhibiting in public or importing into India any copy of the work in violation of the copyright.

The Copyright Act, 1957 provides for both civil and criminal remedies against any violation of a copyright. The remedy may be injunction, damages or share of profits, delivery of violating copies and damages for conversion. According to this Act, courts can punish the copyright violators for criminal offences. The Police are empowered to confiscate, without warrant, pirated copies of the work including the equipment used for making such copies.

Trademark

Definition A trademark (TM) is a word, name, symbol, logo, slogan or other distinctive sign or a combination thereof identifying and distinguishing a product or services from other similar products. It is officially registered at the trademark office, and legally restricted to the use of the owner or manufacturer of the product. A trademark gives protection to the owner of the mark by ensuring exclusive right to use it to identify goods or service. The owner of a trademark can authorize another person to use it in return for money. Trademarks for services such as banking and insurance are called service marks (SM).

***TM, SM and* ®**   The abbreviations "TM" or "SM" are used as notations to inform third parties that the person claims trademark or service mark rights to the word, slogan and phrases. Usually these notations are used as superscripts as in, for example, Philips™. They can be used once the application for registration has been filed. After the grant of the trademark right, the symbol ® can be used, as in, for example, Microsoft®.

Tenure of Trademark   The duration of the exclusive right to keep a trademark is 10 years but the mark can be permanently kept on the Register by renewing it after every ten years.

Design

The design of an article can be registered with the Controller General of Patents, Designs, and Trademarks. The term 'design' refers to two-dimensional and three-dimensional aesthetic aspects of an article. Two-dimensional features may be patterns, lines or colours, and three-dimensional features may be the shape or surface of an article. Design registration does not cover the mechanical feature of the article.

Geographical Indication

A geographical indication (GI) is a name or sign used on certain products that correspond to a specific geographical location or origin (e.g. a town, region or country). The use of GI may serve as a certification that the product has certain qualities, or enjoys certain reputation because of its geographical origin. More importantly, GI also serves to preserve cultural traditions of a locality and to protect community rights.

Every geographical location on earth has its own peculiar environmental features (soil, climate, fauna, flora, etc.). Because of these features a product produced in that region may possess certain peculiar features like flavour, taste, texture, medicinal property, etc. Our country has a rich heritage of products originating from specific regions that are nurtured by knowledge and tradition built by communities over the years. For example, Tea produced in Darjeeling may have distinctive flavour and taste that may not be found in the Tea from other regions. In such cases one would like to protect the use of the name of geographical location from where a product originates. A few other popular examples of GI are: Champagne, Swiss watches, Indian carpets, Kashmir shawl, Kancheepuram silk, and Punjab wheat.

Legal protection for GI   A number of international treaties sponsored by WIPO (World Intellectual Properties Organization) provide for protection of geographical indications. In addition, the agreement on TRIPS (Trade Related Aspects of Intellectual Property Right) deals with the international protection of GI within the framework of the WTO (World Trade Organization).

In India the Geographical Indication Act 2001 defines a geographical indication and according to this definition a GI should have the following characteristics:

i)   It must indicate that a product (e.g. agricultural goods, natural goods, or manufacturing goods) is originating in a territory of country, or region or locality in that territory.

ii)   The quality, reputation or other characteristics of such a product must be due its specific geographic origin.

iii)   If a product is a manufactured one, then at least one of the activities involved in its manufacture should take place in the specific territory.

Applicant of GI   Any association of persons, producers, organization or authority (appointed by the Government) can apply for a GI. The applicant must represent the interest of the producers of the product (e.g. the local community of the territory).

Registration of GI   The application in writing in prescribed form and along with the prescribed fees should be submitted to the Registrar of Geographical Indications. The information required in the application includes:

i)   the class of product (whether it is food, textile, etc.)

ii)   the territory

iii)   particulars of appearance

iv)   particulars of the producers of the product

v)   an affidavit how the applicant claims to represent the interest of the producers of the product

vi)   the specific and general characteristics of the GI

vii) the details of the special characteristics of the GI

viii) detailed description of the boundary of the territory

ix)   certified copies of the map of the territory

x)   number of producers

xi)   particulars of inspection structures, if any, to regulate the use of GI.

On receipt of an application for GI, the Registrar initiates and carries out the following:

i)   allots a number to each application

ii)   scrutinizes it to check whether it meets the requirements of the Act and Rules

iii) constitutes a Consultative Group of experts to ascertain the correctness of the particulars furnished in the application

iv)   issues an Examination Report to the applicant requiring for explanation and clarification for the points raised in the report

v)   considers the explanations given by the applicant

vi)   if he is satisfied with the application in all its aspects, the Registrar advertises the details of the product for which GI is sought in the Geographical Indications Journal, so that anybody concerned can raise objections to granting the specific GI, within a period of four months

vii) if there are no objections, the Registrar registers the GI in Part A of the GI Registry

After the registration of the GI in Part A, any person who produces the product files a separate application to the Registrar to register him in Part B of the Registry, as an authorized user of the GI for his product.

Protection for registered GI   A registered GI is protected by a wide range of laws such as

i)   laws against unfair competition

ii)   consumer protection laws

iii) laws for the protection of the certification marks

iv)   special laws for the protection of GIs or appellations of origin.

According to these laws, unauthorized persons shall not use the GI to mislead the public. Any violation of the GI use will attract judicial sanctions such as court injunctions preventing the unauthorized use, payment of damages, imposition of fine, and, in extremely serious cases, imprisonment.

Assignment and transmission of GI   A geographical indication is a public property belonging to the producers of a product in a specific territory, and therefore, assignment, transmission, licensing, pledge or mortgage is not permissible.

Duration of a GI   The validity of a registered GI is for a period of 10 years. It can be extended by further 10 years by renewing the registration periodically.

Plant Variety and Farmers' Rights Protection

Plants are not patentable in India. However, they are protected under the provisions of the Plant Variety Protection and Farmers' Rights Act. Under this Act, breeders have rights to protect a new variety of plant and crop. The right includes the right to produce, sell, market directly or export, a variety. This Act also defines the farmer's rights. It specifies that a farmer is entitled to save, use, sow, resow, exchange, share, or sell his farm products including a variety of seed protected under this law. However, a farmer is not entitled to sell a branded seed variety.

Trade Secrets

There is a lot of information that is classified as "Confidential Information" such as trade secrets, literary and artistic secrets, personal secrets, and public and Government secrets. Confidential information is meant to be kept as secrets so that the holder of the confidential information will continue to exclusively enjoy the benefits arising out of that secret. Among the various confidential information, trade secrets deserve some detailed discussion.

A trade secret is any information that allows the person who holds the secret to make money because others do not know the information. It could be a formula, computer program, process, method, device, technique, customer list or other non-public information.

One of the famous examples of a trade secret is the formula of Coca-Cola. The formula of this soft drink has a code name "Merchandise 7X". It is known to only a few people and protected in the secret vault of a bank in Atlanta, Georgia, US. The individuals who are aware of the secret formula have signed an agreement not to disclose the secret. Prior to 1991 the law in India required the disclosure of the trade secrets and therefore Coca-Cola could not be marketed in our country. Subsequently, India amended its law on trademarks and trade secrets such that there is no need to disclose trade secrets, and now Coca-Cola is sold.

Trade secrets are quite different from patents, copyrights and trademarks. The patents and copyrights require their holders to disclose their information in the application process and the information ultimately becomes public. Trade secret, on the other hand, requires the holder of the secret to keep the information secret. The duration of trade secret is much longer than those of patents (20 years) and copyrights (60–100 years), it can potentially last forever.

Trade secrets are valid only as long as no one else discovers the information and makes it public. If the trade secret is made public in violation of the non-disclosure agreement, the right holder can sue for damages. However, the secret status of the trade secret is permanently lost.

At present there is no legislation in India to protect trade secrets. However, as a part of its obligations under WTO (World Trade Organization) India has accepted to implement the conditions of the TRIPS (Trade Related Intellectual Property Rights agreement) to protect undisclosed information.

India has a vast repository of knowledge and practices all of which are maintained as secrets by traditional medical practitioners (e.g. Siddha), hakims, artists and artisans. These secrets have high potential commercial value, and therefore the legal protection of these secrets is important for the survival of these systems and practices. One way to protect these secrets is to legislate *sui generi* systems. Such systems will deter illegal transfer of trade secrets by people, especially those who are employed in the concern and have access to the secrets. These legislations along with provisions under Breach of Contract or Non-disclosure Agreement would help to protect the trade secrets and undisclosed information as property of their owners.

OTHER ISSUES RELATING TO IPR

There are several other issues relating to the protection of IPR, which are not strictly covered under law and about which a lot of debate is going on at the national and international levels. We shall briefly consider some of such issues.

Copyleft

A jargon of the term Copyright, the Copyleft is a movement that questions underlying philosophy of IPR. Its claim is based on the belief that creations of the mind are common to everybody on earth and not the right of any individual or organization. According to the proponents of this concept, knowledge multiplies by sharing, and therefore it should be freely available. The Copyleft movement is of the opinion that the IP laws are not in favour of the scientists or authors, but serve the interests of large corporate information monopolies. An example of the Copyleft movement is the open source software, where the source code of computer software is provided in the public domain so that any one can build and improve upon it. Thus the Copyleft movement preaches that the creations of mind, such as software, literary works, etc. should be open to everybody so that anyone can modify or add to it under the condition that the new, improved version also remains open to all.

Traditional Knowledge

Traditional knowledge (TK) is the knowledge that has been continually developed, acquired, used, practised, transmitted and sustained by the communities and/or individuals through generations. Traditional knowledge includes artistic expressions, agricultural methods, and knowledge about the medicinal properties of plants and animals. For instance, there are reports of modern music composers commercially exploiting folk songs of different indigenous communities. Likewise, there are also attempts to patent indigenous medicines, agricultural products, etc. Existing IP laws are inadequate to protect the traditional knowledge. Every effort should be made by both Government and the community concerned for the improvement of the existing knowledge such that they can meet the requirements of the definition of the invention and thus qualify for patenting.

Biodiversity

Most of the traditional knowledge is related to the biodiversity of a region. As defined by the Convention on Biological Diversity (CBD), biodiversity is the variability among the living organisms (animals and plants) from all sources (land, sea and freshwater) and the ecological complexes of which they are a part. Thus biodiversity includes diversity within and between species, and the diversity of the ecosystem.

Developing countries like India and many African and South American countries are rich in biodiversity. The value of the biodiversity is very high because it serves as a feedstock of medicine, industry, healthcare and social development. Therefore there is an imperative need to protect biodiversity against its destruction and unethical exploitation. Thus the biodiversity-rich countries have the responsibility to preserve and sustainably use their biodiversity.

The first initiative for international cooperation to protect biodiversity came in 1992 with the Convention on Biodiversity. In India the Biological Diversity Bill was introduced in 2000 to implement the provisions of the CBD. The important features of the Bill are as follows:

1. Regulation of access to biological resources of the country with the purpose of securing equitable share in benefits arising out of the use of biological resources and associated knowledge relating to biological resources

2. Conservation and sustainable use of biological diversity

3. Respecting and protecting the knowledge of local communities, related to biodiversity

4. Ensuring that the benefits arising out of the biodiversity are shared with local people because they are the conservers of biological resources and holders of knowledge and information relating to the use of biological resources

5. Conservation and development of areas that are important from the standpoint of biological diversity by declaring them as biological diversity heritage sites

6. Protection and rehabilitation of threatened species

7. Involvement of institutions of self-governments in the broad scheme of the implementation of the act through constitution of committees.

In addition to the Biological Diversity Bill, the Protection of Plant Varieties and Farmers' Rights Bill strives to stimulate investments for the growth of the seed industry and ensure the rights of both plant breeders and the farmers. Creation of a Community Biodiversity Register, a database, of local products and their uses would help in regulation of the exploitation of these bioresources by pharmaceutical and other industries with appropriate royalties to the community.

Biopiracy

Most of the traditional knowledge of a geographical territory is based on its biodiversity. Countries rich in biodiversity are also rich in traditional knowledge. In this regard, biodiversity is linked to IPR and TRIPS agreement through the concept of traditional knowledge. Biodiversity-based traditional knowledge is the unregistered and undervalued IP of the indigenous or local communities. This knowledge has in fact contributed a lot for the development of new therapeutic drugs. The issue is that the individuals and companies patent this biodiversity-based traditional knowledge as their own "invention" with an aim to make money. The community or the country from where the knowledge was "stolen" is in no way benefited. This unethical practice of monopolizing a biodiversity-based traditional knowledge is called the Biopiracy.

We shall consider three cases of biodiversity related to traditional knowledge of India, viz., turmeric, neem, and basmati rice. With regard to these three plants, patents were granted in the United States of America. The patenting of the knowledge of these plants is described as Biopiracy, for the following reasons:

i) These plants have been traditionally used for ages in India for various purposes because of their medicinal and agricultural values.

ii) Because of the grant of patent to knowledge relating to these plants, the local communities can be legally prevented from the commercial exploitation of these plants. If they want to use them, they will have to pay royalties to these companies.

iii) It is quite possible, because of the patents, the public may be deprived of cheap traditional medicine, or a pesticide or rice.

There are also strong counter arguments. The defenders of these patents claim the intellectual property protection (patents, or some other form of protection for plant varieties, for example) in these areas as a means of promoting development through research. They also reject some of the assumptions and conclusions of the criticisms—for instance, that patenting means higher costs.

Turmeric   This plant belongs to ginger family and its botanical name is *Curcuma longa*, and vernacular (Tamil) name, *Manjal*. The saffron coloured rhizome of turmeric has several useful properties and therefore used, for example, as

i)   a spice and colouring agent in cooking

ii) an ingredient in medicines (Ayurvedic)

iii) cosmetic and

iv) a colour dye for leather and textile

The turmeric is traditionally used to heal wounds and rashes. Further, it has antibacterial and antifungal properties and, therefore, used in food preservation.

Two Indian emigrants working at the University of Mississippi Medical Centre, USA, were granted patent in 1995 for turmeric to be used to heal wounds. Following protest in India and elsewhere, CSIR (The Indian Council for Scientific and Industrial Research) filed a case with US Patent Office challenging the patent on the grounds that the medicinal healing potency of turmeric is a "prior art", i.e., turmeric has been used for thousands of years for healing wounds and rashes, and therefore its use as a medicine was not a new invention. The CSIR produced an ancient Sanskrit text and a paper published in the Journal of the Indian Medical Association, in support of its case. The objection of the CSIR was upheld by the US Patent Office, and the patent was cancelled. This case is an example of how Biopiracy can be challenged and revoked. The main reason why a patent for a traditional knowledge was awarded at all was probably the absence or inadequacy of the database on such traditional knowledge, which can be checked by anyone interested. It is interesting to learn through the database of the US Patent Office that there are as many as nine patents for using turmeric, including one for treating degenerating musculoskeletal diseases such as rheumatoid arthritis and osteoarthritis.

Neem   Neem (*Azadirachata indica,* vernacular, *Vembu, veppa maram* in Tamil) is a tree native to India. Because of its diverse usefulness, neem is now planted in Africa, Central America, the Caribbean, Hawaii, and Saudi Arabia.

Extracts of neem, especially that of its seed, can be used as a pesticide against as many as 250 pests including whiteflies, aphids, mealy bugs, mites, and termites. Fungal diseases such as rusts and powdery mildew that attack the leaves of ornamental plants and food crops can be controlled by neem extracts.

In Ayurvedic medicine, neem is used as an antimicrobial agent and as a cure for many other infections, including common cold and flu. Neem oil is used in the manufacture of soaps, and offers cheap and easy relief from malaria, skin diseases and even meningitis. Neem leaves are useful as fodder for livestock. Neem oil is used as hair-oil, insect repellent and fertilizer.

Several products of neem have been patented, many of which have been granted to Indian companies. The patented products include a contraceptive (National Institute of Immunology in 1993) and an ecofriendly pesticide (Godrej Soaps, in 1994). However, two patents granted to a US company (WR Grace & Co) have created controversy on the ground that they amount to Biopiracy. The first patent was granted in 1990 for improving the storage stability of neem seed extracts containing azadirachtin, and the other was granted in 1994 for storage of stable insecticidal composition that includes neem seed extract. In both the patents the novelty claimed is increasing the shelf-life stability of azadirachtin solution.

These patents provoked numerous criticisms and protests especially by farmers and NGOs (Non-Governmental Organizations) all over the world that the patents would prevent the farmers from using neem as a source of home-made pesticide, that the cost of pest control in underdeveloped countries would be exorbitantly increased and that the patents are Biopiracy of traditional knowledge.

Heeding to the pressure from various quarters, the Government of India filed a complaint to the US Patent Office accusing the US-based Company of copying an Indian invention. However, the Government of India withdrew the complaint subsequently on the understanding that the patents pertain only to a new invention for the neem extraction process, and that the patents were not based on traditional knowledge. It was also clarified that the farmers can continue to be free to use neem in any traditional way they are used to, and that the use of neem extract, or its seeds or leaves, cannot be patented, because they are traditional knowledge. But any new innovative process by which the properties of neem can be modified can be patented. For example, a process to synthesize a variation of azadirachtin is patentable, because such synthetic product does not exist in nature.

Basmati   Basmati is a superior quality rice grown in the state of Punjab in India and Punjab province of Pakistan. Literally the term 'basmati' means "fragrant earth". This variety of rice with a slender long grain and characteristic flavour is commonly used in the preparation of 'biryani' or 'fried rice' all over the world. Therefore, it is a major export crop for both India and Pakistan.

A patent was granted to a US-based company (RiceTec) by the US Patent Office in 1997 for 'basmati' rice. The patent specifically covers 'novel' varieties of basmati rice, their plants and seeds, a method of breeding them and a method for selecting rice grains according to their starch content so that the cooked rice has the same qualities as the traditional basmati rice. The patent does not cover the use of the name 'basmati', and therefore the US Company does not have any exclusive right to use this word as a trademark.

This patent also has evoked a lot of protests and criticisms from the farmers of India and Pakistan. The main criticisms are that

i)   the collective intellectual and biodiversity heritage of the farmers of India and Pakistan is being unethically usurped,

ii) the patent and/or trademark enables the US Company to take away the market, especially the export market of the farmers, and

iii) consumers are misled by the use of the term basmati for a rice grown in US, and therefore cannot have the same quality as the Punjab-grown rice.

Though the Government of India has protested that the patent could affect its basmati export market and thus adversely affect the livelihood of the farmers, it has not formally challenged the patent in the US Patent Office. Only when such a challenge is made what "novelty" is involved in the US Company's basmati and the related processes, will be known.

The debate on protection of biodiversity and bioresources against piracy or unethical exploitation continues all over the world. There are many complicated issues to be tackled and these are being discussed in a number of international forums, including the WTO's Committee on Trade and Environment, NGOs and activists interested in the safety of our biodiversity. There are also countries that have reached agreements with commercial organizations such that they can carry out research on local biodiversity in order to develop new products and commercially exploit them. However, they must pay royalties commensurate with the commercial value of the products they have developed using the biodiversity of a country. The agreements also insist that there should be some technology transfer so that the local scientist can carry out research on similar lines.

CHAPTER SEVENTEEN

Laboratory Safety

INTRODUCTION

There are several aspects to safety in a biology research laboratory depending on the type of research being carried out and the types of organisms studied there. When we say "laboratory safety" what we mean is the safety of the

    i)   personnel (research students, staff, technical persons, assistants, etc.) working in the laboratory,

    ii) community, and

    iii) biodiversity.

The term 'laboratory' may refer not only to a space within four walls, but also to a field (agricultural, horticultural, aquacultural, industrial, etc.). The term "safety" refers to protection against all hazardous biological, chemical and physical materials, including the harmful radiations. In addition, the safety of the laboratory animals is equally important, especially their safety against "cruelty".

It would be rather an unrealistic task if we try to describe and discuss in detail the various measures that are required for the safety of different laboratories. Therefore, we shall confine our discussion to the principles of safety measures generally needed for most of the biology laboratories. The term "biosafety" is used to denote all aspects of safety measures against hazardous biological materials especially the pathogenic microorganisms.

There are statutory and institutional regulations to how research must be organized in the laboratories and in the field. Apart from the "good laboratory practices" required for any biology laboratory, there are specific guidelines to be strictly adhered to in specialized research laboratories involved in biomedical research, genetic engineering, biotechnology, etc. For instance, WHO (World Health Organization) has published a number of manuals to provide guidelines on matters relating to biosafety. In India, the Department of Biotechnology has issued "Recombinant DNA safety guidelines and regulations". Likewise, every country, university, or research institution has laid down its own guidelines for the conduct of research in biology.

BIOHAZARDOUS AGENTS

Biohazardous agents are generally classified into eight categories:

    i)   Cultured animal cells and the potentially infectious agents these cells may contain

ii) Microorganisms including bacteria, viruses, fungi, protozoans, etc.

iii) Parasites

iv) Allergens

v) Tissues from experimental animals including animal dander (small scales from the hair or feathers of animals that may cause allergy in sensitive individuals)

vi) Plant viruses, bacteria and fungi

vii) Toxins of bacterial or plant origin

RISK GROUPS AND BIOSAFETY LEVELS

The World Health Organization identifies four risk groups with reference to relative hazards of microorganisms and four corresponding Biosafety levels. The classification is on the basis of the microorganisms encountered within the boundaries of a laboratory. The factors taken into account are:

1. The pathogenicity of the organism, i.e., whether the organisms being handled in a laboratory can cause any disease to humans, animals, and plants.

2. The mode of transmission and host range of the organism. These factors may be influenced by the existing levels of immunity in the local population, density and movement of the host population, presence of appropriate vectors, and standards of environmental hygiene.

3. Local availability of effective preventive measures. These may include: prophylaxis by immunization or administration of antisera (passive immunization); sanitary measures such as food and water hygiene; control of animal reservoirs or arthropod vectors.

4. Local availability of effective treatment. This includes passive immunization, post-exposure vaccination and the use of antimicrobials, antivirals and chemotherapeutic agents, and should take into consideration the possibility of the emergence of drug-resistant strains.

The four risk groups and the associated pathogenic agents, safety levels, laboratory practices, and laboratory requirements are summarized in the Table 17.1.

Table 17.1 Risk groups, biosafety levels, practices and requirements.

Risk group	Nature of pathogenic agents	Biosafety level	Examples of laboratory	Laboratory practices	Safety equipment	
					Desirable	Must
1. No or very low individual or community risk	Not associated with disease in healthy adult humans	Biosafety Level 1 - Basic	Basic teaching	Good laboratory practices, microbiological techniques, etc.		On-site autoclave
2. Moderate individual risk, low community risk	Associated with human disease which is rarely serious and for which preventive or therapeutic interventions are *often* available	Biosafety Level 2 - Basic	Primary health services; primary level hospital; diagnostic, teaching and public health	As above plus protective clothing and biohazard sign	Ventilation - inward air flow - mechanical via building system	BSC* I or II, for potential aerosols; on-site autoclave

*BSC , Biological Safety Cabinet

(Contd.)

Table 17.1 (Continued)

Risk group	Nature of pathogenic agents	Biosafety level	Examples of laboratory	Laboratory practices	Safety equipment	
					Desirable	Must
3. High individual risk and low community risk	Associated with serious or lethal human disease for which preventive or therapeutic interventions *may be* available	Biosafety Level 3 - Containment	Special diagnostic	As above plus special clothing, controlled access, directional airflow	Isolation of laboratory; Ventilation – via building system - mechanical, independent BSC Class III	Room sealable for decontamination; Inward air flow ventilation; Filtered air exhaust; Double door entry; Autoclave on site and in laboratory: BSC Class I or II
4. High individual and community risk	Agents that are likely to cause serious or lethal human disease for which preventive or therapeutic interventions *not usually* available	Biosafety Level 4 - Maximum containment	Units handling dangerous pathogens	As above plus airlock entry, shower exit, special waste disposal.	BSC Class I or II	Isolation of laboratory; Room sealable for decontamination; Ventilation (inward airflow, independent mechanical, filtered air exhaust); Double door entry; Airlock; Airlock with shower; Effluent treatment; Autoclave (on site, in laboratory room, double ended); BSC class III.

LABORATORY-ACQUIRED INFECTIONS

Any infection to the personnel by biohazardous agents with which they are working in a laboratory and which are not commonly found in the community in which the personnel live, is described as laboratory-acquired infection. Any infection by such biohazardous agents to any one who is working in the campus in which the laboratory is situated is also a laboratory-acquired infection.

Routes of Exposure

The laboratory personnel may get exposed to biohazardous material through several routes, the most common being

    i)   respiratory

    ii) oral

    iii) skin puncture

    iv) penetration through unbroken skin

    v) conjunctival

    vi) vectors

In laboratory-acquired infections the route of entry may not be the same as when the disease is acquired naturally. The dose or number of organisms required to initiate infection depends on the route.

A laboratory-acquired infection may occur as a result of the following causes:

1. **Accidental inoculation** due to pricking, jabbing or cutting the skin with infected instruments or objects such as hypodermic needles, scalpels, and broken, contaminated glassware.

2. **Accidental ingestion** as a result of mouth pipetting, drinking and smoking in the laboratory.

3. **Splashing** into the face and eyes due to accidental breakage of glassware, etc.

4. Accidental **spillage** of material containing pathogens onto the workbench, floor, or some equipment, noticed or unnoticed, the pathogen getting transmitted by **direct contact**, by fingers to the mouth or the eyes.

5. **Infection by aerosol**   When a liquid is forced under pressure through a small hole, or when a jet of liquid is forced to strike a solid surface, a cloud of very small droplets called aerosol is formed. If the droplets of an aerosol contain bacteria or other pathogenic material they are called **infected airborne particles**. These particles remain suspended in air and are transported for long distances by aircurrent. Infectious airborne particles may also come from lyophilized cultures, dried bacterial colonies, dried materials on stoppers and caps of culture tubes and bottles, fungal spores, which may be released when the containers are opened.

SAFETY MEASURES

Access to the Laboratory

i)   All laboratories in which high-risk organisms (Risk Group 2 and above), especially microorganisms, are handled should display the international biohazard warning symbol (Figure 17.1) with necessary information required.

Figure 17.1   Biohazard warning symbol.

ii) Only authorized persons should have access into the laboratory working area.

iii) The doors of the high-risk laboratories should always be kept closed.

iv) Children should never be permitted to enter the high-risk laboratories.

v) Animal houses associated with the high-risk laboratories also should have restricted admission.

vi) Animals other than the laboratory animals should never be allowed to enter inside the laboratory or animal house.

Personal Safety

1. ***Clothing*** Proper laboratory clothing such as coveralls, gowns, overcoats, face masks or uniforms should be worn by all the personnel working in the laboratory and the animal room during working hours. The laboratory clothing should never be worn outside the laboratory, in places like canteens, offices, libraries, staff rooms and toilets. The protective clothing that is used in the laboratory should not be stored in the same lockers or cupboards that are used for keeping the street clothing.

2. ***Gloves*** Whenever a work involves direct or accidental contact with blood, body fluids or any other infectious material or infected animal, the personnel should wear appropriate gloves. After the work is over, the gloves should be removed aseptically and hands should then be washed using appropriate detergent.

3. ***Protection of face*** In order to protect the eyes and face from splashes, sharp objects and radiation, appropriate glasses, goggles and face shields (visors) should be worn.

4. ***Footwear*** It is not advisable to use open-toed footwear inside the laboratory. Footwear used outside the laboratory should also be avoided.

5. ***Food consumption*** Eating, drinking, smoking or applying cosmetics should be strictly avoided inside the laboratory. Any food meant for human consumption should never be stored in the laboratory.

6. ***Contact lens*** Applying or removing contact lens should never be done inside the laboratory.

Laboratory Practices

1. Pipetting of any solution or water by mouth should be strictly avoided. Aspirator bulb can be used for this purpose.

2. Nothing should be placed in the mouth. For instance, licking labels should never be done.

3. All laboratory procedures must be carried out in such a way that the formation of aerosol or droplets is minimum.

4. Hypodermic syringes and needles should be used only for injection into or aspiration from laboratory animals. That is, they should not be used for any other purpose such as a pipetting device.

5. Any event (e.g. spillage, breakage of culture containers, etc.) that occurs in the laboratory, which is likely to expose infectious materials, should be immediately reported to the laboratory supervisor. It is also desirable to maintain a record of all such accidental events.

6. Appropriate procedure for cleaning any spillage or breakage should be laid down and scrupulously followed.

7. All contaminated solutions must be chemically or physically decontaminated before being discharged in the sewerage.

8. Any material (documents, books, notebooks, etc.) that is likely to be taken out of the laboratory must be protected from contamination while it is inside the laboratory.

Cleanliness of Laboratory

1. All parts of the laboratory should be kept neat, clean and free of unwanted materials.

2. The workbench tops and other work surfaces must be decontaminated every day, preferably at the end of the working day.

3. All contaminated materials like specimens and cultures must be decontaminated before being disposed or cleaned for reuse.

4. If there is need for transporting potentially biohazardous materials, appropriate national or international regulations for packing and transporting such materials should be followed.

5. All windows of the laboratory, especially those that are often opened and closed, should have screens to prevent the entry of insects, spiders, lizards, etc.

Basic Requirements of Laboratory

While designing a research laboratory for specific types of works, appropriate and special attention should be paid to all matters relating to safety problems. The following are a few general suggestions that may be taken into account while designing a laboratory.

1. The laboratory should have ample space to carry out all the work including cleaning and maintenance. There should not be overcrowding of equipment in any one laboratory.

2. The entrance and exit of the laboratory should be so designed that all personnel working in it can easily be evacuated in case of emergency.

3. All surfaces such as those of walls, ceiling, floors, etc., should be smooth, easy to clean, impermeable to liquids and capable of withstanding chemicals and disinfectants generally used in the laboratory. However, the floor should not be slippery.

4. The workbench tops should be resistant to water, disinfectants, acids, alkalis, organic solvents and heat.

5. Adequate illumination, without undesirable reflection and glare, to carry out all types of work should be provided.

6. All furniture used in the laboratory should be sturdy and should not easily topple.

7. Sufficient open spaces between and under the workbenches, cabinets and equipment should be provided for easy cleaning.

8. Sufficient storage spaces for prepared solutions, reagents, etc. should be available in the work area of the laboratory.

9. Storage spaces for personal items (lockers) should be located conveniently, and outside the laboratory.

10. Rest room with facility for eating, drinking water, etc. should be provided outside the laboratory area.

11. Wash basins with facilities like running water, detergent, etc. should be available in every laboratory, preferably close to the exit door.

12. All the doors of a laboratory should be fire-proof, have vision panels, and be self closing.

13. Necessary means of decontamination such as an autoclave should be available close to but separated from the laboratory.

14. Safety systems for protection against fire, electrical emergencies, or any other accident should be readily available.

15. Properly equipped first-aid facility room that can be easily accessed from the laboratory must be provided.

16. Mechanical ventilation systems for an inward airflow without recirculation is desirable. Alternately, the windows should be easily openable, but they should be fitted with appropriate nets to prevent entry of insects and spiders.

17. A constant and assured supply of good quality water is highly essential. The water pipelines of the laboratory should not have any link to the drinking water pipelines.

18. Electricity supply to the laboratory should be reliable and adequate. Emergency lighting must be available. In order to provide uninterrupted supply of power to essential equipment (e.g. incubators, biological safety cabinets, freezers, etc.) and for the ventilation of animal rooms and cages, a stand-by generator or "inverter" must be provided.

19. The gas supply also should reliable and adequate. The gas pipelines should be properly installed and maintained.

20. Security of the laboratories and animal rooms should be strong enough to protect them against theft and any other vandalism.

Basic and Essential Biosafety Equipment

1. Pipetting devices (teats, bulbs, mechanical aspirators, micropipettes, etc.) should be used to avoid mouth pipetting.

2. Biological safety cabinets of appropriate design required for the safety of the laboratory should be used for handling all biohazardous materials, and for preventing production of aerosol and consequent airborne infections. These cabinets are also useful during inoculation of animals and harvesting pathological samples from animals and eggs. The biosafety cabinets may be provided with plastic disposable transfer loops or electric transfer loop incinerators to minimize aerosol formation.

3. As far as possible, screw-capped tubes and bottles should be used.

4. Plastic Pasteur pipettes are preferable to those made of glass.

5. Autoclaves or other means of decontamination (e.g. pressure cooker) should be properly maintained, periodically serviced and readily available.

Disposal of Biohazardous Waste

All laboratory wastes must be segregated into non-biohazardous and biohazardous wastes. Non-biohazardous wastes can be disposed of in the regular garbage.

Biohazardous waste may be a chemical or a material (e.g. culture medium, broken glassware, syringe and needle, carcass of a laboratory animal, etc.) that may contain infectious organisms. In contrast to chemical wastes, biohazardous wastes are capable of replicating and giving rise to a large population in the environment. While "permitted safe levels" can be fixed for chemical wastes to be released into the environment, there can be no such safe level for biohazardous wastes, especially those containing infectious organisms. All biohazardous wastes must be first rendered harmless by either chemical decontamination or heat sterilization before being disposed.

Sterilization is the complete destruction or elimination of the pathogenic, reproductive or infective potential of a biological agent. It is the best method of eliminating the risks associated with biohazardous wastes. Autoclaving (use of saturated steam under pressure) is the sterilization method commonly used in research laboratories and hospitals. It is the most dependable measure for ensuring total destruction of microorganisms. Autoclaving involves heating in a chamber with saturated steam under heavy pressure so that the chamber temperature rises to at least 121°C for a minimum period of 15 minutes (or 3 minutes holding time at 134°C or 10 minutes holding time at 126°C or 25 minutes holding time at 115°C). The duration of the autoclaving is measured after the temperature of the material being sterilized reaches 121°C. All the materials that are being autoclaved must come into contact with steam and heat for complete sterilization. Any air trapped inside the materials may thwart this and, therefore trapping of air should be prevented.

Dry heat may also be used for sterilization, though it is much less effective than wet (steam) heat sterilization. Dry heat sterilization requires very high temperature (160°C to 170°C) and longer durations (2–4 hours).

Decontamination or disinfection is the reduction of many or all disease-causing microorganisms and the destruction of pathogenic microorganisms in or on the surface of an object, so that they are no longer capable of transmitting disease. A few common chemical disinfectants, their use, mode of action, advantages and disadvantages are summarized in Table 17.2.

Table 17.2   Common chemical disinfectants.

Class	Use	Mode of action	Advantages	Disadvantages
Alcohols: Hi-Ethanol, Isopropanol (70–85%)	Cleaning some instruments and surfaces of BSCs	Cell lysis and protein denaturation; Presence of water assists with killing action	Fairly inexpensive, easy to use, not corrosive, effective against most microorganisms	Evaporates quickly, requires long contact time, flammable, inactivated by organic matter
Formaldehydes: 37% Formalin Paraformaldehyde	Surface cleaner; as a gas it is used for decontamination of large spaces (BSCs and rooms)	Denature proteins and require the presence of water vapour	Very effective against all forms of biohazards (including spores) and the gas can penetrate into small cracks and spaces	Requires the use of special personal protective equipment
Phenolics: Phenol-containing detergents	When diluted act as an effective bacteriostatic agent	They have a rapid corrosive action on tissues and cells	Effective against viruses and vegetative bacteria	Corrosive, irritant to skin, sticky and have strong odour

Class	Use	Mode of action	Advantages	Disadvantages
Quaternary Ammonium Compounds - Quats	Good surface cleaners	Affect proteins and cell membrane of microorganisms	Contain detergents to aid in cleansing, rapid action, non-corrosive, and nonstaining	May not be active against some bacteria, spores, viruses and rapidly inactivated by soap and organic matter
Chlorines: Sodium hypochlorite	Spills of human body fluids	Free available chlorine binds with contents within microorganisms; reaction by-products cause death to the cell	Broad spectrum, fast acting, inexpensive	Corrosive, short use life, inactivated by organic matter, irritate skin and eyes
Iodine compounds - Iodophors Tincture iodine	Disinfecting some semi-critical medical equipment	Free iodine enters the microorganism and binds with its cellular components, needs 30–50 ppm	Broad spectrum, cleansing action, built-in colour indicator, inexpensive and few health or disposal problems	Inactivated by hard water, may stain, weakly corrosive, and react with organic matter

ADDITIONAL HAZARDS

In addition to the biohazards in a biological research laboratory, there are other hazards about which researchers and other laboratory personnel should be aware of and for which they must have necessary safety measures. These hazards are chemicals, fire, electricity, radiation and noise.

Chemical Hazards

All those working in a laboratory should be aware of the possible hazards from exposure to the chemicals used in the laboratory. The hazards may produce toxic effects to humans and other organisms (e.g. laboratory animals, plants, etc.) or may cause serious physical injury due to faulty handling and storage. Exposure to hazardous chemicals can occur by any one or more of the following ways:

- i)   Inhalation
- ii) Contact
- iii) Needle pricks
- iv) Through open skin

The toxic effects of chemicals may produce harmful effects in the skin, respiratory system, blood, lungs, liver, kidneys, gastrointestinal system and other organs and tissues. Some chemicals are carcinogenic or teratogenic.

All the chemicals should be kept in a protected storeroom. Only small amounts of chemicals necessary for use should be kept in the laboratory. Every care should be taken to avoid incompatible chemicals coming into contact with each other because such a contact may lead to fire and/or explosion. For example, alkali metals (sodium, potassium, cesium, lithium) should not come into contact with carbon dioxide, chlorinated hydrocarbons, and water. Halogens are incompatible with ammonia, acetylene, and hydrocarbons. Acetic acid, hydrogen sulphide, aniline, hydrocarbons and sulphuric acid should not come into contact with oxidizing agents such as chromic acid, nitric acid, peroxides and permanganates.

The following are general guidelines to counter the spillage of any toxic chemical in the laboratory.

1. Appropriate spillage charts (methods describing how to deal with spills) should be displayed in a prominent position in the laboratory.

2. Spillage kits should be available in an easily accessible place.

3. Protective clothing (e.g. heavy-duty rubber gloves, overshoes or rubber boots, respirators) should be used while cleaning the spillage.

4. Cleaning accessories such as scoops and dustpans, forceps for picking broken glass, mops, cloths, papers towels, buckets, etc. should be available.

5. Sodium carbonate (soda ash) or sodium bicarbonate are useful for neutralizing acids and corrosive materials.

6. Sand may be used to cover alkali spills.

7. Only non-flammable detergents should be used for cleaning the spillage area.

8. In case of emergency, all non-essential personnel of the laboratory should be evacuated.

9. Immediate first-aid and subsequent medical attention should be given to all those persons who may have been contaminated by the spilled chemical.

10. If the spilled chemical is inflammable, all open flames must be immediately put off, gas supply must be closed, windows opened and any electrical gadget/appliance that may produce spark must be switched off.

11. Appropriate procedure should be followed to clean the spilled chemical.

Fire Hazards

Fire accidents may occur in a laboratory due to the following causes;

i) Overloading of electrical circuit

ii) Poor electrical maintenance

iii) Loose electrical connections producing sparks and leakage of gas

iv) Leaving equipment unnecessarily switched on

v) Open flames

vi) Spoiled gas tubing

vii) Improper handling and storage of inflammable or explosive materials

viii) Insufficient and improper ventilation

To counter any accidental fire in the laboratory, fire extinguishers should be available at convenient places. Fire alarms should be installed and maintained in working conditions. Fire escape routes should be displayed prominently in all parts of the building in which the laboratory is situated. In case of emergency the help of the fire service should be immediately summoned.

Each laboratory must have fire-fighting equipment including hoses, buckets of sand or water and fire extinguisher. Water should be used to extinguish fire in paper, wood, or cloth. It should not be used to put out electrical fires and burning liquids and metals. Carbon dioxide is useful to put out fire of inflammable liquids and gases and electrical fires. It should not be used to extinguish fire of alkali metals and paper. Foam is suitable to counter fire in inflammable liquids but not for electrical fires.

All gas cylinders (e.g. LPG) should be properly fixed to some solid support in such a way that they are not easily toppled or moved about unnecessarily.

Electrical Hazards

The chief electrical hazards are fire due to spark or overloading of electrical wiring, and electric shock to personnel. To counter such hazards all electrical installations and equipment, including earthing/grounding systems, should be maintained properly. Circuit breakers and earth-fault-interrupters should be installed in all the laboratories. Circuit-breakers are necessary to protect the electric wiring from being overloaded and thus prevent fires. They cannot protect people from receiving electric shocks. Earth-fault-interrupters are useful for protection from this electric hazard. All electrical equipment should be earthed/grounded, preferably through three-prong plugs. All equipment and materials used in wiring should conform to the national electrical safety standards and codes (e.g. ISI certification).

Noise

Prolonged exposure to excessive noise may cause health problems. Therefore, research personnel who work with instruments that produce significant noise should have appropriate protection (e.g. ear plugs). Further, such personnel should be medically monitored for any possible ill effects due to consistent exposure to noise.

Radiation Hazard

Sources Many biology research laboratories are potential sources of ionizing radiation. Some of the common radiations and their sources in a biology laboratory are:

i)   UV-radiation with a wavelength less than 315 nm (e.g. UV-transilluminator)

ii) Lasers with a classification of 3R, 3B or 4 (e.g. confocal laser scanning microscope)

iii) Microwave (e.g. microwave ovens)

iv) Ultrasound (e.g. ultrasonic disintegrators, sonicators)

v)   Isotopes (e.g. phosphorus-32, ^{32}P; iodine-125, ^{125}I)

vi) X-ray generators

vii) cellphones

Harmful effects Generally, the harmful effects of ionizing radiation may be somatic or genetic. Somatic effects of radiation are those that can be observed as clinical symptoms in persons who were exposed to radiation. The effect may be less severe as in minor skin damage, hair loss, blood deficiencies, gastrointestinal damage or cataract formation. At the other extreme, ionizing radiations may cause cancers in blood (leukemia), bone, lung and skin. The onset of such cancerous effect may occur after several years after the exposure to the radiation.

Ultraviolet (UV) radiation where the wavelength is shorter than 315 nm has the greatest potential to produce burns (sunburn and "arc-eye") and long-term somatic effects such as skin cancer.

Lasers of particular intensities can permanently damage the eyes.

Low frequency electromagnetic radiation (e.g. cellphones) can cause malfunctions in heart pacemakers, which can lead to the death of the person. High intensity microwave radiation can increase the core temperature of people beyond their physiological capacity to cope. Ultrasonic noise can permanently impair hearing.

Genetic (i.e., hereditary) effects are those that are observed in the descendants of the persons who were exposed to radiation. The genetic effect is due to the harmful effects (e.g. chromosome damage or gene mutation) of the radiation on the reproductive cells of the exposed individual. Irradiation of gonads with high doses may lead to cell death and consequent infertility of the exposed person. It may cause disturbance in the menstrual cycle. If an early foetus (about 8–15 weeks old) is exposed to radiation, it will run the risk of having congenital malformations, mental impairment or even cancer at a later period of life.

Protection   Protection from ionizing radiation involves:

i)   Controlled use of all radioactive materials (e.g. isotopes) and compliance with all relevant national standards.

ii)   *Minimizing the radiation exposure time*   The radiation dose a person receives from a radioactive source is directly proportional to the duration of exposure. The relationship is as follows:

$$\text{Dose} = \text{Dose rate} \times \text{Duration of exposure}$$

Therefore, the duration a person spends during handling radioactive materials should be minimal. It can be minimized by:

a)   using dummies (non-radioactive materials) for practising new and unfamiliar techniques

b)   planning the work schedule in such a way that the time of working with radioactive materials is minimal

c)   returning all radioactive materials to storage immediately after use

d)   removing radioactive wastes from the laboratory as frequently as possible

e)   minimizing the stay in the radiation area or laboratory.

iii)   *Maximizing the distance from the radioactive materials*   The radiation dose a person receives from a radioactive source is inversely proportional to the distance between the person and the source, i.e., longer the distance, lesser the radiation dose. The relationship between the distance and dose is as follows:

$$\text{Dose rate} = \text{Constant} \times (1/\text{distance}^2)$$

Therefore, as far as possible, radioactive materials should be handled away from the body.

iv)   *Shielding the radiation source*   Barriers made of lead, concrete, or water provide effective protection from radiations such as gamma rays and neutrons. For this reason radioactive materials are stored or handled under water or by remote control in lead–lined concrete. There are special plastic shields that can block beta particles. Air is capable of stopping alpha particles. Therefore, inserting a proper and appropriate shield between the person and the radiation source will minimize the radiation dose to a great extent.

In addition, use of Perspex screens, face shields, safety spectacles, gloves, and long-sleeved coats will protect the eyes and body. UV transilluminators should be shielded with Perspex while in use.

Persons who are working in open fields can protect themselves from the harmful effects of cosmic radiation by wearing light clothes made of natural fibres (e.g. cotton), and wide-brimmed hats, and by applying sunscreen lotions on exposed skin.

SAFETY IN GENETIC ENGINEERING

Genetic engineering is the use of *in vitro* techniques such recombinant DNA technology for deliberate manipulation of genes within or between species, for the purpose of genetic analysis, product improvement or creation of genetically modified organisms (GMOs) like transgenic and 'knockout' animals and plants. The products of genetic engineering, especially the GMOs, are something that may never have existed in nature before. Such products might have unpredictable and undesirable properties, which may be biohazardous to the environment and biodiversity. For instance, cross fertilization between a GMO and an indigenous species may lead to the loss of the latter due to competition in the ecosystem. If the GMO happens to be a pathogen, that too a more virulent one, its effect on biodiversity could be devastating.

On the other hand, the benefits of genetic engineering, especially the rDNA technology, are numerous. For instance, with the availability of the nucleotide sequence of the entire human genome and their functions, gene therapy may become a routine treatment for certain diseases. Transgenic plants and animals are also likely to play an increasingly significant role in the fields of medicine and agriculture in the days to come.

Biosafety risk assessment In order to minimize the potential dangers and still enjoy the benefits, the genetic engineering experiments involving the construction and use of GMOs should be performed only after carrying out a biosafety risk assessment. Since the pathogenic properties and potential hazards of a GMO may be novel and not well-characterized, the risk evaluation should take into consideration

    i)   the properties of the donor organism,

ii) the nature of the DNA sequence that is to be transferred,

iii) the properties of the recipient organism and

iv) the characteristics of the environment into which the GMO is to be released.

The risk assessment would help to determine the biosafety level that is required for the safe handling of the GMO, and to decide the biological and physical containment systems necessary for the experiment to be carried out.

The Department of Biotechnology, Government of India, has recognized three categories of research activity in the field of genetic engineering involving GMOs and transgenic crops, on the basis of the level of the associated risk and requirement for the approval of competent authority.

Category I   Routine cloning work of defined genes, defined non-coding stretches of DNA and open reading frames in defined genes in *E. coli* or other bacterial and fungal hosts that are generally considered safe to humans, animals and plants.

Category II   Experiments carried out in laboratory and greenhouse or net house using defined DNA fragments that are non-pathogenic to humans and animals for genetic transformation of plants, both as model species and crop species.

Category III Experiments having high risk, that is, if the experimental transgenic trait escapes from the laboratory (including greenhouses, net houses and open fields) into the environment, it could bring about significant alterations in the biosphere, the ecosystem, plants and animals by dispersing the new genetic traits, the effects of which cannot be evaluated precisely.

Safety of viral vectors   Transfer of desired genes to the recipient cell or organism is effected through viral vectors such as adenovirus vector. Such viral vectors used in laboratories lack certain virus replication genes and they can be propagated in special cell lines that complement the deficiency. The stock of replication-deficient virus should be protected from contamination by replication-competent viruses. Such contamination may occur due to rare spontaneous recombination events in the propagating cell lines. It may also be due to insufficient purification. Therefore, the vectors should be handled at the same biosafety level as the parent stock.

Safety of transgenic animals   Transgenic animals are those carrying foreign genetic material. Such animals should be handled in containment levels appropriate to the characteristics of the products of the inserted genes. For example, a transgenic animal (e.g. mice) can be created for the purpose of using it as a model for studying a disease. Normally, the animal is not infected by the pathogen (e.g. human poliovirus) that causes the disease in humans. The gene for the receptors for the particular pathogen is inserted into the genome of the animal to make it a transgenic animal. The expression of the transgene now makes the recipient animal susceptible to infection by the specific pathogen, and therefore can be used as a model to study and to find a cure for the disease. The risk is when the transgenic animal escapes from the laboratory into the wild. It is theoretically possible that such an escaped transgenic animal can transmit the transgene to the wild population of that species, and thus create a reservoir for the pathogen. Therefore, all measures should be taken to ensure strict containment of the transgenic animals.

Safety of transgenic plants   Transgenic plants are created to give them tolerance to herbicides or resistance to pests. They are also created using animal or human genes to develop medicinal and nutritional products. The food-safety of such transgenic plants, and the long-term ecological consequences of their cultivation need to be thoroughly investigated before they are allowed to be grown in open and made available for human consumption. Appropriate risk assessment should be made to determine the biosafety level for the production of these plants.

In India, the Department of Biotechnology and the Ministry of Environment have formulated the Indian recombinant DNA safety guidelines and regulation. These guidelines stipulate the necessary safety measures required for the research works at academic level and at industrial level.

SAFETY OF LABORATORY ANIMALS

Biological Model System

A wide range of animals, from protozoa to apes, is used in laboratories for experimental purposes. The aim of most of the biological and biomedical researches is to understand all aspects of life and to improve the quality of life of human beings and animals. The necessity for the use of animals in research is to have a biological model system as an experimental subject to learn about the human and animal systems.

A biological model system can be defined as a system that can be observed instead of the original system (human or animal), which is of ultimate interest to the investigator. The use of model system helps to answer questions which could not be answered using the original system with the available technology. Further, it enables the isolation and study of certain features which would be too complex to study or impossible to isolate in the original system. Generally, four types of models are used in biology and biomedical researches. They are:

i) whole, living animals (including humans),

ii) living systems composed of samples from the original system (e.g. tissue culture),

iii) non-living, mechanical or molecular systems and

iv) mathematical models such as computer simulation.

Animal model systems involving simple organisms (e.g. invertebrates), which have very short lifespan (life cycle), are useful to study many basic biological processes because they are easy to grow and observe. Vertebrates are useful as experimental animals because of their closer phylogenetic relationship with man. Mammals among vertebrates are much more suited for biomedical research because humans are also mammals. Further, many diseases that affect human beings also affect other mammals, but they do not occur in plants, insects or bacteria. More importantly, as required in most of the biological and biomedical experiments, it is possible to have genetically, morphologically and physiologically more uniform animals.

Animal Welfare

While using animal model systems, especially those involving mammals, it is very essential to ensure that only healthy animals that are well adapted to their housing conditions are used. Only such properly maintained animal models would yield reliable and valid results. Healthy animals are a prerequisite not only for good science but also for good animal welfare.

The welfare of animals used for experimental purposes can be assessed from the following facilities available to them:

i) Ready access to water and food to maintain health and vigour, i.e., the animals should not suffer from hunger and thirst

ii) Suitable environment to make the animals comfortable

iii) Prevention and treatment of any pain, injury and disease the animals might suffer from

iv) Conditions to avoid mental suffering (fear, distress, etc.)

v) Sufficient space and adequate facilities for the animals to express their natural behaviour

Drinking water and diet are the basic and important factors that are to be monitored carefully to maintain the health of the laboratory animals. Since water is a vehicle for microorganisms, the water system should be protected from being contaminated. The quality of water should be checked daily. The diet should be so formulated to satisfy the nutritional requirements of the animals. All food hoppers and utensils should be cleaned and sterilized frequently to prevent contamination.

The temperature of the animal room is another important factor that should be carefully controlled and monitored continuously. Different species of animals would require different ambient temperature. If room temperature is high, the temperature of the cages in which the animals are housed would be much higher. Fluctuations in temperature might cause stress to the animals and, therefore, should be avoided. Appropriate ventilation is important in regulating the environment of the animal room.

In India, the welfare of animals in general is statutorily protected by Prevention of Cruelty to Animals Act 1960, Chapter 14 of which deals specifically with the animals used for experimental purposes (Appendix I). A statutory body, the Committee for the Purpose of Control and Supervision of Experimental Animals (CPCSEA), formed in 1964 and revived in 1998, enforces the provisions of the PCA Act. CPCSEA has formulated specific guidelines for laboratory animals facilities (Appendix II). The CPCSEA follows the "Operational guidelines for observance of good practices in CPCSEA" (Appendix III) in enforcing its guidelines.

According to the provisions of CPCSEA, all research laboratories that are using or intending to use animals for experimental purposes are required to get themselves registered with CPCSEA, get their animal house facilities inspected, and also get specific projects for research cleared before commencing the research. The breeding and trade of experimental animals is also under the control of the provisions of CPCSEA. Every research institute must constitute an "Institutional Animal Ethics Committee (IAEC)" and this committee is empowered to approve research project proposals that intend to use small experimental animals such as rats, mice, guinea pigs or rabbits. All research activities involving larger animals such as dogs, pigs, cattle and monkeys can be approved by a panel of scientific experts constituted for this purpose.

Apart from the guidelines on laboratory animal care and practice, the CPCSEA has formulated a protocol for the production of immunobiologicals from horses. It is also working out alternatives to live animal models in biology research and teaching.

In addition to CPCSEA, other organizations such as Indian National Science Academy (Appendix IV) and Indian Council of Medical Research have also formulated guidelines for care and use of animals in scientific research and teaching.

Appendices

APPENDIX   I

The Prevention of Cruelty to Animals Act, 1960
(59 of 1960)
As amended by Central Act 26 of 1982

The Prevention of Cruelty To Animals Act, 1960
(59 of 1960)
(26 December, 1960)

An act to prevent the infliction of unnecessary pain or suffering on animals and for that purpose to amend the law relating to the prevention of cruelty to animals.

Be it enacted by Parliament in the Eleventh year of the Republic of India as follows:

CHAPTER 1   PRELIMINARY

Short title, extent and commencement

1.   (1) This Act may be called the Prevention of Cruelty to Animals Act, 1960.

      (2) It extends to the whole of India except the State of Jammu and Kashmir.

      (3) It shall come into force on such date as the Central Government may, by notification in the official Gazette, appoint, and different dates may be appointed for different States and for the different provisions contained in this Act.

Definitions

2.   In this Act, unless the context otherwise requires:

      (a) "animal" means any living creature other than a human being;

      *[(b)] "Board" means the Board established under Section 4, and as reconstituted from time to time under Section 5A]

      (c) "captive animal" means any animal (not being a domestic animal) which is in captivity or confinement, whether permanent or temporary, or which is subjected to any appliance of contrivance for the purpose of hindering or preventing its escape from captivity or confinement or which is pinioned or which is or appears to be maimed;

      (d) "domestic animal" means any animal which is tamed or which has been or is being sufficiently tamed to serve some purpose for the use of man or which, although it neither has been nor is intended to be so tamed, is or has become in fact wholly or partly tamed;

---

*See Annexure for notifications under section 1 (3)

*Subs. by Act. 26 of 1982 which came into force on 30th July 1982.

(e)   "local authority" means a municipal committee, district board or other authority for the time being invested by law with the control and administration of any matters within a specified local area;

(f)   "owner" used with reference to an animal, includes not only the owner but also any other person for the time being in possession or custody of the animal, whether with or without the consent of the owner;

(g)   "phooka" or "doom dev" includes any process of introducing air or any substance into the female organ of a milch animal with the object of drawing off from the animal any secretion of milk;

(h)   "prescribed" means prescribed by Rules made under this Act;

(i)   "street" includes any way, road, lane, square, court, alley, passage or open space, whether a thorough fare or not to which the public have access.

Duties of persons having Charge of animals

3.   It shall be the duty of every person having the care or charge of any animal to take all reasonable measures to ensure the well-being of such animal and to prevent the infliction upon such animal of unnecessary pain or suffering

CHAPTER II   *(ANIMAL WELFARE BOARD OF INDIA)

Establishment of Animal Welfare Board of India

4.   (1)   For the promotion of animal welfare generally for the purpose of protecting animals from being subjected to unnecessary pain or suffering, in particular, there shall be established by the Central Government, as soon as may be after the commencement of this Act, a Board to be called the **(Animal Welfare Board of India)

(2)   The Board shall be a body corporate having perpetual succession and a common seal with power, subject to the provisions of this Act, to acquire, hold and dispose of property and may by its name sue and be sued

Constitution of the Board

5.   (1)   The Board shall consist of the following persons, namely:

---

*Subs. Act 26 of 1982; S.3, for the words "Animal Welfare Board".

**Sub-ibid, S.4 for the words "Animal Welfare Board".

(a) the Inspector General of Forests, Government of India, *ex-officio*;

(b) the Animal Husbandry Commissioner to the Government of India, *ex-officio*;

*(ba) two persons to represent respectively the Ministries of the Central Government dealing with Home Affairs and Education, to be appointed by the Central Government;

(bb) one person to represent the Indian Board for Wildlife, to be appointed by the Central Government;

(bc) three persons who, in the option of the Central Government are or have been actively engaged in animal welfare work and are well-known humanitarians, to be nominated by the Central Government;

(c) one person to represent such association of veterinary practitioners as in the opinion of the Central Government ought to be represented on the Board, to be elected by that association in the prescribed manner;

(d) two persons to represent practitioners of modern and indigenous systems of medicine, to be nominated by the Central Government;

**[(e) one person to represent each of such two municipal corporations as in the opinion of the Central Government ought to be represented on the Board, to be elected by each of the said corporations in the prescribed manner];

(f) one person to represent each of such three organizations actively interested in animal welfare as in the opinion of the Central Government ought to be represented on the Board, to be chosen by each of the said organizations in the prescribed manner;

(g) one person to represent of such three societies dealing with prevention of cruelty to animals as in the opinion of the Central Government ought to be represented on the Board, to be chosen in the prescribed manner;

(h) three persons to be nominated by the Central Government

---

*inserted by Act 26 of 1982; S. 5 (a) (i)

**Subs. - ibid S. 5 (a) (ii) for the original clause.

(i)   six Members of Parliament, four to be elected by the House of the People (Lok Sabha) and two by the Council of States (Rajya Sabha)

(2)   Any of the persons referred to in clause 9 (a) or *clause (b) or *clause (ba) or clause (bb) of sub-section (1) may be depute any other person to attend any of the meetings of the Board.

**[(3)The Central Government shall nominate one of the Members of the Board to be its Chairman and another member of the Board to be its Vice-Chairman]

@[5A (1) In order that the Chairman and other members of the Board hold office till the same date and that their terms of office come to an end on the same date, the Central Government may, by notification in the Official Gazette, reconstitute, as soon as may be after the Prevention of Cruelty to Animals (Amendment) Act, 1982 comes into force, the Board].

(2)   The Board as reconstituted under sub-section (1) shall be reconstituted from time to time on the expiration of every third year, from the date of its reconstitution under sub-section (1).

(3)   There shall be included amongst the members of the Board reconstituted under sub-section (1), all persons who immediately before the date on which such reconstitution is to take effect, are Members of the Board but such persons shall hold office only for the unexpired portion of the term for which they would have held office if such reconstitution had not been made and the vacancies arising as a result of their ceasing to be Members of the Board shall be filled up as casual vacancies for the remaining period of the term of the Board as so reconstituted;

Provided that nothing in this sub-section shall apply in relation to any person who ceases to be member of the Board by virtue of the amendment made in sub-section (1) of section 5 by sub clause (ii) of clause (a) of section 5 of the Prevention of Cruelty to Animals (Amendment) Act, 1982.

---

*Subs. by Act 26 of 1982: S. 5 (b),for the word, brackets and letter "Clause(b)

**Subs. ibid, S. 5 (c) for the original clause.

@ins. ibid, S.6.

Terms of Office and conditions of service of members of the Board

6.   *(1) The term for which the Board may be reconstituted under section 5 A shall be three years from the date of the reconstitution and the Chairman and other Members of the Board as so reconstituted shall hold office till the expiry of the term for which the Board has been so reconstituted

(2)   Not withstanding anything contained in sub-section (1):-

    (a)   the term of office of any ex-officio Member shall continue so long as he holds the office by virtue of which he is such a Member;

    (b)   the term of office of a Member elected or chosen under clause (c), clause (e), clause (g), clause (h) or clause (i) of section 5 to represent any body of persons shall come to an end as soon as he ceases to be a Member of the body which elected him or in respect of which he was chosen;

    (c)   the term of office of a Member appointed, nominated, elected or chosen to fill a casual vacancy shall continue for the remainder of the term of office of the Member in whose place he is appointed, nominated, elected or chosen;

    (d)   the Central Government may, at any time, remove for reasons to be recorded in writing a member from office after giving him a reasonable opportunity of showing cause against the proposed removal and any vacancy caused by such removal shall be treated as casual vacancy for the purpose of clause (c)

(3)   The members of the Board shall receive such allowance, if any, as the Board may, subject to the previous approval of the Central Government, provided by regulations made in this behalf;

(4)   No act done or proceeding taken by the Board shall be questioned on the ground merely of the existence of any vacancy in, or defect in the constitution of the Board and in particular, and without prejudice to the generality of the foregoing, during the period intervening between the expiry of the term for which the Board has been reconstituted under section 5A and its further reconstitution under that section, the ex-officio members of the Board shall discharge all the powers and function of the Board

---

*Subs., by Act 26 of 1982, S.7, for the original Section.

Secretary and other employees of the Board

7.   (1)   The Central Government shall appoint *xxxxx the Secretary of the Board

   (2)   Subject to such rules as may be made by the Central Government in this behalf, the Board may appoint such number of other officers and employees as may be necessary for the exercise of its powers and the discharge of its functions and may determine the terms and conditions of service of such officers and other employees by regulations made by it with the previous approval of the Central Government.

Funds of the Board

8.   The funds of the Board shall consist of grants made to it from time to time by the Government and of contributions, subscriptions, bequests, gifts and the like made to it by any local authority or by any other person.

Functions of the Board

9.   The functions of the Board shall be:-

   a)   To keep the law in force in India for the Prevention of Cruelty to Animals under constant study and to advise the Government on the amendments to be undertaken in any such law from time to time;

   b)   To advise the Central Government on the making of rules under the Act with a view to preventing unnecessary pain or suffering to animals generally, and more particularly when they are being transported from one place to another or when they are used as performing animals or when they are kept in captivity or confinement;

   c)   To advise the Government or any local authority or other person on improvements in the design of vehicles so as to lessen the burden on draught animals;

   d)   To take all such steps as the Board may think fit for **(amelioration of animals) by encouraging, or providing for the construction of sheds, water troughs and the like and by providing for veterinary assistance to animals;

---

*The words "one of its officers to be" omitted by Act 26 of 1982,

**Subs, by Act 26 of 1982, S.9 (a) for the word "ameliorating the condition of beasts of burden.

e) To advise the Government or any local authority or other person in the design of slaughter houses or the maintenance of slaughter houses or in connection with slaughter of animals so that unnecessary pain or suffering, whether physical or mental, is eliminated in the pre-slaughter stages as far as possible, and animals are killed, wherever necessary, in as humane a manner as possible;

f) To take all such steps as the Board may think fit to ensure that unwanted animals are destroyed by local authorities, whenever it is necessary to do so, either instantaneously or after being rendered insensible to pain or suffering;

g) To encourage by the grant of financial assistance or otherwise, *(the formation or establishment of pinjrapoles, rescue homes, animal shelters, sanctuaries and the like), where animals and birds may find a shelter when they have become old and useless or when they need protection;

h) To co-operate with, and co-ordinate the work of associations or bodies established for the purpose of preventing unnecessary pain or suffering to animals or for the protection of animals and birds;

i) To give financial assistance and other assistance to Animal Welfare Organisations functioning in any local area or to encourage the formation of Animal Welfare Organisations in any local area which shall work under the general supervision and guidance of the Board;

j) To advise the Government on matters relating to the medical care and attention which may be provided in animal hospitals, and to give financial and other assistance to animal hospitals whenever the Board think it is necessary to do so;

k) To impart education in relation to the humane treatment of animals and to encourage the formation of public opinion against the infliction of unnecessary pain or suffering to animals and for the promotion of animal welfare by means of lectures, books, posters, cinematographic exhibitions and the like;

l) To advise the Government on any matter connected with animal welfare or the prevention of infliction of unnecessary pain or suffering on animals;

---

*Subs, ibid, S, 9 (b) for the words "the formation of pinjrapoles, sanctuaries and the like"

Power of the Board to make regulations

10.   The Board may, subject to the previous approval of the Central Government, make such regulations as it may think fit for the administration of its affairs and for carrying out its functions.

CHAPTER III   CRUELTY TO ANIMALS GENERALLY

Treating animals cruelly.

11.   (1)   If any person

    (a)   beats, kicks, over-rides, over-drives, over-loads, tortures or otherwise treats any animal so as to subject it to unnecessary pain or suffering or causes, or being the owner permits, any animal to be so treated; or

    (b)   *(employs in any work or labour or for any purpose any animal which, by reason of its age or any disease) infirmity, wound, sore or other cause, is unfit to be so employed or, being the owner, permits any such unfit animal to be employed; or

    (c)   wilfully and unreasonably administers any injurious drug or injurious substance to **(any animal) or wilfully and unreasonably causes or attempts to cause any such drug or substance to be taken by ***(any animal;) or

    (d)   conveys or carries whether in or upon any vehicle or not, any animal in such a manner or position as to subject it to unnecessary pain or suffering; or

    (e)   keeps or confines any animal in any cage or other receptacle which does not measure sufficiently in height, length and breadth to permit the animal a reasonable opportunity for movement; or

    (f)   keeps for an unreasonable time any animal chained or tethered upon an unreasonably short or unreasonably heavy chain or cord; or

---

*Subs. by Act 26 of 1982, S. 10 (a) (i) for the words "employ in any work or labour any animal which, by reason of any disease";

**Subs. ibid S.10 (a) (ii) for the words "any domestic or captive animal".

***Subs. ibid S. 10 (a) (ii) for the words "any captive animal".

(g)   being the owner, neglects to exercise or cause to be exercised reasonably any dog habitually chained up or kept in close confinement; or

(h)   being the owner of (any animal) fails to provide such animal with sufficient food, drink or shelter; or

(i)   without reasonable cause, abandons any animal in circumstances which tender it likely that it will suffer pain by reason of starvation of thirst; or

(j)   willfuly permits any animal, of which he is the owner, to go at large in any street, while the animal is affected with contagious or infectious disease or, without reasonable excuse permits any diseased or disabled animal, of which he is the owner, to die in any street; or

(k)   offers for sale or without reasonable cause, has in his possession any animal which is suffering pain by reasons of mutilation, starvation, thirst, overcrowding or other ill-treatment; or

*[(l)   mutilates any animal or kills any animal (including stray dogs) by using the method of strychnine injections in the heart or in any other unnecessarily cruel manner; or)]

**(m) solely with a view to providing entertainment;

(i)   confines or causes to be confined any animal (including tying of an animal as a bait in a tiger or other sanctuary) so as to make it an object or prey for any other animal; or

(n)   ***(XXXX) organizes, keeps, uses or acts in the management or, any place for animal fighting or for the purpose of baiting any animal or permits or offers any place to be so used or receives money for the admission of any other person to any place kept or used for any such purposes; or

(o)   promotes or takes part in any shooting match or competition wherein animals are released from captivity for the purpose of such shooting;

---

*Subs. by Act 26 of 1982. S. 10 (a) (iv) for the original clause.

**Subs. ibid S. 10 (a) for the original clause.

***The words "for the purposes of his business" omitted by Act 26 of 1982, S. 10 (a) (iv) he shall be punishable *(in the case of a first offence, with fine which shall not be less than ten rupees but which may extend to fifty rupees and in the case of a second or subsequent offence committed within three years of the previous offence, with fine which shall not be less than twenty five rupees but which may extend, to one hundred rupees or with imprisonment for a term which may extend, to three months, or with both.)

(2) For the purposes of section (1) an owner shall be deemed to have committed an offence if he has failed to exercise reasonable care and supervision with a view to the prevention of such offence; Provided that where an owner is convicted permitting cruelty by reason only of having failed to exercise such care and supervision, he shall not be liable to imprisonment without the option of a fine

(3) Nothing in this section shall apply to:-

   (a) the dehorning of cattle, or the castration or branding or noseroping of any animal in the prescribed manner, or

   (b) the destruction of stray dogs in lethal chambers **(by such other methods as may be prescribed) or

   (c) the extermination or destruction of any animal under the authority of any law for the time being in force; or

   (d) any matter dealt with in Chapter IV; or

   (e) the commission or omission of any act in the course of the destruction or the preparation for destruction of any animal as food for mankind unless such destruction or preparation was accompanied by the infliction of unnecessary pain or suffering.

Penalty for practicing phooka or doom dev.

12. If any persons practices upon any cow or other milch animal the operation called phooka or ***[doom dev or any other operation (including injection of any substance) to improve lactation which is injurious to the health of the animal] or permits such operation being performed upon any such animal in his possession or under his

---

*Subs. ibid S. 10 (a) (vii) for the portion beginning with the words "in the case of a first offence" and ending with words "or with both".

**Subs. by Act 26 of 1982, S. 10 (b), for the words "by the other methods with a minimum of suffering".

***Subs. ibid S. 11, for the words, "doom dev"

control, he shall be punishable with fine which may extend to one thousand rupees, or with imprisonment for a term which may extend to two years, or with both, and the animal on which the operation was performed shall be forfeited to the Government.

Destructions of suffering animals

13. (1) Where the owner of an animal is convicted of an offence under section 11, it shall be lawful for the court, if the court is satisfied that it would be cruel to keep the animal alive, to direct that the animal be destroyed and to assign the animal to any suitable person for that purpose, and the person to whom such animal is so assigned shall as soon as possible, destroy such animal or cause such animal to be destroyed in his presence without unnecessary suffering; and any reasonable expense incurred in destroying the animal may be ordered by the court, (if the court is satisfied that it would be cruel to keep the animal alive, to direct that the animal be destroyed and to assign the animal to any reasonable expense incurred in destroying the animal may be ordered by the court) to be recovered from the owner as if it were a fine; Provided that unless the owner assents thereto, no order shall be made under this section except upon the evidence of a veterinary officer in charge of the area.

    (2) When any magistrate, commissioner of police or district superintendent of police has reason to believe that an offence under section 11 has been committed in respect of any animal, he may direct the immediate destruction of the animal, if in his opinion, it would be cruel to keep the animal alive.

    (3) Any police officer above the rank of a constable or any person authorized by the State Government in this behalf who finds any animal so diseased or so severely injured or in such a physical condition that in his opinion it cannot be removed without cruelty, may, if the owner is absent or refuses his consent to the destruction of the animal, forth with summon to the veterinary officer incharge of the area in which the animal is found, and if the veterinary officer certifies that the animal is mortally injured or so severely injured or in such a physical condition that it would be cruel to keep it alive, the police officer or the person authorizes, as the case may be, may, after obtaining orders from a magistrate, destroy the animal injured or cause it to be destroyed; *(in such manner as may be prescribed).

---

*Ins. By Act 26 of 1982, S. 12

(4) No appeal shall lie from any order of a magistrate for the destruction of an animal.

CHAPTER IV   EXPERIMENTATION OF ANIMALS

Experiments on animals

14. Nothing contained in this Act shall render unlawful the performance of experiments (including) experiments involving operations) on animals for the purpose of advancement by new discovery of physiological knowledge or of knowledge which will be useful for saving or for prolonging life or alleviating suffering or for combating any disease, whether of human beings, animals or plants.

Committee for the Purpose of Control and Supervision of Experiments on Animals

15. (1) If at any time, on the advice of the Board, the Central Government is of opinion that it is necessary so to do for the purpose of controlling and supervising experiments on animals, it may by notification in the Official Gazette constitute a Committee consisting of such number of officials and non-officials, as it may think fit to appoint thereto.

(2) The Central Government shall nominate one of the Member of the Committee to be its Chairman.

(3) The Committee shall have power to regulate its own Procedure in relation to the performance of its duties.

(4) The funds of the Committee shall consist of grants made to it from time to time by the Government and of contributions, donations, subscriptions, bequests, gifts and the like made to it by any person.

Sub-committee

*[15A (1) The Committee may constitute as many sub-committees as it thinks fit for exercising any power or discharging any duty of the Committee or for inquiring into or reporting and advising on any matter which the Committee may refer.

(2) A sub-committee shall consist exclusively of the Members of the Committee].

---

*Ins. by Act 26 of 1982, S. 13.

Staff of the Committee

16. Subject to the control of the Central Government, the Committee may appoint such number of officers and other employees as may be necessary to enable it to exercise its powers and perform its duties and may determine the remuneration and other terms and conditions of service of such officers and other employees.

Duties of the Committee and power of the Committee to make rules relating to experiments on animals

17. (1) It shall be the duty of the Committee to take all such measures as may be necessary to ensure that animals are not subjected to unnecessary pain or suffering before, during or after the performance of experiments on them, and for the purpose it may, by notification in the Gazette of India and subject to the condition of previous publication, make such rules as it may think fit in relation to the conduct of such experiments.

*[(1A) In particular, and without prejudice to the generality to the foregoing power, such rules may provide for the following matters namely:

  (a) the registration of persons or institutions carrying on experiments on animals;

  (b) the reports and other information which shall be forwarded to the Committee by persons and institutions carrying on experiments on animals.]

  (2) In particular, and without prejudice to the generality of the foregoing power, rules made by the Committee shall be designed to secure the following objects, namely:

  (a) that in cases where experiments are performed in any institution, the responsibility therefore is placed on the person in charge of the institution and that, in cases where experiments are performed outside an institution by individuals, the individuals, are qualified in that behalf and the experiments are performed on their full responsibility;

  (b) that experiments are performed with due care and humanity and that as far as possible experiments involving operations are performed under the influence of some anaesthetic of sufficient power to prevent the animals feeling pain;

---

*Ins. by Act 26 of 1982, S. 14.

(c)   that animals which, in the course of experiments under the influence of anesthetics, are so injured that their recovery would involve serious suffering, are ordinarily destroyed while still insensible;

(d)   that experiments on animals are avoided wherever it is possible to do so; as for example; in medical schools, hospitals, colleges and the like, if other teaching devices such as books, models, films and the like, may equally suffice;

(e)   that experiments on larger animals are avoided when it is possible to achieve the same results by experiments upon small laboratory animals like guinea pigs, rabbits, frogs and rats;

(f)   that, as far as possible, experiments are not performed merely for the purpose of acquiring manual skill;

(g)   that animals intended for the performance of experiments are properly looked after both before and after experiments;

(h)   that suitable records are maintained with respect to experiments performed on animals.

(3)   In making any rules under this section, the Committee shall be guided by such directions as the Central Government (consistently with the objects for which the Committee is set up) may give to it, and the Central Government is hereby authorized to give such direction.

(4)   All rules made by the Committee shall be binding on all individuals performing experiments outside institutions and on persons in-charge of institutions in which experiments are performed.

Power of entry and inspection

18.   For the purpose of ensuring that the rules made by it are being complied with the Committee may authorize any of its officers or any other person in writing to inspect any institution or place where experiments are being carried on and report to it as a result of such inspection, and any officer or person so authorized may-

(a)   enter at any time considered reasonable by him and inspect any institution or place in which experiments on animals are being carried on; and

    (b)   require any person to produce any record kept by him with respect to experiments on animals

Power to prohibit experiments on animals

19.   If the Committee is satisfied, on the report of any officer or other person made to it as a result of any inspection under section 18 or otherwise that the rules made by it under section 17 are not being animals, the Committee may, after giving an opportunity to the person or institution carrying on experiments on animals, of being heard in the matter, by order, prohibit the person or institution from carrying on any such experiments either for a specified period or indefinitely, or may allow the person or institution to carry on such experiments subject to such special conditions as the Committee may think fit to impose.

Penalties

20.   If any person:

    (a)   contravenes any order made by the Committee under Section 19; or

    (b)   commits a breach of any condition imposed by the Committee under that section:

he shall be punishable with fine which may extend to two hundred rupees, and, when the contravention or breach of condition has taken place in any institution the person in-charge of the institution shall be deemed to be guilty of the offence and shall be punishable accordingly.

CHAPTER V   PERFORMING ANIMALS

"Exhibit" and "Train" defined

21.   In this chapter, "exhibit" means exhibit or any entertainment to which the public are admitted through sale of tickets, and "train" means train for the purpose of any such exhibition, and the expressions "exhibitor" and "trainer" have respectively the corresponding meanings

Restriction on exhibition and training of performing animals

22.   No person shall exhibit or train

    (i)   any performing animal unless he is registered in accordance with the provisions of this chapter;

    (ii)   as a performing animal, any animal which the Central Government may, by notification in the official gazette, specify as an animal which shall not be exhibited or trained as a performing animal.

Procedure of registration

23.   (1)   Every person desirous of exhibiting or training any performing animal shall, on making an application in the prescribed form to the prescribed authority and on payment of the prescribed fee, be registered under this Act unless he is a person who, by reason of an order made by the court under this Chapter, is not entitled to be so registered.

    (2)   An application for registration under this Chapter shall contain such particulars as to the animals and as to the general nature of the performance in which the animals are to be exhibited or for which they are to be trained as may be prescribed, and the particulars so given shall be entered in the register maintained by the prescribed authority;

    (3)   The prescribed authority shall give to every person whose name appears on the register kept by them, a certificate of registration in the prescribed form containing the particulars entered in the register.

    (4)   Every register kept under this Chapter shall at all reasonable times be open for inspection on payment of the prescribed fee, and any person shall on payment of the prescribed fee, be entitled to obtain copies thereof or make extracts therefrom.

    (5)   Any person whose name is entered in the register shall, subject to the provisions of any order made under this Act by any court, be entitled, on making an application for the purpose, to have the particulars entered in the register with respect to him varied and where any such particulars are so varied, the existing certificate shall be cancelled and a new certificate issued.

Power of court to prohibit or restrict exhibition and training of performing animals

24.   (1)   Where it is proved to the satisfaction of any magistrate on a complaint made by a police officer or an officer authorized in writing by the prescribed authority referred to in Section 23, that the training or exhibition of any performing animals has been accompanied by unnecessary pain or suffering and should

be prohibited or allowed only subject to conditions, the court may make an order against the person in respect of whom the complaint is made, prohibiting the training or exhibition or imposing such conditions in relation thereto, as may be specified by the order.

(2) Any court by which an order is made under this section, shall cause a copy of the order to be sent, as soon as may be after the order is made, to the prescribed authority by which the person against whom the order is made is registered and shall cause the particulars of the order to be endorsed upon the certificate held by the person, and that person shall produce his certificate on being so required by the court for the purposes of endorsement, and the prescribed authority to which a coy of an order is sent under this section shall enter the particulars of the order in that register.

Power to enter premises

25. (1) Any person authorized in writing by the prescribed authority referred to in section 23 and any police officer not below the rank of a sub-inspector may

   (a) enter at all reasonable times and inspect any premises in which any performing animals are being trained or exhibited or kept for training or exhibition, and any such animals found therein; and

   (b) require any person who, he has reason to believe is a trainer or exhibitor of performing animals to produce his certificate of registration.

(2) No person or police officer referred to in sub section (1) shall be entitled under this section to go on or behind the stage during a public performance of performing animals.

Offences

26. If any person

   (a) not being registered under this chapter, exhibits or trains any performing animal; or

   (b) being registered under the Act, exhibits or trains any performing animal with respect to which or in a manner with respect to which, he is not registered; or

(c)   exhibits or trains as a performing animal, any animal which is not to be used for the purpose by reason of a notification issued under clause(ii) of section 22; or

(d)   obstructs or willfully delays any person or police officer referred to in section 25 in the exercise of powers under this Act as to entry and inspection; or

(e)   conceals any animal with a view to avoiding such inspection; or

(f)   being a person registered under the Act, on being duly required in pursuance of this Act to produce his certificate under this Act, fails without reasonable excuse so to do; or

(g)   applies to be registered under this Act when not entitled to be so registered

he shall be punishable on conviction with fine which may extend to five hundred rupees or with imprisonment which may extend to three months, or with both.

Exemptions

27.   Nothing contained in this Chapter shall apply to-

(a)   the training of animals for bona fide military or police purpose or the exhibition of any animals so trained; or

(b)   any animals kept in any zoological garden or by any society or association which has for its principal object the exhibition of animals for educational or scientific purposes.

CHAPTER VI   MISCELLANEOUS

Saving as respects manner or killing prescribed by religion

28.   Nothing contained in this Act shall render it an offence to kill any animal in a manner required by the religion of any community.

Power of court to deprive person convicted of ownership of animal

29.   (1)   If the owner of any animal is found guilty of any offence under this Act, the court upon his conviction thereof, may, if it thinks fit, in addition to any other punishment make an order that the animal with respect to which the offence was committed shall be forfeited to Government and may, further, make such order as to the disposal of the animal as it thinks fit under the circumstances.

(2)   No order under sub section (1) shall be made unless it is shown by evidence as to a previous conviction under this Act or as to

the character of the owner or otherwise as to the treatment of the animal that the animal if left with the owner, is likely to be exposed to further cruelty.

(3) Without prejudice to the provisions contained in sub-section (1), the court may also order that a person convicted of an offence under this Act shall, either permanently or during such period as is fixed by the order, be prohibited from having the custody of any animal of any kind whatsoever, or as the court thinks fit of any animal of any kind or species specified in the order.

(4) No order under sub-section (3) shall be made unless-

   (a) it is shown by evidence as to a previous conviction or as to the character of the said person or otherwise as to the treatment of the animal in relation to which he has been convicted that an animal in the custody of the said person is likely to be exposed to cruelty;

   (b) it is stated in the complaint upon which the conviction was made that it is the intention of the complaint upon the conviction of the accused to request that an order be made as aforesaid; and

   (c) the offence for which the conviction was made was committed in an area in which under the law for the time being in force a license is necessary for the keeping of any such animal as that in respect of which the conviction was made.

(5) Notwithstanding anything to the contrary contained in any law for the time being in force, any person in respect of whom an order is made under sub-section (3) shall have no right to the custody of any animal contrary to the provisions of the order, and if he contravenes the provisions of any order, he shall be punishable with fine which may extend to one hundred rupees, or with imprisonment for a term which may extend to three months, or with both.

(6) Any court which has made an order under sub-section (3) may at any time, either on its own motion or on application made to it in this behalf, rescind or modify such order.

Presumption as to guilt in certain cases

30. If any person is charged with the offences of killing a goat, cow or its progeny contrary to the provisions of clause (1) of sub-section (1) or

section 11, and it is proved that such person had in his possession, at the time the offence is alleged to have been committed, the skin of any such animal as is referred to in this section with any part of the skin of the head attached thereto, it shall be presumed until the contrary is proved that such animal was killed in a cruel manner.

Cognizability of offences

31.　Notwithstanding anything contained in the code or criminal procedure, 1898, (5 of 1898) an offence punishable under clause (1) or clause (n) or clause (o) of sub-section (1) of section 11 or under section 12 shall be a cognizable offence within the meaning of that code

Powers of search and seizure

32.　(1)　If a police officer not below the rank of sub-inspector, or any person authorized by the State Government in this behalf has reason to believe that an offence under clause (l) of sub-section (1) of section 11 in respect of any such animal as is referred to in section 30 is being, or that any person has in his possession the skin of any such animal with any part of the skin of the head attached thereto, he may enter and search such place or any place in which he has reason to believe any such skin to be, and may seize such skin or any article or thing used or intended to be used in the commission of such offence.

　　(2)　If a police officer not below the rank of sub-inspector, or any person authorized by the State Government in this behalf, has reason to believe that phooka or *(doom dev or any other operation of the nature referred to in section 12) has just been or is being, performed on any animal within the limits of his jurisdiction, he may enter any place in which he has reason to believe such animal to be, and may seize the animal and produce it for examination by the Veterinary Officer in charge of the area in which the animal is seized.

Search Warrants

33.　(1)　If a Magistrate of the first or second class or a Presidency Magistrate or a Commissioner of Police or District Superintendent of Police, upon information in writing; and after such inquiry as he thinks necessary, has reason to believe that an offence under this Act is being, or is about to be, or has been

---

*Subs. by Act 26 of 1982 S. 15 for the words "doom dev"

committed in any place, he may either himself enter and search or by his warrant authorize any police officer not below the rank of Sub-Inspector to enter and search the place.

(2) The provisions of the code of criminal procedure, 1898, relating to searches shall so far as those provision can be made applicable, apply to searches under this Act.

General power of seizure for examination

34. Any police officer above the rank of a constable or any person authorized by the State Government in this behalf, who has reason to believe that an offence against this Act has been or is being, committed in respect of any animal, may, if in his opinion the circumstances so require, seize the animal and produce the same for examination by the nearest Magistrate or by such Veterinary Officer as may be prescribed; and such police officer or authorized person may, when seizing the animal, require the person in charge thereof to accompany it to the place of examination.

Treatment and care of animals

35. (1) The State Government may be general or special order appoint infirmaries for the treatment and care of animals in respect of which offences against this Act have been committed, and may authorize the detention therein of any animal pending its production before a Magistrate.

(2) The Magistrate before whom a prosecution for an offence against this Act has been instituted may direct that the animals concerned shall be treated and cared for in an infirmary, until it is fit to perform its usual work or is otherwise fit for discharge, or that it shall be sent to a pinjrapole, or if the veterinary officer in charge of the area in which the animal is found or such a veterinary officer may be authorized in this behalf by rules made under this Act certified that it is incurable or cannot be removed without cruelty, that it shall be destroyed.

(3) An animal sent for care and treatment to any infirmary shall not, unless the magistrate directs that it shall be sent to a pinjrapole or that it shall be destroyed, be released from such place except upon a certificate of its fitness for discharge issued by the veterinary officer in charge of the area in which the infirmary is situated or such other veterinary officer as may be authorized in this behalf by rules made under this Act.

(4)   The cost of transporting the animal to an infirmary or pinjrapole and of its maintenance and treatment in an infirmary, shall be payable by the District Magistrate, or, in presidency towns, by the Commissioner of Police.

Provided that when the magistrate so orders on account of the poverty of the owner of the animal, no charge shall be payable for the treatment of the animal.

(5)   Any amount payable by an owner of an animal under sub-section (4) may be recovered in the same manner as an arrear of land revenue.

(6)   If the owner refuses or neglects to remove the animal within such time as a Magistrate may specify, the magistrate may direct that the animal be sold and that the proceeds of the same be applied to the payment of such cost.

(7)   The surplus, if any, of the proceeds of such sale shall, on application made by the owner within two months from the date of the sale be paid to him.

Limitation of prosecutions

36.   A prosecution for an offence against this Act shall not be instituted after the expiration of three months from the date of the commission of the offence.

Delegation of powers

37.   The Central Government may, by notification in the Official Gazette, direct that all or any of the powers exercisable by it under this Act, may, subject, to such conditions as it may think fit to impose, be also exercised by any State Government.

Power to make rules

38.   (1)   The Central Government may, by notification in the Official Gazette and subject to the condition of previous publication, make rules to carry out the purposes of this Act.

(2)   In particular, and without prejudice to the generality of the foregoing power, the Central Government may make rules providing for all or any of the following matters, namely:

(a)   the *(xxxx) conditions of service of members of the Board, the allowances payable to them and the manner in which they may exercise their powers and discharge their functions

---

*The Words "terms and" omitted by Act 26 of 1982, S. 16 (a) (i).

*[(aa) the manner in which the persons to represent municipal corporation are to be elected under clause (e) of sub-section (1) of section 5;)]

(b)   the maximum load (including any load occasioned by the weight of passengers) to be carried or drawn by any animal;

(c)   the conditions to be observed for preventing the overcrowding of animals;

(d)   the period during which, and the hours between which, any class of animals shall not be used for draught purposes;

(e)   prohibiting the use of any bit or harness involving cruelty to animals;

**[(ea) the other methods of destruction of stray dogs referred to in clause (b) of sub-section (3) of section 11;

(eb) the methods by which any animal which cannot be removed without cruelty may be destroyed under sub-section (3) of section 13]

(f)   requiring persons carrying on the business of a farrier to be licensed and registered by such authority as may be prescribed and levying a fee for the purpose;

(g)   the precautions to be taken in the capture of animals for purposes of sale, export or for any other purpose, and the different appliances or devices that may alone be used for the purpose; and the licensing of such capture and the levying of fees for such licenses;

(h)   the precautions to be taken in the transport of animals whether by rail, road, inland waterway, sea or air and the manner in which and the cages or other receptacles in which they may be so transported;

(i)   requiring person owning or in charge of premises in which animals are kept or milked to register such premises, to comply with such conditions as may be laid down in relation to the boundary walls or surroundings of such premises, to permit their inspection for the purpose of ascertaining whether any offence under this Act is being, or has been

---

*Ins ibid S. 16 (a) (ii).

**Ins by Act 26 of 1982 S. 16 (a) (iii)

committed therein, and to expose in such premises copies of section 12 in a language or languages commonly understood in the locality;

(j) the form in which applications for registration under Chapter V may be made, the particulars to be contained therein the fees payable for such registration and the authorities to whom such applications may be made;

*[(ja) the fees which may be charged by the Committee constituted under section 15 for the registration of persons or institutions carrying on experiments on animals or for any other purpose;]

(k) the purposes to which fines realized under the Act may be applied, including such purposes as the maintenance of infirmaries, pinjrapole and veterinary hospitals;

(l) any other matter which has to be, or may be prescribed

(3) If any person contravenes, or abets the contravention of, any rules made under this section, he shall be punishable with fine which may extend to one hundred rupees, or with imprisonment for a term which may extend to three months, or with both

** [xxxx]

Rules and Regulations to be laid before Parliament

38A. ***[Every rule made by the Central Government or by the Committee constituted under section 15 and every regulation made by the Board shall be laid, as soon as may be after it is made, before each House of Parliament, while it is in session, for a total period of thirty days which may be comprised in one session or in two or more successive sessions, and if, before the expiry of the session immediately following the session or the successive sessions aforesaid, both Houses agree in making any modification in the rule or regulation, as the case may be, should not be made the rule or regulation shall there after have effect only in such modified form or be of no effect, as the case may be; so, however, that any such modification or annulment shall be without prejudice to the validity of anything previously done under that rule or regulation.]

---

*Ins. by Act 26 of 1982. S. 16 (a) (iv)

**Sub-section (4) of the Principal Act omitted by Act 26 of 1982. S. 16(b).

*** Ins. Ibid S, 17.

Persons authorized under Section 34 to be public servants

39. Every person authorized by the State Government under Section 34 shall be deemed to be a public servant within the meaning of section 21 of the Indian Penal Code.

Indemnity

40. No suit, prosecution or other legal proceeding shall lie against any person who is, or who is deemed to be a public servant within the meaning of section 21 of the Indian Penal Code in respect of anything in good faith done or intended to be done under this Act.

Repeal of Act 11 of 1890

41. Where in pursuance of a notification under sub-section (3) of section 1 any provision of this Act comes into force in any State, any provision of the Prevention of Cruelty to Animals Act, 1890, which corresponds to the provision so coming into force, shall thereupon stand repealed.

APPENDIX II

CPCSEA GUIDELINES FOR LABORATORY ANIMAL FACILITY

Committee for the Purpose of Control and Supervision on Experiments on Animals

Good Laboratory Practices (GLP) for animal facilities is intended to assure quality maintenance and safety of animals used in laboratory studies while conducting biomedical and behavioural research and testing of products.

GOAL

The goal of these guidelines is to promote the humane care of animals used in biomedical and behavioural research and testing with the basic objective of providing specifications that will enhance animal well-being, quality in the pursuit of advancement of biological knowledge that is relevant to humans and animals.

VETERINARY CARE

Adequate veterinary care must be provided and is the responsibility of a veterinarian or a person who has training or experience in laboratory animal sciences and medicine.

Daily observation of animals can be accomplished by someone other than a veterinarian; however, a mechanism of direct and frequent communication should be adopted so that timely and accurate information on problems in animal health, behaviour, and well-being is conveyed to the attending veterinarian.

The veterinarian can also contribute to the establishment of appropriate policies and procedures for ancillary aspects of veterinary care, such as reviewing protocols and proposals, animal husbandry and animal welfare; monitoring occupational health hazards, containment, and zoonosis control programs; and supervising animal nutrition and sanitation. Institutional requirements will determine the need for full-time or part-time or consultative veterinary services.

ANIMAL PROCUREMENT

All animals must be acquired lawfully as per the CPCSEA guidelines.

A health surveillance program for screening incoming animals should be carried out to assess animal quality. Methods of transportation should also be taken into account (Annexure 4).

Each consignment of animals should be inspected for compliance with procurement specifications, and the animals should be quarantined and stabilized according to procedures appropriate for the species and circumstances.

QUARANTINE, STABILIZATION AND SEPARATION

Quarantine is the separation of newly received animals from those already in the facility until the health and possibly the microbial status of the newly received animals have been determined. An effective quarantine minimizes the chance for introduction of pathogens into an established colony. A minimum duration of quarantine for small lab animals is one week and larger animals is 6 weeks (cat, dog and monkey).

Effective quarantine procedures should be used for non-human primates to help limit exposure of humans to zoonotic infections.

Regardless of the duration of quarantine, newly received animals should be given a period for physiologic, psychologic and nutritional stabilization before their use. The length of time stabilization will depend on the type and duration of animal transportation, the species involved and the intended use of the animals.

Physical separation of animals by species is recommended to prevent interspecies disease transmission and to eliminate anxiety and possible physiological and behavioural changes due to interspecies conflict.

Such separation is usually accomplished by housing different species in separate rooms; however, cubicles, laminar-flow units, cages that have filtered air or separate ventilation, and isolators shall be suitable alternatives.

In some instances, it shall be acceptable to house different species in the same room, for example, if two species have a similar pathogen status and are behaviourally compatible.

SURVEILLANCE, DIAGNOSIS, TREATMENT AND CONTROL OF DISEASE

All animals should be observed for signs of illness, injury, or abnormal behaviour by animal house staff. As a rule, this should occur daily, but more-frequent observations might be warranted, such as during post-operative recovery or when animals are ill or have a physical deficit. It is imperative that appropriate methods be in place for disease surveillance and diagnosis (Annexure 1 and 2).

Unexpected deaths and signs of illness, distress, or other deviations from normal health condition in animals should be reported promptly to ensure appropriate and timely delivery of veterinary medical care. Animals that show signs of a contagious disease should be isolated from healthy animals in the colony. If an entire room of animals is known or believed to be exposed to an infectious agent (e.g. *Mycobacterium tuberculosis* in non-human primates), the group should be kept intact and isolated during the process of diagnosis, treatment, and control. Diagnostic clinical laboratory may be made available.

ANIMAL CARE AND TECHNICAL PERSONNEL

Animal care programs require technical and husbandry support. Institutions should employ people trained in laboratory animal science or provide for both formal and on-the-job training to ensure effective implementation of the program (Annexure 7).

PERSONAL HYGIENE

It is essential that the animal care staff maintain a high standard of personal cleanliness. Facilities and supplies for meeting this obligation should be provided e.g. showers, change of uniforms, footwears etc. Clothing suitable for use in the animal facility should be supplied and laundered by the institution. A commercial laundering service is acceptable in many situations; however, institutional facilities should be used to decontaminate clothing exposed to potentially hazardous microbial agents or toxic substances. In some circumstances, it is acceptable to use disposable wear such as gloves, masks, head covers, coats, coveralls and shoe covers. Personnel should change clothing as often as is necessary to maintain personal hygiene. Outer garments worn in the animal rooms should not be worn outside the animal facility.

Washing and showering facilities appropriate to the program should be available. Personnel should not be permitted to eat, drink, smoke or apply cosmetics in animal rooms. A separate area or room should be made available for these purposes.

ANIMAL EXPERIMENTATION INVOLVING HAZARDOUS AGENTS

Institutions should have policies governing experimentation with hazardous agents. Institutional Biosafety Committee whose members are knowledgeable about hazardous agents are in place in most of the higher level education, research institutes and in many pharmaceutical industries for safety issues. This committee shall also examine the proposal on animal experiments involving hazardous agents in addition to its existing functions (Annexure 8).

Since the use of animals in such studies requires special consideration, the procedures and the facilities to be used must be reviewed by both the Institutional Biosafety Committee and Institutional Animal Ethics Committee (IAEC).

MULTIPLE SURGICAL PROCEDURES ON SINGLE ANIMAL

Multiple surgical procedures on a single animal for any testing or experiment are not to be practiced unless specified in a protocol only approved by the IAEC.

DURATIONS OF EXPERIMENTS

No animal should be used for experimentation for more than 3 years unless adequate justification is provided.

PHYSICAL RESTRAINT

Brief physical restraint of animals for examination, collection of samples, and a variety of other clinical and experimental manipulations can be accomplished manually or with devices being suitable in size and design for the animal being held and operated properly to minimize stress and avoid injury to the animal.

Prolonged restraint of any animal, including the chairing of non-human primates, should be avoided unless essential to research objectives. Less restrictive systems, such as the tether system or the pole and collar system, should be used when compatible with research objectives.

The following are important guidelines for the use of restraint equipments:

Restraint devices cannot be used simply as a convenience in handling or managing animals.

The period of restraint should be the minimum required to accomplish the research objectives.

Animals to be placed in restraint devices should be given training to adapt to the equipment.

Provision should be made for observation of the animal at appropriate intervals. Veterinary care should be provided if lesions or illness associated with restraint are observed. The presence of lesions, illness, or severe behavioural change should be dealt with by the temporary or permanent removal of the animal from restraint.

PHYSICAL PLANT

The physical condition and design of animal facility determine, to a great extent, the efficiency and economy of their operation. The design and size of an animal facility depend on the scope of institutional research activities, animals to be housed, physical relationship to the rest of the institution, and geographic location. A well-planned, properly maintained facility is an important element in good animal care.

PHYSICAL RELATIONSHIP OF ANIMAL FACILITIES TO LABORATORIES

Good animal husbandry and human comfort and health protection require separation of animal facilities from personnel areas such as offices, conference rooms, and most laboratories.

- Laboratory animals are very sensitive to their living conditions. It is important that they shall be housed in an isolated building located as far away from human habitations as possible and not exposed to dust, smoke, noise, wild rodents, insects and birds. The building, cages and environment of animal rooms are the major factors, which affect the quality of animals.

- This separation can be accomplished by having the animal quarters in a separate building, wing, floor, or room. Careful planning should make it possible to place animal housing areas adjacent to or near laboratories, but separated from them by barriers such as entry locks, corridors, or floors.

- In planning an animal facility the space should be well divided for various activities. The animal rooms should occupy about 50–60% of the total constructed area and the remaining area should be utilized for services such as stores, washing, office and staff, machine rooms, quarantine and corridors. The environment of animal room (macroenvironment) and animal cage (microenvironment) are factors on which the production and experimental efficiency of the animal depends. Since animals are very sensitive to environmental changes, sharp fluctuations in temperature, humidity, light, sound and ventilation should be avoided. The recommended space requirements for animal rooms, for different species are given in (Annexure 3).

FUNCTIONAL AREAS

The size and nature of a facility will determine whether areas for separate service functions are possible or necessary. Sufficient animal area is required to:

- ensure separation of species or isolation of individual projects when necessary
- receive, quarantine, and isolate animals and
- provide for animal housing

In facilities that are small, maintain few animals or maintain animals under special conditions (e.g. facilities exclusively used for housing germ-free colonies or animals in runs and pens) some functional areas listed below could be unnecessary or included in a multipurpose area. Professional judegment must be exercised when developing a practical system for animal care.

- Specialized laboratories
- Individual areas contiguous with or near animal housing areas for such activities as surgery, intensive care, necropsy, radiography, preparation of special diets, experimental manipulation, treatment, and diagnostic laboratory procedures containment facilities or
- Equipment, if hazardous biological, physical, or chemical agents are to be used
- Receiving and storage areas for food, bedding
- Pharmaceuticals and biologics and supplies
- Space for administration, supervision and direction of the facility
- Showers, sinks, lockers and toilets for personnel
- An area for washing and sterilization of equipment and supplies
- An autoclave for equipment
- Food and bedding and separate areas
- For holding soiled and unclean equipment
- An area for repairing cages and equipment
- An area to store wastes prior to incineration or removal

PHYSICAL FACILITIES

a) **Building materials** should be selected to facilitate efficient and hygienic operation of animal facilities. Durable, moisture-proof, fire-resistant, seamless materials are most desirable for interior surfaces including vermin and pest resistance.

b) **Corridor(s)** should be wide enough to facilitate the movement of personnel as well as equipments and should be kept clean.

c) **Utilities** such as water lines, drain pipes and electrical connections should preferably be accessible through service panels or shafts in corridors outside the animal rooms.

d) **Animal room doors**   Doors should be rust, vermin and dust proof. They should fit properly within their frames and provided with an observation window. Door closures may also be provided. Rodent barriers can be provided in the doors of the small animal facilities.

e) **Exterior windows**   Windows are not recommended for small animal facilities. However, where power failures are frequent and backup power is not available, they may be necessary to provide alternate source of light and ventilation. In primate rooms, windows can be provided.

f) **Floors**   Floors should be smooth, moisture proof, non-absorbent, skid-proof, resistant to wear, acid, solvents, adverse effects of detergents and disinfectants.

They should be capable of supporting racks, equipment, and stored items without becoming gouged, cracked, or pitted, with minimum number of joints.

A continuous moisture-proof membrane might be needed. If sills are installed at the entrance to a room, they should be designed to allow for convenient passage of equipment.

g) **Drains**   Floor drains are not essential in all rooms used exclusively for housing rodents. Floor in such rooms can be maintained satisfactorily by wet vacuuming or mopping with appropriate disinfectants or cleaning compounds. Where floor drains are used, the floors should be sloped and drain taps kept filled with water or corrosion free mesh. To prevent high humidity, drainage must be adequate to allow rapid removal of water and drying of surfaces.

h) **Walls and ceilings**   Walls should be free of cracks, unsealed utility penetrations, or imperfect junctions with doors, ceilings, floors and corners.

Surface materials should be capable of withstanding scrubbing with detergents and disinfectants and the impact of water under high pressure.

i) **Storage areas**   Separate storage areas should be designed for feed, bedding, cages and materials not in use.

Refrigerated storage, separated from other cold storage, is essential for storage of dead animals and animal tissue waste.

j) **Facilities for sanitizing equipment and supplies**   An area for sanitizing cages and ancillary equipment is essential with adequate water supply.

k) **Experimental area** All experimental procedures in small animals should be carried out in a separate area away from the place where animals are housed. For larger animal functional areas for aseptic surgery should include a separate surgical support area, a preparation area, the operating room or rooms, and an area for intensive care and supportive treatment of animals.

ENVIRONMENT

a) **Temperature and humidity control**

Air conditioning is an effective means of regulating these environmental parameters for laboratory animals. Temperature and humidity control prevents variations due to changing climatic conditions or differences in the number and kind of room occupants. Ideally, capability should be provided to allow variations within the range of approximately 18 to 29°C (64.4 to 84.2 °F), which includes the temperature ranges usually recommended for common laboratory animals.

The relative humidity should be controllable within the range of 30% to 70% throughout the year. For larger animals a comfortable zone (18 to 37°C) should be maintained during extreme summer by appropriate methods for cooling.

b) **Ventilation**

In renovating existing or in building new animal facilities, consideration should be given to the ventilation of the animals' primary enclosures.

Heating, ventilating, and air-conditioning systems should be designed so that operation can be continued with a standby system. The animal facility and human occupancy areas should be ventilated separately.

c) **Power and lighting**

The electrical system should be safe and provide appropriate lighting and a sufficient number of power outlets. It is suggested that a lighting system be installed that provides adequate illumination while people are working in the animal rooms and a lowered intensity of light for the animals.

Fluorescent lights are efficient and available in a variety of acceptable fixtures.

A time-controlled lighting system should be used to ensure a regular diurnal lighting cycle wherever required. Emergency power should be available in the event of power failure.

d) Noise control

The facility should be provided with noise free environment. Noise control is an important consideration in designing an animal facility. Concrete walls are more effective than metal or plaster walls in containing noise because their density reduces sound transmission.

ANIMAL HUSBANDRY

Caging or Housing System

The caging or housing system is one of the most important elements in the physical and social environment of research animals. It should be designed carefully to facilitate animal well being, meet research requirements, and minimize experimental variables.

The housing system should:

- provide space that is adequate, permit freedom of movement and normal postural adjustments, and have a resting place appropriate to the species; (Annexure 3)

- provide a comfortable environment

- provide an escape proof enclosure that confines animal safety

- provide easy access to food and water

- provide adequate ventilation

- meet the biological needs of the animals, e.g. maintenance of body temperature, urination, defecation and reproduction. Keep the animals dry and clean, consistent with species requirements

- facilitate research while maintaining good health of the animals.

They should be constructed of sturdy, durable materials and designed to minimize cross-infection between adjoining units. Polypropylene, polycarbonate and stainless steel cages should be used to house small lab animals, monkeys should be housed in cages made of steel or painted mild steel and for other animals such as sheep, horses, the details can be seen in Annexure 3.

To simplify servicing and sanitation, cages should have smooth, impervious surfaces that neither attract nor retain dirt and a minimum number of ledges, angles, and corners in which dirt or water can accumulate.

The design should allow inspection of cage occupants without disturbing them. Feeding and watering devices should be easily accessible for filling, changing, cleaning and servicing.

Cages, runs and pens must be kept in good condition to prevent injuries to animals, promote physical comfort, and facilitate sanitation and servicing. Particular attention must be given to eliminate sharp edges and broken wires, keeping cage floors in good condition.

Sheltered or Outdoor Housing

When animals are maintained in outdoor runs, pens, or other large enclosures, there must be protection from extreme temperature or other harsh weather conditions and adequate protective and escape mechanism for submissive animals, as in case of monkeys by way of an indoor portion of a run, should be provided.

Shelter should be accessible to all animals, have sufficient ventilation, and be designed to prevent build up of waste materials and excessive moisture.

Houses, dens, boxes, shelves, perches, and other furnishings should be constructed in a manner and made of materials that allow cleaning or replacement in accordance with generally accepted husbandry practices when the furnishings are soiled or wornout.

Ground-level surfaces of outdoor housing facilities can be covered with absorbent bedding, sand, gravel, grass, or similar material that can be removed or replaced when needed to ensure appropriate sanitation.

Build up of animal waste and stagnant water should be avoided, for example, by using contoured or drained surface. Other surfaces should be able to withstand the elements and be easily maintained.

Social Environment

The social environment includes all interactions among individuals of a group or among those able to communicate. The effects of social environment on caged animals vary with the species and experience of the animals.

In selecting a suitable social environment, attention should be given to whether the animals are naturally territorial or communal and whether they will be housed singly or in groups.

When appropriate, group housing should be considered for communal animals. In grouping animals, it is important to take into account population density and ability to disperse; initial familiarity among animals; and age, sex, and social rank.

Population density can affect reproduction, metabolism, immune responses, and behaviour. Group composition should be held as stable as possible, particularly for canine, non-human primates, and other highly social mammals, because mixing of groups or introducing new members can alter behavioural and physiological functions.

Non-human primates should have a run for free ranging activities.

ACTIVITY

Provision should be made for animals with specialized locomotor pattern to express these patterns, especially when the animals are held for long periods. For e.g. ropes, bars, and perches are appropriate for branching non-human primates.

Cages are often used for short-term (up to 3 months) housing of dogs and may be necessary for post surgical care, isolation of sick dogs, and metabolic studies.

Pens, runs, or other out-of-cage space provide more opportunity for exercise, and their use is encouraged when holding dogs for long periods.

FOOD

Animals should be fed palatable, non-contaminated, and nutritionally adequate food daily unless the experimental protocol requires otherwise.

Feeders should allow easy access to food, while avoiding contamination by urine and faeces.

Food should be available in amounts sufficient to ensure normal growth in immature animals and maintenance of normal body weight, reproduction, and lactation in adults.

Food should contain adequate nutrition, including formulation and preparation; free from chemical and microbial contaminants; bio-availability of nutrients should be at par with the nutritional requirement of the animal.

Laboratory animal diets should not be manufactured or stored in facilities used for farm feeds or any products containing additives such as rodenticides, insecticides, hormones, antibiotics, fumigants, or other potential toxicants.

Areas in which diets are processed or stored should be kept clean and enclosed to prevent entry of insects or other animals.

Precautions should be taken if perishable items such as meats, fruits, and vegetables are fed, because these are potential sources of biological and chemical contamination and can also lead to variation in the amount of nutrients consumed.

Diet should be free from heavy metals (e.g. lead, arsenic, cadmium, nickel, mercury), naturally occurring toxins and other contaminants.

Exposure to extremes in relative humidity, unsanitary conditions, light, oxygen, and insects hasten the deterioration of food.

Meats, fruits, vegetables, and other perishable items should be refrigerated if required to be stored. Unused, open food should be stored in vermin-proof condition to minimize contamination and to avoid potential spread of disease agents.

Food hoppers should not be transferred from room to room unless cleaned and sanitized. The animal feed should contain moisture, crude fibre, crude protein, essential vitamins, minerals crude fat and carbohydrate for providing appropriate nutrition.

BEDDING

Bedding should be absorbent, free of toxic chemicals or other substances that could injure animals or personnel, and of a type not readily eaten by animals. Bedding should be used in amounts sufficient to keep animals dry between cage changes without coming into contact with watering tubes.

Bedding should be removed and replaced with fresh materials as often as necessary to keep the animals clean and dry. The frequency is a matter of professional judgement of the animal care personnel in consultation with the investigation depending on the number of animals and size of cages. However it is ideal to change the bedding twice a week.

The desirable criteria for rodent contact bedding is ammonia binding, sterilizable, deleterious products not formed as a result of sterilization, easily stored, non-desiccating to the animal, uncontaminated, unlikely to be chewed or mouthed, non-toxic, non-malodorous, nestable, disposable by incineration, readily available, remains chemically stable during use, manifests batch uniformity, optimizes normal animal behaviour, non-deleterious to cage-washers, non-injurious and non-hazardous to personnel, non-nutritious and non-palatable.

Nesting materials for newly delivered pups wherever can be provided (e.g. paper, tissue paper and cotton).

WATER

Ordinarily animals should have continuous access to fresh, potable, uncontaminated drinking water, according to their particular requirements. Periodic monitoring of microbial contamination in water is necessary.

Watering devices, such as drinking tubes and automatic waterers if used should be examined routinely to ensure their proper operation. Sometimes it is necessary to train animals to use automatic watering devices.

It is better to replace water bottles than to refill them, however, if bottles are refilled, care should be taken that each bottle is replaced on the cage which it was removed.

SANITATION AND CLEANLINESS

Sanitation is essential in an animal facility. Animal rooms, corridors, storage spaces, and other areas should be cleaned with appropriate detergents and disinfectants as often as necessary to keep them free of dirt, debris, and harmful contamination.

Cleaning utensils, such as mops, pails, and brooms, should not be transported between animal rooms.

Where animal waste is removed by hosing or flushing, this should be done at least twice a day. Animals should be kept dry during such procedures. For larger animals, such as dogs, cats, and non-human primates, soiled litter material should be removed twice daily.

Cages should be sanitized before animals are placed in them. Animal cages, racks, and accessory equipments, such as feeders and watering devices, should be washed and sanitized frequently to keep them clean and contamination free. Ordinarily this can be achieved by washing solid bottom rodent cages and accessories once or twice a week and cages, racks at least monthly.

Wire-bottom rodent cages for all other animals should be washed at least every 2 weeks. It is good practice to have extra cages available at all times so that a systematic cage-washing schedule can be maintained. Cages can be disinfected by rinsing at a temperature of 82.2°C (180 °F) or higher for a period long enough to ensure the destruction of vegetative pathogenic organisms.

Disinfection can also be accomplished with appropriate chemicals; equipments should be rinsed free of chemicals prior to use. Periodic microbiologic monitoring is useful to determine the efficacy of disinfection or sterilization procedures.

Rabbits and some rodents, such as guinea pigs and hamsters, produce urine with high concentration of proteins and minerals. Minerals and organic compounds in the urine from these animals often adhere to cage surfaces and necessitate treatment with acid solutions before washing.

Water bottles, sipper tubes, stoppers, and other watering equipment should be washed and then sanitized by rinsing with water of at least 82.2°C (180 °F) or appropriate chemicals agents (e.g. hyperchlorite) to destroy pathogenic organisms. If bottles are washed by hand, powered rotating brushes at the washing sink are useful, and provision should be made for dipping or soaking the water bottles in detergents and disinfectant solutions. A two compartment sink or tub is adequate for this purpose.

Some means for sterilizing equipments and supplies, such as an autoclave or gas sterilizer, is essential when pathogenic organisms are present. Routine sterilization of cages, food and bedding is not considered essential if care is taken to use clean materials from reliable sources. Where hazardous biological, chemical, or physical agents are used, a system of equipment monitoring might be appropriate.

Deodorizers or chemical agents other than germicidal should not be used to mask animal odours. Such products are not a substitute for good sanitation.

ASSESSING THE EFFECTIVENESS OF SANITATION

Monitoring of sanitation practices should be appropriate to the process and materials being cleaned; it can include visual inspection of the materials, monitoring of water temperatures, or microbiologic monitoring.

The intensity of animal odours, particularly that of ammonia, should not be used as the sole means of assessing the effectiveness of the sanitation program.

A decision to alter the frequency of cage bedding changes or cage-washing should be based on factors such as the concentration of ammonia, the appearance of the cage, the condition of the bedding and the number and size of animals housed in the cage.

WASTE DISPOSAL

Wastes should be removed regularly and frequently. All waste should be collected and disposed of in a safe and sanitary manner. The most preferred method of waste disposal is incineration. Incinerators should be in compliance with all central, state, and local regulations.

Waste cans containing animal tissues, carcasses, and hazardous wastes should be lined with leak-proof, disposable liners. If wastes must be stored before removal, the waste storage area should be separated from other storage facilities and free of flies, cockroaches, rodents, and other vermin. Cold storage might be necessary to prevent decomposition of biological wastes. Hazardous wastes should be rendered safe by sterilization, contamination, or other appropriate means before they are removed from an animal facility for disposal.

PEST CONTROL

Programs designed to prevent, control, or eliminate the presence of or infestations by pests are essential in an animal environment.

EMERGENCY, WEEKEND AND HOLIDAY CARE

Animals should be cared for by qualified personnel every day, including weekends and holidays, to safeguard their well-being including emergency veterinary care. In the event of an emergency, institutional security personnel and fire or police officials should be able to reach people responsible for the animals. This can be enhanced by prominently posting emergency procedures, names, or telephone numbers in animals facilities or by placing them in the security department or telephone centre. A disaster plan that takes into account both personnel and animals should be prepared as part of the overall safety plan for the animal facility.

RECORD KEEPING

The animal house should maintain the following records:

- Animal house plans, which includes typical floor plan, all fixtures etc.
- Animal house staff record—both technical and non-technical
- Health record of staff/ animals
- All standard operating procedures (SOPs) relevant to the animals
- Breeding, stock, purchase and sales records
- Minutes of Institute Animals Ethics Committee meetings
- Records of experiments conducted with the number of animals used (copy of Form D)
- Death Record
- Clinical record of sick animals
- Training record of staff involved in animal activities
- Water analysis report

STANDARD OPERATING PROCEDURES (SOPs)/GUIDELINES

The Institute shall maintain SOPs describing procedures/methods adapted with regard to animal husbandry, maintenance, breeding, animal house microbial analysis and experimentation records.

A SOP should contain the following items:

- Name of the Author
- Title of the SOP
- Date of preparation
- Reference of previous SOP on the same subject and date (Issue No. and Date)
- Location and distribution of SOPs with sign of each recipient
- Objectives
- Detailed information of the instruments used in relation with animals with methodology (Model No., Serial No. and Date of commissioning)
- The name of the manufacturer of the reagents and the methodology of the analysis pertaining to animals
- Normal value of all parameters
- Hazard identification and risk assessment

PERSONNEL AND TRAINING

The selection of animal facility staff, particularly the staff working in animal rooms or involved in transportation, is a critical component in the management of an animal facility.

The staff must be provided with all required protective clothing (masks, aprons, gloves and gumboots and other footwear) while working in animal rooms. Facilities should be provided for change over with lockers, wash basin, toilets and bathrooms to maintain personal hygiene. It is also important a regular medical check-up is arranged for the workers to ensure that they have not picked up any zoonotic infection and also that they are not acting as a source of transmission of infection to the animals. The animal house in-charge should ensure that persons working in animal house do not eat, drink, smoke in animal room and have all required vaccination, particularly against tetanus and other zoonotic diseases.

Initial in-house training of staff at all levels is essential. A few weeks must be spent on the training of the newly recruited staff, teaching them the animal handling techniques, cleaning of cages and importance of hygiene, disinfection and sterilization. They should also be made familiar with the activities of normal healthy and sick animals so that they are able to spot the sick animal during their daily routine check up of cages (Annexure 7).

TRANSPORT OF LABORATORY ANIMALS

The transport of animals from one place to another is very important and must be undertaken with care. The main considerations for transport of animals are, the mode of transport, the containers, the animal density in cages, food and water during transit, protection from transit infections, injuries and stress.

The mode of transport of animals depends on the distance, seasonal and climatic conditions and the species of animals. Animals can be transported by road, rail or air taking into consideration of above factors. In any case the transport stress should be avoided and the containers should be of an appropriate size so as to enable these animals to have a comfortable, free movement and protection from possible injuries. The food and water should be provided in suitable containers or in suitable form so as to ensure that they get adequate food and more particularly water during transit. The transport containers (cages or crates) should be of appropriate size and only a permissible number of animals should be accommodated in each container to avoid overcrowding and infighting (Annexure 4).

ANAESTHESIA AND EUTHANASIA

The scientists should ensure that the procedures, which are considered painful, are conducted under appropriate anaesthesia as recommended for each species of animals.

It must also be ensured that the anaesthesia is given for the full duration of experiment and at no stage the animal is conscious to perceive pain during the experiment. If at any stage during the experiment the investigator feels that he has to abandon the experiment or he has inflicted irreparable injury, the animal should be sacrificed. Neuromuscular blocking agents must not be used without adequate general anaesthesia (Annexure 5).

In the event of a decision to sacrifice an animal on termination of an experiment or otherwise, an approved method of euthanasia should be adopted (Annexure 6) and the investigator must ensure that the animal is clinically dead before it is sent for disposal. The data about large animals, which have been euthanised, should be maintained.

Anaesthesia

Unless contrary to the achievement of the results of study, sedatives, analgesics and anaesthetics should be used to control pain or distress under experiment. Anaesthetic agents generally affect cardiovascular, respiratory and thermo-regulatory mechanism in addition to central nervous system.

Before using actual anaesthetics the animal is prepared for anaesthesia by overnight fasting and using pre-anaesthetics, which block parasympathetic stimulation of cardio-pulmonary system and reduce salivary secretion. Atropine is the most commonly used anticholinergic agent. Local or general anaesthesia may be used, depending on the type of surgical procedure.

Local anaesthetics are used to block the nerve supply to a limited area and are used only for minor and rapid procedures. This should be carried out under expert supervision for regional infiltration of surgical site, nerve blocks and for epidural and spinal anaesthesia.

A number of general anaesthetic agents are used in the form of inhalants. General anaesthetics are also used in the form of intravenous or intramuscular injections such as barbiturates. Species characteristics and variation must be kept in mind while using an anaesthetic. Side effects such as excessive salivation, convulsions, excitement and disorientation should be suitably prevented and controlled. The animal should remain under veterinary care till it completely recovers from anaesthesia and post-operative stress.

Euthanasia

Euthanasia is resorted to events where an animal is required to be sacrificed on termination of an experiment or otherwise for ethical reasons. The procedure should be carried out quickly and painlessly in an atmosphere free from fear or anxiety. For accepting an euthanasia method as humane it should have an initial depressive action on the central nervous system for immediate insensitivity to pain. The choice of a method will depend on the nature of study, the species of animal to be killed (Annexure 6). The method should in all cases meet the following requirements:

a) Death, without causing anxiety, pain or distress with minimum time lag phase.

b) Minimum physiological and psychological disturbances.

c) Compatibility with the purpose of study and minimum emotional effect on the operator.

d) Location should be separate from animal rooms and free from environmental contaminants.

Tranquilizers have to be administered to larger species such as monkeys, dogs and cats before an euthanasia procedure.

LABORATORY ANIMAL ETHICS

All scientists working with laboratory animals must have a deep ethical consideration for the animals they are dealing with. From the ethical point of view it is important that such considerations are taken care at the individual level, at institutional level and finally at the national level.

TRANSGENIC ANIMALS

Transgenic animals are those animals, into whose germ line foreign gene(s) have been engineered, whereas knockout animals are those whose specific gene(s) have been disrupted leading to loss of function. These animals can be bred to establish transgenic animal strains. Transgenic animals are used to study the biological functions of specific genes, to develop animal models for diseases of humans or animals, to produce therapeutic products, vaccines and for biological screening. These can be either developed in the laboratory or produced for R&D purpose from registered scientific/ academic institutions or commercial firms, and generally from abroad with approval from appropriate authorities.

MAINTENANCE

Housing, feeding, ventilation, lighting, sanitation and routine management practices for such animals are similar to those for the other animals of the species as given in guidelines. However, special care has to be taken with transgenic/gene knockout animals where the animals can become susceptible to diseases where special conditions of maintenance are required due to the altered metabolic activities. The transgenic and knockout animals carry additional genes or lack genes compared to the wild population. To avoid the spread of the genes in wild population care should be taken to ensure that these are not inadvertently released in the wild to prevent cross breeding with other animals. The transgenic and knockout animals should be maintained in clean room environment or in animal isolators.

DISPOSAL

The transgenic and knockout animals should be first euthanized and then disposed off as prescribed elsewhere in the guidelines. A record of disposal and the manner of disposal should be kept as a matter of routine.

BREEDING AND GENETICS

For initiating a colony, the breeding stock must be procured from CPCSEA registered breeders or suppliers ensuring that genetic makeup and health status of animal is known. In case of an inbred strain, the characters of the strain with their gene distribution and the number of inbred generation must be known for further propagation. The health status should indicate their origin, e.g. conventional, specific pathogen free or transgenic gnotobiotic or knockout stock.

ANNEXURE 1

Haematological data of commonly used laboratory animals.

	Mouse	Rat	Hamster	G. pig	Rabbit	Cat	Dog (Beagle)	Monkey (Rhesus)
RBC (X10 mm^3)	7–25	7–10	6–10	4.5–7	4–7	5–10	55–95	3.56–6.98
PCV (%)	39–49	36–18	36–55	37–48	38–48	30–15	37–55	26–48
Hb (g/dl)	10.2–16.6	11–18	10–16	11–15	10–15.5	8–15	12–18	8.8–16.5
WBC ($\times$ 10^3/mm^3)	6–15	6–17	3–11	7–18	9–11	5.5–19.5	6–17	2.5–26.7
Neutrophils (%)	10–40	9–34	10–42	28–44	20–75	35–75	60–70	5–88
Lymphocytes (%)	55–95	85–85	50–95	39–72	30–85	20–55	12–30	8–92
Eosinophils (%)	0–4	0–6	0–4.5	1–5	0–4	2–12	2–10	0–14
Monocytes (%)	0.1–3.5	0–5	0–3	3–12	1–4	1–4	3–10	0–11
Basophils (%)	0–0.3	0–1.5	0–1	0–3	2–7	Rare	Rare	0–6
Platelets ($\times$ 10^3/mm^3)	160–410	500–1300	200–500	250–850	250–858	300–700	200–900	109–597

*Neutrophils often resemble eosinophils due to granules.

(**Note:** The range of normal values may vary in a laboratory using specific species strain or substrain of these animals. Any major deviation on higher or lower side may be considered as a condition and not a disease per se).

ANNEXURE 2

Biochemical data of commonly used laboratory animals.

	Mouse	Rat	Hamster	G. pig	Rabbit	Cat	Dog	Monkey
Protein	3.5–7.2	5.6–7.6	4.5–7.5	4.6–6.2	5.4–7.5	6–7.5	6–7.5	4.9–9.3
Albumin (g/dl)	2.5–4.8	2.8–4.8	2.6–4.1	2.1–3.9	2.7–4.6	2.5–4.0	3–4	2.8–5.2
Globulin (g/dl)	0.6	1.8–3	2.7–4.2	1.7–2.6	1.5–2.8	2.5–3.8	2.4–3.7	1.2–5.8
Glucose (mg/dl)	62–175	50–135	60–150	60–125	75–150	81–108	54–99	46–178
Urea nitrogen	12–28	15–21	12–25	9–31.5	17–23.5	3.5–8.0	3.5–7.5	8–40
Creatinine (mg/dl)	0.3–1	0.2–0.8	0.91–0.99	0.6–2.2	0.8–1.8	<180	<20 (n mol/l)	0.1–2.8 (n mol/l)
Bilirubin (mg/dl)	0.1–0.9	0.2–0.55	0.25–0.8	0.3–0.9	0.25–0.74	<4.0	<5.0 (n mol/l)	0.1–2 (n mol/l)
Cholesterol (mg/dl)	26–82	40–130	25–135	20–43	35–53	2–4	4–7 (n mol/l)	108–263 (n mol/l)

The range of normal values may vary in a laboratory using specific species, strain or sub strain of these animals. Any major deviation on higher or lower side may be considered as a condition and not a disease per se.

ANNEXURE 3A

Minimum floor area recommended for laboratory animals (based on their weight/size and behavioural activity).

Animal	Weight in grams	Floor area/animal (cm^2)	Cage height (cm^2)
Mice	<10	38.7	
	up to 15	51.6	
	up to 25	77.4	
	>25	96.7	12
Rats	<100	109.6	
	up to 200	148.3	
	up to 300	187.0	
	up to 400	258.0	
	up to 500	387.0	
	>500	>=451.5	14
Hamsters/Gerbills/Mastomys/Cotton rats	>60	64.5	
	up to 80	83.8	
	up to 100	103.2	
	>100	122.5	12
Guinea pigs	<350	387.0	
	>350	>=651.4	18

		Floor area		Height
		(Sq.ft)	(Sq.metre)	(inches)
Rabbits	<2000	1.5	0.135	14
	up to 4000	3.0	0.27	14
	up to 5400	4.0	0.36	14
	>5400	5.0	0.45	14
	Mother with kids	4.5	0.40	14

ANNEXURE 3B

Example for calculating the number of mice to be kept per cage, based on floor area recommended for animal according to their weight (size) and size of the cage.

Recommended floor area/animal (cm^2)	38.7	51.6	77.4	96.7
Weight of animals (g)	<10	up to 15	up to 25	>25
Example 1:				
Cage size 24 × 14 cm i.e., floor area of 336 cm^2				
Maximum number of animals	8	7	4	3
Example 2:				
Cage size 32.5 × 21 cm i.e., floor area of 682 cm^2				
Maximum number of animals	17	14	9	7

Note: Cage size, specially length and breadth may vary. However, the minimum floor area and cage height recommended for group housing has to be taken into consideration. Thus, the number of animals which can be housed in a particular cage (of different sizes) can be calculated on the basis of a) floor area of the cage, b) recommended floor area per animal and c) weight of animal.

In case of breeding pairs, three adults (i.e., 1 male and 2 females) along with the pups from delivery up to weaning stage are permitted.

ANNEXURE 3C

Example for calculating the number of rats to be kept for cage, based on floor area recommended per animal according to their weight (size) and size of the cage.

Recommended floor area/animal (cm^2)	109.6	148.3	187.0	258.0	387.0	>451.5
Weight of animals (g)	<100	up to 200	up to 300	up to 400	up to 500	>500
Example: Cage size 32.5 × 21 cm, i.e., floor area of 682 cm^2 maximum number of animals	6	5	4	3	2	1

Note: Cage size, specially length and breadth may vary. However the minimum floor area and cage height recommended for group housing has to be taken into consideration. Thus, the number of animals which can be housed in a particular cage (of different sizes) can be calculated on the basis of (a) floor area of the cage, (b) recommended floor area per animal and (c) weight of animal.

ANNEXURE 3D

Example for calculating the number of Hamster/Gerbils/Mastomys/Cotton rats to be kept per cage based on floor area recommended per animal according to their weight (size) and size of the cage.

Recommended floor area/animal (cm^2)	64.5	83.8	103.2	122.5
Weight of animals (g)	<60	up to 80	up to 100	>100
Example: Cage size 32.5 × 21 cm, i.e., floor area of 682 cm^2				
Maximum number of animals	10	8	6	5

Note: Cage size, specially length and breadth may vary. However the minimum floor area and cage height recommended for group housing has to be taken into consideration. Thus, the number of animal which can be housed in a particular cage (of different sizes) can be calculated on the basis of (a) floor area of the cage, (b) recommended floor area per animal and (c) weight of animal.

ANNEXURE 3E

Minimum floor area and height recommended for monkeys (rhesus and bonnet) based on their weight (size) and behavioural activity (for langurs, the recommended space is in the foot note).

Weight (in kg)	Floor area		Height (cm)
	ft²	cm²	
Up to 1	1.6	1440	50
Up to 3	3.0	2700	72
Up to 10–12	4.3	3870	72
Up to 12–15	6.0	5400	72
Up to 15–25	8.0	7200	90

Note:

a) The height of the cage should be sufficient for the animals to stand erect with their feet on the floor, whereas the minimum height of the cage for langurs has to be 90 cm as mentioned in INSA guidelines.

b) The floor area for langurs up to 6 kg weight, 5000 cm² and above 6 kg weight. 6000–9000 cm² is recommended. The height of the case in either case remains the same, i.e., 90 cm as mentioned in INSA guidelines.

c) If the experimental protocol demands caging for more than 6 months, animals should be provided with double the floor space mentioned above.

d) All primate facilities should have one or more runs as big as possible with minimum floor space of 150 sq.ft and height not less than 2 metres for free ranging activities.

ANNXURE 3F

Recommended space for cats, dogs and birds.

Animals	Weight, kg	Floor area/animal (ft^2)	Height inches
Cats	<4	3.0	24
	<4	>4.0	24
	<15	8.0	–
	up to 30	12.0	–
	>30	24.0	–
Pigeons	–	0.8	–
Chicken	<0.25	0.25	–
	up to 0.5	0.50	–
	up to 1.5	1.00	–
	up to 3.0	2.00	–
	>3.0	>3.00	–

ANNEXURE 3G

Recommended space for commonly used farm animals.

Animal/enclosure	Weight (kg^2)	Floor area/animal (ft^2)	Height (ft)
Sheep and goats			8
	<25	10.0	
	up to 50	15.0	
2–5	>50	20.0	
	<25	8.5	
	up to 50	12.5	
>5	>50	17.0	
	<25	7.5	
	up to 50	11.3	
	>50	15.0	

(Contd.)

Animal/enclosure	Weight (kg^2)	Floor area/animal (ft^2)	Height (ft)
Swine			8
	up to 25	12.0	
	up to 50	15.0	
	up to 100	24.0	
	up to 200	48.0	
	>200	>60.0	
2–5	<25	6.0	
	up to 50	10.0	
	up to 100	20.0	
	up to 200	40.0	
	>200	>52.0	
>5	<25	6.0	
	up to 50	9.0	
	up to 100	18.0	
	up to 200	36.0	
	>200	>48.0	
Cattle			8
1	<75	24.0	
	up to 200	48.0	
	up to 350	72.0	
	up to 500	98.0	
	up to 650	124.0	
	>650	144.0	
2–5	<75	20.0	
	up to 200	40.0	
	up to 350	60.0	
	up to 500	80.0	
	up to 650	105.0	
	>650	120.0	
>5	<75	18.0	
	up to 200	36.0	
	up to 350	54.0	
	up to 500	72.0	
	up to 650	93.0	
	>650	>108.0	
		144.0	

(Contd.)

Animal/enclosure	Weight (kg^2)	Floor area/animal (ft^2)	Height (ft)
Horses/ponies			
1–4	<200	72.0	8
>4	>200	60.0	10
		>72.0	

*To convert kilograms to pounds multiply by 2.2; To convert square feet to square metres multiply by 0.09 larger animals might require more space.

ANNEXURE 4

Requirements for transport of laboratory animals by road, rail and air.

	Mouse	Rat	Hamster	G. pig	Rabbit	Cat	Dog	Monkey
Maximum no. of animals per cage	25	25	25	12	2	1 or 2	1 or 2	1
Material used in transport box	Metal card board, synthetic material	Metal card board, synthetic material	Metal card board, synthetic material	Metal card board, synthetic material	Metal card board, synthetic material	Metal	Metal	Bamboo/ wood/metal
Space per animal (cm^2)	20–25	80–100	80–100	160–180	1000–1200	1400–1500	3000	2000–4000
Minimum height of box (cm)	12	14	12	15	30	40	50	48

ANNEXURE 5

Commonly used anesthetic drugs for laboratory animals.

Drugs (mg/kg)	Mouse	Rat	Hamster	G. pig	Rabbit	Cat	Dog	Monkey
Ketamine HCl	22–24 i/m	22–24 i/m	–	22–24 i/m	22–24 i/m	30 i/m	30 i/m	15–40
Pentobarbitone sodium	35 i/v	25 i/v	35 i/v	30 i/v	30 i/v	25 i/v	20–30 i/v	35 i/v
"	50 i/p	50 i/p	–	40 i/p	40 i/p	–	–	–
Thio pentone sodium	25 i/v	20 i/v	20 i/v	20 i/v				25 i/v
"	50 i/p	40 i/p	40 i/p	55 i/p	20 i/v	25 i/v	25 i/v	60 i/p
Urethane	–	0.75 i/p	–	1.5 i/p	25 i/v, i/v	1.25 i/v	1.00 i/v	1.0 i/v
"						1.50 i/p		

Atropine: Dose 0.02 – 0.05 mg/kg for all species by s/c or i/m or i/v routes used to reduce salivary and bronchial secretions and protect heart from vagal inhibition, given prior to anaesthesia.

i/m = intramuscular, i/v = intravenous, i/p = intraperitoneal, s/c = subcutaneous

ANNEXURE 6

Euthanasaia of laboratory animals.

[A-Methods Acceptable for species of animals indicated NR-Not Recommended]

Species		Mouse	Rat	Hamster	Guinea pig	Rabbit	Cat	Dog	Monkey
a)	Physical methods								
	Electrocution	NR	NR	NR	NR	NR	NR	NR	NR
	Exsanguination	A	A	A	A	A	A	NR	NR
	Decapitation (for analysis of stress)	A	A	NR	NR	NR	NR	NR	NR
	Cervical dislocation	A	A	A	NR	NR	NR	NR	NR
b)	Inhalation of gases								
	Carbon monoxide	A	A	A	A	A	A	A	A
	Carbon dioxide	A	A	A	A	A	A	NR	NR
	Carbon dioxide plus	A	A	A	A	A	A	NR	NR
	Chloroform/halothane	A	A	A	A	A	A	A	A

c) Drug administration

Barbiturate overdose (route)	A(IP)	A(IP)	A(IP)	A(IP)	A(IV, IP)	A(IV, IP)	A(IV, IP)	A(IV, IP)
Chloral hydrate overdose (route)	NR	NR	NR	NR	A(IV)	A(IV)	A(IV)	A(IV)
Ketamine overdose (route)	A(IM/IP)	A(IM/IP)	A(IM/IP)	A(IM/IP)	A(IM/IV)	A(IM/IV)	A(IM/IV)	A(IM/IV)
Sodium pentothol [overdose (route)]	IP	IP	IP	IP	IV	IV	IV	IV

IP = intraperitoneal   IV = intravenous   IM = intramuscular

Methods not acceptable for any species of animals

a) Physical methods
    i)   Decompression
    ii) Stunning
b) Inhalation of gases
    i)   Nitrogen Flushing
    ii) Argon Flushing

c) Drug administration
    i)   Curariform drugs
    ii)   Nicotine sulphate
    iii) Magnesium sulphate
    iv)   Potassium chloride
    v)   Strychnine
    vi)   Paraquat
    vii) Dichloryos
    viii) Air Embolism

ANNEXURE 7

Certificate course for Laboratary attendant
(Basic education: 8th standard)

Introduction—Definition of plants and animals—types of animals—animals without backbones(invertbrates) and those with backbones (chordates/ vertebrates)—animals that live in water (aquatic), air (aerial), land (terrestrial)—wild animals and domesticated animals—poisonous and non-poisonous animals—laboratory bred and non-laboratory bred animals— diurnal and nocturnal animals (suitable and relevant Indian examples to be given).

Animal rooms—animal chambers/cages—sizes of animal chambers general dimensions for monkey and rat cages stocking density—need for light (LD cycles), air water and feed—cleaning animal chambers,animal runs, aquana and animal rooms—frequency of feeding—frequency of cleaning.

Handling of animals—precautions while handling animals—common injuries and ailments in animals—litters—weaning—maintenance—record keeping.

Personal hygiene—need to use apron, gloves, mask, handling of detergents and other cleaning substances—zoonoses—need of safety handling—antidotes for specific poisons if handling poisonous animals like venomous snakes—first aid.

Emergency situations: escaping animals—use of fire extinguishers— emergency lamps—sirens.

ANNEXURE 8

Institutional Biosafety Committee (IBSC)

Institutional Biosafety Committee (IBSC)is to be constituted in all centres engaged in genetic engineering research and production activities.The committee will constitute the following.

i)   Head of the institution or his nominee.

ii) Three or more scientists engaged in DNA work or molecular biology with an outside expert in the relevant discipline.

iii) A member with medical qualification—Biosafety officer (in case of work with pathogenic agents/large scale used).

iv. One member nominated by DBT.

The Institutional Biosafety Committee shall be the point for interaction within institution for implementation of the guidelines. Any research project which is likely to have biohazard potential(as envisaged by the guidelines) during the execution stage or which involve the production of either microorganisms or biologically active molecules that might cause biohazard should be notified to ISBC. ISBC will allow genetic engineering activity on classified organisms only at places where such work should be performed as per guidelines. Provision of suitable safe storage facility of donor, vectors, reciepients and other materials involved in experimental work should be made and may be subjected to inspection on accountability.

The biosafety functions and activity include the following:

i) Registration of Biosafety Committee membership composition with RCGM and submission of report.

IBSC will provide half yearly reports on the ongoing projects to RCGM regarding the observance of the safety guidelines on accidents, risks and on deviations if any. A computerized Central Registry for collation of periodic reports on approved subjects will be setup with RCGM to monitor compliance on safeguards as stipulated in the guidelines.

ii) Review and clearance of project proposals falling under restricted category that meets the requirements under the guidelines.

IBSC would make efforts to issue clearance certificates quickly on receiving the research proposals from investigators.

iii) Tailoring biosafety program to the level of risk assessment.

iv) Training of personnel on biosafety.

v) Instituting health monitoring program for laboratory personnel.

Complete medical check up of personnel working in projects involving work with potentially dangerous microorganism should be done prior to starting such projects. Follow up medical check ups including pathological test should be done periodically, annually for scientific workers involved in such projects. Their medical record should be accessible to the RCGM. It will provide half yearly reports on the ongoing projects to RCGM regarding the observance of the safety guidelines on accidents, risks and on deviations if any.

vi) Adopting emergency plans.

So far biosafety committee have been already set up in 24 institutions. The other institutions will be asked to take similar action.

APPENDIX III

OPERATIONAL GUIDELINES FOR OBSERVANCE OF GOOD PRACTICES IN CPCSEA

Experimentation on animals in course of medical research and education is covered by provisions of the Prevention of Cruelty to Animals Act, 1960 and the Rules under the Act of 1998 and 2001. This is enforced by the Committee for the Purpose of Control and Supervision of Experiments on Animals (CPCSEA), a statutory body under the Prevention of Cruelty to Animals Act, 1960. Under these provisions, the concerned establishments are required to get themselves registered with CPCSEA, get their Animal House Facilities inspected, and also get specific projects for research cleared by CPCSEA before commencing the research. Further, breeding and trade of animals for such experimentation are also regulated under these Rules.

GUIDELINES

1. **Main Activities**

    (a) *Registration of establishments for Breeding of animals*   Under Rule 3 of Breeding of and Experiments on Animals (Control and Supervision) Rules 1998.

    (b) *Registration of establishments for experiments on animals*   Under Rule 4 of Breeding of and Experiments on Animals (Control and Supervision) Rules 1998.

    (c) *Approval of Animal House Facilities*   Rule 5 (b) of Breeding of and Experiments on Animals (Control and Supervision) Rules 1998 prescribes for inspections of such facilities before registration. Sub Rule (c) of Rule 5 as amended in 2001, provides for stipulating conditions at the time of registration. A specific condition regarding obtaining approval of Animal House Facilities, after detailed inspection before conduct of experiment may be imposed by the Member-Secretary at the time of registration.

    (d) *Permission of Committee for Conducting Experiments*   Under Rule 8 of Breeding of and Experiments on Animals (Control and Supervision) Rules 1998.

(e) Recommendation for import under Rule 10 (e) of Breeding of and Experiments on Animals (Control and Supervision) Rules 1998 as amended in 2001.

2. Regulatory Mechanism

2.1 *Committee for the Purpose of Control and Supervision of Experiments on Animals (CPCSEA)* Constituted under the Provision of Section 15 of the Prevention of Cruelty to Animals (PCA) Act, 1960. The main mandate of the Committee is to ensure that animals are not subjected to unnecessary pain or suffering before, during or after performance of experiments on them. The CPCSEA functions within the ambit of the Act and the Rules for Breeding of and Experiments on Animals (Control & Supervision), 1998 as amended in February, 2001.

2.2 *Sub-Committees of the CPCSEA* In exercise of the powers conferred under Section 15(A) of the Prevention of Cruelty to Animals Act 1960, which provides for constitution of Sub-Committees for exercising any power or discharging any duty of the Committee, the CPCSEA has constituted following 2 sub-committees for exercising some of its specific powers.

2.2.1 *Sub-Committee on Large Animals (SCLA)*

All projects involving experiments on large animals have to be approved by the Sub-Committee on Large Animals (SCLA). The Sub-Committee was re-constituted vide order No. 25/20/2003-AWD dated 2nd June, 2003.

2.2.2 *Sub-Committee for selection of CPCSEA Nominees on Institutional Animals Ethics Committees*

The CPCSEA in its meeting held on 11th July, 2003 decided to constitute a Sub-Committee for selection of CPCSEA Nominees on Institutional Animals Ethics Committees. (This has been communicated vide order No.25/39/2003-AWD dated 21st August, 2003). The scope of this Sub-Committee is proposed to be enlarged to include selection of panel for conduct of Inspection Teams for Animal House Facilities. The selected panel would comprise of veterinarians, persons engaged in activities of Animal Welfare, and persons dealing with animal experimentations. The panels would be made zone wise. Besides, the Sub-Committee would also draw a panel of persons in-charge of some existing good animal house facilities to work in an advisory capacity.

2.3 *Institutional Animal Ethics Committee (IAEC)*

Every establishment constituted and operated in accordance with the procedures specified by the CPCSEA is required to constitute an Institutional Animal Ethics Committee.

As per Rule 13 of the Breeding of and Experiments on Animals (Control and Supervision) Rules, 1998, every IAEC shall include; a biological scientist, two scientists from different biological disciplines, a veterinarian involved in the care of animals, the scientist in charge of animals facility of the establishment concerned, a scientist from outside the institute, a non-scientific socially aware member and a representative or nominee of the CPCSEA. A specialist may be co-opted while reviewing special projects using hazardous agents such as radioactive substances and deadly microorganisms.

To each IAEC, a CPCSEA Nominee would be nominated and his presence would be necessary for every meeting. In order to ensure smooth functioning of the IAEC, there will be a link CPCSEA nominee, nominated for each IAEC who will participate in the meeting in event of inability of the regular CPCSEA nominee to attend such meeting.

The IAEC will scrutinize all project proposals for experimentation on animals. In case of small animals, it would give the final approval. In case of large animals, it would make its recommendation to SCLA which will be the final clearance authority.

2.4 *Member-Secretary, CPCSEA*   As per Rules, Member-Secretary, is responsible for registration of establishment engaged in Breeding of Animals and also in Experimentation on Animals. He is also responsible for causing the application for permission for conducting experiments to be brought before the CPCSEA and SCLA. Besides, he has also been authorized to constitute the Inspection Teams for inspection of Animal House Facilities, from the approved panel.

3.   Standing time tables of meetings of the regulator/expert body

(a) *Committee for the Purpose of Control and Supervision of Experiments on Animals (CPCSEA)*   The meetings are to take place quarterly. The schedule for the meetings will be 20th December, 20th March, 20th June, and 20th September every year. If the

20th happens to be a holiday, the meeting will be held the next working day.

(b) *Sub-Committee on Large Animals (SCLA)*   The meetings are to be held on 10th of every alternate month. If the 10th happens to be a holiday, the meeting will be held the next working day.

(c) *Sub-Committee for selection of CPCSEA Nominees on Institutional Animals Ethics Committee*   The meeting is proposed to be convened once in six months say in January and July.

(d) *Institutional Animal Ethics Committee (IAEC)*   This committee functions within the research establishments and the standing timetable of each IAEC would vary as per its needs.

(e) *Member Secretary, CPCSEA*   No requirement of standing timetable.

Besides, in emergent situations, and to clear backlog, special meetings of the above regulatory bodies, may be convened with the approval of the Chair.

4. **Nomination of Co-chair/Vice chair of the regulator/expert body for presiding over the meeting in the absence of the chair**

(a) *Committee for the Purpose of Control and Supervision of Experiments on Animals (CPCSEA)*

As a standing arrangement, it is proposed to nominate JS or an equivalent in charge of Animal Welfare as Vice-Chair of the Committee. In the absence of JS or equivalent, the members present could nominate any member from amongst themselves to preside over proceedings of a meeting.

(b) *Sub-Committee on Large Animals (SCLA) and Sub-Committee for selection of CPCSEA Nominees*

The CPCSEA could make a provision that in absence of the Chair, in the above two Sub-Committees, the members present, would nominate any member from amongst themselves to preside over the proceedings of a meeting.

5. **Creation of separate Web page**   A web page of CPCSEA within the MOEF web site will be created with the help of NIC, displaying the following minimum information.

(a) Ground Rules.

(b) Names of the Chair/Vice-chair of CPCSEA/ its Sub-Committees and the provision of nomination of any member among themselves to preside over the proceedings of the meeting in the absence of Chairman and Vice-Chairman.

(c) Standing timetable of the Committee meetings.

(d) A notice that, if necessary to consider all the listed cases the meeting would continue the following working day.

(e) A notice that project proponents, if they wish to, may remain present at the time of the meeting to make a personal presentation.

(f) Agenda for the next meeting.

(g) Status of all cases received in the Ministry.

6. **Ground Rules**

(i) Registration of Establishment for Breeding of Animals and for Experiments of Animals:

The Member-Secretary would take a decision on any application for Registration within two months of its receipt. In case the application cannot be decided within two months, the Member-Secretary would record speaking reasons for the same and bring the case to the notice of CPCSEA at its ensuing meeting.

(i) The application should be accompanied by a bank draft of Rs. 1,000/- in favour of CPCSEA as registration fee.

(ii) The establishment should have constituted an IAEC before applying for registration.

(iii) Before taking the decision regarding Registration, a spot inspection of the organization would be necessary. The spot inspection would be carried out in the following manner:

(a) In case of Registration of an Establishment for Experiments on Animals (under Rule 4 of Breeding Rules), the initial inspection would be preliminary in nature, and carried out by one inspecting member. However, before conduct of experiments on animals specific approval regarding Animal House Facility will be necessary through inspection by a team constituted under para 6 (ii).

(b) In case of Registration of establishments for Breeding purposes (under Rule 3 of Breeding Rules), the spot inspection would be carried out by the inspection team as constituted under para 6 (ii).

(ii) Approval of Animal House Facilities:

(a) Detailed inspection by a Inspection team. As per the existing practice the following is the composition of inspection team:

(i) CPCSEA's Nominee.

(ii) Expert Consultant, CPCSEA.

(iii) Person in charge of another animal house in the same State, where the institution carrying out experiment is located. In the event of any constraint, from the adjoining state in the same region.

To ensure representation from all concerned sectors, the composition of the inspection team is now proposed as below:

(i) Expert Consultant, CPCSEA-Coordinator

(ii) CPCSEA Nominee

(iii) CPCSEA Nominee

The two CPCSEA Nominees should be so selected as to ensure balanced representation to address concerns both of veterinarians as well as those engaged in activities of Animal Welfare. Besides, a person in charge of some good Animal House Facility will be co-opted to assist the Inspection team.

The selection of members of the Inspection team including the co-opted member to assist the team, will be made from the approved panel.

Approval of Animal House Facility will be accorded by CPCSEA or by the Sub-Committee on Large Animals (SCLA), if so authorized.

(b) The time limit prescribed by the regulator should be adhered to even if one or two members of the team are not in a position to undertake the visit on scheduled date. In such case the Member-Secretary could substitute the members in the Inspection team from the approved panel, preferably from the same zone.

(iii) Permission of Committee for Conducting Experiments.

(a) No proposal for Animal Experimentation would be entertained, unless the Animal House Facility of the concerned establishment has been approved by competent authority.

(b) The approval for experimentation on small animals will be accorded by IAEC.

(c) Approval for experimentation on Large Animals will be considered by SCLA on the recommendation of IAEC.

(d) The decision in the IAEC would normally be taken by consensus. If divergent views are expressed by the members these may be recorded in minutes and a broad consensus be recorded as per understanding of the Chair. However, in case of dissent by the CPCSEA nominee, the proposal with the report of IAEC and the dissenting note of the CPCSEA nominee, would be submitted to SCLA for taking a final decision.

(e) The decision in the SCLA would be taken by consensus. If divergent views are expressed by the members, these may be recorded in the minutes and a broad consensus be recorded as per understanding of the Chair.

7. Check-list for Standardized information

(i) Registration of establishment for Breeding of animals and for experiments of animals:

  (i) Name and address of the Institute with telephone/fax numbers, as well as e-mail ID if available

  (ii) Name and address of Head of the Institution

  (iii) Nature of activities

  (a) Breeding purpose

  (b) Experiment purpose (please tick the portion applicable)

  (iv) Layout plan of the establishment

  (v) Whether it is in line with the prescribed guidelines

  (vi) Whether there is any deviation from the prescribed Guidelines (Please mention)

  (vii) Present stock of animals (Species-wise)

  Male No.

  Female No.

  (viii) Source of procurement of Animals—Address of the supplier

  (ix) Whether animals have been supplied to other institute. If yes, details of Institutes for the last three years.

  (x) Availability of manpower—trained/untrained

  (xi) Purpose of registration

   (ii)　Permission of Committee for Conducting Experiments:

      1.　Name and address of the Institute with telephone/fax numbers, as well as e-mail ID if available

      2.　Name and address of Head of the Institution

      3.　Registration number and date of registration

      4.　Project Title

      5.　Nature, purpose and justification of experiment to be undertaken

      6.　Source for procurement of animals

         (a)　Bred by establishment

         (b)　Procured from Breeder—mention Name, address and Registration number of breeder

         (c)　Procured from other legal sources

         (d)　Are the records duly maintained

7.　Place where the animals are presently kept (or proposed to be kept)

8.　Place where the experiment is to be performed

9.　Date on which the experiment is to commence and duration of experiment

10.　Chief Investigator

   (i)　Name

   (ii)　Designation

   (iii)　Dept./Div./Lab

   (iv)　Telephone number

11.　List of names of all individuals authorized to conduct procedures under this proposal

12.　Funding Source

13.　Duration of the project

   (i)　Number of months

   (ii)　Date of initiation

   (iii)　Date of completion

14.　Study Objectives [the aims of study (and why they are important) to be explained briefly using non-technical terms as far as possible]

15. Animals required

    (i)   Species

    (ii) Age/Weight/Size

    (iii) Gender

    (iv) Numbers to be used (Year-wise breakups and total figures needed to be given)

    (v)   Number of days each animal will be housed

16. Rationale for animal usage

17. Why is animal usage necessary for these studies?

18. Why are the particular species selected required?

19. Why are the estimated number of animals essential?

20. Similar experiments conducted in the past. If so, the number of animals used and results obtained in brief.

    (i)   If yes, why new experiment is required?

21. Have similar experiment(s) been made by any other organization/ agency? If so, their results in your knowledge.

22. Description of procedures to be used

23. Does the protocol prohibit use of anaesthetic or analgesic for the conduct of painful procedures (any which cause more pain than that associated with routine injection, or blood withdrawal)? If yes, explanation and justification

24. Will survival surgery be done? If Yes, the following to be described

    (i)   List and description of all such surgical procedures (including methods of asepsis)

    (ii) Names, qualifications and experience levels of operators

    (iii) Description of post-operative care

    (iv) Justification if major survival surgery is to be performed more than once on a single individual animal

25. Care of animals post-experimentation

26. Rehabilitation/Euthanasia (In case of euthanasia, justification for not undertaking rehabilitation and drug dosage and route for anesthesia, where appropriate, as well as methods of carcass disposal post experimentation)

27. Animal transportation methods if extra-institutional transport is envisaged

28. Use of hazardous agents (Copy of IBC approval to be attached in case hazardous agents are used)

    (iii) Format for Inspecting Teams for approval of Animal House Facilities for breeding of animals/conduct of experimentation on animals.

    (a)   Details of animals, Species wise, kept at the time of Inspection in the Animal House

    (b)   Veterinary Care of animals

    (c)   Health status of animals

    (d)   Animal Procurement

    (e)   Quarantine, Stabilization and Separation

    (f)   Physical Facilities

    (i)   Building materials

    (ii)   Corridor(s)

    (iii) Utilities

    (iv)   Animal Room Doors

    (v)   Exterior windows

    (vi)   Floors

    (vii) Drains

    (viii) Walls and ceilings

    (ix)   Storage areas

    (x)   Facilities for sanitizing equipment and supplies

    (xi)   Experimental Area

    (xii) Environment

    (xiii) Temperature and Humidity control

    (xiv) Ventilation

    (xv)   Power and lighting

    (xvi) Noise control

    (g) Animal Husbandry

    (i)   Caging or housing system

      (ii)   Sheltered or outdoor housing

      (iii) Social environment

(h)   Food

(i)   Bedding

(j)   Water

(k)   Sanitation and Cleanliness

(l)   Waste Disposal

(m)   Pest Control

(n)   Emergency, weekend and holiday care

(o)   Record keeping

(p)   Personnel and Training

(q)   No. of technical staff, supporting staff, details of the training of the supporting staff

(r)   Transport of laboratory animals

(s)   Anaesthesia and Euthanasia

(t)   Laboratory animal ethics

(u)   Transgenic animals

(v)   Maintenance

(w)   Disposal

(x)   Details of rehabilitation facilities

(y)   Overall assessment

(z)   Recommendation (__________________________________)

    1.   Recommended for approval (without any stipulations)

    2.   Recommended for approval with suggestions for improvement
               (please specify here)

    3.   Recommended for fulfillment of stipulated conditions before consideration for approval
               (please specify here)

    4.   Recommended for rejection with specific grounds
               (please specify here)

________________	________________	________________
Member's Signature	Member's Signature	Member's Signature

Note: Please tick whichever is applicable.

8. Check Memo

(I) *Registration*

Date of action

Initiated   Completed

(a) Processing of application by Expert Consultant

(b) Inspection of the organization

(c) Approval of Member-Secretary

(d) Conveying of decision to organization

(II) *Clearance of proposal for conducting the experiment on large animals*

Date of action

Initiated   Completed

(a) Approval of Animal House Facility

    (i) Inspection of animal house facility

    (ii) Placing of Inspection Report
      before CPCSEA

    (iii) Conveying the approval of facility to organization

(b) Approval of the Project Proposal

    (i) After approval of facility, placing of project proposal
      before SCLA

    (ii) Conveying the decision of SCLA to institutes

9. Where specific guidelines have not been provided in forgoing provisions, the guidelines issued vide MOEF letter No. 20011/3/2002-GC dated 15th September, 2004 regarding observance of Good Practices in Environmental Regulation will be applicable (copy enclosed).

APPENDIX IV

GUIDELINES FOR CARE AND USE OF
ANIMALS IN SCIENTIFIC RESEARCH

**This report was submitted by the Expert Committee
Constituted by the Indian National Science Academy,
New Delhi**

1. INTRODUCTION

The use of laboratory Animals in Scientific Research has been a subject of debate for over a century. Though the animals were first used in research in second century AD their systematic use in research began about 100 years ago, when vaccines for polio and rabies came up for production. Since then, the animals have been used in research investigations and production of biologicals and have played an important role in unfolding vital information about the human and animal life processes. This has helped in the advancement of medicine, development of drugs, diagnostics and production of biologicals for alleviating sufferings of both humans and animals.

It must be emphasized that use of animals in research is inevitable and cannot be abandoned in the interest of human and animal welfare. *In vitro* alternate methods cannot replace animal experimentation totally, but can work only as adjuncts and reduce the number of animals in some cases. However, efforts to develop *in vitro* models should continuously be made.

The scientists are deeply concerned about the rational and humane use of animals in research. Ethics committees are functional in many institutes. They are concerned about avoiding unnecessary pain or suffering or injury to animals during holding, experimentation and post-experimental period by monitoring and improving their housing, environment, feeding and veterinary care. The Government of India has authorized the National Accreditation Board of Testing and Calibration Laboratories (NABTCL), promoted by the Department of Science and Technology, to provide accredition [sic] services to laboratories covering a wide range of subjects including biological and clinical laboratories. The NABTCL is a full member of the International Laboratory Accreditation Cooperation and the

Asia Pacific Laboratory Cooperation. Such accreditation of animal facilities would demonstrate their commitment to responsible animal care and use and good science since such an accreditation is an indicator of an institution's ability to comply with its assurances.

In India, the need to develop guidelines for the use of animals in research has been discussed at various forums. Unfortunately no standard document was available for reference till 1992 when the Indian National Science Academy developed the guidelines for use of animals in scientific research. Considering the knowledge generated internationally over the years and the guidelines of WHO, NIH associated NRC, USA and European Union, the INSA guidelines have been updated.

(a)   *The Need*

The biomedical scientists generally work to unfold the complicated processes of life and to provide new measures for the health and welfare of the society, i.e., the humans, the animals and the environment. There is, therefore, need to provide them certain degree of freedom and adequate facilities to use animals wherever necessary. It is evident that certain life processes can not be investigated without involving whole animal system. The *in vitro* alternatives can only provide limited information. These cannot totally replace the animals in experiments. This is why the use of animals continues to be mandatory to meet the statutory regulatory requirements. At the same time, it is an obligation of the scientists to ensure that the experiments conducted on animals are rational and unavoidable, and no unnecessary pain or injury is inflicted on them and they are maintained in best possible environmental conditions. It is, therefore, necessary to have well-defined guidelines which will safeguard the pursuit of knowledge, the interest of society and the welfare of animals.

(b)   *The Objectives*

To provide guidelines for

1.   housing, care, breeding and maintenance of experimental animals to keep them in physical comfort and good health and to permit them to grow, reproduce and behave normally;

2.   sources of experimental animals of known genetic, health and nutritional status;

3. development of training facilities for scientists, technicians and other supportive staff for the care of animals and their use in experiments;

4. acceptable experimental techniques and procedures for anaesthesia and euthanasia;

5. developing alternate *in vitro* systems to replace animal experiments;

6. the constitution of institutional ethics committees, their functions and the legal and ethical obligations to ensure minimal and ethical use of animals.

(c) Current Status

The INSA guidelines are the only ones available at the National level and adopted by well established institutions in India for care and use of their laboratory animals. However, there is need to adopt these in all the research animal facilities. It is therefore essential that these guidelines are accepted as the national guidelines.

2. SOURCES OF EXPERIMENTAL ANIMALS

Animals for experiments should be procured by scientists from recognized animal facilities. The animals trapped from the wild, e.g. the monkeys, feral dogs and cats are also used in research as they are readily available and less expensive compared to colony bred animals. These wild and feral animals are generally quarantined and stabilized in animal facility before use in experiments. The health and genetic status of these animals are not known and therefore a careful screening during quarantine is necessary. The wild and feral animals should be acquired after due clearance from Institutional Animal Ethics Committees and through certified suppliers.

The only authentic source of getting right type of animals for research should be from recognized scientific animal facilities where the animal colonies of known genetic and health status are available. Such animals only can provide reliable results. The scientists should therefore insist upon getting defined animals through organized colonies eliminating unscrupulous traders, which not only supply poor stock of animals but also maintain these animals under most unethical and unhygienic conditions. A list of Scientific Institutions which maintain recognized animal strains is given in Annexure 1. A directory of animal species and strains available with each of these institutions should be prepared and circulated. A list of 'Physiological Norms of Commonly Used Laboratory Animals' and 'Reproductive Data of Commonly Used Laboratory Animals' is given in Annexure 2 and 3 respectively.

3. LABORATORY ANIMAL HUSBANDRY AND MANAGEMENT

(i) *Housing and Environment*

Laboratory animals are very sensitive to their living conditions. It is important that they are housed in an isolated building located as far away from human habitations as possible and not exposed to dust, smoke, noise, wild rodents, insects and birds. The building, cages and environment of animal rooms are the major factors which affect the quality of animals.

In planning an animal facility the space should be well divided for various activities. The animal rooms should occupy about 50–60% of the total constructed area and the remaining area should be utilized for services such as stores (8–10%), washing (8–10%), office and staff (8–10%), machine rooms (4–5%) quarantine and corridors (12–15%).

The cages should be made of suitable metal (stainless steel, galvanized iron sheet/rods) or synthetic material (polypropylene/ polycarbonate). They should be of suitable size for each species of animal and should have adequate arrangement for feeding and watering. They must be free from crevices, corners and sharp edges for easy cleaning and to avoid injury. The bedding should be of right material and sterilized before use. Common bedding materials used in India are paddy husk, saw dust, paper cuttings, dry grass and crushed corn cobs.

The environment of animal room (macro-environment) and animal cage (micro-environment) is an important factor on which the production and experimental efficiency of the animal depends. Since animals are very sensitive to environmental changes, sharp fluctuations in temperature, humidity, light, sound and ventilation should be avoided. The recommended environmental requirements for animal rooms, for different species are given in Annexure 4.

A constant room temperature is essential, because variation in room temperature causes change in food and water intake. A change in temperature of 4°C can cause 10-fold alteration in biological responses. The temperature also affects fertility and lactation. Coupled with high humidity the increase in temperature causes ammonia build up. If the ventilation is not proper the high ammonia concentration causes respiratory irritation to both animals and attendants, predisposing them to infection by lowering their resistance. An effective ventilation system with 10–12 air changes per hour of 100% fresh air must be provided for animal rooms.

Light and sound are other important factors. The light intensity, the wave length and the photo cycle affect the health and behaviour of the animals. Sudden and sharp sounds in the animal rooms disturb the health and behaviour of animals and may give rise to ear damage, hypertension, cannibalism, etc.

(ii) *Breeding and Genetics*

For initiating a colony, the breeding stock must be procured from a reliable/accredited source, ensuring that genetic make-up and health status of animals is known. In case of an inbred strain, the characters of the strain with their gene distribution and the number of inbred generation must be known for further propagation. The health status should indicate their origin, e.g. conventional, specific pathogen free or transgenic gnotobiotic or knock-out stock. The known nutritional status and feeding habits of the stock are also of advantage.

The animal colonies may be inbred or outbred. In the case of inbred colonies the number of generation of brother × sister mating and latest genetic monitoring parameters for various markers should be known. Mutations or genetic contamination can be detected by using screening methods, such as histocompatability (skin grafting), biochemical markers, coat colour studies, mandibular biometry or immunological studies. Phenotypic visual characters may also sometime provide clue of genetic contamination (Annexure 5).

(iii) *Transgenic/Knock-out Animals*

Transgenic animals are those animals into whose germ line foreign gene(s) have been engineered, whereas knock-out animals are those whose specific gene(s) have been disrupted leading to loss of function. These animals can be bred to establish transgenic animal strains. Transgenic animals are used to study the biological functions of specific genes, to develop animal models for diseases of humans or animals, to produce therapeutic products, vaccines and for biological screening, etc. These can be either developed in the laboratory or procured for R & D purpose from scientific/academic institution or commercial firms, generally from abroad.

Those laboratories developing transgenic animals should pay special attention to the following points:

1. The photoperiod is a critical regulator of reproductive behaviour of many species of animals. Inadvertent light exposure during the dark cycle should be minimized or avoided. A time controlled lighting system should be used to ensure regular diurnal cycle.

2. Embryo transfer has to be carried out using anaesthetics.

3. Pseudopregnant females which receive embryos should be kept in separate rooms where there is no disturbance.

4. Bedding, feed, water or cage should not be changed for about 3–4 days after embryo transfer as at this stage there is high risk of embryo resorption and termination of pregnancy.

5. The bedding changes and handling of the female should be carried out by skilled caretaker till delivery and weaning is over.

6. A high protein diet should be given to the lactating mother.

Maintenance   Housing, feeding, ventilation, lighting, sanitation and routine management practices for such animals are similar to those for the other animals of the species as given in the guidelines. However, special care has to be taken with transgenic/ gene knock-out animals where the animals can become susceptible to diseases or where special conditions of maintenance are required due to the altered metabolic activities. The transgenic and knock out animals carry additional genes or lack genes compared to the wild population of the species, and therefore to avoid the spread of the genes in wild population, neither they should inadvertently cross breed with other animals or be released in the wild. Special care should be taken to maintain those animals which have been genetically manipulated to produce models for diseases of humans or other animals. The transgenic and knockout animals should be maintained in clean room environment or in animal isolators.

Disposal The transgenic and knock-out animals should be first euthanasised and then disposed off as prescribed elsewhere in the guidelines. A record of disposal and the manner of disposal should be kept as a matter of routine.

The transgenic and knock-out animals need greater level of monitoring than other animals as transgenes might have unexpected effect on the phenotype and its interaction with the environment.

(iv) *Nutrition and Feeding*

The results of an experiment are likely to be influenced by co-existence of nutritional deficiencies and imbalance. It is, therefore, essential that laboratory animals are maintained on a balanced diet based on nutritional requirements of each species. Special care is needed on nutritional elements, ingredients used in diet, and feeding practices. A balanced diet should contain protein, carbohydrates, fat, minerals, vitamins, roughage and water in required proportions for each species of animal. These requirements for commonly used species are given in Annexure 6.

Only quality ingredients should be used in a diet and they should be free from dust, moulds, fungi and other contaminants. Each animal must get required quantity of feed, based on animal maintenance and production requirements.

The feed should be palatable so that it is consumed in adequate quantity by the animals. Any undesirable odour always causes under consumption resulting in nutritional deficiency in the animals.

No drug, hormone or antibiotic should be added in the feed as these are likely to disturb the normal metabolism of the animals and produce biased results.

The ingredients and the prepared feed must be stored and handled carefully so as to avoid any contamination. The food must be of right consistency and should be presented to animals in proper type of hoppers to avoid wastage. In some cases the feed may be divided in 2–3 meals during the day.

Pelleted feeds balanced for different species of animals are now available commercially. These are easy to procure and use without wastage. However one has to be careful on quality of the feed from batch to batch. It should be obligatory on manufacturer to mark each bag with the type of food, date of manufacture, the batch number, the ingredients used and chemical composition. Random chemical analysis must be carried to major nutrients to monitor the quality of food from time to time.

Clean, chlorinated water should be available to the animals *ad-lib*.

(v) *Hygience and Disease Control*

The building for housing the animals should be provided with barriers to control the entry of contamination into the building through men, material and wild animals. Strict barriers should be provided to avoid the entry of wild rodents, birds, insects and pests. Visitors and service staff should be allowed entry with care and when necessary.

On the exit side an efficient monitoring service should be established to monitor the prevalence of any infection in the colony. A regular medical checkup of the staff, postmortem of dead and sacrificed animals and screening of waste material of the rooms are essential.

(vi) *Personnel and Training*

The selection of animal facility staff, particularly the staff working in animal rooms or involved in transportation, is a critical component in the management of an animal facility.

The staff must be provided with all required protective clothing (masks, aprons, gloves, gumboots, etc.) while working in animal rooms. Facilities should be provided for change over with lockers, wash basins, toilets and bathrooms to maintain personal hygiene. It is also important that a regular medical checkup is arranged for the workers to ensure that they have not picked up any zoonotic infection and also that they are not acting as a source of transmission of infection to the animals. He should ensure that persons working in animal house don't eat, drink, smoke in animal room and have all required vaccination, particularly against tetanus and other zoonoses.

Initial in-house training of staff at all levels is essential. A few weeks must be spent on the training of the newly recruited staff, teaching them the animal handling techniques, cleaning of cages and importance of hygiene, disinfection and sterilization. They should also be made familiar with the activities of normal healthy and sick animals so that they are able to spot the sick animal during their daily routine checkup of the cages.

At national level suitable training programmes should be organized by the National Centres to provide training in care, breeding, management, handling of animals for the staff working in animal breeding and holding units. Orientation training programmes should also be initiated for the investigators working in different areas to acquaint themselves with various experimental techniques. Such a course should address the undermentioned topics:

a)   biology and husbandry of laboratory animals

b)   genetic make-up

c)   microbiology and diseases

d)   health-hazards in the animal house

e)   anaesthesia, analysis and experimental procedures

f)   alternatives to animal use

g)   ethical aspects and legislation

The national level training programmes of following type are essentially required.

Training level	Qualification	Duration	Course contents
1. Technician level	Matriculate	6–12 weeks	Basics
2. Supervisory level	Graduate	12–24 weeks	Comprehensive
3. Scientist level	Veterinary or Medical Graduate/ Post-Graduate in Natural Sciences	8–12 weeks	Specialised

The training courses and workshops may also be organized for the senior level biological scientists to evoke awareness among them about the use of animals in research, alternatives available and the ethical and legal provisions in regard to use of animals. This is particularly important care and management for the supervisory staff and veterinarians exists in the country.

(vii) *Records and Evaluation*

Good quality animals are those which are free from disease. Animal of a specified strain should also have all the characteristics of that strain, i.e., they should be genetically anthesised. The results of regular monitoring of parameters of genetic purity must be scrupulously recorded.

Proper record-keeping is extremely important and vital for an animal facility. The forms should be simple but complete and preferably computer compatible. Too exhaustive and unnecessary recording should be avoided as these are not useful. Records of breeding and experimentation and deaths of all experimental animals at various stages are essential.

Receipt and issue of food and other stores should be recorded. Log books of various machines such as incinerator, boilers, air-conditioning plant should be maintained. Monthly and annual reports of the activities should be prepared and reviewed for evaluation of work and future planning.

(viii) *Experimentation and Veterinary Care*

The experimental animal units should generally be looked after by qualified investigators. These units must have adequate housing and technical facilities for experiment and post-operative care. The equipment provided in the experimental unit should be appropriate for the needs of the experiments. No technique should be used which may cause avoidable discomfort to the animals. The post-operative holding rooms and cages should be comfortable and such animals should remain under the care and supervision of an experienced scientist or a qualified veterinarian.

The person actually incharge of animal facility should preferably be a veterinarian or a person qualified in laboratory animal sciences. In any case an experienced veterinarian must be readily available in an animal holding for health care, monitoring, diagnosis and treatment of diseases and injuries. A veterinarian could also be helpful to investigators in animal anaesthesia and surgery.

4. TRANSPORT OF LABORATORY ANIMALS

The transport of animals from one place to another is very important and must be undertaken with care. The main considerations for transport of animals are, the mode of transport, the containers, the animal density in cages, food and water during transit, protection from transit infections, injuries and stress.

The mode of transport of animals depends on the distance, seasonal and climatic conditions and the species of animals. Animals can be transported by road, rail or air taking into consideration above factors. In any case the transport stress should be avoided and the containers should be of an appropriate size so as to enable these animals to have a comfortable, free movement and protection from possible injuries. The food and water should be provided in suitable containers or in suitable form so as to ensure that they get adequate food and more particularly water during transit. The transport containers (cages or crates) should be of appropriate size and only a permissible number of animals should be accommodated in each container to avoid overcrowding and infighting. Requirements of space in transport cages for each species of animals are given in Annexure 7.

5. ANAESTHESIA AND EUTHANASIA

The scientists should ensure that the procedures which are considered painful are conducted under appropriate anaesthesia as recommended for each species of animals (Annexure 8). It must also be ensured that the anaesthesia is given for the full duration of experiment and at no stage the animal is conscious to perceive pain during the experiment. If at any stage during the experiment the investigator feels that he has to abandon the experiment or he has inflicted irreparable injury, the animal should be sacrificed by overdose of anaesthetic. Neuromuscular blocking agents must not be used without adequate general anaesthesia.

In the event of a decision to sacrifice an animal on termination of an experiment or otherwise an approved method of euthanasia (Annexure 9) should be adopted and the investigator must ensure that the animal is clinically dead before it is sent for disposal.

(a) *Anaesthesia*

Unless contrary to the achievement of the results of study sedatives, analgesics and anaesthetics should be used to control pain or distress under experiment. Anaesthetic agents generally affect cardiovascular, respiratory and thermo-regulatory mechanism in addition to central nervous system.

Before using actual anaesthetic the animal is prepared for anaesthesia by over night fasting and using pre-anaesthetics, which block parasympathetic stimulation of cardio-pulmonary system and reduce salivary secretion. Atropine is the most commonly used anti-cholinergic agent. Local or general anaesthesia may be used, depending on the type of surgical procedure. Anaesthetic agents used for common species of laboratory animals and their doses are given in Annexure 8.

Local Anaesthesia   Local anaesthetics are used to block the nerve supply to a limited area and are used only for minor and rapid procedures. This should be carried out under expert supervision for regional infiltration of a surgical site, nerve blocks and for epidural and spinal anaesthesia.

General Anaesthesia   A number of agents are used as inhalants. General anaesthetics are also used in the form of intravenous or intra-muscular injections such as barbiturates. Species characteristics and variation must be kept in mind where using an anaesthetic. Side-effects such as excessive salivation, convulsions, excitement and disorientation should be suitably prevented and controlled.

The animal should remain under veterinary care till it completely recovers from anaesthesia and post-operative stress.

(b) *Euthanasia*

Euthanasia means "easy death" and is resorted to in events where an animal is required to be sacrificed on termination of an experiment or otherwise for ethical reasons. The procedure should be carried out quickly and painlessly in an atmosphere free from fear or anxiety. For accepting of an euthanasia method as humane it should have an initial depressive action on the central nervous system for immediate insensitivity to pain. The choice of a method will depend on the nature of study, the species of animals and number of animals to be sacrificed. The method should in all cases meet the following requirements

(a) Death, without causing anxiety, pain or distress with minimum time lag phase.

(b) Minimum physiological and psychological disturbances.

(c) Compatibility with the purpose of study and minimum emotional effect on the observer and operator.

(d) Location should be separate from animal rooms and free from environmental contaminations.

(e) Method should be reliable, reproducible and safe to the personnel involved.

(f) Simple and economical.

It is recommended that tranquilizers be administered to larger species such as monkeys, dogs and cats before an euthanasia procedure.

A number of euthanasia methods have been recognized humane which could be physical, use of inhalant gases, injectable drugs and general anaesthetics in heavy dose. The methods recognized as appropriate for a commonly used species of animal have been listed in Annexure 9.

6. DISPOSAL OF ANIMAL CARCASSES

All animal carcasses whether healthy, infectious or radioactive, must be packed in polythene bags before sending them for disposal. All healthy or infectious animals may be buried deep in the ground covered with lime and disinfectants or burnt in an incinerator. Animals with radioactive material should be packed in double polythene bags and dumped in a special pit meant for this purpose. Details of the pit can be obtained from Bhabha Atomic Research Centre, Mumbai. Strict precautions should be taken to safeguard the health of the personnel handling infectious and radioactive material and in no case these should be brought out in open containers for disposal.

The investigators working with infectious and radioactive material must ensure that the animals are properly disposed off on termination of their experiment and that infection is not transmitted to other animals in the room or to the personnel involved in handling such animals during the experiment and also at the time of disposal. The staff handling such animals for disposal must be apprised of the hazards involved in their job and protective clothing, gloves and mask must be provided to them for their personal safety.

7.   LABORATORY ANIMAL ETHICS

All scientists working with laboratory animals must have a deep ethical consideration for the animals they are dealing with. From the ethical point of view it is important that such considerations are taken care at the individual level, at institutional level and finally at the national level.

Individually each investigator has an obligation to abide by all the ethical guidelines laid down in this regard at institutional level. The Head of the Institution, maintaining animals for scientific experiments, should constitute an Animal Ethics Committee for experimentation to ensure that all experiments conducted on animals are rational, do not cause undue pain or suffering to the animals and only minimum number of animals are used. The constitution and terms of reference of the Animal Ethics Committee should be well-defined.

An Animal Ethics Committee should include a senior biological scientist of the institute, two scientists from different biological disciplines, a veterinarian involved in care of animals, the scientist incharge of animal facility, a scientist from outside the institute, a non-scientific socially aware member and a member or nominee of appropriate regulatory authority of Government of India. A specialist may be co-opted while reviewing special projects using hazardous agents such as radioactive substances and deadly microorganisms etc. The investigator may also be called in for any clarification, if required.

The Animal Ethics Committee has to examine all projects involving use of animals before implementation, to ensure that minimum number of animals is used in the project and the ethical guidelines are strictly adhered to. It will also examine that the scientists and technicians handling animals possess adequate skill to perform the experiment. All animals will be maintained under standard living conditions and experiments will be conducted with care. All invasive experiments will be conducted under proper anaesthesia and on termination of an experiment, the animal will be humanely sacrificed under anaesthesia. Before disposal it must be ensured that the animal is clinically dead.

(a)   *Ethical Guidelines for Use of Animals in Scientific Research*

1.   Animal experiments should be undertaken only after due consideration of their relevance for human or animal health and the advancement of knowledge.

2. The animals selected for an experiment should be of an appropriate species and quality, and minimum number should be used to obtain scientifically and statistically valid results.

3. Investigators and other personnel should treat animals with kindness and should take proper care by avoiding or minimizing discomfort, distress or pain.

4. Investigators should assume that all procedures which would cause pain in human beings may cause pain in other vertebrate species also (although more needs to be known about the perception of pain in animals).

5. Procedures that may cause more than momentary pain or distress should be performed with appropriate sedation, analgesia or anaesthesia in accordance with accepted veterinary practice. Surgical or other painful procedures should not be performed on unanesthetized animals.

6. At the end of, or when appropriate during an experiment, the animal that would otherwise suffer severe chronic pain, distress, discomfort, or disablement that cannot be relieved or repaired should be painlessly killed under anaesthesia.

7. The best possible living condition should be provided to animals used for research purpose. Normally the care of animals should be under the supervision of a veterinarian or a person having adequate experience in laboratory animal care.

8. It is the responsibility of the investigator to ensure that personnel conducting experiment on animals possess appropriate qualifications or experience for conducting the required procedures. Adequate opportunities have to be provided by the institution for inservice training for scientific and technical staff in this respect.

9. *In vitro* systems to replace or reduce the number of animals should be used wherever possible.

(b) *In vitro Systems to Replace Animals*

A number of *in vitro* systems can be used to reduce/replace animals in experimentation. These systems could be the living or the non-living systems. The living systems are tissue and organ culture, lower animals and microorganisms and human volunteers in restricted cases. The non-living systems could also be used in place of animals in certain areas and these include chemicals, mechanical models, mathematical models, computer simulation, DNA recombinant technology and synthetic substances.

8. LEGAL PROVISION

The Prevention of Cruelty to Animal Act of 1960 has provided the constitution of a committee under the Animal Welfare Board to control and supervise experiments on animals. It has also provided inspection of all animal holdings through designated inspectors and the institutions not abiding by the standard requirements can be prosecuted. [See Appendix I, The Prevention of Cruelty to Animals Act 1960]

ANNEXURE 1

List of Major Institutions Maintaining Animal Strains

(A) National Level Facilities of Laboratory Animals:
1. National Laboratory Animal Centre, Central Drug Research Institute, Lucknow.
2. National Centre for Laboratory Animal Sciences (NCLAS) National Institute of Nutrition, Hyderabad.

(B) Institutional Laboratory Animal Facilities at:
(i) Research/Academic Institutions:

All India Institute of Medical Sciences, New Delhi

Cancer Research Institute, Mumbai

Central Food Technology Research Institute, Mysore

Central Research Institute, Kassauli (H.P.)

Centre for Cellular and Molecular Biology, Hyderabad

Haffkine Bio-Pharmaceutical Corporation, Mumbai

Haryana Agricultural University, Hissar

Indian Institute of Science, Bangalore

Indian Veterinary Research Institute, Izatnagar

Institute of Microbial Technology, Chandigarh

Institute for Research in Reproduction, Parel, Mumbai

Indian Institute of Chemical Biology, Calcutta

National Institute of Communicable Diseases, New Delhi

National Institute of Immunology, New Delhi

National Institute of Virology, Pune

Post Graduate Institute of Medical Education and Research, Chandigarh

Regional Research Laboratory, Jammu Tawi

National Centre for Cell Science, Pune

(ii) Industrial Institutions :

Indian Drugs and Pharmaceuticals Ltd., Hyderabad

Indian Drugs and Pharmaceuticals Ltd, Rishikesh

Indian Immunologicals, Hyderabad

Hoechst Pharmaceuticals Ltd, Bombay

Sarabhai Chemicals Research Centre, Baroda

Ranbaxy Research Laboratory, New Delhi

Reddy Research Laboratory, Hyderabad

ANNEXURE 2

Physiological norms of commonly used laboratory animals.

	Mouse	Rat	Hamster	G. pig	Rabbit	Cat	Dog (Beagle)	Monkey (Rhesus)
Weight at birth (grams)	1–2	4–5	2–3	80–100	40–60	100–30	400–500	460–500
Age at weaning (weeks)	3	3	3	3	8	4–6	6–8	20–24
Wt. at weaning (grams)	9–12	40–50	30–40	250–300	800–900	400–700	–	400–700
Age at maturity W = weeks Y = years	6–8w	10–12w	6–8w	16–20w	24–32w	30–35w	1–1.2y	4–5y
Wt. at maturity	18–22g	150–200g	80–90g	250–400g	1.5–2.0kg	4–6kg	15–25kg	9–10kg
Adult weight	25–30g	200–300g	80–100g	400–500g	2.0–2.5kg	3–5kg	12–15kg	10–12kg
Rectal Temp °C(average)	37.4	37.5	37.6	38.6	38.7	39.5	38.6	38.4
Respiratory rate per minute	90–180	80–150	40–120	60–110	35–56	20–30	14–28	30–54
Pulse rate per minute (average)	600	300	450	150	133	110	95	200
Life span (years)	1.5–2.0	2.5–3.0	1.5–2.0	4–5	4–5	8–12	10–15	15–20
Diploid chromosome number 2n=	40	42	44	64	44	38	78	42

ANNEXURE 3

Reproductive date of commonly used laboratory animals.

	Mouse	Rat	Hamster	G. Pig	Rabbit	Cat	Dog (Beagle)	Monkey (Rhesus)
Oestrus cycle (days)	4–5	4–5	4–5	16		14	Biannual	28
Duration of oestrus	10h	13–15h	20h	6–11h	–	3–6d	14–21d	–
Time of ovulation h = hour da = day	2–3 h after Est. (Spont.)	8–10h (Spont.)	8–10h (Spont.)	10h (Spont.)	Induced 10–11h after mating	Induced 25–26h after mating	1–3 d (Spont.)	11–14d after onset of menstruation
Gestation period (days) average	21	21	16	68	30	63	62	164
Litter size	6–10	8–12	5–8	1–4	4–6	3–6	4–8	1
Oestrus after parturition	Post partum	Post partum	1–8d	Post partum	35d	4th week or lactation	Next neat reason	After weaning of young ones
Reproductive life span (years)	1	1	1	3–4	2–3	6	6–8	12–15
Mating system M:F ratio	Pair/Trio/ Harem	Pair/ Harem	Pair Harem	Harem	Hand Mating	Harem	Pair Harem	Pair Harem
Max. number of females per male	5	5	5	6	10	6	6	10
Mammary Glands (T.A.P.) no. of pairs	3, 1, 1 Five	3, 12 Six	–, 5, 1 Six	–, –, 1 One	1, 2, 1 Four	2, 2, – Four	2, 2, 1 Five	1, –, – One

A = Abdominal, P = Pelvic, T = Thoracic, h = hours, d = days = Menstrual cycle.

ANNEXURE 4

Housing and environment requirement for commonly used laboratory animals.

	Mouse	Rat	Hamster	G. Pig	Rabbit	Cat	Dog (Beagle)	Monkey (Rhesus)	Mouse	Rat
Avg. adult weight	25–30g	200–300g	30–100g	400–500g	2.0–2.5kg	3–5kg	12–15kg	3.5–8kg	3.5–8kg	6–15kg
Type of housing	Cage	Cage	Cage	Cage Pan	Cage Run	Run	Cage	Cage	Cage	Cage
Floor area per annual (sq. cm)	65–100	100–150	90–120	300–600	3700–4600	2500–3500	7000–12000	4000–6000	4000–6000	6000–9000
Cage height minimum (cm)	12	14	12	18	36	36	–	72	72	90
Room temp°C	22–24	22–24	22–24	22–24	22–24	Air dried	Air dried	Air dried	Air dried	Air dried

Relative humidity (%)	45–60	50–60	45–60	45–60	45–60	45–60	45–60	45–60	45–60	45–60
Suitable bedding material	Paddy husk saw dust	Paddy husk saw dusk	Paddy husk shredded paper	Paddy husk saw dusk	–	–	–	–	–	–
Nesting material	Paper cutting	Paper cutting	Paper cutting	Wood shavings dry grass	Cotton paper cutting dry grass	Cotton paper cutting	Cotton paper cutting	Fine jute paper cutting	–	–
Ventilation air changes per hour	10–12	10–12	10–12	10–12	10–12	Continuous	Continuous	Continuous	Continuous	Continuous
Light intensity (LUX)	300–400	300–400	300–400	300–400	300–400	300–400	300–400	300–400	300–400	300–400
Photocycle (Light dark)	12:12	12:12	12:12	12:12	12:12	12:12	12:12	12:12	12:12	12:12

ANNEXURE 5

Genetic Monitoring

Genetic monitoring of the animals is necessary particularly among the inbred and special strains to ensure genetic homogenicity of desired characters in a particular strain. The inbred strains particularly need such monitoring at regular time interval to ensure freedom from genetic contamination. The different methods, used in detecting the status of inbred strains for their homogenicity are:

1. ***Histocompatibility or Skin grafting***   It is a very convenient and reliable method for routine testing of all inbred strain against genetic contamination. Skin grafts exchanged among members of the same inbred strain or F1 hybrid or from either parent to an F1 hybrid should be accepted.

2. ***Electrophoresis or Biochemical Markers***   Inbred strains differ at many genetic loci. Some of these loci code for proteins and enzymes that may be identified by a number of biochemical techniques, the most important of which is electrophoresis. Typically, the samples of body fluids or organ homogenates are electrophorosed on starch cellulose acetate or polyacrilamide gels with specified buffer systems and under controlled conditions. After an appropriate time the gels are stained.

3. ***Immunological Markers***   Inbred strains differ in the alloantigens that they carry and these immunological markers can be used effectively in routine genetic quality control. For immunological markers, there are two basic methods for demonstrating allo-antibodies. The first is the haemagglutination technique in which the erythrocytes are the test cells. The second is the cytotoxic test in which the lymphocytes are typically the test cells.

4. ***Coat Colour Studies***   The coat colour genes carried by most inbred strains of mice and many inbred strains of rats are known. Some pigmented strains have an unusual or unique coat colour, which may be sufficient both to type the strain and to serve as an indicator in the event of genetic contamination. A cross with virtually any other strain would result in offspring that would not have the parental coat colour, and the occurrence of coat colour variants would immediately suggest genetic contamination.

5. ***Mandibular biometry***   Strain differences in the morphology of the skeleton have been known for many years. One method of genetic quality control using skeleton morphology is based on mandible shape. The technique is highly sensitive. All strains seem to differ in mandible shape and even closely related sublines can be distinguished, by measuring different parameters on the mandibles of 10–15 random animal samples taken from the group.

ANNEXURE 6

Nurtritional requirements of common laboratory animals.

	Mouse	Rat	Hamster	Guinea pig	Rabbit	Cat	Dog	Monkey
Protein (%)	18.00	12.00	15.00	18.00	17.00	30.00	20.00	15.00
Fat (%)	5.00	-	5.00	1.00	2.00	9.00	4.50	-
Linoleic acid (%)	0.30	0.60	-	-	-	1.00	0.90	1.00
Fibre (%)	5.00	-	-	10.00	10.12	-	-	-
Digestible energy (kcal)	3000	3800	4200	3000	2500	-	-	100
Vitamins								
A(IU/kg)	500	4000	3636	23333	580	25000	4500	10000–15000
D(IU/kg)	150	1000(e)	2484	1000	-	1000	450	2000
E(IU/kg)	20	30(f)	3	50	40	120	45	50
KI(IU/kg)	3000	50(g)	4000	2500	-	-	-	t
C(mg/kg)	t	t	t	200	t	t	t	100
Biotin (mg/kg)	0.20	t	0.60	0.30	-	-	0.90	0.10
Choline (mg/kg)	600	1000	2000	1000	1200	3000	1100	-
Folic acid (mg/kg)	0.50	1.00	2.00	4.00	-	1.00	0.16	0.20
Niacin (mg/kg)	10.00	20.00	90.00	10.00	180.00	4.00	10.30	50.00

Pantothenic acid	10.00	8.00	40.00	20.00	–	5.00	9.00	15.00
Riboflavin	7.00	3.00	15.00	3.00	–	4.00	2.00	5.00
Thaimine	5.00	4.00	20.00	2.00	–	5.50	0.90	–
VitaminB6	1.00	6.00	6.00	3.00	39.00	4.00	0.90	2.50
VitaminB12 (mg/kg)	10.00	50.00	10.00	10.00	–	–	20.00	t
Minerals								
Calcium (%)	0.40	0.50	0.60	0.8	0.75	1.00	1.00	0.50
Chloride (%)	t	0.50	–	–	0.30	1.00	1.00	0.2–0.5
Magnesium (%)	0.05	0.04	0.06	0.10–0.30	0.04	0.054	0.036	0.15
Phosphorus (%)	0.40	0.40	0.30	0.40–0.70	0.50	0.80	0.80	0.40
Potassium (%)	0.20	0.36	0.61	0.50–1.40	0.60	0.60	0.50	0.80
Sodium (%)	t	0.05	0.15	–	0.20	–	1.00	0.20–0.40
Sulphur (%)	–	0.03	–	–	–	–	–	–
Copper (mg/kg)	4.50	5.00	1.60	6.00	3.00	7.20	6.50	–
Iodine (mg/kg)	0.25	0.15	1.60	1.00	0.20	0.54	1.39	2.00
Iron	25.00	35.00	140.00	50.00	100.00	65.00	54.00	180.00
Manganese (mg/kg)	45.00	50.00	3.65	40.00	8.50	5.00	4.50	40.00
Selenium (mg/kg)	t	0.10	0.10	0.10	–	0.15	0.10	–
Zinc (mg/kg)	30.00	12.00	9.20	20.00	1.00	54.00	45.00	10.00

(Contd.)

ANNEXURE 6 (CONTINUED)

	Mouse	Rat	Hamster	Guinea pig	Rabbit	Cat	Dog	Monkey
L-Amino Acids								
Arginese (%)	0.30	0.60	0.76	–	0.60	–	–	–
Histidine (%)	0.20	0.30	0.40	–	0.30	–	–	–
Isoleucine (%)	0.40	0.50	0.89	–	0.60	–	–	–
Leucine (%)	0.70	0.80	1.39	–	1.10	–	–	–
Lysine (%)	0.40	0.70	1.20	–	0.65	–	–	–
Methionine (%)	0.50	0.60	0.32	–	0.60	–	–	–
Phenylalanine–								
Tyrosine (%)	0.40	0.80	0.83	–	1.10	–	–	–
Niasine (%)	0.50	0.80	0.83	–	0.20	–	–	–
Trytophan (%)	0.10	0.15	0.34	–	0.20	–	–	–
Water consumption (ml/day)	3–7	20–45	8–12	12–15	80–100	100–200	25–35	350–950
Food consumption (g/per day)	3–6	10–20	7–15	20–35	75–100	110–225	250–1200	350–50

-Not known/No data availbale

ANNEXURE 7

Requirements for transport of laboratory animals by road, rail and air

Nutritional requirements of common laboratory animals.

	Mouse	Rat	Hamster	Guinea pig	Rabbit	Cat	Dog	Monkey
Maximum no of animals per cage	25	25	25	12	2	1 or 2	1 or 2	1
Material used in transport	Metal Cardboard Synthetic Material	Metal Cardboard Synthetic Material	Metal Cardboard Synthetic Material	Metal Cardboard Synthetic Material	Metal Cardboard Synthetic Material	Metal	Metal	Metal
Space per animal (sq.)	20–25	80–100	80–100	160–180	1000–1200	1400–1500	3000	4000
Minimum height of box (cm)	12	14	12	15	30	40	50	48

ANNEXURE 8

Commonly used anaesthetic drugs for laboratory animals.

Drugs (mg/kg)	Mouse	Rat	Hamster	Guinea pig	Rabbit	Cat	Dog	Monkey
Ketamine HCl	22–24i/m	22–24i/m	-	22–24	22–24	30i/m	30i/m	15–40
Pentobarbitone	35i/v	25i/v	35i/p	30–i/v	30i/v	25i/v	20–30i/v	35i/v
Sodium	50i/p	50i/p		40i/p	40i/v			
Thiopentone	25i/v	20i/v	20i/v	20i/v	20i/v	25i/v	25i/v	25i/v
	50i/p	40i/p	55i/p					60i/p
Urethane	-	0.75i/p	-	1.5i/p	1.0i/p,i/v	1.25i/v 1.00i/v 150i/p	1.00i/v	1.0i/v

Atropine: Dose 0.02–0.05mg/kg for all species by s/c or i/m or i/v routes used to reduce salivary and bronchial secretions and protect heart from vagal inhibition, given prior to anaesthesia

i/m = intramuscular, i/v = intravenous, i/p = intraperitoneal, s/c = subcutaneous

ANNEXURE 9

Euthanasia of laboratory animals.

A.   Methods acceptable for species of animals indicated NR=Not Recommended

Species	Mouse	Rat	Hamster	Guinea pig	Rabbit	Cat	Dog	Monkey
a) Physical Methods								
Electrocution	NR	NR	NR	NR	NR	NR	A	NR
Exsanguination	NR	A	NR	A	A	A	A	A
Decapitation Cervical	A	A	A	NR	A	NR	NR	A
Dislocation	A	A	A	A	A	NR	NR	A
b) Inhalation of Gases								
Carbon Monoxide	A	A	A	A	A	A	A	A
Carbon Dioxide	A	A	A	A	A	A	NR	NR
Carbon Dioxide-Chloroform	A	A	A	A	A	A	NR	NR
c) Drug Administration								
Barbiturate Overdose (route)	A(IP)	A(IP)	A(IP)	A(IP)	A(IV, IP)	A(IV, IP)	A(IV, IP)	A(IV, IP)
Chloral hydrate Overdose (route)	NR	NR	NR	NR	A(IV)	A(IV)	A(IV)	A(IV)
Ketamine Overdose (route)	A(IM)	A(IM)	A(IM)	A(IM)	A(IM)	A(IM)	A(IM)	A(IM)

B.   Methods not acceptable for any species of animals

a)   Physical Methods

   (i)   Decompression

   (ii)   Stunning

b)   Inhalation of Gases

   (i)   Nitrogen Flushing

   (ii)   Argon Fushing

c)   Drug Administration

   (i)   Curariform drugs

   (ii)   Nicotine Sulphate

   (iii)   Magnesium Sulphate

   (iv)   Potassium chloride

   (v)   Strychnine

   (vi)   Paraquat

   (vii)   Dichlorvos

   (viii) Air emboism

APPENDIX   V

SI UNITS OF MEASUREMENTS

The International System of Units, or SI (*Système Internationale* [Fr.]) units, is the standard system of measurement used since October 1960 in all scientific works. It is based on seven principal units. These base units can combine with different prefixes to form derived SI units defining smaller and larger quantities.

CONVENTIONS IN THE USAGE OF SI UNITS

The following are some of the important conventions followed in the correct use of the SI units.

i) A base unit may take only one prefix. For example 'millimillimetre' is incorrect and should be written as 'micrometre'.

ii) Most prefixes that make a unit bigger are written in capital letters (M, G, T, etc.), and that make a unit smaller are lower cased (m, n, p, etc.). Exceptions to this are the kilo [k] to avoid any possible confusion with kelvin [K]; hecto [h]; and deca [da] or [dk].

iii) The names of many SI units are eponymous, that is they are named after persons, who were associated with and made significant contributions to the fields in which the units are used. All such units are written in lower case (newton, volt, pascal, etc.) when expressed in full, but starting with a capital letter (N, V, Pa, etc.) when abbreviated. An exception to this rule is the litre which, if written as a lower case 'l' could be mistaken for a '1' (one) and so a capital 'L' is allowed as an alternative. It is proposed that a single letter (l or L) will be decided upon in the future when it becomes clear which letter is being favoured most in use.

iv) Units written in abbreviated form are never pluralized. Therefore, 'm' could always be either 'metre' or 'metres'. An 'ms' represents 'millisecond'. Use of 'gms' for grams, 'kms' for kilometers, etc., is not permitted.

v) An abbreviation (such as J, N, g, Pa, etc.) is never followed by a fullstop unless it is the end of a sentence.

vi) In order to make reading of the numbers easier, the numbers are divided into groups of 3 separated by spaces (or half-spaces) but not commas.

vii) A decimal fraction is expressed with a comma (e.g. 7743,6750,) to separate the whole number from its fractional part. However, the practice of using a point is acceptable provided that the point is placed on the line of the bottom edge of the numbers as in 7743.6750 and not in the middle as in 17.456

Table 1   Definitions of the SI base units.

Physical quantity	Name of SI base unit	Symbol for SI unit	Definition
Length	metre	m	Meter is the distance travelled by light, in vacuum, in 1/299792458th of a second.
Mass	kilogram	kg	Kilogram is the mass of an international prototype in the form of a platinum-iridium cylinder kept at Sevres in France. *It is now the only basic unit still defined in terms of a material object, and also the only one with a prefix [kilo] already in place.*
Time	second	s	Second is the length of time taken for 9192631770 periods of vibration of the caesium-133 atom to occur.
Electric current	ampere	A	The ampere is that current which produces a specified force between two parallel wires that are 1 metre apart in vacuum. *It is named after the French physicist Andre Ampere (1775–1836).*
Thermodynamic temperature	kelvin	K	The Kelvin is 1/273.16*th* of the thermodynamic temperature of the triple point of water. *It is named after the Scottish mathematician and physicist William Thomson 1st Lord Kelvin (1824–1907).*

(Contd.)

Table 1   (Continued)

Physical quantity	Name of SI base unit	Symbol for SI unit	Definition
Amount of substance	mole	mol	The mole is the amount of substance that contains as many elementary units as there are atoms in 0.012 kg of carbon-12.
Luminous intensity	candela	cd	The candela is the intensity of a source of light of a *specified frequency*, which gives a specified amount of power in a given direction.

Table 2   Derived SI units with special names.

Physical quantity	Name of SI unit	Symbol	Equivalent in SI base units	Definition
Frequency	hertz	Hz	S^{-1}	The hertz is the SI unit of the frequency of a periodic phenomenon. One hertz indicates that 1 cycle of the phenomenon occurs every second. For most work much higher frequencies are needed such as the kilohertz [kHz] and megahertz [MHz]. *It is named after the German physicist Heinrich Rudolph Hertz (1857–94).*
Force	newton	N	$m\ kg\ S^{-2}$	One newton is the force required to give a mass of 1 kilogram an acceleration of 1 metre per second. *It is named after the English mathematician and physicist Sir Isaac Newton (1642–1727).*

(Contd.)

Table 2 (Continued)

Physical quantity	Name of SI unit	Symbol	Equivalent in SI base units	Definition
Pressure	pascal	Pa	Nm^{-2}	One pascal is the pressure generated by a force of 1 newton acting on an area of 1 square metre. It is a rather small unit as defined and is more often used as a kilopascal [kPa]. *It is named after the French mathematician, physicist and philosopher Blaise Pascal (1623–62).*
Energy, work	joule	J	Nm	One joule is the amount of work done when an applied force of 1 newton moves through a distance of 1 metre in the direction of the force. *It is named after the English physicist James Prescott Joule (1818–89).*
Power or rate of doing work	watt	W		One watt is a power of 1 joule per second. *It is named after the Scottish engineer James Watt (1736–1819).*
Electrical charge or quantity of electricity	coulomb	C	As	One coulomb is the quantity of electricity transported in 1s by a current of 1 A. *It is named after Charles Augustin de Coulomb (1736–1806).*
Electrical potential	volt	V	$J\,C^{-1}$	One volt is the difference of potential between two points of an electrical conductor when a current of 1 ampere flowing between those points dissipates a power of 1 watt. *It is named after the Italian physicist Count Alessandro Giuseppe Anastasio Volta (1745–1827).*

(Contd.)

Table 2 (Continued)

Physical quantity	Name of SI unit	Symbol	Equivalent in SI base units	Definition
Electrical resistance	ohm	Ω	VA^{-1}	The ohm is the SI unit of resistance of an electrical conductor. Its symbol, is the capital Greek letter 'omega'. *It is named after the German physicist Georg Simon Ohm (1789–1854).*
Electrical conductance	siemens	S	AV^{-1}	It is equal to the ratio of current (measured in amps) divided by voltage (measured in volts). It also equals 1/R. 1 picosiemen equals 10^{-12} siemens and is convenient to use for ion channels. For example, the ion channel gramicidin has a conductance of 30 pS for cations. This is equivalent to 6.28×10^6 ions per second per applied volt of conductance of cations thru the pore when the concentration of cations is equal on both sides of the membrane. *It is named after Ernst Werner von Siemens, German electrical engineer (1816–1892).*
Electrical capacitance	farad	F	CV^{-1}	The farad is the SI unit of the capacitance of an electrical system, that is, its capacity to store electricity. It is a large unit as defined and so often used as a microfarad. *It is named after the English chemist and physicist Michael Faraday (1791–1867).*

(Contd.)

Table 2   (Continued)

Physical quantity	Name of SI unit	Symbol	Equivalent in SI base units	Definition
Luminous flux	lumen	lm	cd sr	One lumen is the luminous flux emitted in a solid angle of 1 steradian by a uniform point source having an intensity of 1 candela. The steradian (sr) is the solid angle that, having its vertex in the centre of a sphere, cuts off an area of the surface of the sphere equal to that of a square with sides of length equal to the radius of the sphere.
Illuminance	lux	Ix	cd sr m^{-2}	One lux is lumens per square meter (cd sr m^{-2}). It is the total amount of visible light illuminating (incident upon) a point on a surface from all directions above the surface. This "surface" can be a physical surface or an imaginary plane. Therefore illuminance is equivalent to *irradiance* weighted with the response curve of the human eye.
Radio-activity	becquerel	Bq	S^{-1}	A becquerel is "one disintegration per second". One Bq of radioactive material is that amount of material in which one atom is transformed or undergoes one disintegration in every second. It can express a concentration of radioactivity. Bq replaces the earlier measure "Curies". *It is named after Antoine Henri Becquerel (1852-1908), French physicist who discovered that rays emitted by uranium salts affect photographic plates.*

(Contd.)

Table 2   (Continued)

Physical quantity	Name of SI unit	Symbol	Equivalent in SI base units	Definition
Absorbed dose of radiation	gray	Gy	$J\ kg^{-1}$	It is a measure of the amount of energy absorbed by the body. The rad is the traditional unit of absorbed dose. It is being replaced by the unit gray (Gy), which is equivalent to 100 rad. One rad equals the dose delivered to an object of 100 ergs of energy per gram of material. *It is named for the British physician L. Harold Gray (1905–1965), an authority on the use of radiation in the treatment of cancer.*
Celsius temperature	degree Celsius	°C		The Celsius temperature scale was designed so that the freezing point of water is 0 degrees, and the boiling point is 100 degrees at standard atmospheric pressure. A change in temperature of 1°C is equal to a change in temperature of 1 K. *It is named after the Swedish astronomer Anders Celsius (1701 to 1744), who first proposed a similar system in 1742.*

Table 3   Prefixes used with SI units.

Prefix	Symbol	Value	Expression
yotta	Y	1 000 000 000 000 000 000 000 000	10^{24}
zetta	Z	1 000 000 000 000 000 000 000	10^{21}
exa	E	1 000 000 000 000 000 000	10^{18}
peta	E	1 000 000 000 000 000	10^{15}
tera	T	1 000 000 000 000	10^{12}
giga	G	1 000 000 000 (a billion)	10^{9}
mega	M	1 000 000 (a million)	10^{6}
kilo	k	1 000 (a thousand)	10^{3}
hecto	h	100 (a hundred)	10^{2}
deca	da	10 (ten)	10^{1}
		1 (one)	10^{0}
deci	d	0.1 (a tenth)	10^{-1}
centi	c	0.01 (a hundredth)	10^{-2}
milli	m	0.001 (a thousandth)	10^{-3}
micro	μ	0.000 001 (a millionth)	10^{-6}
nano	n	0.000 000 001 (a billionth)	10^{-9}
pico	p	0.000 000 000 001	10^{-12}
femto	f	0.000 000 000 000 001	10^{-15}
atto	a	0.000 000 000 000 000 001	10^{-18}
yocto	y	0.000 000 000 000 000 000 000 001	10^{-24}

Table 4 Derived SI units for other quantities.

Physical quantity	Coherent[*] SI Unit	Other SI Units
area	m^2	cm^2, km^2
volume	m^3	cm^3, dm^3
speed, velocity	$m\ s^{-1}$	$km\ s^{-1}$
acceleration	$m\ s^{-2}$	$cm\ s^{-2}$
wave number	m^{-1}	cm^{-1}
density, mass density	$kg\ m^{-3}$	$g\ cm^{-3}$
specific volume	$m^3\ kg^{-1}$	$cm^3\ g$
amount concentration	$mol\ m^{-3}$	$mol\ dm^{-3}$, $mol\ 1^{-1}$
molar volume	$m^3\ mol^{-1}$	$1\ mol^{-1}$, $cm^3\ mol^{-1}$
molar energy	$J\ mol^{-1}$	$kJ\ mol^{-1}$
molar heat capacity	$J\ K^{-1}\ mol$	

Coherent SI units are units expressed in terms of the base units (Table 1) and/or the derived units with special names (Table 2), with the numerical factor always unity, e.g. the coherent SI unit for volume is m^3, i.e., $1\ m^3$.

Table 5   Some acceptable non-SI units and symbols.

Unit	Symbol	SI equivalent
minute	min	60 s
hour	h	3600 s
day	d	86 400 s
week	wk	604 800 s
year	Yr	($\approx$365.25 d)
degree	°	(π/180) rad
minute	'	(π/10.800) rad
second	"	(π/648.000) rad
ångstrom	Å	10^{-10} m, 0.1 nm
liter	l or L	1 dm^3, 10^{-3} m^3
bar[a]	bar	10^5 Pa
curie	Ci	3.7×10^{10} Bq
svedberg	Sv	10^{-13} s
Electronvolt[*]	Ev	$\approx 1.60218 \times 10^{-19}$ J
Uamu[**], dalton	u, Da = ma (^{12}C)/12	$\approx 1.66054 \times 10^{-27}$ kg

[*]The bar is recommended as the "metric atmosphere" to replace the non-SI unit "standard atmosphere; 1 atm = 760 mmHg = 1.01325 bar; 1 bar = 750.062 mmHg (= 750.0 mmHg for most purposes (<0.01% error)).

[**]1 Da, or u, is one-twelfth the mass or one atom or ^{12}C. Unified atomic mass unit ("uamu") and symbol u are the terms recognized by the General Conference on Weights and Measures (CGPM), the organization responsible, by treaty, for units internationally; dalton and Da have not been approved by CGPM but are widely used by chemists and biochemists. Both the dalton and the electronvolt must be related to SI units through measurements, the uncertainty or which limits the precision of their equivalents in SI units.

Table 6   Not acceptable units and symbols.

Not acceptable units and symbol	Recommended units and symbols
micron, μ	micrometer, um, μm
microgram, γ	microgram, ug, μg
microliter, λ	microliter, ul, μl, μL
Calorie[*], cal	joule, J
kilocalorie, kcal	kilojoule, kJ
atmosphere, atm	bar
mmHg, Torr	mbar, ubar, Pa

*For data related directly to nutrition, the calorie and the kilocalorie (= I Calorie) may be used; otherwise, joule (kilojoule, microjoule) should be used. The current practice in thermodynamics is to use joule and kilojoule rather than calorie.

Table 7   Some biophysical quantities and symbols.

Quantity	Symbol
absorbance	A
acceleration of gravity	g
equilibrium constant	K
Michaelis constant	K_m
relative molar mass	M_r
retardation factor	R_F
specific rotation	$[\alpha]^t$
sedimentation coefficient	s
sedimentation coefficient in water at 20°C, extrapolated to c = 0	$S^\circ_{20,w}$
diffusion coefficient	D
Gibbs function (Gibbs energy)	G
change in Gibbs function	ΔG
enthalpy, change in enthalpy	H, ΔH
change in enthalpy at standard state	$\Delta H\theta$, ΔH°
entropy, change in entropy	S, ΔS
energy	U, (E)
heat capacity	C
heat capacity at const.p, V	C_p C_V

Table 8   Metric system of measurements.

Length		Area	
10 millimetres	= 1 centimetre	100 sq. mm	= 1 sq. cm
10 centimetres	= 1 decimeter	10 000 sq. cm	= 1 sq. metre
10 decimetres	= 1 metre	100 sq. metres	= 1 acre
10 metres	= 1 decametre	100 acres	= 1 hectare
10 decametres	= 1 hectometre	10 000 sq. metres	= 1 hectare
10 hectometres	= 1 kilometre	100 hectares	= 1 sq. kilometre
1000 metres	= 1 kilometre	1 000 000 sq. metres	= 1 sq. kilometre

Volume		Capacity	
1000 cu. mm	= 1 cu. cm	10 millilitres	= 1 centilitre
1000 cu. cm	= 1 cu. decimetre	10 centilitres	= 1 decilitre
1000 cu. dm	= 1 cu. metre	10 decilitres	= 1 litre
1 million cu. cm	= 1 cu. metre	1000 litres	= 1 cu. metre

Mass	
1000 grams	= 1 kilogram
1000 kilograms	= 1 tonne

Notes   There is no real distinction between "volume" and "capacity"
1 millilitre = 1 cubic centimeter 1 litre = 1 cubic decimetre

APPENDIX VI

ABBREVIATIONS AND ACRONYMS

Abbreviation is intentional shortening of a word or phrase, while writing. The shortening is done arbitrarily and conveniently (convenient to read and pronounce) by trimming letters of a word. The trimmed letters may be in the middle or at the end of a word. For example "Mister" is abbreviated as "Mr.", "Mistress" as "Mrs.", "Doctor" as "Dr.", "Limited" as "Ltd.", "foot or feet" as "ft", etc. by trimming some or all the letters between the first and the last letters. Or, they may be at the end of a word as in "Gen." for "General", "in." for "inches". "Co." for "company", etc. While abbreviating a phrase, letters are trimmed from each word in the phrase. For example, "deoxyribose nucleic acid' is abbreviated as D.N.A., "United States of America" as U.S.A.

A period (full stop) is used to denote an abbreviation of each word that is abbreviated. However, this practice is becoming obsolete as in, for instance, DNA, USA, NATO, etc. Some abbreviations use an apostrophe mark to represent the missing letters as in "can't" for "cannot". A period is never used in abbreviations having apostrophes.

An abbreviated word is pronounced as its expanded form or as its abbreviated form. For instance, "Mr." is read as "mister", "Dr." as "doctor" and "Prof." as "professor". On the other hand, DNA, USA, etc. are pronounced using the letters.

Many abbreviations are used as symbols for units of measurements (see Appendix on SI Units), and such abbreviations may be pronounced in original form or in abbreviated form. For example, cm may be pronounced as centimeter or as cm.

Acronyms are also abbreviations formed by a combination of first letters or syllables in a group of words such that the new group of letters serves as a "short name" representing the original phrase. Acronyms are made such that they can be easily pronounced and used as a noun. A few examples of acronyms are: Radar—Radio detection and ranging; Laser—Light Amplification by Stimulated Emission of Radiation; Maser—Microwave Amplification by Stimulated Emission of Radiation. The main difference between an abbreviation and an acronym is that the latter is pronounceable.

There are many non-English phrases (French, Latin, etc.) commonly used in abbreviated forms. A few examples are: *e.g. et al., ad-lib, i.e., etc.* The meaning and usage of a few of these abbreviations are given in Appendix.

In biology, the creation and use of abbreviations and acronyms are ever-increasing especially in the fields of molecular biology, biotechnology, and bioinformatics. It would be an unwieldy task if we try to list all the abbreviations that are currently used in the different fields of biology. The following list contains common abbreviations used in biological sciences.

A.D.	*anno Domini* [in the year of the Lord]
A.M.	*ante meridiem* [before noon]; *Artium Magister* [Master of Arts]
aa	amino acid(s)
AAQ	Ambient Air Quality
AAQS	Ambient Air Quality Standards
Ab	antibody(s)
abbr.	abbreviation(s), abbreviated
ABER	Annual Blood Examination Rate
ac	alternating current
Acad.	Academy
ACTH	adrenocorticotropin
ADCC	Antibody Dependent Cellular Cytotoxicity
ADDC	Antibody-Dependent Cell-mediated Cytotoxicity
ADP	adenosine diphosphate
AES	Atomic Emission Spectrometer
af	audio frequency
AFP	Alfa-Feto Protein
Ag	Antigen(s)

(Contd.)

AIDS	Acquired Immuno Deficiency Syndrome
AIIMS	All India Institute of Medical Sciences
ALARA	As Low As Reasonably Achievable
ALMPGIBMS	A. Lakshmanaswamy Mudaliar Postgraduate Institute of Basic Medical Sciences
alt.	altitude
AM	amplitude modulation
AMP	adenosine monophosphate
AMU	Aligarh Muslim University
amu	atomic mass unit
ANGIS	The Australian National Genomic Information Service
ANOVA	Analysis of Variance
AP	alkaline phosphatase
API	Annual Parasite Index
AR	Analytical Reagent
ARS	Acute Radiation Syndrome
ASI	Archaeological Survey of India
Assn.	Association
ASV	Anti snake venom
at. no.	atomic number
at. wt	atomic weight
ATP	adenosine triphosphate
ATPase	adenosine triphosphatase
B cell	bursa-derived lymphocyte
B.C.	Before Christ
b.p.	boiling point
BALT	Bronchus-Associated Lymphoid Tissue
BAPMON	Background Air Pollution Monitoring

(Contd.)

BARC	Baba Atomic Research Centre
BBC	British Broadcasting Corporation
BEIR	Biological Effects of Ionizing Radiation
BHU	Banaras Hindu University
BIS	Bureau of Indian Standards
BLAST	Basic Local Alignment Search Tool
BLOSUM	BLOcks Substitution Matrix
BOD	Biological oxygen demand
BP	Blood Pressure; Biologically Pure
bp	base pair; boiling point
BSA	Bovine Serum Albumin
BSC	Biological Safety Cabinet
BSM	Biomolecular Structure and Modelling
Bt	*Bacillus thuringiensis*
BTIC	Biotechnology Information Centre
Btu	British thermal unit(s)
BWC	Biological Weapons Convention
BWR	Boiling Water Reactor
c.	*circa* [about]
CAM	Cell Adhesion Molecule
cAMP	cyclic AMP
CARI	Central Agricultural Research Institute
CATH database	Class, Architecture, Topology, Homology database
CBD	Convention on Biological Diversity
CBS	Centre for Biological Sequence Analysis
CBT	Centre for Biotechnology

(Contd.)

CCMB	Centre for Cellular and Molecular Biology
CCRI	Central Coffee Research Institute
CD	Circular Dichroism
CDFD	Centre for DNA Fingerprinting and Diagnostics
cDNA	complementary DNA
CDRI	Central Drug Research Institute
CD-ROM	Compact Disc-Read Only Memory
CECRI	Central Electrochemical Research Institute
CESE	Centre for Environmental Science and Engineering
CFC	Chlorofluorocarbon
CFTRI	Central Food Technological Research Institute
CGHS	Central Government Health Scheme
CGPM	General Conference on Weights and Measures
CHG	Centre for Human Genetics
CHO	Chinese hamster ovary
CICR	Central Institute of Cotton Research
CIFA	Central Institute of Freshwater Aquaculture
CIFE	Central Institute of Fisheries Education
CIFT	Central Institute of Fisheries Technology
CIMAP	Central Institute of Medicinal and Aromatic Plants
CIRG	Central Institute of Research on Goats
CJIL	Central JALMA Institute for Leprosy
CLRI	Central Leather Research Institute
CMC	Christian Medical College
CMI	Cell Mediated Immunity

(Contd.)

CNS	Central Nervous System
CoA	Coenzyme A
Coag.	Coagulation
COD	Chemical Oxygen Demand
conc.	concentration; concentrated
const.	constant
CP	Chemically Pure; Coat protein
CPCB	Central Pollution Control Board
CPCSEA	Committee for the Purpose of Control and Supervision of Experiments on Animals
CPE	cytopatheic effect
CPM	counts per minute
cRNA	complementary RNA
CRP	C-reactive protein
CSE	Centre for Science and Environment
CSF	Cerebrospinal fluid; Colony-stimulating Factor
CSI	Computer Society of India
CSIR	Council for Scientific and Industrial Research
CSIRO	Commonwealth Scientific and Industrial Research Organization
CSRTI	Central Sericultural Research and Training Institute
CT	Computerized Tomography
CTL	Control; Cytotoxic T Lymphocyte
CTRI	Central Tobacco Research Institute
CTX	Cholera toxin
CVP	Children Vaccine Programme
CWMF	Central Waste Management Facility
cyclic AMP	adenosine monophosphate 3', 5'-cyclic

(Contd.)

DAE	Department of Atomic Energy
DBT	Department of Biotechnology
dc	direct current
DDBJ	DNA Data Bank of Japan
Dept.	Department
DFRL	The Defence Food Research Laboratory
Dist.	District
Div.	Division
DNA	Deoxyribose nucleic acid
DNase	Deoxyribonuclease
DoE	Department of Energy
DPM or dpm	Disintegration per minute
DPS or dps	Disintegrations per second
Dr.	Doctor
DRDO	Defence Research Development Organization
ds	double stranded
DTH	Delayed Type Hypersensitivity
Dx	Diagnosis
Dz	Disease
EAA	Essential amino acid
EB	Electron Beam
EBI	European Bioinformatics Institute
ECL	Enhanced Chemiluminescence
ECM	Extracellular matrix
E-Commerce	Electronic Commerce
ed.	edited, edition, editor(s)
ED_{50}	effective dose, 50%
EDE	effective dose equivalent

(Contd.)

EDI	Electronic Data Interface
EDTA	Ethylenediamine tetra acetic acid
EGF	Epidermal growth factor
ELISA	Enzyme linked immunosorbent assay
EM	electron micrography; electron microscopy; electron microscope
EMBnet	The European Molecular Biology network
EMCP	Enhanced Malaria Control Programme
emf	electromotive force
EMU	Electric Multiple Unit
EPA	Environmental Protection Act
EPCA	Environmental Pollution Control Authority
EPD	Eukaryotic Promoter Database
epidem.	epidemiology
EPR	Electron Paramagnetic Resonance
equiv. wt.	equivalent weight
ER	Endoplasmic Reticulum
ES	Embryonic stem
ESL	Environmental Survey Laboratory
ESR	Electron Spin Resonance; Erythrocyte Sedimentation Rate
EST	Expressed Sequence Tag
et al.	*et alii* [and others]
ETP	Effluent treatment plant
EU	European Union
Exp. or exp.	Experiment; Experimental
F	Fahrenheit
FAD	Flavin-adenine dinucleotide
FAO	Food and Agriculture Organization

(Contd.)

FAQ	Frequently Asked Questions
FBP	Final Boiling Point
FBR	Fast Breeder Reactor
FBS	Foetal Bovine Serum
FBTR	Fast Breeder Test Reactor
FCI	Food Corporation of India
FCS	Foetal Calf Serum
FDA	Food and Drug Administration
fl oz	fluid ounce(s)
FM	Frequency Modulation
FMN	Flavin mononucleotide
fp	freezing point
FRCP	Fellow of the Royal College of Physicians
FRI	Forest Research Institute
FSH	Follicle stimulating hormone
FSI	Forest Survey of India
GI	Gastrointestinal
GAC	Genome Annotation Consortium
GALT	Gut-Associated Lymphoid Tissue
GAVI	Global AIDS vaccine initiative
Gaz.	Gazette
GBPIHED	G.B. Pant Institute of Himalayan Environment and Development
GC	Gas chromatography
GDB	Genome database
GDP	Gross Domestic Product
GEAC	Genetic Engineering Approval Committee
GEF	Global Environment Facility

(Contd.)

Gen.	General, Genesis
GIS	Geographical Information System
GLC	gas/liquid chromatography
GLP	Good laboratory practice
GMO	Genetically Modified Organism
GMP	Good manufacturing practice
GMT	Greenwich mean time; Good microbiological techniques
GSI	Geological Survey of India
gp	gene product
GPS	Global Protection System
GSDB	The Genome Sequence Database
GSETT	Group of Scientific Experiments Technical Test
HA	Haemagglutinin
Hb	Haemoglobin
HbCO	carbon monoxide haemoglobin
HbO_2	oxyhaemoglobin
HBSS	Hanks' balanced salt solution
HDI	Human Development Index
HEPA filter	High-Efficiency Particulate Air filter
HEV	Hepatitis E virus
hf	high frequency
HGI	Human Gene Index
HGMP-RC	Human Genome Project—Resource Centre
HGP	Human Genome project
HIV	Human Immuno Virus; Human Immunodeficiency Virus
HLA	Human Leukocyte Antigen
HPLC	High Performance (or Pressure) Liquid Chromatography
HPTLC	High performance thin layer chromatography
HPU	Health Physics Unit

(Contd.)

HRC	Human Rights Commission
HRP	Horseradish peroxidase
HSADL	High Security Animal Disease Laboratory
HTLV	Human T Lymphocyte Virus
HTML	Hypertext Markup Language
HTTP	Hypertext Transfer Protocol
HVAC	Heating, Ventilation and Air-Conditioning
HYV	High Yielding Varieties
i.d. or o.d.	Inside diameter, or outside diameter
i.e.	*id est* [that is]
i.l.	intraluminal(ly)
i.m.	intramuscular(ly)
i.p.	intraperitoneal(ly)
i.v.	intravenous(ly)
IAEA	International Atomic Energy Agency
IAEC	Institutional Animal Ethics Committee
IARI	Indian Agricultural Research Institute
IBRC	Insect Biopesticide Research Centre
IBSC	Institutional Biosafety Committees
IBSD	Institute of Bioresources and Sustainable Development
IC_{50}	Inhibitory Concentration, 50%
ICAR	Indian Council for Agricultural Research
ICGEB	International Centre for Genetic Engineering and Biotechnology
ICJ	International Court of Justice
ICMR	Indian Council of Medical Research
ICN	International Computer Network

(Contd.)

ICPO	Institute of Cytology and Preventive Oncology
ICRISAT	International Crops Research Institute for the Semi Arid Tropics
ICRP	International Commission on Radiological Protection
ID	identify or identification
ID_{50}	Infective dose, 50%
IF	Immunofluorescence
IFAD	International Fund for Agricultural Development
IFGTB	Institute of Forest Genetics and Tree Breeding
IFLA	International Federation of Library Association
IFN	Interferon
IgG, IgM	Immunoglobulin G, M
IGIB	Institute of Genomics and Integrative Biology
IHBT	Institute of Himalayan Bioresource Technology
IICB	Indian Institute of Chemical Biology
IIF	Indirect immunofluorescence
IIFM	Indian Institute of Forestry Management
IIHR	Indian Institute of Horticultural Research
IIIT	Indian Institute of Information Technology
IIPR	Indian Institute of Pulses Research
IIPS	International Institute of Population Studies
IISc	Indian Institute of Science
IISR	Indian Institute of Spices Research
IIT	Indian Institute of Technology
IITM	Indian Institute of Tropical Meteorology
IIVR	Indian Institute of Vegetables Research
ILRI	Indian Lac Research Institute
IMA	Indian Medical Association

(Contd.)

IMTECH	Institute of Microbial Technology
inc.	incorporated
INDOEX	Indian Ocean Experiment
INM	Integrated nutrient management
INPM	Integrated nutrient and pest management
INS	Indian Newspapers Society
INSA	Indian National Science Academy
Inst.	Institute, Institution
INTERNET	International Computer Network
IP	Intellectual Property; Internet Protocol
IPM	Integrated pest management
IPR	Intellectual Property Right
IR	Infrared; immune response
IRDP	Integrated Rural Development Programme
IREDA	Indian Renewable Energy Development Agency
IRMS	Infrared Mass Spectrometer
IRR	Institute for Research in Reproduction, (now renamed as National Institute of Research in Reproduction)
IRRI	International Rice Research Institute
IS	Indian Standard
ISI	Indian Statistical Institute
ISO	International Organization for Standardization; Indian Space Organisation
ISP	Internet Service Provider
ISRO	Indian Space Research Organisation
IT	Information Technology
ITLCV	Indian Tomato Leaf Curl Virus
ITRC	Industrial Toxicology Research Center
IU	International Unit

(Contd.)

IVCOL	Indian Vaccines Corporation Limited
IVD	*in vitro* diagnostics
IVRI	Indian Veterinary Research Institute
IWRM	Integrated Water Resources Management
JFM	Joint Forest Management
JIPID	Japan International Protein Information Database
JNCASR	Jawaharlal Nehru Centre for Advanced Scientific Research
JNU	Jawaharlal Nehru University
Jr.	Junior
Kb or kb	kilobase(s); 1000 nucleotide residues (ss nucleic acids)
Kbp	kilobase pair(s);1000 base pairs (ds nucleic acids)
Kbps	kilobits or thousands of bits per second
KFRI	Kerala Forest Research Institute
KSSDI	Karnataka State Sericulture Development Institute
LR	Laboratory Reagent
lat.	latitude
LD_{50}	Lethal dose, median
LET	Linear Energy Transfer
LF	Leaf folder
Lib.	Library
LLW	Low-level waste
LNG	Liquefied Natural Gas
Long.	longitude
LPG	Liquefied Petroleum Gas
LRI	Local Radiation Injury
LSC	Liquid Scintillation Counter
Ltd.	Limited

(Contd.)

M.D.	*Medicinae Doctor* [Doctor of Medicine]
mAb	monoclonal antibody
MALT	Mucosa-Associated Lymphoid Tissue
MCRC	Murugappa Chettiar Research Centre
MEM	Eagle's Minimum Essential Medium
MGE	Mobile genetic element
MHC	Major Histocompatibility Complex
MIPS	Martinsried Institute for Protein Sequences
MKU	Madurai Kamaraj University
MLR	Mixed Lymphocyte Reaction
MMDB	Molecular Modeling Database
MNES	Ministry of Non-conventional Energy Sources
MO	Monocyte
MoEF	Ministry of Environment and Forests
MoHFW	Ministry of Health and Family Welfare
MORGAN	Multiframe Optimal Rule–based Gene Analyzer
MoU	Memorandum of Understanding
mp	melting point
mph	miles per hour
MRI	Magnetic Resonance Imaging
mRNA	messenger RNA
MS	Mass spectroscopy; mass spectrometry; manuscript
MSL	Mean Sea Level
MSSRF	M S Swaminathan Research Foundation
mt.	Mount, Mountain
MTCC	Microbial Type Culture Collection
MtDNA	mitochondrial DNA
MVI	Malaria vaccine initiative
N	North
NAAQS	National Ambient Air Quality Standards

(Contd.)

NABL	National Accreditation Board of Testing and Calibration Laboratories
NAMP	National Anti-Malaria Programme; National Air Quality Monitoring Programme
NAPS	Narora Atomic Power Station
NASA	National Aeronautics and Space Administration
NBAGR	National Bureau of Animal Genetic Resources
NBP	Net Biome Production
NBPGR	National Bureau of Plant Genetic Resources
NBRC	National Brain Research Centre
NBRF	National Biomedical Research Foundation
NBRI	National Botanical Research Institute
NCA	National Commission on Agriculture
NCAR	National Centre for Atmospheric Research
NCBI	NCL Centre for Biodiversity Informatics
NCBI	National Centre for Biotechnology Information
NCCS	National Centre for Cell Science
NCL	National Chemical Laboratory
NCSA	National Centre for Supercomputing Applications
NDDB	National Dairy Development Board
NDRI	National Dairy Research Institute
NDT	Non-Destructive Technique
NE	North-east
NEERI	National Environmental Engineering Research Institute
NEHU	North-Eastern Hill University
NGF	Nerve growth factor
NIAID	National Institute for Allergy and Infectious Diseases
NICED	National Institute of Cholera and Enteric Diseases
NIH	National Institute of Health
NII	National Institute of Immunology

(Contd.)

NIMHANS	National Institute of Mental Health and Neurosciences
NIP	National Immunization Programme
NIRR	National Institute of Research in Reproduction, (formerly Institute for Research in Reproduction)
NLN	National Library of Medicine
NMEP	National Malaria Eradication Programme
NMR	Nuclear Magnetic Resonance
No. or no.	Number
NRC	Nuclear Regulatory Commission
NRC Grapes	National Research Centre for Grapes
NRDB	Non-Redundant DataBase
NSAID	Non-Steroid Anti-Inflammatory Drug
NSF	National Science Foundation
OD	Optical density
OHP	Overhead Projector
OMIM	Online Mendelian Inheritance in Man
ONGC	Oil and Natural Gas Commission
Op.	*Opus* [work]
OPV	Oral Polio Vaccine
ORD	Optical Rotatory Dispersion
ORF	Open Reading Frame
OS	Operating System
OSH	Occupational Safety and Health
PAGE	Polyacrylamide gel electrophoresis
PAM	Point Accepted Mutation
PBS	Phosphate-buffered saline
PCA	Perchloric acid
PCR	Polymerase Chain Reaction
PCS	Process Control System

(Contd.)

PDB	Protein Data Base; Protein Databank
PDF	Portable Document Format; Portable Digital Format
PETA	People for Ethical Treatment of Animals
PFA Act	Prevention of Food Adulteration Act
PGD	Plasmid Genome Database
PGDB	Plant Genome Data Bank
PGIMER	Postgraduate Institute of Medical Education and Research
Ph. D.	Doctor of Philosophy
PHCs	Primary Health Centres
PIR	Protein Information Resource
PM	Particulate Matter
pmf	proton-motive force
PMN	Polymorphonuclear leukocyte
PMT	Photo Multiplier Tube
POSSUM	Picture Of Standard Syndromes and Undiagnosed Malformations
ppt	Precipitate; parts per thousand
PPV	Protection of Plant Variety
PSI-BLAST	Position Specific Iterated BLAST
PUC	Pollution Under Control
PVC	Polyvinyl chloride
QAQC	Quality Assurance Quality Control
QED	Quantum Electrodynamics
R&D	Research and Development
R&M	Renovation & Maintenance
RAPS	Rajasthan Atomic Power Station
RBC	Red Blood Cell
RBD	Randomized Block Design

(Contd.)

RBE	Relative Biological Efficiency (radiation)
RCGM	Review Committee on Genetic Manipulation
RDAC	Recombinant DNA Advisory Committee
REBASE	Restriction Enzyme Database
REE	Rare Earth Elements
REGP	Rural Employment Generation Programme
Rem	Roentgen equivalent, man
RER	Rough Endoplasmic Reticulum
rev.	revised
rf	radio frequency
RGCB	Rajiv Gandhi Centre for Biotechnology
RHdb	Radiation Hybrid database
RIA	Radioimmunoassay
RIS	Research Information System
RMRC	Regional Medical Research Centre
RNA	Ribonucleic acid
RNase	Ribonuclease
RO	Reverse Osmosis
RPM or rpm	Revolutions Per Minute
RRL	Regional Research Laboratory
rRNA	ribosomal RNA
RSA	Recurrent spontaneous abortions
RT	Room temperature
RT-PCR	Reverse transcription PCR
Rx	Therapeutic drugs, care, etc.
S	south
S&T	Science and Technology

(Contd.)

s.c.	subcutaneous(ly)
SALT,	Skin-Associated Lymphoid Tissue
SARS	Severe Acute Respiratory Syndrome
SBCC	State Biosafety Coordination Committees
SCI	Science Citation Index
SCLA	Sub-Committee on Large Animals
SCOP	Structural Classification of Proteins
SD	Standard Deviation
SDS	sodium dodecyl sulfate
SDS-PAGE	Sodium dodecyl sulphate-polyacrylamide gel electrophoresis
SEM	Standard error of the mean
SFPP	Special Foodgrains Production Programme
SI	*Le Systeme International d'Unites*
SIB	Swiss Institute of Bioinformatics
SIIL	Serum Institute of India Ltd
SMD	Soil Moisture Deficit Ratio
SMI	Soil Moisture Index
SNP	Single nucleotide polymorphism
SOPs	Standard operating procedures
SPCB	State Pollution Control Board
SPV	Solar Photo Voltaic
Sr.	Senior
SRBC	Sheep red blood cell
SRS	Sequence Retrieval System
ss	single stranded
SSLP	Simple sequence length polymorphism
SST	Sea Surface Temperature

(Contd.)

SSTL	Silkworm Seed Technology Laboratory
STD	Sexually Transmitted Disease
STS	Sequence Tag Site
SV40	simian virus 40
SWAT	Soil and Water Assessment Tool
Sx	Symptoms
T cell	Thymus-derived lymphocyte; many subtypes
$t_{1/2}$	half-life (half-time)
TANUVAS	Tamil Nadu University of Veterinary and Animal Sciences
TB	Tuberculosis
TBS	Tris-buffered saline
TC	Tissue culture
TCA	Trichloroacetic acid
TCP	Transmission Control Protocol
TCR	T Cell Receptor
TDF	Transcription Factor Database
TDS	Total dissolved solids
TEDE	The total effective dose equivalent
TERI	The Energy and Resources Institute, (formerly Tata Energy Research Institute)
TFR	Total Fertility Rate
TIGR	The Institute for Genomic Research
TLC	Thin-layer chromatography
TNAU	Tamil Nadu Agricultural University
TrEMBL	Translated EMBL
TRIPS	Trade Related Intellectual Property Rights
tRNA	transfer RNA

(Contd.)

TRRD	Transcription Regulatory Region Database
Tx	treatment
U	unit
uamu	Unified atomic mass unit
UCL	University College of London
UGC	University Grants Commission
UHF	Ultra High Frequency
UK	United Kingdom
UK MRC	United Kingdom Medical Research Council
UN	United Nations
UNCED	United Nations Conference on Environment and Development
UNCETDG	United Nations Committee of Experts on the Transport of Dangerous Goods
UNCLOS	United Nations Conference of Law of the Sea
UNDCP	United Nations Drug Central Programme
UNDP	United Nations Development Programme
UNEP	United Nations Environment Programme
UNESCO	United Nations Educational, Scientific and CulturalOrganisation
UNFCCC	United Nations Framework Convention on Climate Change
UNICEF	United Nations Children's Fund (United Nations International Children's Emergency Fund)
Univ.	University
UNO	United Nations Organization
UPOV	Union for Protection Of new Varieties of plants
URL	Universal resource locator; Uniform Resource Locator
USA	United States of America

(Contd.)

USG	Ultrasonography
UV	Ultraviolet
VHF	Very High Frequency
vol.	Volume(s)
vs.	*versus*
W	West
WAN	Wide Area Network
WB	World Bank
WCD	World Council of Dams
WEF	World Economic Forum
WEP	World Food Programme
WEU	Western European Union
WGA	Wheat Germ Agglutinin
WHO	World Health Organization
WIPO	World Intellectual Property Organization
wk	Week
WMO	World Meteorological Organization
WOFOST	World Food Study Programme
WSF	Water Soluble Fraction
WSSD	World Summit on Sustainable Development; White spot shrimp disease
WSSV	White spot shrimp Virus
wt.	weight
WTO	World Trade Organization
WWF	World Wildlife Fund; World Wide Fund for Nature
WWW or www or W3	World Wide Web
XML	Extensible Markup Language
XRF	X-ray Radio Frequency

APPENDIX   VII

CODES FOR AMINO ACIDS

Amino acid	Three letter code	Single letter code
Nonpolar amino acids (Hydrophobic)		
glycine	Gly	G
alanine	Ala	A
valine	Val	V
leucine	Leu	L
isoleucine	Ile	I
methionine	Met	M
phenylalanine	Phe	F
tryptophan	Trp	W
proline	Pro	P
Polar (Hydrophilic)		
serine	Ser	S
threonine	Thr	T
cysteine	Cys	C
tyrosine	Tyr	Y
asparagine	Asn	N
glutamine	Gln	Q
Electrically charged (Negative)		
aspartic acid	Asp	D
glutamic acid	Glu	E
Electrically charged (Positive)		
lysine	Lys	K
arginine	Arg	R
histidine	His	H

APPENDIX VIII

THE GREEK ALPHABET

Letter	Uppercase type	Lowercase type	English equivalent
alpha	A	α	a
beta	B	β	b
gamma	Γ	γ	g
delta	Δ	δ	d
epsilon	E	ε	e
zeta	Z	ζ	z
eta	H	η	ē
Theta	Θ	θ	th
iota	I	ι	i
kappa	K	κ	k
lambda	Λ	λ	l
mu	M	μ	m
nu	N	ν	n
xi	Ξ	ξ	x
omicron	O	o	o
pi	Π	π	p
rho	P	ρ	r, rh
sigma	Σ	σ	s
tau	T	τ	t
upsilon	Y	υ	u
phi	Φ	φ	ph
chi	X	χ	kh
psi	Ψ	ψ	ps
omega	Ω	ω	ō

ACCENTS AND DIACRITICAL MARKS

Mark #	Name #	Example
´	acute	é
°	bolle	å
˘	brene	?
¸	cedilla	Ç
^	circumflex	ê
¨	umlant/diaeresis	ö
`	grave	è
ˇ	háček	Č
?	macron	ō
/	sterg	Ø
~	tilde	ñ

APPENDIX   IX

SOME FRENCH AND LATIN PHRASES COMMONLY USED IN ENGLISH

French phrases

à la	in the manner of
à la carte	according to the menu; a selected meal; each available separately
à la mode	In the manner; in the fashion; fashionable
au contraire	to the contrary
au courant	up-to-date, abreast of current affairs
avant garde	before the guard (vanguard); applied to cutting-edge or radically innovative movements in art and literature
bête noire	black beast; special object of dislike; someone or something which is detested or avoided
carte blanche	blank card; full power; unlimited authority; free hand
coup de grâce	blow of mercy; finishing stroke; a death-blow
coup d'état	takeover of state; a sudden change in government by force; a swift stroke of policy
cul-de-sac	a blind alley; a dead-end street
déjà vu	the impression or illusion of having seen or experienced something before
détente	relaxation of strained relations between countries; easing of diplomatic tension
double entendre	double meaning, often improper in inference
en bloc	in a lump; taken all together
en masse	in a body; all together

(Contd.)

French phrases

en passant	in passing
en route	on the way to
encore	again; a request to repeat a performance
entente	diplomatic agreement or cooperation
expose	a published exposure of a fraud or scandal
fait accompli	something that has happened and is unlikely to be reversed
faux pas	a false step; A social blunder; An embarrassing social error
milieu	environment, setting
objet d'art	a piece of art
outré	bizarre, eccentric
par excellence	by excellence; pre-eminently; quintessential.
précis	a concise summary
raison d'etre	reason for existence
rapprochement	the establishment of cordial relations
rendezvous	a meeting, appointment, or date; usually written rendez-vous in French and sometimes in English
répondez s'il vous plaît. (RSVP)	please reply; a form of words appended to an invitation
résumé	a summary or abstract; a document listing one's qualifications for employment
tête-à-tête	head-to-head; face to face; a private or tense meeting; a private conversation between two people
touché	acknowledgment of an effective counterpoint
vis-à-vis	face to face; opposite; in comparison with or in relation to; also used to refer to the opposite corner of an intersection
volte-face	a complete reversal of opinion or position

Latin phrases

A contrario	On the contrary
A posteriori	From the latter; from effect to cause; based on observation; inductive; relating to or derived by reasoning from observed facts; the opposite of *a priori*; Used in probability, philosophy and logic to denote something that is known after a proof has been carried out.
A priori	From the former; from cause to effect; presupposed, the oppsite of *a posteriori* deductive; relating to or derived by reasoning from self-evident propositions; presupposed by experience; being without examination or analysis; presumptive; formed or conceived beforehand; used in probability, philosophy and logic to denote something that is known or postulated before a proof has been carried out
Ab initio	From the beginning or from the start
Ab origine	From the origin
Ab ovo	From the egg; from the very beginning
Ad absurdum	To absurdity; taken to an absurd extreme (in logic); short version of *reductio ad absurdum*
Ad fundum	To the bottom or to the end
Ad hoc	For this (purpose); *i.e.,* improvised, made up on the spot for a specific use, purpose, etc.
Ad infinitum	To infinity; going on forever; without end or limit; used to designate a property which repeats in all cases in mathematical proof
Ad interim	In the meantime; for the time being
Ad libitum (ad lib)	At pleasure; as much as you like; according to one's pleasure; often used to indicate the liberty to "improvise", "just ramble on"; especially in music, theatrical scripts, etc.; to indicate the frequency of feeding or drinking allowed for the experimental animals

(Contd.)

Latin phrases

Ad nauseam	To sea-sickness; to the point of disgust; to a sickening or excessive degree
Ad referendum	Subject to reference
Ad valorem	By the value; *e.g. ad valorem* tax
Addendum	An item to be added; an appendix
Alias	Otherwise; applied to a alternate name of a person
Alibi	Elsewhere; the legal defence that a person was elsewhere when the crime was committed
Alma mater	Nourishing or fostering mother; applied by the students to the school or university where they studied; the word "matriculation" is derived from "mater". The term suggests that the students are "fed" knowledge and taken care of by the university; also used for a university's traditional school anthem
Alumnus/Alumna	A graduate or former student of a school, college, or university
Anno Domini (A.D.)	In the year of the lord; indicates a year counted from the traditional date of birth of Jesus; also called the *Common Era (C.E.)* to remove religious implications
Ante cibum (a.c.)	Before meals (medical shorthand)
Ante meridiem (a.m.)	Before noon; in the period from midnight to noon
Ante prandium (a.p.)	Before lunch; *i.e.,* before a meal; used on pharmaceutical prescriptions
Aqua fortis	Strong water; nitric acid
Aqua vitæ	Water of life; spirits of wine, brandy
Camera lucida	Light chamber; an instrument to trace the outline of a specimen through microscope; opposite of *camera obscura*

(Contd.)

Latin phrases

Camera obscura	Dark chamber; a darkened enclosure having an aperture usually provided with a lens through which light from external objects enters to form an image of the objects on the opposite surface
Caveat	Beware! A warning enjoining one from certain acts or practices; an explanation to prevent misinterpretation; a legal warning to a judicial officer to suspend a proceeding until the opposition has a hearing
*Circa (*also *ca.* or *c.)*	Around; used in the sense of "approximately, about"; usually of a date, magnification of photomicrograph, etc.
Confer (cf.)	Bring together; compare; used as an abbreviation in text to recommend a comparison with another thing
De facto	From the fact; in reality; said of something that actually *is* the case, in contrast to a legal or official rule or status or version, which is described as *de jure*. In some contexts *de facto* refers to the "way things really are" rather than what is "officially" presented as the fact. When speak about a *de facto* Government, we mean the one that is actually in power whether legal or otherwise
De jure	By law; official; rightful; opposite of *de facto*
De novo	Again; anew
e.g.	See *Exempli gratia*
Emeritus	From merit; often used to denote a position held at the point of retirement, as an honor. *e.g. Professor emeritus.* This does not necessarily mean that the honoree is no longer active
Et alibi (et al.)	And elsewhere; used (albeit uncommonly) like "etc." to stand for a list of places
Et alii (et al.)	And others; used like "etc." to stand for a list of names; the abbreviation et al. is used in reference citation in the text, to stand for more than two names

(Contd.)

Latin phrases

Et cetera (etc.)	And the rest; and others especially of the same kind: and so forth; and more
Et sequens (et seq.)	And the following
Ex gratia	From kindness or from grace; referring to someone performing an act out of kindness as opposed to being forced to do it; something given out of kindness, as in *ex gratia* payment
Ex officio	From the office; by virtue of one's office; when someone holds one position by virtue of holding another, *e.g.* the vice president is the *ex officio* Chairman of Rajya Sabha
Ex parte	From part; from one side; from a partisan; by (or for) one party; a legal term, as in *ex parte* judgement; in the interest of one side only
Ex tempore	This instant; right away; immediately; without preparation; *ex tempore* measure, one hastily provided; *ex tempore speech*
Exempli gratia (e.g. or eg)	For the sake of example; used as "for example", "example given"
Experimentum crucis	Critical experiment; a decisive test of a scientific theory
Facsimile	Do the like; make a similar one; a perfect copy; origin of the word fax
Habeas corpus	"You must have the body", *i.e.,* you must justify an imprisonment; first two words of the writ to bring a prisoner to court, and commonly used as the general term for a prisoner's legal right to have the charge against them specifically identified
i.e.	See *Id est*
Ibidem (ibid.)	In the same place; usually in bibliographic citations, for subsequent references to the same place in the same authority

(Contd.)

Latin phrases

Id est (i.e.)	that is (to say), in this case; used in the explanatory sense; it is never equivalent to *e.g.*
In absentia	"In the absence", *e.g.* of a trial carried out in the absence of the accused
In camera	In a room or chamber; used to mean "in secret" or "in private"
In situ	In its original place, position, or arrangement; in medical contexts it implies that the condition is "still" in its original place and has not spread
In toto	In all; entirely; totally; completely
In vacuo	In empty space; in a vacuum; in isolation from other things
In vitro	In glass; an experiment or process performed in a non-natural laboratory setting, for example in a test tube
In vivo	In (a) living (organism); an experiment or process performed in a living specimen, as opposed to *in vitro*
inst.	abbreviation for instant; used in formal correspondence to refer to the current month, as opposed to last or next month, as in "Thank-you for your kind letter of the 7th inst"
Inter alia	Among other things
Ipso facto	By the fact itself; virtually; as an inevitable result
Lingua franca	Frankish language, a common language, a mixture of Italian French, Spanish, Greek, and Arabic that was formerly spoken in Mediterranean ports; any of various languages used as common or commercial tongues among peoples of diverse speech; something resembling a common language
Magnum opus	Great work; the greatest achievement of an artist or writer

(Contd.)

Latin phrases

Mala fide	In bad faith; said of an act done with knowledge of its illegality, or with intention to defraud or mislead someone
Memento	Something that serves to warn or remind; something presented to a guest to remember an event
Modus operandi (M. O.)	Method of operation; the way of working; usually used to describe a criminal's methods
Nihil per os (n.p.o.)	Nothing by mouth (medical shorthand)
Nota bene (NB)	Note it well; mark well; please note; important note; used to call attention to something important
Oculus dexter (O.D.)	Right eye (ophthalmologist shorthand)
Oculus sinister (O.S.)	Left eye (ophthalmologist shorthand)
Opera omnia	All works; the collected works of an author
Opera posthuma	Posthumous works; i.e., published after the author's death
Opere citato (op. cit.)	In work (already) cited; used in research reports when referring again to the last source mentioned or used
Passim	Throughout; here and there; frequently; of a word that occurs several times in a cited texts; also, in proof reading, of a change that is to be repeated everywhere needed
Per annum	Per year
Per capita	By heads; per head; per person; equally to each individual; per unit of population: by or for each person
Per definitionem	By definition
Per diem	Per day; by the day; for each day use or service by the day; daily; paid by the day; a specific amount of money an organization allows an individual to spend per day; travel allowance; *per diems,* a daily allowance

(Contd.)

Latin phrases

Per os (p.o.)	By mouth (medical shorthand)
Per se	By itself; in itself; by its own nature; without referring to anything else, intrinsically, taken without qualifications, etc.; for instance, negligence *per se*
Persona grata	a pleasing person; personally acceptable or welcome; a welcome visitor
Persona non grata	Person not wanted; an unwanted or undesirable person. In diplomatic contexts, a person rejected by the host government; unacceptable; unwelcome; banned
Post cibum (p.c.)	After meals (medical shorthand)
Post meridiem (p.m.)	After noon; in the period from noon to midnight
Post mortem	After death; occuring or done after death; pertaining to a post-mortem examination, an examination of the body of the deceased; an autopsy
Post partum	After birth; of or occuring in the period shortly after childbirth
Post scriptum (p.s.)	Post script; used to mark additions to a letter, after the signature
Prima facie	At first sight; at first view; at first appearance on the first appearance; true, valid, or sufficient at first impression; apparent; self-evident; used to designate evidence in a trial which is suggestive, but not conclusive, of something (e.g. a person's guilt);legally sufficient to establish a fact or a case unless disproved
Pro forma	For form; as a matter of form; made or carried out in a perfunctory manner or as a formality; a standard document or form
Pro tempore	For the time (being); temporary
q. v.	See *Quod vide*

(Contd.)

Latin phrases

Quaque die (qd)	Every day (medical shorthand)
Quaque hora (qh)	Every hour (medical shorthand)
Quater in die (qid)	Four times a day (medical shorthand)
Quid pro quo	Tit for tat; this for that; a thing for a thing; in English, a favour for a favour
Quo vadis	Where are you going?
Quod vide (q.v.)	"Which see"; used after a term or phrase that should be looked up elsewhere in the current document or book
Sic	Thus; just so; states that the preceding quoted material appears exactly that way in the source, usually despite errors of spelling, grammar, usage, or fact
Signetur (sig or S/)	Let it be labelled (medical shorthand)
Sine die	Without a (set) day; indefinitely; without any future date being designated (as for resumption)
Sine qua non	Without which not; used to denote something that is an essential part of the whole; something absolutely indispensable or essential
Statim (stat)	Immediately (medical shorthand)
Status quo (ante)	The state that was (before); as things were before; the status of affairs or situation prior to some upsetting event
Stet	Let it stand; marginal mark in proofreading to indicate that something previously deleted or marked for deletion should be retained
Sub judice	Under a judge; under judicial consideration; not yet decided; said of a case that cannot be publicly discussed until it is finished
Sub poena (subpoena)	Under penalty; writ commanding one to attend a court of law or be punished

(Contd.)

Latin phrases

Sui generis	Of its (own) kind; peculiar to itself; in a class of its own; unique
Suo moto *[alt. Suo motu]*	Upon one's own initiative; usually used when a court of law, upon its own initiative, (i.e., no petition has been filed) proceeds against a person or authority that it deems has committed an illegal act
Ultimo (ult.)	Used in formal correspondence to refer to the previous month
Ultra vires	Without authority; beyond the powers possessed
Ut infra	As below
Ut retro	As backwards; as on the back side; i.e., "as above" or "as on the previous page"
Ut supra	As above
Vade mecum	Go with me; a *vade-mecum* or *vademecum* is an item one carries around, constantly consulted aid, a handbook
Versus (vs or v.)	Against; as in "Good *versus* Evil"
Veto	I forbid; a right to unilaterally stop a certain piece of legislation
Via	By way of; through; by means of
Via media	Middle path; middle course
Vice-versa	With places exchanged; the order being reversed; conversely
Vide	See
Vide infra (v.i.)	See below
Vide supra (v.s. or supra)	See above; see earlier in this writing
Videlicet (viz.) Videre licet	One may see; to wit; namely; used to introduce examples or a listing of something just named
Vox populi	Voice of the people

APPENDIX   X

PROOFREADING

A necessary part of producing a research report is careful and repeated proofreading and editing of the text before submission for publication or evaluation. The text should be free from errors in content, style, form, grammar, and spelling. The document should be consistent throughout, especially with regard to formatting such as paragraph indents, headings at different levels, numbering at different levels, fonts used, etc.

Proofreading skill can be mastered only by practice. The following are a few hints for successful proofreading:

1. We must develop a sense of doubting the correctness of every word, sentence, etc. we have written in the report. We might be aware that we tend to commit certain types of errors more frequently (e.g. spelling errors). We should double-check for the occurrence of such errors.

2. While proofreading a text, we should read slowly and loudly, one word at a time. The intention is to read what is actually written and not what we think we have written. This practice of reading loudly word by word what we have written is rather difficult, and therefore, it would be prudent if we take the help of a third person for this job.

3. It is useful to proofread twice, once for spelling and grammar and once for correctness of the content, formatting, etc.

4. A break between document preparation and proofreading always helps to do the proofreading effectively.

5. We must choose the right time of day for proofreading, i.e., the time when we are mentally alert and when we are free from external disturbances. Generally, early morning is better suited for proofreading.

6. Good lighting is essential. We can try natural, incandescent and fluorescent lights and identify the one that suits us best.

7. It is much easier to proofread from hard copy (printout, larger font and double spaced) than from soft copy (on-screen).

8. Though it is very useful, we must be very careful with electronic spelling and grammar checks because they may not catch all errors. It is often handy to use the "find" function to identify the common spelling and grammar mistakes we tend to commit.

9. Double or, even, treble checking of numerical data, especially in tables, is necessary. Use of a ruler (scale) to isolate lines of numerical data is helpful.

10. All research students should learn the meaning and usage of proofreading marks not only to understand what their supervisors suggested on the manuscripts of the research report but also to provide proper guidelines to the typists.

11. If more than one person is involved in proofreading, use of different colours is advisable. All proof copies must be saved to compare with the final copy.

Given under are a few proofreading marks and abbreviations used in proof correction.

Proofreading marks

Proofreading Symbols	Meaning	Example
⋀	Insert what is shown in the margin	See sample proofread page below
ℰ or ℘	Delete	See sample proofread page below
⊙	Add a period	See sample proofread page below
⋃ or ⋂	Transpose	See sample proofread page below
⌣	Close up	See sample proofread page below
⌣	Delete and close up	
¶	New paragraph	
⊔⊔⌿	Italicize (italics)	
≡	Capitalize	See sample proofread page below

(Contd.)

Proofreading symbols	Meaning	Example
≢	No capitals; lower case	See sample proofread page below
═	Small capitals	
𝖫𝖥𝖩	No italics; change to Roman upright type	See sample proofread page below
Y	Insert space, between words, lines, etc.	See sample proofread page below
↑	Reduce space, between words, lines, etc.	
⊏	Take over	
⊐	Take back	
⊐	Move to the right or indent	See sample proofread page below
⊐	Move to the left	
⤶	Run on; no new paragraph	See sample proofread page below
⊏___	Move to the left	
___⊐	Move to the right	
⌊___⌋	Move up	
⌉⌈	Move centre	
⌈___⌉	Move down	
∿∿∿	Change to bold	
∿∤∿	Unbold	See sample proofread page below
∿∿∿ (underlined)	Change to bold italic	See sample proofread page below
⟨ ⟩	Add parenthesis	
⟦ ⟧	Add square brackets	

(Contd.)

Proofreading symbols	Meaning	Example
∧, ∧̦, ❝❞, ❜	Insert, comma, double quotation marks, apostrophe, etc.	See sample proofread page below
# more #, less #	space; leave more space; reduce space	See sample proofread page below
‖	Correct vertical alignment	
(wf)	Wrong font	
- - - - -	Leave unchanged; let it stand	See sample proofread page below
(sp)	Spell out	
(=)	Insert hyphen	See sample proofread page below

A sample page with proofreading marks

APPLICATION PEPTIDE MAPPING AND ADVANTAGES OF HUNTER SYSTEM OVER OTHER TECHNIQUES

Peptide mapping is a powerful technique to help determine peptide structure and composition of proteins. Peptide maps fingerprints of proteolysed proteinsare usually obtained by resolution on either one-dimensional SDS- PAGE analysis (cleveland), reverse-phase hplc or by Two-dimensional separation on thin layer plates. Perhaps the most common applications of peptide mapping are: 1) to reveal identities between proteins suspected to be encoded by the same or related genes,

2) to prepare individual peptides to determine amino acid composition and sequence, and 3) to determine the precise location of amino acid residues that are posttransitionally modified by either fatty acid acylation glycosylation methylation, acetylation, or phosphorylation. Because the biochemist is often faced with the reality of obtaining only vanishingly small amounts of protein for analysis, it is often difficult or impossible to perform crusial experiments which reveal some of these important characteristics Two-dimensional separation of proteolytic digests using the Hunter system and

chromatography on thin layer cellulose plates is a technique that is well-suited to solve at least some of these problems and certain advantages over reverse-phase hplc and **SDS-PAGE** First, it is an extremely sensitive technique that requires only small amounts of metabolically labeled product (*only a few dpm*). Second, because digests are resolved in two-dimensions, a variety of information is derived that often yields subtle but-important clues

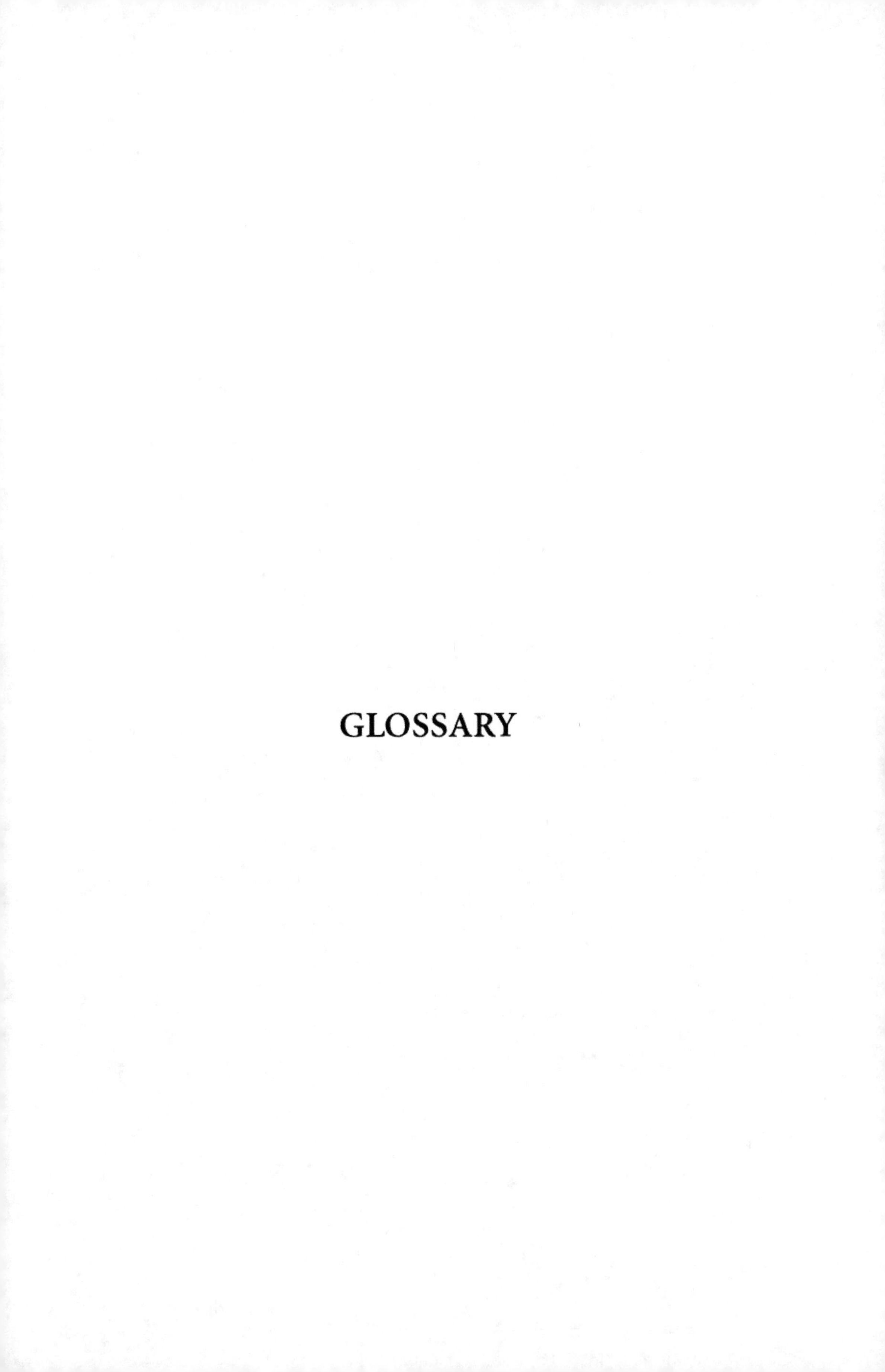

GLOSSARY

absorbance When light passes through a solution, light of certain wavelength is absorbed by the coloured chemical substance and it is referred to as the absorbance, and the remaining light is transmitted which is referred to as transmittance; absorbance is also known as Optical Density (OD) or Extinction.

absorbance spectrum Absorbance spectrum (absorption spectrum) of a substance of specific concentration is the plot of its absorbance of radiation at various wavelengths; often described as the "fingerprint" of that substance and necessary for the quantitative determination of that substance. The spectrum would show that specific wavelength at which the absorbance is more and this specific wavelength where there is maximum absorption is referred to as *absorption maxima* (A_{max} or λ_{max}).

abstract A summary (in about 250 words) of the information reported in a research article; gives the salient features of the main sections (Introduction, Materials and Methods, Results and Discussion) of a paper; the abstracts of research articles of standard Journals are published or made available on the Internet by indexing services such as *Biological Abstracts*, *Medline*, etc.

accession number A term used in Bioinformatics and is an identifying number assigned by the curators of the major biological databases to any submission of a new entry of data like base sequence of a gene, amino acid sequence of a protein, etc.

acclimatization The adaptation of a living organism (plant, animal or microorganism) to a changed environment (e.g. laboratory conditions, new habitat, etc.), which might cause physiological stress to it.

active site The amino acid residues at the catalytic site of an enzyme. These residues provide the binding and activation energy needed to place the substrate into its transition state and bridge the energy barrier of the reaction undergoing catalysis.

adenovirus DNA viruses causing diseases in animals and in man, they produce acute respiratory tract infections with symptoms resembling common cold. They are used in gene cloning, as vectors for expressing large amounts of recombinant proteins in animal cells and to make live-virus vaccines against pathogens.

ADEPT (Antibody-directed enzyme pro-drug therapy) A method to target a drug to a specific tissue. The drug is administered as an inactive pro-drug, and then converted into an active drug by an enzyme administered with a second injection. The enzyme is coupled to an antibody that concentrates it in the target tissue. When the enzyme arrives at the target tissue, the pro-drug is activated to form the active drug, while elsewhere it remains inactive.

adsorption chromatography A chromatographic technique in which the components of a sample are separated on the basis of the differences in their adsorption–desorption behaviour between the mobile and the stationary phases.

affinity chromatography A column chromatographic technique which uses column packing, called the affinity matrix (e.g. agarose), that has been chemically altered by attaching a compound with a specific affinity for the desired molecules (e.g. biological compounds such as an enzyme). The ligands that have been genetically engineered to have specific affinity are inserted into the matrix. The desired molecules in the mobile phase adsorb to the ligands because of their specific interaction (e.g. antigen–antibody reaction). A solution of high salt concentration is then passed through the column to cause desorption of the molecules from the ligands, and to elute from the column. There are several types; (i) In immuno-affinity chromatography, an antibody is used as the immobilized molecule to "capture" its antigen; (ii) In pseudo-affinity chromatography, a compound similar to a biological ligand is immobilized on a solid material to bind with an enzyme or other proteins; (iii) Metal affinity chromatography uses a metal ion which is immobilized on a solid support; the metal ions bind tightly and specifically to many biomolecules.

affinity tag, purification tag   A sequence of amino acids that has been engineered into a protein to make its purification easier. The incorporated tag can work in a number of ways: (i) The tag could be another protein, which binds to some other material very tightly and thus allow the protein to be purified by affinity chromatography; (ii) The tag could be a short amino acid sequence, which is recognized by an antibody. The antibody would then bind to the protein whereas it would not have done so before. One such short peptide called FLAG, has been designed so that it is particularly easy to make antibodies against it; (iii) The tag could be a few amino acids, which are then used as a chemical tag on the protein. For example, a string of

positively charged amino acids will bind very strongly to a negatively charged filter: this could be used as the basis of a separation system. (iv) Some amino acids bind metals very strongly, especially in pairs; this chemical property can be exploited by using a filter with metal atoms chemically linked onto it to pull a protein out of a mixture of proteins.

aflatoxin   Toxic substance produced by fungi of the *Aspergillus flavus* group. It is found in stored food or feed contaminated with these fungi. The toxin binds to DNA and prevents replication and transcription, and causes acute liver damage and cancer.

AFLP (Amplified fragment length polymorphism)   A type of DNA marker generated by (i) digestion of genomic DNA with two restriction enzymes to create many DNA fragments, (ii) ligation of specific sequences of DNA (called adaptors) to the ends of these fragments, (iii) amplification of the fragments via PCR (using a set of primers with sequences corresponding to the adapters, plus various random combinations of three additional bases at the end), and (iv) visualization of fragments via gel electrophoresis. The PCR will amplify any fragment whose sequence happens to start with any of the three-base sequences in the set of primers. AFLPs have the important advantage that many markers can be generated with relatively little effort. They are a very useful means of quantifying the extent of genetic diversity within and between populations. Their major disadvantage is that they are not specific to a particular locus and, because they are scored as the presence or absence of a band, heterozygotes cannot be distinguished from homozygotes.

agar A gel used as supporting medium in electrophoresis. It is a compound of two galactose-based polymers, agarose and agaropectin that dissolves in aqueous buffer at 40°C and solidifies to form a gel at 38°C. Low concentration of agar yields gels with larger pore size, which virtually has no molecular sieving effect; therefore, they have low diffusion resistance and are useful in immunoelectrophoresis. They are also useful in the electrophoresis of high molecular weight biomolecules such as proteins and nucleic acids. The main disadvantage is the severe electroosmosis produced as a result of the sulphated nature of the agaropectin.

agarose A gel used as supporting medium in electrophoresis. It is a natural colloid that is extracted from seaweed (red algae). It is very brittle and easily damaged during handling; since the pore size is relatively large, this gel is mainly used to separate large molecules which have molecular mass of about 200 kdal or more; though faster than the polyacrylamide gel, the resolution is poor; the bands formed are blurred and spread far apart.

agarose gel electrophoresis An electrophoretic technique in which the stationary phase matrix is a highly purified form of agar. It is used to separate larger DNA and RNA molecules and is used as preparative gel that has applications in Northern blots, Southern blots, apoptosis "ladder" assays and DNA mapping. It is prepared by dissolving agarose powder in appropriate buffer (Tris-acetate, Tris-borate, Tris-phosphate, and Alkaline buffers), by heating. Higher the agarose concentration better is the resolution of the DNA fragments. The advantages are its easy and quick set-up and adequate resolutions, though not as superior as that of the PAGE.

AIDS (Acquired immunodeficiency syndrome) The human disease in which the immune system is destroyed by the Human Immunodeficiency Virus (HIV) which is a retrovirus. The virus destroys helper T-cells, which are essential for developing immunity against infections.

algorithm Well-defined series of steps of a procedure or formula for solving a problem. It can be coded into a programming language and executed. In bioinformatics it is used to process, store, analyse, visualize and make predictions from biological data.

alignment A term used in bioinformatics that refers to the outcome of a comparison of two or more genes or protein sequences in order to determine their degree of base or amino acid similarity. Sequence alignments are used to determine the similarity, homology, function or other degree of relationship between two or more genes or gene products.

analytical centrifugation A centrifugation process used in analytical ultracentrifuges, to characterize the properties of the particles in pure sample. The design of the rotor of an analytical ultracentrifuge incorporates an optical system and special centrifuge tub holders with which the concentration of the particles can be measured as a function of the radius from the axis of rotation. It is used in the determination of molecular weights of molecules such a nucleic acids and proteins, the purity of DNA preparations, proteins and microorganisms including viruses, and conformational changes in DNA and protein.

animal cell immobilization Entrapment of animal cells in some solid material in order to produce some natural product or genetically engineered protein. The solid material may be the hollow fibre membrane bioreactors, or porous carriers made of polysaccharide, protein, plastic or ceramic materials with microscopic holes inside which the cells grow.

animal model systems Model systems involving invertebrate and vertebrate animals. Simple organisms (e.g. invertebrates) with very short lifespan (life cycle) are useful to study many basic biological processes because they are easy to grow and observe; vertebrates are useful as experimental animals because of their closer phylogenetic relationship with man; mammals among vertebrates are suited for biomedical research because humans are also mammals.

antibody An immunological protein, called an immunoglobulin (Ig) produced by lymphocytes of the immune system of an organism in response to an antigen; it has the ability of specifically binding with the antigen and rendering it harmless. Each antibody has two identical "light" chains and two identical "heavy" chains. Each chain comprises a constant region, i.e., a region that is the same between antibodies of the same class and sub-class, and a variable region that differs between. The antigen-binding region or binding site—*complementarity determining region*—is in the variable region. The antibody can be cut by proteases into several fragments, known as Fab, Fab', and Fc.

antigen/immunogen A substance that elicits an immune response by stimulating the production of antibodies. The antigen, usually a protein, when introduced into a vertebrate is bound by the antibody or a T cell receptor.

antigenic determinant A surface feature of a microorganism or macromolecule, such as a glycoprotein, that elicits an immune response. *See* epitope.

antiseptic A compound that inhibits the growth and development of microorganisms without necessarily killing them. Antiseptics are usually applied to body surfaces.

apoptosis The process of cell death, occurring naturally as a part of normal development, maintenance and renewal of tissue in an organism. It differs from necrosis, in which cell death is caused by a toxic substance.

appendix A section of a research report to present supplementary material, which is helpful but not essential to be included in the text; e.g. raw data, detailed protocols of procedures, questionnaires, statistical tables, glossary of terms, and extensive quotations.

applied research Research, in which the main aim is "practical application" of the results, such as cure of a disease, increased crop production, development of biological weapon, etc. A problem for applied research is selected from a going concern, and is analysed and investigated with well-designed scientific methods.

aquaculture A biotechnological process of controlled culturing of water plants (sea weeds) and animals (oysters, prawns, fish, etc), which normally grow in rivers or seas. It is "aquaculture" when freshwater is used and "mariculture" if sea water is used. It involves growing of organisms in large volumes of clean, well-aerated water for the animals to grow in. Food, such as krill or powdered synthetic food and food additives, such as astaxanthins are provided to ensure that fish and prawns have the right colour. It is also used to mass-culture macro- and micro-algae for chemicals, vitamins and pigments. Genetic methods to produce triploid and tetraploid organisms, and hybrid algae through plant cell fusions are employed to improve the yield; for example, triploid trout are sterile, and can be used for biocontrol of weeds without the threat of their being able to breed themselves.

astigmatism A microscopic optical aberration in which the off-axis image of the observed spot of specimen appears as a line or ellipse instead of discrete point. The contrast is also decreased as the distance of the image from the centre increases. Astigmatism in microscopes is due to (i) asymmetric lens curvature, a manufacturing defect; (ii) improper mounting of lens on its frame; or (iii) incorrect orientation of the lenses within the objective barrel. It can be corrected by proper designing of objectives with precise spacing of individual lens elements, appropriate lens shapes and refractive indices.

atomic force microscope (AFM) A type of scanning probe microscope that images almost any type of surface, (polymers, ceramics, composites, glass and biological samples) by moving a sharp metal probe (with a tip diameter of about a few nm) over the specimen surface at a constant distance; the interatomic forces (the electrons in the probe are repelled by the electrons of the atoms in the sample) provide interaction mechanism, which depends on the nature of the sample, the probe tip and the distance between them; variations in the force cause the tip to move up and down as it scans the surface. It is calculated by measuring the deflection by a cantilever mechanism attached to the probe and an image is created.

attenuated vaccine A virulent organism that has been changed to produce a less virulent form, but has the ability to elicit antibodies against the virulent form.

Bacillus thuringiensis (Bt) A bacterium that kills insects; a major component of the microbial pesticide industry.

bacteriophage A virus infecting a bacteria; simply called a phage.

baculoviruses A class of insect viruses that are used to make DNA cloning vectors for gene expression in eukaryotic cells. They are non-infective and non-pathogenic to vertebrates.

base pair (bp) Two nucleotide bases on different strands of the nucleic acid molecule that bond together: adenine (A) with thymine (T) in DNA, or adenine (A) with uracil (U) in RNA and guanine (G) with cytosine (C). The size of a nucleic acid molecule is often described in terms of the number of base pairs (symbol: bp) or thousand base pairs (kilobase pairs; symbol: kb; a more convenient unit) it contains.

basic research/pure research The main objective of such research is expansion of the existing knowledge. It does not envisage its results to have any social or economic relevance. A problem for basic research is selected from any source and tackled with appropriately planned scientific methodology.

Beer–Lambert law Gives the relationship between absorbance of light by a solution, the concentration of the solute and the path length. It states that for a given path length, the concentration of a substance in solution is directly proportional to the absorbance of light by that solution, provided that no other solute is absorbing light at the specific wavelength. It is valid only for monochromatic radiation (radiation of a single wavelength), and only when there is no change in the physical or chemical state of the solute with changes in concentration; written as $A = kcl$, where, A is the absorbance (no units, since $A = log(I_0/I)$), k is the molar absorptivity with units of mol^{-1} L cm^{-1}, c is the concentration of the compound in solution expressed in mol L^{-1}, and l is the path length of the sample, i.e., the diameter of the cuvette in which the sample is contained, in centimetres.

bibliography A reading list and a list of works, consulted at different stages of the research.

bioaccumulation Progressive accumulation and concentration of materials (metals or other compounds like pesticides, e.g. DDT) in an organism, which are not needed for survival of the organism. Many organisms (e.g. plants, fungi, protists, bacteria, etc.) accumulate metals when grown in a solution of these compounds, either as part of their defence mechanism against their poisonous effect, or as a side effect of the chemistry of their cell walls. This is important as part of the microbial mining cycle, removing toxic metals from waste water, as a purification (bioremediation) process, etc.

bioassay A method for assessing the activity of a substance by measuring its effect in living cells or on organisms. Laboratory animals have been used widely in drug research in bioassays for the pharmacological efficacy of drugs. However, bioassays are now usually made using prokaryotic (bacteria) or eukaryotic (animals and plants) cells in culture, as these are easier to handle, cheaper to make and keep, and free from the ethical problems associated with testing of animals. Bioassay is sometimes used to detect minute amounts of substances that influence or are essential to growth.

bioaugmentation Addition of pre-grown microbial cultures to an environment to perform a specific function.

biocide Any agent that is capable of killing organisms.

biocontrol/biological control The control of living organisms (especially pests) by biological means. Biocontrol is any process using deliberately introduced living organisms to restrain the growth and development of other pathogenic organisms. For instance, spider mites are used to control cassava mealy bugs. Biocontrol also refers to use of disease-resistant crop cultivars. Biotechnology approaches biocontrol in various ways—using fungi, viruses or bacteria, which are known to attack an insect or weed pest.

bioconversion Conversion of one chemical compound into another by living organisms, as opposed to its conversion by enzymes (which is biotransformation) or by chemical processes. The process is specific and can work in moderate conditions. It is useful for introducing chemical changes at specific points in large, complex molecules, and is applied in the manufacture of steroids.

biodegradation The breakdown of a compound to its chemical constituents by living organisms. Substances that can be easily biodegraded are described as *biodegradable* and those that are not, *nonbiodegradable* (e.g. plastic).

biodiversity The variety of species (species diversity) or other taxa of animals, microorganisms and plants in a natural community or habitat, or of communities in a particular environment (ecological diversity), or of genetic variation in a species (genetic diversity). It is the variability among the living organisms (animals and plants) from all sources (land, sea and freshwater) and the ecological complexes of which they are a part, and includes diversity within and between species, and the diversity of the ecosystem.

bioengineering The use of artificial tissues, organs and parts of an organ to replace parts of the body that are damaged, lost or malfunctioning.

bioenrichment The process of adding nutrients or oxygen to increase microbial breakdown of pollutants.

bioethics The branch of ethics that deals with biological research, especially of medical techniques such as organ transplantation and of the life sciences, and their potential impact on society. They are useful in focusing attention on problems that need to be rectified.

biofilms Layers of colonies of microorganisms growing on surfaces of inanimate materials (e.g. catheters, medical prosthetic devices, ship hulls, water pipes, cooling towers). They are organized and have complex structural and functional characteristics and physical and chemical gradients that influence microbial metabolic processes.

biofuel An energy-rich gaseous, liquid or solid fuel derived from a biological source. For example, rapeseed oil or fish liver oil can be used in place of diesel fuel in modified engines. Biodiesel is a vegetable oil-based fuel that runs diesel engines—cars, buses, trucks, construction equipment, boats, generators, and home heating units. It is usually made from any vegetable oil (jatropha; sunflower; rapeseed) and can also be made from recycled fryer oil. It can be blended with regular diesel or used as 100% biodiesel.

biogas An alternative energy source; a mixture of methane and carbon dioxide produced from the anaerobic decomposition of waste such as domestic, industrial and agricultural sewage. Gobar biogas (methane) is generated by the decomposition of organic cow dung.

biohazard Any hazard of biological origin; generally refers to those microbial or antigenic entities presenting a risk or potential risk to the well-being of man, other animals, or plants, either directly through infection or indirectly through disruption of the environment. Any substance that contains or potentially contains biohazardous agents is referred to as biohazardous material; any set of equipment and procedures used to prevent or minimize the exposure of workers and their environment to biohazardous agents or materials is called biohazard control.

biohazardous agents Any agent of biological origin causing hazard including: (i) cultured animal cells and the potentially infectious agents these cells may contain, (ii) microorganisms including bacteria, viruses, fungi, protozoa, etc., (iii) parasites, (iv) allergens, (v) tissues from experimental animals including animal dander (small scales from the hair or feathers of animals that may cause allergy in sensitive individuals), (vi) plant viruses, bacteria and fungi, and (vii) toxins of bacterial or plant origin.

bioinformatics   Organized storage, retrieval, and analysis of information of biological interest such as gene structure and function, protein structure and function, etc; also concerned with biomedical experimental results, patient statistics, and scientific literature. The field is highly interdisciplinary employing techniques and concepts from computer science, statistics, mathematics, chemistry, biochemistry, biophysics, etc. and has wide range of applications in biology and medicine.

biolistics   Acronym of **bio**logical and bal**listics**; a technique to insert DNA into cells. The technique involves the mixing of DNA with minute metal particles (usually tungsten or gold) having diameter in nanometer and firing into a cell at very high speed. The particles puncture the cell and carry the DNA into the cell. It has an advantage over other techniques like transfection, transduction, etc., because it can apply to any cell, or to parts of a cell. It has been used to insert DNA into animal, plant and fungal cells, and into mitochondria, inside cells.

biological containment   Restricting the movement of potential biohazardous organisms (e.g. genetically modified organism) by creating barriers to prevent them from growing outside the laboratory. It can be achieved by two forms, either making the organism unable to survive in the outside environment, or making the outside environment inhospitable to the organism. For bacteria and yeasts, which can survive anywhere in any environment, the favoured containment method is to mutate the genes in the organism so that they require a supply of a specific nutrient that is usually available only in the laboratory; it may also involve the use of vector molecules and host organisms which have been genetically disabled such that they can survive only in the peculiar conditions provided by the experimenter and which are unavailable outside the laboratory.

biological model system   A system that can be observed instead of the original system (human or animal), which is of ultimate interest to the investigator. It helps to answer questions that could not be answered using the original system with the available technology. It enables the isolation and study of certain features which would be too complex to study or impossible to isolate in the original system. There are four types of models used in biology and biomedical researches: (i) whole, living animals (including humans), (ii) living systems composed of samples from the original system (e.g. tissue culture), (iii) non-living, mechanical or molecular systems and (iv) mathematical models such as computer simulation.

bioluminescence   The emission of light by live cells; light produced through oxidation of molecules by an enzyme luciferase.

biomass   It may refer to (i) the cell mass produced by a population of living organisms; (ii) the organic mass that can be used either as a source of energy or for its chemical components; or (iii) the total organic matter that derives from the photosynthetic conversion of solar energy. The amount of biological material in a specific volume is called biomass concentration.

biopesticide A biodegradable agent that kills organisms by virtue of its specific biological effects rather than as a chemical poison. Includes bio-insecticides and bio-fungicides and is different from biocontrol agents in that the biopesticides are passive agents, whereas biocontrol agents are predatory, hunting the pest to be killed. An example is *Bacillus thuringiensis* (Bt) toxin that specifically interferes with the absorption of food from the guts of some insect pests but is harmless to mammals.

biopiracy The unethical practice of monopolizing a biodiversity-based traditional knowledge.

biopolymer Any large polymeric molecule such as protein, nucleic acid, polysaccharide, lipid, etc. produced by microorganisms, animals and plants.

bioprocess Any method that uses entire living cells or their organelles (e.g. chloroplasts) or their products (e.g. enzymes) to produce desired physical or chemical changes.

bioreactor A container in which cells, cell extracts or enzymes carry out a biological reaction. It refers to a growth chamber (fermenter, fermentation vessel) for cells or microorganisms.

bioremediation A practice that uses living organisms to remove contaminants, pollutants or unwanted substances from soil or water.

biosensor A device that employs an immobilized biological agent (e.g. enzyme, antibiotic, organelle or whole cell) to detect or measure a chemical compound. The reaction between the immobilized agent and the molecule being analysed is transduced into an electric signal.

biotechnology Technology for the use of biological processes or organisms for the production of materials and services of benefit to humankind—for the improvement of the characteristics of economically important plants and animals and for the development of microorganisms to act on the environment. It also involves the scientific manipulation of living organisms, especially at the molecular genetic level, to produce new products, such as hormones, vaccines or monoclonal antibodies.

biotransformation The transformation of one chemical or substance into another using an enzyme, or a whole, dead microorganism that contains an enzyme or several enzymes.

blot, dot blot, colony blot A method to transfer DNA, RNA or protein to an immobilizing matrix, or the autoradiograph produced during the blotting techniques. Dot blots are DNA, RNA or protein that were dotted directly onto the membrane support, on which they form discrete spots. Colony blots are the molecules (usually DNA) which are from colonies of bacteria or yeast growing on a bacteriological plate.

blotting technique Any of the techniques of transferring DNA, RNA, or protein to an immobilizing matrix (cellulose acetate membrane) for analysis. Differences in blotting techniques depend on the molecules that are transferred—Southern blot for DNA molecules, northern blot for RNA, western blot for protein with a labelled antibody probe and southwestern blot for protein, with DNA probe.

bright field microscope   Microscope in which the specimen is viewed as dark objects in a bright background. It is the common compound microscope used in biological and clinical laboratories.

botulism   A food poisoning caused by neurotoxin (botulin) synthesized by *Clostridium botulinum* serotypes A–G; found in contaminated canned or preserved food.

CAAT box   A conserved sequence found within the promoter region of the protein-encoding genes of many eukaryotes. It has the consensus sequence GGCCAATCT and occurs around 75 bases prior to the transcription initiation site. It is one of several sites for recognition and binding of regulatory proteins called transcription factors.

camera lucida   An equipment that helps to draw images of the objects that are viewed under a microscope. It is basically a four-sided reflecting prism with specific angles so that the light from the object is reflected through an eye lens. Camera lucida is mounted above the eyepiece of the microscope in such a way that the large mirror and small prism bring the object and the paper into the same view. It permits the observer to see the specimen along with the paper, through the eyepiece. The observer can then trace the image while looking through the microscope.

capillary electrochromatography   A combination of liquid chromatography and capillary electrophoresis; the capillary is packed with stationary phase through which an electric field is applied, inducing an electroosmotic flow which moves the mobile phase through the packed column; the separation of the sample molecules depends on the partition between the stationary phase and the mobile phase. It yields high degree of resolution of neutral compounds in a very short duration. The main advantages are separation of closely related compounds, shorter analysis time, and increased sample throughput.

capillary electrophoresis   An electrophoretic technique which uses capillary columns to separate the sample molecules at high voltages (10–30 kV). It has very high efficiencies (handling more than 10^5 sample molecules at a time) and very high order mass sensitivities (up to 10^{-18} to 10^{-20} moles). The major features are: versatility of application ranging from inorganic ions to large DNA fragments; different separation modes with different selectivity; extremely low volume of sample; negligible operating cost; use of different types of highly sensitive detection systems; easy automation; quantification; ruggedness; and simplicity of the instrumentation.

capillary gel electrophoresis (CGE)   An adaptation of the conventional slab gel electrophoresis to the capillary format. The capillary is filled with a non-crosslinked polymer (e.g. linear polyacrylamide, polyethylene glycol or cellulose derivatives) or crosslinked polymers (e.g. polyacrylamide gel, agarose gel). The sample molecules are separated according to their size in a nonconvective medium. It is useful for the electrophoretic separation of large biological molecules like proteins and DNAs, which have similar electrophoretic mobility in free buffer solution because of their similar charge-to-mass ratio. The advantages of CGE over slab-gel electrophoresis are a wider range of gel matrix and composition, online detection, improved quantification and automation.

capillary isoelectric focusing (CIEF) A capillary electrophoretic technique used for the separation of amphoteric molecules such as proteins, peptides, amino acids, and pharmaceuticals. The basis of the separation is the differences in the isoelectric point (pI) of sample components rather than differences in apparent velocity. It is performed by filling the capillary with a mixture of ampholytes and the sample; the ampholytes (zwitterions) are used to generate a pH gradient inside the focusing capillary; ampholytes form a stable pH gradient; a molecule in the sample with a net positive charge migrates towards the cathode and when it encounters a pH at which it has a zero net charge, it stops its migration.

capillary isotachophoresis (CITP) An electrophoresis technique in which the sample molecules are separated by displacement process in a multizonal medium, performed in a capillary system. In CITP, a sample is introduced between a leading electrolyte and a trailing electrolyte without electroosmotic flow. The leading electrolyte possesses higher mobility while the trailing electrolyte has lesser mobility when compared to those ions in the sample zone; the separation is on the basis of the differences in the velocities of the ions in the sample zone.

capillary zone electrophoresis (CZE) A capillary electrophoresis working on the basis of ionic properties of the sample molecules and electroosmotic flow (EOF). It is used for the separation of both anionic and cationic molecules in a single electrophoresis run. The sample is applied as narrow zone (band), which is surrounded by the separation buffer. When a strong electric field is applied, each component in the sample zone migrates according to its electrophoretic mobility; the cations and anions migrate in different directions, but they are all swept toward the cathode by the EOF, which is higher than the velocity of the migrating molecules. Eventually, all sample component molecules separate from each other to form individual zones of pure material. Cations elute first because the direction of their migration is the same as that of the EOF. Neutral solutes, which would remain unresolved, will elute next because they lack electrophoretic mobility and are carried by the EOF; anions are the last to elute because they migrate in the opposite direction of the EOF.

capillary-gas chromatography A gas chromatography with capillary column. The capillary wall is made up of glass or fused silica, and is coated with an adsorbent. Because of the small amount of the stationary phase, the column has only a limited capacity but yields quick separation of mixtures.

case study Observation and description of one or a few occurrences of an event, occurring anywhere and among humans, other animals, plants, in the atmosphere, freshwater bodies, oceans and so on.

cDNA, Complementary DNA Double-stranded DNA complement of an mRNA sequence. It is synthesized *in vitro* from a mature RNA template using reverse transcriptase (to create a single strand of DNA from the RNA template) and DNA polymerase (to create the double-stranded DNA). cDNA is the first step in cloning of desired DNA sequences. It is used as specific and sensitive probes in hybridization studies, because cDNAs usually do not include regulatory or other controlling sequences, and so they can be used to identify (probe) and isolate genes and their associated sequences from genomic DNA.

cDNA clone A double-stranded DNA molecule that is carried in a vector and was synthesized *in vitro* from an mRNA sequence by using reverse transcriptase and DNA polymerase.

cDNA cloning A method of cloning the coding sequence of a gene, starting with its mRNA transcript. It is used to clone a DNA copy of a eukaryotic mRNA. The cDNA copy, being a copy of a mature messenger molecule, will not contain any intron sequences and may be readily expressed in any host organism if attached to a suitable promoter sequence within the cloning vector.

cDNA library A collection of cDNA clones that were produced *in vitro* from the mRNA sequences isolated from an organism or a specific tissue or cell type or population of an organism.

cell cycle The stages a cell passes through between one division and the next. It consists of G, S, and G_2 phases; G phase, high rate of biosynthesis and growth; S phase, doubling of the DNA content as a result of chromosome replication; G_2 phase, the final preparations for cell division (cytokinesis).

cell differentiation Transformation of a pluripotent embryonic cell into specialized adult cell; results from the controlled activation and de-activation of genes.

cell fusion *In vitro* formation of a single hybrid cell from two cells of different species; cells fuse and coalesce, but their nuclei may remain separated. During subsequent cell division, a single spindle is formed so that each daughter cell has a single nucleus containing sets of chromosomes from each parental line. Subsequent divisions often result in the loss of chromosomes and therefore of genes; used to determine the control of specific genes and their assignment to chromosomes.

cellulose acetate A supporting medium of electrophoresis that has several advantages like no adsorption property, chemical purity with no impurities such as lignins, hemicellulose or nitrogen, translucency, so that it can be subjected to direct photoelectric scanning of the bands, low glucose content, therefore suitable for electrophoresis of polysaccharides and subsequent PAS (periodic acid Schiff) staining. It is used in the electrophoretic separation of glycoproteins, lipoproteins, and haemoglobin of blood samples; also useful in immunoelectrophoresis and immunodiffusion.

central dogma The basic concept that genetic information generally can flow only from DNA to RNA to protein. In retroviruses, the information contained in RNA molecules can also flow back to DNA.

centrifugal force A force acting on objects moving in a circular path, tending to pull the object out, away from the centre; a force which equals and thus balances the centripetal force that pulls the object towards the centre. It is directly proportional to the speed of rotation (angular velocity) and the radius of the rotation; centrifugal force (F) = (angular velocity, ω)^2 × radius; angular velocity is related to revolutions (rotations) per minute.

centrifugation A method for separation of molecules, cell organelles, microorganisms, etc., by size or density, using the action of centrifugal force to accelerate sedimentation in a solid–liquid mixture. Centrifugal force is generated by a spinning rotor; G-forces of several hundred thousand times gravity are generated in ultracentrifugation.

centrifuge An instrument to separate solid or liquid particles of different densities by rotating them in a tube in a horizontal, slanting or vertical circle; the denser particles tend to move along the length of the tube to a greater radius of rotation, displacing the lighter particles to the other end.

centripetal force The force that maintains an object in circular motion (force seeking to move towards the centre).

chain terminator Codons that do not code for an amino acid but give signal to ribosomes to terminate protein synthesis. Any two of the triplets UAA, UAG and UGA, named ochre, amber and opal, respectively, are found together at the end of a coding sequence of RNA. The term also refers to di-deoxynucleoside triphosphates that are added as chain terminators, in the synthesis of a complementary DNA strand by the Sanger method of DNA sequencing. They are also known as stop codons and termination codons.

chimera or chimaera Literally chimaera is a mythical monster having the head of a lion, the body of a goat and the tail of a serpent, i.e., an organism whose cells are derived from different animals. In biology it may refer to (i) an animal having two or more genotypes in patches derived from two or more embryos; an individual derived from two embryos by experimental intervention. (ii) part of a plant with a genetically different constitution as compared with other parts of the same plant. It may result from different zygotes that grow together, or from artificial fusion (grafting). It may either be periclinal chimera, in which one tissue lies over another as a glove fits a hand; mericlinal chimera, where the outer tissue does not completely cover the inner tissue; and sectoral chimera, in which the tissues lie side by side; (iii) a recombinant DNA molecule that contains sequences from different organisms.

chimeric DNA A recombinant DNA molecule containing unrelated genes.

chimeric gene A semi-synthetic gene consisting of the coding sequence from one organism fused to promoter and other sequences derived from a different gene; used in transformation.

chimeric protein *See* fusion protein.

chromatic aberration A common optical aberration in a microscope. It occurs because the lens refracts the various colours present in the illuminating white light at different angles according to the wavelengths of the colours. For example, red rays are not refracted at the same angle as green or blue rays. The effect is the focal points of these colours vary in the optical axis of the lens. Red is focused farther away when compared to green and blue. Blue is focused closest to the lens. Chromatic aberration, also referred to as *dispersion*, results in the formation of a *colour fringe* or *halo* surrounding the image of the specimen. Chromatic aberrations can be reduced to a great extent, or even eliminated, by designing compound lenses that consist of individual elements having different colour properties. Refinement of the optical system with more sophisticated glass formulas and shapes has helped to almost eliminate chromatic aberrations in modern microscopes.

chromatography   A wide range of physical methods used for the separation and analysis of complex mixture, on the basis of the differences in the physical, chemical, molecular, immunological, etc. properties of the components of a mixture. The basic principle of all types of chromatography is the differential movement of a mixture over or through an adsorptive material. The differential movement is related to the differences in the physical, chemical, and other properties of the components of the mixture. The mixture itself does not move but is carried by another material which may be a liquid or a gas. The material carrying the mixture is referred to as the mobile phase. The material over or through which the mobile phase is flushed is called the stationary phase. The stationary phase may be solid, gel, liquid, or a mixture of liquid and solid. This phase is immobilized, hence called the stationary phase. The stationary and mobile phases are mutually immiscible.

chromosome jumping   A method for cloning together two segments of duplex DNA that are separated by thousands of base pairs (about 200 kb). After sub-cloning, each segment can be used as a probe to identify cloned DNA sequences that, at the chromosome level, are roughly 200 kb apart.

chromosome walking   A method for identifying overlapping cloned DNA fragments that form one continuous segment of a chromosome. These fragments can be generated either by random shearing or by partial digestion with a four-base-pair cutter such as *Sau*3A. A series of colony hybridizations is then carried out, starting with some cloned fragment which has already been identified and which is known to be in the region encompassed by the overlapping clones. This identified fragment is used as a probe to pick out clones containing adjacent sequences. These are then used as probes themselves to identify clones carrying sequences adjacent to them and so on. At each round of hybridization one "walks" further along the chromosome from the initial fragment.

citation index   An index based on the principle that there is some meaningful relationship between one paper and some other paper that it cites or that cites it, and thus between the work of the two authors or two groups of authors who published the papers. The entire SCI (Science Citation Index) database is amenable to extensive manipulation and analysis. In the case of authors, it is possible to identify the frequency with which they and their papers are cited in the literature, over any chosen time period. Counts of this sort are strictly quantitative and objective. An author's or a paper's frequency of citation has been found to correlate well with professional standing. It provides a useful objective criterion for judging the quality of paper. The same principle is used in evaluation of Journals; when a Journal's articles are cited more frequently, it implies that the Journal is a carrier of useful information.

citation of reference Acknowledging in the text of a research report the source of information that was used while writing the different sections of the report.

clonal   propagation Asexual propagation of many new plants from an individual; all have the same genotype.

clonal selection The production of a population of plasma cells, all producing the same antibody in response to the interaction between a B lymphocyte producing that specific antibody and the antigen bound by that antibody.

clone Refers to (i) a group of cells or organisms that are genetically identical as a result of asexual reproduction, breeding of completely inbred organisms, or forming genetically identical organisms by nuclear transplantation; (ii) a population of cells that all carry a cloning vehicle with the same insert DNA molecule.

clone bank *See* gene bank.

cloning Refers to a process involving (i) the mitotic division of a progenitor cell to give rise to a population of identical daughter cells that are called clones; (ii) incorporation of a DNA molecule into a chromosomal site or a cloning vector; (iii) the creation of a whole animal by mitotic divisions from a single diploid somatic cell, typically by the process of nuclear transfer.

cloning vector/cloning vehicle A small, self-replicating DNA molecule, usually a plasmid or viral DNA, into which foreign DNA is inserted in the process of cloning genes or other DNA sequences of interest. The vector can carry the inserted DNA and be perpetuated in a host cell.

colorimetry Estimation of the concentration of a substance in a solution on the basis of the relationship between the concentration of the substance in that solution and the colour intensity of that solution, the colour intensity being proportional to the concentration.

column chromatography A chromatographic technique in which the stationary phase of the chromatographic system is contained in a tube called the column.

coma An optical aberration due to off-axis light rays, and mainly occurs in microscopes with improper alignment. It causes an aberration in which the point of focus, especially one at the periphery of the viewfield, exhibits a streak of light that resembles the tail of a comet, hence the name coma. When the angle of incidence of the light rays on different parts of the lens becomes more and more oblique (off-axis), the refraction differences become more pronounced, and results in coma. The degree of coma aberration is directly related to the objective aperture, it being greater for lenses with wider apertures. It can be partially corrected by reducing the aperture size to accommodate the diameter of the object field for a given objective and eyepiece combination.

conclusions   *See* summary (and conclusions).

confocal microscope   A modification of an epifocal fluorescence microscope; a design in which a pinhole aperture-screen combination is added at the point where the image is formed at the back of the objective. An image of a spot of the specimen is formed at the minute aperture (pinhole/slit) of the screen. Thus the two points, focal point and the pinhole, conjugate, and hence the name "confocal" microscope. The image is formed only by the light rays that pass through the pinhole because the rays from other areas of the specimen are blocked by the screen. Consequently, the image of the fluorescing spot is free from the "out-of focus" flare and has higher resolution.

confocal scanning laser microscope (CSLM)   A light microscope in which a monochromatic laser scans across the specimen at a specific level and illuminates one area at a time to form an image. Stray light from other parts of the specimen is blocked out to give an image with good contrast and resolution.

control   In a biological experiment when a sample unit receives a "treatment" by one of several methods, it is possible that the effect observed might be due to factors other than the one that the treatment contained. Therefore, in order to demonstrate that the effect is in fact due to the treatment factor and not due to any others including the method of treatment itself (e.g. handling of the experimental animals, medium in which the treatment factor is prepared, etc.) an appropriate control is included in the experimental design. A control is a sample unit that is "identical", at least theoretically, to the sample unit that is treated with an experimental factor (e.g. addition or removal of a test factor) except for that experimental factor. If to the experimental unit is added a factor, the control unit is deprived of that factor, hence the control is a negative control. On the other hand, if the experimental unit is deprived of a factor, the control is added with that factor, and therefore referred to as positive control. Both positive and negative controls are together known as baseline controls, which are basically opposite of experimental units, with reference to only one specific condition, all the other conditions being the same. Normal, healthy, untreated animals, plants, etc. may also serve as baseline controls. If the observer (researcher or evaluator) is intentionally not aware as to which sample unit is experimental and which is control, then the control is described as a blind control. Double blind control refers to conditions in which both the observer and the subjects (usually, human) are unaware about the treatment.

copyleft   A jargon of the term Copyright, the Copyleft is a movement that questions underlying philosophy of IPR, based on the belief that creations of the mind are common to everybody on earth and not the right of any individual or organization; knowledge multiplies by sharing, and therefore it should be freely available.

copyright   An intellectual property protection granted under law to the creators of original works such as literary, dramatic, musical, artistic and cinematographic works and of films, and sound recordings.

cosmid A plasmid vector that contains the two cos (cohesive) ends of phage lambda and one or more selectable markers such as an antibiotic resistance gene. Cosmids exploit certain properties of phage lambda to enable large, 40–50 kb, DNA fragments to be cloned at high efficiency. Cosmids and cosmid recombinants replicate as plasmids.

cross-sectional study The study of a defined population by observing samples collected from that population, and in which the variables are not manipulated. A possible correlation, association, etc. among the variables is investigated.

cryoprotectant Compound preventing cell damage during freezing and thawing processes. They are agents with high water solubility and low toxicity. There are two types, permeating (glycerol) and non-permeating (sugars, dextran, ethylene glycol, polyvinyl pyrolidone and hydroxyethyl starch).

CTAB (Cetyltrimethylammonium bromide) A common cationic surfactant used in electrophoresis and used for the electrophoresis of samples that pose difficulties for SDS-PAGE. For example, it is used in the electrophoresis of extremely acidic or extremely basic samples, because the extremely acidic samples may not bind with SDS, and the extremely basic samples may precipitate when SDS is added. These problems are overcome by CTAB, which produces a uniform charge-to-mass ratio among the sample molecules.

dark-field microscope Microscope in which living, unstained, transparent specimens are viewed as bright objects against a black background; also referred to as dark-ground microscope. Illumination involves focusing a hollow cone of light on the specimen so that direct light (unreflected and unrefracted) does not enter into the objective lens, because the objective lies in the hollow dark portion of the cone. However, light reflected and refracted by the specimen will enter the objective and form an image. The clear portions of the specimen will not reflect or refract any light and therefore will appear dark; the medium in which the specimen is mounted also does not reflect or refract light, and therefore will appear black, resulting in the formation of brightly illuminated specimen against a dark background.

databank In life sciences, a databank (or data bank) is an organized set of raw data, like DNA sequences from sequencing projects (e.g. the EMBL and GenBank databases), protein structures, etc.

database A collection of data that is organized and structured so that its contents can easily be accessed, managed, and modified by a computer. In Bioinformatics, a database is a curated storehouse of raw data containing annotations, further analysis, and links to other databases. For example SWISSPROT is a database for annotated protein sequences, and the FlyBase is a database of genetic and molecular data for *Drosophila melanogaster.*

decontamination Any process for removing and/or killing microorganisms and removing or neutralizing hazardous chemicals and radioactive materials. It also refers to the reduction of many or all disease-causing microorganisms and the destruction of pathogenic microorganisms in or on the surface of an object, so that they are no longer capable of transmitting disease.

denaturing gradient electrophoresis (DGGE)   A gel electrophoresis in which the gel is incorporated with a denaturing agent. When a small sample of DNA or RNA is applied to such a gel and electrophoresed, the nucleic acid is subjected to increasingly extreme denaturing conditions, and therefore melts into single stranded fragments which may be as small as 200–700 base pairs. It is used to understand the differences in DNA sequences or mutations in various genes. Since small differences of few base pairs in the sequence of otherwise identical fragments cause them to partially melt at different positions in the gel, by comparing the melting behaviour of the polymorphic DNA fragaments side by side on denaturing gradient gel, it is possible to detect fragments that have mutations. By parallel running of two samples that were allowed to denature together, it is possible to identify even the minutest differences in the two fragments.

dendrogram   A tree-like diagram that is drawn to graphically summarize the mutual similarities and relationships between organisms.

denitrification   A process in which nitrates in the soil are reduced to molecular nitrogen, which is released into the atmosphere.

densitometers   Devices used to record results from almost all types of electrophoresis support media, including cellulose acetate, polyacrylamide, agar, and starch cells. They are spectrophotometers fitted with gel scanning accessories; a monochromatic beam of light passing through a stained gel scans the different bands, recording the percentage of transmission of the light, which is inversely proportional to the density of the bands.

density gradient centrifugation   High-speed centrifugation in which molecules move and get suspended at a point where their density equals that in a gradient medium (e.g. caesium chloride or sucrose). The density gradient may either be formed before centrifugation by mixing two solutions of different density (as in sucrose density gradients) or it can be formed by the process of centrifugation itself (as in CsCl and Cs_2SO_4 density gradients).

depth-of-field   In photography, the range in which all objects appear in adequate focus; also referred to as the zone of focus.

desiccation   The process of drying, exhausting or depriving of water or moisture. Any chemical used to desiccate is called a desiccant and the apparatus for drying and preventing desiccated hygroscopic samples from rehydrating is a desiccator.

design   Refers to two-dimensional and three-dimensional aesthetic aspects of an article. Two-dimensional features may be patterns, lines or colours, and three-dimensional features may be shape or surface of an article. Design registration does not cover the mechanical feature of the article.

detergent   A compound which lowers the surface tension of a solution, improving its cleaning properties (e.g. Tween-20TM, a surfactant and wetting agent).

differential centrifugation   A centrifugation method of separating sub-cellular particles according to their sedimentation coefficients, which are proportional to their size; cell extract is subjected to successive centrifugal runs at progressively increasing rpms; larger organelles, such as nuclei or mitochondria, precipitate at relatively slow speeds; higher G force is required to sediment minute particles such as ribosomes.

differential interference contrast microscope (DIC) A microscopic design very similar to the phase-contrast microscope and that can produce increased contrast and three-dimensional effect to the image. It also creates an image by detecting differences in the refractive indices of the different component parts of the specimen.

diffraction grating Used for the dispersion of polychromatic radiation in the monochromator of a spectrophotometer; a grating is a highly polished reflecting surface on which many thousands (6000–50000 per cm) of narrow parallel grooves have been etched; the dimensions of the grooves are of the same order as the wavelength of the light that is to be dispersed; a beam of radiation directed on to the grating is scattered, or diffracted, in all directions; radiation waves reinforce each other in certain directions and cancel out in other directions.

diffusion The migration of molecules from a region of higher concentration to a region of lower concentration.

digital photography Capturing images of objects in a digital format achieved either by scanning film-based photographs or by capturing images by a digital camera. A digital photograph is computer based and is made up of minute squares called picture elements or pixels,

numbering from several thousands to millions. A computer screen is divided into a grid of pixels and the values stored in a digital photograph are translated to specify the brightness and colour of each pixel. The process of controlling individual pixels in terms of digital values is called bit mapping and the images thus formed are called bit-maps. The same principle applies to printing of digital images. The resolution of the digital image, both as seen on a computer monitor and printed on a paper, depends on the number of pixels used to make the image. Larger number of pixels gives better quality to the image. If a digital image is enlarged past a certain point, the individual pixels of the image will become visible, an effect called pixelization.

discussion Section of a research report that brings out the principles, relationships and generalizations shown by the results of an investigation. It interprets the results obtained and discusses them in the light of what is already known (information available in the literature) and states the significance of and conclusions arrived out of the results.

disinfectant A compound used to kill microorganisms, but not necessarily spores; usually applied to inanimate surfaces or objects.

disinfection A physical or chemical method of killing microorganisms, but not necessarily spores; the elimination or inhibition of the activity of surface-adhering microorganisms.

displacement chromatography A preparative liquid chromatography for the separation of protein components of a sample. It uses a displacer compound, which has a higher affinity for the stationary phase than the solutes in the sample. First, a large volume of the sample is loaded onto the column. Next, a displacer solution is pumped constantly into the column. The displacement process is based on the competition of solutes for the adsorption sites on the stationary phase according to their relative binding affinities and the displacer solution concentration. The action of the displacer solution, which has a higher affinity to the stationary phase than the solutes, causes the sample components to migrate faster through the column. The sample components elute from the column as adjacent zones of highly concentrated pure solute, in the order of increasing affinity of adsorption.

disposal   The process of getting rid of laboratory wastes, both hazardous and non-hazardous, which is governed by various regional, national and international regulations. Biohazardous wastes are made non-hazardous by processes like autoclaving and disposed by processes like off-site incineration or in licensed landfill sites.

dissecting microscope   A microscope with a low magnifying power of about 2.5–50×, used to examine or dissect small animals, etc.

dissociating buffer system   A type of buffer system used in the PAGE (e.g. SDS-PAGE) and includes agents which break the sample molecules down into subunits. Urea is used to break the hydrogen bonds; heating helps in denaturation of the proteins; BME (b-mercaptoethanol) or DTT (dithiothreitol) are added to break the sulphide bonds; however, its denaturing effect makes it unsuitable for the

separation of enzymes with their biological activity intact.

distillation   The process of heating a mixture to separate the more volatile from the less volatile parts, and then condensing the resulting vapour by cooling so as to obtain a pure or refined substance; e.g. distillation of water, alcohol, etc.

DNA amplification   *In vitro* multiplication of a piece of DNA into many thousands of millions of copies by processes like polymerase chain reaction (PCR) system and ligase chain reaction (LCR).

DNA chip   *See* microarray.

DNA fingerprinting   The use of restriction enzymes to measure the genetic variation of individuals. This technology is often used as a forensic tool to detect similarities in blood and tissue samples at crime scenes. In the cattle business, this technology is used to verify the parentage of individual animals. DNA fingerprint is the unique pattern of DNA fragments identified either by Southern hybridization or PCR.

DNA hybridization   The pairing of two DNA molecules, often from different sources, by hydrogen bonding between complementary nucleotides; a technique used to detect the presence of a specific nucleotide sequence in a DNA sample.

DNA ligase   An enzyme that catalyses a reaction linking two DNA molecules by the formation of a phosphodiester bond between the 3′ hydroxyl and 5′ phosphate of adjacent nucleotides. It plays an important role in DNA repair and replication. One of the essential tools of recombinant DNA technology for the incorporation of foreign DNA into vectors. The DNA ligase encoded by phage T4 is used in gene-cloning experiments. It requires ATP as a cofactor. T4 is used *in vitro* to join the vector and insert DNAs.

DNA polymerase An enzyme that catalyses the synthesis of double-stranded DNA, using single-stranded DNA as a template.

DNA polymorphism The presence of two or more alternative forms (alleles) of a chromosomal locus that differ in nucleotide sequence or have variable numbers of repeated nucleotide units.

DNA probe A labelled (tagged) fragment of DNA that is able, after a DNA hybridization reaction, to detect a specific DNA sequence in a mixed collection of sequences. If the tagged sequence is complementary to any one in the collection, the two sequences will form a double helix. This will be identified by the label, i.e., the tag, either by radioactivity or fluorescence.

DNA sequencing Method for determining the nucleotide sequence of a DNA fragment. There are two methods for doing this: (i) the Maxam and Gilbert technique (chemical degradation), that uses different chemicals to break the DNA into fragments at specific bases; (ii) the Sanger technique (called the di-deoxy or chain-terminating method) that uses DNA polymerase to make new DNA chains, with di-deoxy nucleotides (chain terminators) to stop the chain randomly as it grows. In both the methods, the DNA fragments are separated according to length by polyacrylamide gel electrophoresis, enabling the sequence to be read directly from the gel.

Dolly Name of the first mammal to be created by cloning a cell from an adult

animal in 1997. The cell came from the mammary tissue of an adult ewe (sheep). The cloning was achieved by nuclear transfer.

domestic animal diversity (DAD) The variety of genetic differences within each breed, and across all breeds within each domestic animal species, together with the species differences, all of which are available for the sustainable intensification of food and agriculture production.

double-beam spectrophotometer A design of the spectrophotometer that eliminates the problem of drift by measuring the absorbance of the blank and the sample virtually at the same time. A "chopper" alternately transmits and reflects the radiation so that it travels down the blank and the sample optical paths to a single detector. The chopper causes the radiation to switch paths at about 50 Hz enabling the detector to receive signals from blank and the sample light paths, alternately, in short successions. These signals are processed electronically to calculate the ratio of the reference and sample signals and to give either transmittance or absorbance as output.

double diffusion agar assay Ouchterloney's technique of identifying antigens in samples, based on the principle that diffusion of both antibody and antigen (hence, double diffusion) through agar can form stable and easily observable immune complexes. Unknown sample solutions of antigen and known antibodies are loaded into the separate wells punched in agar. The solutions diffuse outward, and when antigen and the appropriate antibody meet, they react and precipitate at the equivalence zone, producing an indicator line or lines; the visible line of precipitation enables comparison of antigens for identity (same antigenic determinants), partial identity (cross-reactivity) or non-identity against a given selected antibody.

DPT (Diphtheria-pertussis-tetanus) vaccine   A vaccine containing three antigens and that is used to immunize against diphtheria, pertussis or whooping cough and tetanus.

drug delivery   Procedure by which a therapeutic agent is delivered to its site of action. A range of new therapeutic-agent delivery systems, such as liposomes and other encapsulation techniques, and a range of mechanisms that target a therapeutic agent to a particular cell or tissue have been developed.

drum micrometer   A specialized eyepiece micrometer functioning on the same principle as a standard eyepiece-reticle combination, but also includes a mechanism to move the ocular micrometer (OM) scale by means of a precision screw mechanism that is operated by rotating an external drum. The drum micrometer should also be calibrated for the ocular micrometer using a stage micrometer. The drum of the micrometer screw is also divided into 100 intervals, so that one interval of the drum division corresponds to 0.1 interval of the eyepiece scale. Full rotation of the drum translates the measuring rule (line) across one interval of the eyepiece scale. The drum micrometer eliminates the necessity to estimate values of fractions of an OM division. It is also known as filar micrometer.

ECG (Electrocardiogram)   A test that records the electrical activity of the heart and used to measure the rate and regularity of heartbeats as well as the size and position of the cardiac chambers, the presence of any damage to the heart, and the effects of drugs or devices (e.g. pacemaker) used to regulate the heart.

electroblotting   A transfer technique used in western blotting. A sandwich of gel and nitrocellulose is encased in a cassette, which is immersed in a buffer between two parallel electrodes. When electric current passes through the cassette submerged in buffer, the protein bands in the gel get detached and deposited onto the nitrocellulose membrane in identical positions. The nitrocellulose paper with blotted protein bands is referred to as a blot. The blot is used for further analysis.

electrochemical sensor   Type of biosensor in which a biological process is linked to an electrical sensor system (e.g. enzyme, oxygen or pH electrode).

electromagnetic radiation   Energy waves produced by the oscillation or acceleration of an electric charge. Electromagnetic waves have both electric and magnetic properties. Electromagnetic radiation is usually arranged in a spectrum that extends from waves of extremely high frequency and short wavelength to extremely low frequency and long wavelength. Visible light is a small part of the electromagnetic spectrum. In descending order of frequency, the electromagnetic spectrum consists of gamma rays, hard and soft X rays, ultraviolet radiation, visible light, infrared radiation, microwaves, and radio waves. Electromagnetic radiation is used to produce mutant cells or organisms, or, in the case of UV, disinfestation and sterilization, in tissue culture.

electromagnetic spectrum The range of wavelengths or frequencies over which electromagnetic radiation extends.

electron microscope A microscope that uses an electron beam as the source of illumination, and magnetic condensers instead of glass lenses; broadly two types, scanning electron microscope (SEM) and transmission electron microscope (TEM).

electronic flash Artificial lighting used for indoor photography. An accessory fitted to the camera and the flash and the exposure can be synchronized. A glass quartz tube is filled with an inert gas, usually xenon. When electricity is applied to the electrodes that are sealed to the ends of the tube, the inert gas inside produces bright light for a very short span of time ranging from 1/1000 to 1/5000 of a second. The burst of intense light and the exposure are synchronized at a shutter speed of 1/125 second or less; also used in outdoor photography as fill-flash, i.e., to fill in a shadowy foreground in front of the object.

electrophoresis An analytical method in molecular biology for the separation and characterization of biological macromolecules such as proteins and nucleic acids. It is also used for the isolation and analysis of minute particles like viruses and small organelles. It works on the principle that charged particles of a sample migrate in an applied electrical field. If conducted in a solution, the molecules (particles) are separated according to their net charge. It also uses paper or, most frequently, gel as a support medium. Gel (agarose, polyacrylamide) matrix provides a sieving effect so that molecules are characterized by both charge and size. Addition of a charged detergent (e.g. SDS, sodium dodecyl sulphate) to the gel equalizes the surface charge and, therefore, permits differential migration of protein molecules only according to their sizes, thus helping in the estimation of molecular weight of proteins.

electrophoretic mobility During electrophoresis molecules in a sample migrate according to voltage density (V in cm^{-1}) of the gel. The velocity of migration ($cm\ sec^{-1}$) of a molecule at $1\ V\ cm^{-1}$ is referred to as the electrophoretic mobility that is expressed as $cm^{-1}V^{-1}sec^{-1}$. The electrophoretic mobility of a molecule is also influenced by its net electrical charge and its frictional coefficient (determined by the shape and molecular weight of the molecule).

electroporation An electrical treatment of cells that induces transient pores, through which DNA can enter the cell. It is used for the introduction of DNA or RNA into protoplasts or other cells by the momentary disruption of the cell membrane through exposure to an intense electric field. Pores form by the local polarization of the cell membrane when it is exposed to a high electric potential and persist for varying periods, depending upon the temperature at which the cell is treated. Macromolecules enter through these pores by diffusion or by electrophoretic movement; the membrane pores then re-seal, trapping the introduced macromolecule and preventing leak of the cell contents.

ELISA (Enzyme-linked immuno-sorbent assay) A method for accurate determination of specific molecules in a sample. The amount of protein or other antigen in a given sample is estimated by means of an enzyme-catalysed colour change. Pimary antibody specific to the test protein is adsorbed onto a solid substrate, and a known amount of the sample is added; all molecules of the test protein in the sample are bound by the antibody. A second antibody specific for a second site on the test protein is added; this secondary antibody is conjugated with an enzyme that catalyses a colour change in the fourth reagent, added finally; the colour change is measured photometrically and compared against a standard curve to give the concentration of protein in the sample. This method is widely used for diagnostic and other purposes.

elution chromatography A preparative liquid chromatographic method for the isolation of sample components in sufficient quantities for further purification and study or for a specific use. It is carried out under one of the following conditions (i) isocratic, i.e., with constant mobile phase composition, (ii) gradient, i.e., continuous change of mobile phase composition, and (iii) step elution. Under these conditions, the sample is injected into the column as finite volume pulse. The solutes in sample feed migrate through the column at different speeds, as dictated by the mobile phase velocity and the distribution of the solutes between the mobile and stationary phases.

embryonic stem cells Cells of the early embryo that can give rise to any type of differentiated cells, including germ line cells.

endangered species A plant or animal species in imminent danger of extinction because its population size has reached a critical level or its habitats have been significantly reduced.

endemic Describing a plant or animal species whose distribution is restricted to one or a few geographical areas; describing a disease or a pest that is always present in a geographical area.

enzyme commission number Systematic name and number that identify an enzyme in technical literature. Assigned by the Enzyme Commission (EC), it consists of four numbers separated by dots: the first classifies the enzyme into one of the six broad groups: (i) Oxidoreductases; (ii) Transferases; (iii) Hydrolases; (iv) Lyases; (v) Isomerases; (v) Ligases. Each of these groups is subdivided into sub-groups, each sub-group into sub-sub-groups, and the last number is specific for the enzyme, e.g. EC 3.1.21.1 is deoxyribonuclease I.

epidemiology The study of the factors determining and influencing the frequency and distribution of disease, injury and other health-related events and their causes in defined human populations.

epitope An area of the antigen molecule that stimulates the production of, and combines with, specific antibodies; also known as antigenic determinant site.

***ex situ* conservation** A conservation method by removing germplasm resources (seeds, pollen, sperm, individual organisms) from the original habitat or natural environment. For instance, *ex situ* conservation of farm animal genetic diversity involves conservation of genetic material *in vivo*, but out of the environment in which it developed, and *in vitro* including, *inter alia,* the cryoconservation of semen, oocytes, embryos, cells or tissues. *Ex situ* conservation and *ex situ* preservation are considered synonymous.

***ex vivo* gene therapy**   The delivery of a gene or genes to the isolated cells of an individual. Following culturing, the transformed cells are introduced back into the individual by transfusion, infusion or injection, to ease a genetic disorder.

experimental error   The variations among observations made on the experimental units, which were treated alike.

experimental unit   Material used, such as a laboratory animal, cage, culture plate, agriculture field, a plot in the garden, a pot in a greenhouse, etc. A treatment is applied to an experimental unit.

exposure   In photography, "exposure" means allowing specific quantity of light rays from the focused object to fall on the film for a specific duration. Control of the quantity and duration of light is effected by the aperture of the diaphragm and time setter, both adjusted in conjunction with each other.

expressivity   Degree of expression of a trait controlled by a gene. A particular gene may show different degrees of expression in different individuals.

extinction   The irreversible condition of a species or other group of organisms of having no living representatives in the wild, which follows the death of the last surviving individual of that species or group. Extinction may occur on a local or global level. It can result from various human activities, including the destruction of habitats or the overexploitation of species that are hunted or harvested as a resource.

factorial design   *See* two-way classification design.

farm animal genetic resources (AnGR)   Animal species that are used, or may be used, for food and agriculture, and the populations within each of them. Populations within each species can be classified as wild and feral populations, landraces and primary populations, standardized breeds, selected lines, and any conserved genetic material.

feedback inhibition   The negative feedback mechanism by which the end product of a biochemical pathway stops synthesis of that product. A late metabolite of a synthetic pathway regulates synthesis at an earlier step of the pathway. When the end product accumulates in excess, it inhibits its own synthesis. It is also known as end product inhibition.

field curvature   Also referred to as *curvature of field*. It is caused by lenses with curved surfaces, a common feature in most microscopes. When light rays are focused through a lens with curved surfaces, the image plane will be curved. The image can be focused over a certain range in the centre or the periphery of the viewfield. The entire image cannot be simultaneously focused especially onto a flat surface such as a film or digital imager. Field curvature is corrected by adding corrective lens elements to the objective. Such corrected objectives are called flat-field or plane- or plano-objectives.

film speed   A characteristic of colour as well as black-and-white photographic films; a numerical measure of the film's sensitivity to light. It determines the amount of exposure required to photograph an object under specific lighting conditions.

filter sterilization Method of sterilizing a liquid by forcing the liquid through a filter with pores so small that they prevent the passage of micro-organisms and microbial spores.

filter In photography, it is a coloured or white transparent or ground gelatin or glass and is fitted to the front of the camera lens. In microphotography, it is inserted in between the condenser of the microscope and the light source. It is also used in enlargers while making prints of negatives. It functions to alter the quality and quantity of the light, modifying the colour balance of the light, contrast, brightness, and haze. Colour filters (red, blue, green, yellow, etc.) transmit light of one colour only. They are commonly used in black-and-white photography to enhance and reduce the tone of specific colour.

filtration Separation of solids from liquids by using a porous material that permits passage of only the liquid or solid of size smaller than the pore size of the filter. The material passing the filter forms the filtrate. It also refers to the removal of cell aggregates to obtain a filtrate of single cells that can be utilized as plating inocula.

first aid The skilled application of accepted principles of medical treatment at the time and place of an accident. It is the approved method of treating a casualty until he or she is placed in the care of a doctor for definitive treatment of the injury.

first-aid box A box constructed of materials that will keep the contents dust- and damp-free. It is always kept in a prominent place; is easily recognized by the symbol a white cross on a green background. The first-aid box should contain: (i) Instruction sheet giving general guidance; (ii) Individually-wrapped sterile adhesive dressings in a variety of sizes; (iii) Sterile eye-pads with attachment bandages; (iv) Triangular bandages; (v) Sterile wound coverings; (vi) Safety pins; (vii) A selection of sterile but unmedicated wound dressings; (viii) An authoritative first-aid manual, e.g. one issued by the International Red Cross. Protective equipment for the person rendering first aid includes: (i) Mouthpiece for mouth-to-mouth resuscitation; (ii) Gloves and other barrier protections against blood exposures; (iii) Clean-up kit for blood spills.

flow cytometry A technique for counting, examining and sorting microscopic particles suspended in a stream of fluid. It allows simultaneous multiparametric analysis of the physical and/or chemical characteristics of single cells flowing through an optical/electronic detection apparatus. A beam of light (e.g. laser light) of a single frequency is directed onto a hydrodynamically focused stream of fluid. A number of detectors are aimed at the point where the stream passes through the light beam. Each suspended particle (e.g. RBC, lymphocytes, etc.) passing through the beam scatters the light in some way, and fluorescent chemicals in the particle may be excited into emitting light at a lower frequency than the light source. This combination of scattered and fluorescent light is picked up by the detectors, and by analysing fluctuations in brightness at each detector it is possible to deduce various facts about the physical and chemical structure of each individual particle. Measurable parameters are: (i) volume and morphological complexity of cells; (ii) cell pigments; (iii) DNA (cell cycle analysis, cell kinetics, proliferation, etc.); (iv) RNA; (v) chromosome analysis and sorting (library construction, chromosome paint); (vi) proteins; (vii) cell surface antigens; (viii) intracellular antigens (various cytokines, secondary mediators, etc.); (ix) nuclear antigens; (x) enzymatic activity; (xi) pH; (xii) intracellular ionized calcium, magnesium; (xiii) membrane potential; (xiv) membrane fluidity; (xv) apoptosis (quantification, measurement of DNA degradation, mitochondrial membrane potential, permeability changes); (xvi) cell viability; (xvii) monitoring electropermeabilization of cells; (xix) oxidative burst; (xx) characterizing multi-drug resistance (MDR) in cancer cells; (xxi) glutathione; (xxii) various combinations (DNA/surface antigens, etc.).

fluorescence   Defined as luminescence of a material when it is excited. The material may be a whole organism like bacteria, specific cells in a tissue or even molecules in a cell. Some biological materials (e.g. chlorophyll) have primary fluorescence and therefore auto-fluoresce, i.e., emit light of their own accord without any addition of fluorescent molecules. Biological materials which do not have autofluorescence property are linked with some exogenous fluorescent molecules, usually called as fluorochromes or fluorescence dyes, in order to create the required fluorescent probe.

fluorescence microscope   A microscopic design based on the principle that certain substances are capable of emitting light if they are excited by radiation and that the emitted light forms an image of the substance. While other light microscopes are used to study the structural details of or an organism or cell, a fluorescence microscope is used as a probing tool to ascertain the presence of a specific organism, cell, organelle, or molecule in the specimen.

fluorescence-activated cell sorting (FACS)   Separation of cells which have been pre-treated with a fluorescent dye that binds to DNA. It is separated according to the quantity of fluorescence detected by a laser beam and used in sperm sexing, based on the principle that X-bearing sperm contains more DNA than Y-bearing sperm.

fluorochromes   Acidic and basic dyes that are mostly yellow or orange in colour and fluoresce intensely when exposed to lights of shorter wavelengths such as UV, violet, and blue lights. Fluorescein is used as fluorescence probe for free protein amino groups. It emits green light when struck with blue excitation light. Acridine-orange, a basic fluorochrome, is used to discriminate between living and dead cells or microorganisms. A number of basic fluorochromes such as auramine O, acridine yellow, and acriflavine are used to create probes for nucleic acids, especially, DNA.

footnotes   Essential information that could not be accommodated in the body of a Table and given beneath the Table and used for three purposes: Source notes: to indicate the source of the data shown in the Table; General notes: information pertaining to the entire Table; Specific notes: necessary comments on specific items, indicated by some symbol as superscript to the specific item.

founder animal   An organism that carries a transgene in its germ line and can be used in matings to establish a pure-breeding transgenic line, or one that acts as a breeding stock for transgenic animals.

frame shift mutation   A mutation that changes the reading frame of an mRNA, either by inserting or deleting nucleotides.

free-living conditions   Natural or greenhouse conditions where the plantlets are transferred from *in vitro* conditions to soil mixtures, where the plantlets must manufacture their own food and survive.

fusion biopharmaceuticals   Biopharmaceutical proteins created as a result of fusion proteins. They have certain advantages such as: (i) synergistic activities in one molecule; thus when the molecule binds to a cell, it does two things at once; (ii) the adverse effect or poor stability of one molecule may be offset by the properties of the other; (iii) one molecule acts as a targeting mechanism to bring the other to the site where it is meant to act.

fusion gene   A hybrid gene produced by joining portions of two different genes, to produce a new protein; also made by joining a gene to a different promoter to alter or regulate gene transcription.

fusion protein   A polypeptide made from a recombinant gene that contains portions of two or more different genes. The different genes are joined so that their coding sequences are in the same reading frame. The cellular genetic system reads the gene fusion as a single gene, and therefore produces a fusion protein. It is also known as hybrid protein or chimeric protein. Such proteins are used: (i) to add an affinity tag to a protein; (ii) to produce a peptide as part of a larger protein, which is then cut up after it has been made by cloning; (iii) to produce a protein with the combined characteristics of two natural proteins; (iv) to produce a protein where two different activities are physically linked.

gas chromatography (GC)   A chromatographic technique which employs a gaseous fluid as the mobile phase, called the carrier gas, an inert gas. The stationary phase is a solid adsorbent or a liquid distributed over the surface of a porous, inert support. Depending on the nature of the stationary phase of a gas chromatographic system, it may be gas–solid chromatography or gas–liquid chromatography. The efficiency of a GC depends on the selection of appropriate column packing/coating and temperature programming.

gas–liquid chromatography (GLC)   A gas chromatography with a column that contains an inert porous solid, the stationary phase, coated with a viscous liquid. Diatomaceous earth is often used as the solid. Sample injected into the column dissolves into the liquid phase and eventually vaporizes, and the separation of the sample components is based on the relative volatilities.

gel   A supporting medium in most of the modern electrophoresis techniques. There are several types of gels such starch gel, agar gel, agarose gel and polyacrylamide gel.

gel electrophoresis   An analytical method for separating molecules according to their size. An electric field across the gel forces the molecules loaded at one end of a slab of polymer gel to migrate. The smaller molecules pass more easily and so move towards the other end faster; the various molecules end up at different positions according to their size. Gels are made from different materials, such as agarose, polyacrylamide, starch, etc. Various chemicals can be included in the gel to assist effective separation, such as the detergent sodium dodecyl sulphate (SDS) in protein gels to unfold proteins, or urea in DNA sequencing gels, which unfolds DNA.

gel-filtration chromatography *See* molecular exclusion chromatography.

gel-permeation chromatography *See* molecular exclusion chromatography.

gene A segment of chromosome. At the molecular level, it is a segment of nucleic acid that encodes peptide or RNA; the unit of heredity transmitted from generation to generation during sexual or asexual reproduction. While some genes direct synthesis of proteins, others have regulatory functions.

gene bank The term is used in the following senses: (i) The storage of collections of genetic material in the form of seeds, tissues or reproductive cells of plants or animals. (ii) A facility, referred to as field gene bank, established for the *ex situ* storage and maintenance of individual plants, using horticultural techniques; (iii) A collection of cloned DNA fragments from a single genome; contains cloned representatives of all the DNA sequences in the genome; (iv) A population of microorganisms, each of which carries a DNA molecule that was inserted into a cloning vector. Also called gene library or clone bank.

gene cloning The practice of synthesizing multiple copies of a particular DNA sequence using a bacterial cell or another organism as a host. The desired gene is isolated and inserted into a self-replicating DNA molecule (DNA vector, a plasmid) and the resulting recombinant DNA molecule is amplified in an appropriate host cell. It is used in genetic engineering and is also referred to as molecular cloning.

gene conservation The conservation of species, populations, individuals or parts of individuals, by *in situ* or *ex situ* methods, to provide a diversity of genetic materials for present and future generations.

gene library Collection of cloned DNA fragments that includes all the genetic information of an organism. If the original source of the DNA is the genomic DNA from an organism, then it is called a genomic library; if the DNA is from cDNA made by enzymatic copying of RNA, then the library includes representative fragments from all genes that were being expressed at the time the RNA was sampled, and would be called a cDNA library.

gene mapping Determination of the relative positions of genes on a chromosome.

gene pool The total genetic information in all the genes in a breeding population at a given time. In Plant Genetic Resource (PGR), the term may refer to primary, secondary or tertiary gene pool. The primary gene pool includes plants that are inter-fertile; the secondary gene pool includes those which can cross with the members of the primary pool; the members of the tertiary gene pool require extreme techniques to achieve crossing.

gene probe   A single-stranded DNA or RNA fragment used in genetic engineering to search for a particular gene or other DNA sequence; it has a base sequence complementary to the target sequence and will thus bind to it by base pairing; labelled probes help identification of target sequence after subsequent separation and purification.

gene sequencing   The process of determination of the sequence of nucleotide bases in a strand of DNA.

gene splicing   A step in the processing of mRNA, taking place only in eukaryotic cells. Intervening base sequences (introns) are removed from the primary RNA transcript (hnRNA), and the coding exons are joined together to form the mature mRNA molecule.

gene therapy   The treatment of inherited diseases by inserting the wild-type copies of the defective gene causing the disorder, into the cells of affected individuals. If the genes are introduced into reproductive cells, the procedure is called germ-line or heritable gene therapy; if introduced into cells other than reproductive cells, the procedure is called somatic cell or non-inheritable gene therapy.

generalization   The aim of a biological investigation is to make an inference about the population using the sample data, i.e., estimation of population parameters using sample statistics. In other words, what is observed in a small sample is generalized for the entire population from which the sample was obtained.

genetic code   The mechanism by which genetic information is stored in living organisms. It uses sets of three nucleotide bases (codons) to make the amino acids, which in turn constitute proteins. The set of 64 nucleotide triplets (codons) specify the 20 amino acids and termination codons (UAA, UAG, UGA).

genetic counselling   Counselling given to prospective parents about the chances of genetic disorders in a future child, given on the basis of the family (genetic) history of the two concerned people.

genetic diversity   The heritable variation within and among populations which is created, enhanced or maintained by evolutionary forces.

genetic drift   Fluctuation in the relative frequency of different genotypes in a small population owing to the chance of disappearance of particular genes as individuals die or do not reproduce.

genetic engineering   The manipulation (introduction or elimination of specific genes) of genetic material in order to alter genes and hence the characteristics of the organism concerned. Involves the use of a vector for transferring useful genetic information from a donor organism into a cell or organism that does not possess it. It also refers to selective breeding and other means of artificial selection.

genetic fingerprint   A set of genetic characteristics derived from the tissues or secretions of an individual and used to identify the person.

genetic fingerprinting   A technique in which a person's DNA is analysed to reveal the pattern of repetition of particular nucleotide sequences throughout the genome.

genetic immunization Delivery of a cloned gene that encodes an antigen to a host organism. After the cloned gene is expressed, it elicits an antibody response that protects the organism from infection by a virus, bacterium or other disease-causing organism.

genetic profile A description enumerating the significant genetic characteristics of an organism. It is used for identification, prediction of inherited disorders, etc.

genetic resources conservation *See* gene conservation.

genetically engineered organism (GEO) *See* genetically modified organism.

genetically modified organism (GMO) An organism the genome of which has been modified by the application of recombinant DNA technology by addition (transgenic) or deletion (knock out) of gene sequence.

genome The gene complement of an organism. It comprises the information of the entire genetic material of an organism.

genomic DNA library, genomic library A collection of clones containing the genomic DNA sequences of an organism; sequences propagated in bacteria or phage; tool used in the process of isolating desired genes.

genomics, functional genomics, structural genomics Study of all aspects of the genome of an organism. Functional genomics (also referred to as functional proteomics) is the study of the function of the proteome (the protein complement encoded by an organism's entire genome). It is the systematic study using large-scale experimental methodologies combined with statistical analysis of the results. Structural genomics is the study of the complete structural description of a defined set of molecules, ultimately for an organism's entire proteome. It applies to X-ray crystallography and NMR spectroscopy.

geographical indication (GI) A name or sign used on certain products, which corresponds to a specific geographical location or origin (e.g. a town, region or country). The use may serve as a certification that the product has certain qualities, or enjoys certain reputation because of its geographical origin. It also serves to preserve cultural traditions of a locality and to protect community rights.

geometric distortion An aberration common in stereomicroscopy, involving geometric distortion of the image of the specimen. It causes changes in the shape of the image but does not affect the sharpness or colour spectrum. Positive and negative geometric distortions, also called as pincushion and barrel respectively, are the common aberrations of this type.

germ-line gene therapy The introduction of a gene or genes to a fertilized egg or an early embryonic stage. The inserted gene(s) may be present in all or some of the nuclei of the cells of the adult individual, including the reproductive cells. This therapy modifies the phenotype of the individual.

GLP (Good Laboratory Practice), GMP (Good Manufacturing Practice) Codes of practice designed to reduce accidents that could affect a research project or a manufactured product. The essential point of both GLP and GMP is that everything is recorded, and that only established procedures are used and by people who have been trained to use them.

GM food (Genetically Modified food)   One that has been modified, in whole or in part, by the application of recombinant DNA technology.

G-proteins   Proteins which relay signals in mammalian cells. They occur on the inner surface of the plasma membrane and transmit signals from outside the membrane, via transmembrane receptors, to adenylate cyclase, which catalyses the formation of the second messenger, cyclic AMP, inside the cell. The name is derived from their ability to bind to guanine nucleotides, namely GTP and GDP. The GTP-protein complex is able to activate adenylate cyclase, whereas the GDP-protein complex is not. G-proteins are activated when the signalling molecule (e.g. a hormone) binds to the transmembrane receptor.

GRAS   (Generally Regarded as Safe)   Certification given to foods, drugs, and other materials that have been used for a considerable period of time and have a history of not causing any illness to humans, even though far-reaching toxicity testing has not been conducted. More recently, certain host organisms for recombinant DNA experimentation are given this status.

green revolution   Name given by William Goud to describe the dramatic increase in crop productivity during the third quarter of the 20th century, as a result of integrated advances in genetics and plant breeding, agronomy, and pest and disease control.

HEPA filter (High Efficiency Particulate Air filter)   A filter capable of screening out particles larger than 0.3 μm and used in laminar air flow cabinets (hoods) for sterile transfer work.

herbicide   Any natural or synthetic substance that is toxic to plants; agrochemicals used to kill specific unwanted plants, such as weeds.

herbicide resistance   The ability of a plant to withstand herbicide. It is the target of plant genetic engineering, so that when a herbicide is sprayed onto a field planted with herbicide-resistant crops, all the plants except the crop would be killed, thus providing an effective method of weed control without having to develop herbicides specific to each weed type.

high performance liquid chromatography (HPLC)   The most common type of liquid chromatography. It is called "high performance" because it has a very effective separation and high resolution. The column is shorter and has a smaller diameter, but the column is equivalent to having large number of equilibrium stages, i.e., plates. The stationary phase (silica particles with long-chain alkanes) is held in a column and the mobile phase (e.g. methanol) is forced through under pressure. Separation of sample solutes is much faster.

high voltage paper electrophoresis   A paper electrophoresis in which the disadvantage (marked diffusion effects and the consequent poor resolution) of low voltage (2 to 50 V) used in order to minimize heat generation, is overcome. It involves the application of high voltage in the range of 2500 to 5000 V and a current of about 500 mA with a cooling system that covers the electrophoresis paper and keeps it in cool, constant temperature.

histocompatibility The degree to which tissue from one organism (donor) will be tolerated by the immune system of another organism (recipient).

histocompatibility complex, histocompatibility system The collection of genes coding for peptides present on the surface of nucleated cells; these peptides are responsible for the differences between genetically non-identical individuals that cause rejection of tissue grafts between such individuals. Originally called histocompatibility antigens, now called histoglobulins, reflecting their structural similarity to immunoglobulins.

homogenotization A genetic method used to replace a gene, or a DNA sequence within a genome, with an altered copy of that sequence. It involves cloning of the DNA sequence, its alteration (e.g. insertion of a transposon, which is encoded to antibiotic resistance), and replacement of the original sequence by the mutated sequence by recombination *in vivo*.

homologous proteins Two proteins with related folds and related sequences. They are of two types—orthologous and paralogous proteins; orthologous proteins evolved from a common ancestral gene; paralogous proteins were created by gene duplication.

humoral immune response Antibody-mediated immune response; the synthesis of antibodies by B cells in response to an encounter of the cells of the immune system with a foreign immunogen.

hybrid vigour, heterosis The superior performance of a hybrid in one or more traits is better than the average performance of the two parental populations.

hybridization It may refer to (i) interbreeding of species, races, varieties, etc., among plants or animals; a process of forming a hybrid by cross-pollination of plants or by mating animals of different types; (ii) the production of offspring of genetically different parents, normally from sexual reproduction, but also asexually by the fusion of protoplasts or by transformation; (iii) the pairing of two polynucleotide strands, often from different sources, by hydrogen bonding between complementary nucleotides.

hybridoma A hybrid cell created by the fusion of a specific antibody-producing B (bone marrow) lymphocyte and a cancerous myeloma cell, which grows indefinitely in tissue culture and is selected for the secretion of the specific antibody produced by that B cell.

hybridoma technology A technology for producing monoclonal antibodies (MAbs). It involves fusion of specific antibody producing B cells and plasma cells (lymphocytes, isolated from mice immunized with specific antigen) with myeloma cells (mutant cancerous plasma cells incapable of producing immunoglobulins, but can be cultured indefinitely). The fused cells, the hybridomas (hybrid of two cells), are cultured in HAT medium (a combination of hypoxanthine, amino-pterin and thymidine). Aminopterin is a poison that blocks a specific metabolic pathway in cells. Myeloma cells do not have the necessary enzyme that would allow them to grow in the presence of this poison; but, the lymphocytes (from the immunized mice) have this enzyme and therefore can continue to grow and multiply provided the intermediate metabolites hypoxanthine and thymidine are available. Thus, the hybridomas have the properties (i) to synthesize specific antibody, (ii) to survive even in the presence of a poison (aminopterin), and (iii) of immortality of cancerous myeloma cells. As a result only the hybridomas grow in the HAT medium, but myeloma cells die. The lymphocytes also die naturally after some time, because they are not immortal as the cancerous myeloma cells are. The hybridoma cells are tested individually for the production of the desired antibody, selected for further cloning (monoclone) and used for mass-producing the antibody.

hypothesis   Tentative solution to a problem or an explanation for an observed event. A statement making a prediction that an event will occur under stated, specific conditions; a tentative theory or supposition provisionally adopted to explain certain facts and to guide in the investigation of other facts; a testable theory; one difficult to "prove" beyond any doubt. Once proven by rigorous scientific investigation, it becomes a theory or a law.

immobilized cells   Cells entrapped in matrices such as alginate, polyacrylamide and agarose designed for use in membrane and filter bioreactors.

immortalization   The genetic transformation of a cell type into a cell line that can proliferate indefinitely.

immunity   The protection of an animal or plant against infection by disease-producing organisms or to the harmful effects of their poisons.

immunization   The creation of immunity in an individual by artificial means. Active immunization involves the introduction, either orally or by injection, of specially treated bacteria, viruses or their toxins so as to stimulate the production of antibodies.

immuno-affinity chromatography A type of affinity chromatography; a purification technique in which an antibody is bound to a matrix and is used to bind and separate a protein from a complex mixture.

Immunoassay, immunodiagnostics An assay method that detects proteins by using an antibody specific to that protein; a positive result is seen as a precipitate of an antibody–protein complex; positive reaction can also be detected if the antibody is linked to a radioactive atom or to an enzyme which catalyses an easily monitored reaction such as a change in colour.

immunoblotting   An immunological technique used in clinical microbiology and molecular biology. It involves PAGE separation of proteins in a sample, followed by the transfer of the separated proteins to nitrocellulose sheets. The transferred bands are visualized by treating the nitrocellulose sheets with enzyme-tagged antibodies. It is used to identify the presence of common and specific proteins among different strains of microorganisms and as a diagnostic tool.

immunoelectrophoresis   An electrophoresis working on the principle of antigen–antibody specificity (immuno-precipitation) in combination with molecular sieving of the electrophoresis gel and gel diffusion. The agar gel is prepared on a glass plate with a circular sample well in the middle, which is loaded with known antibodies, and the gel is subjected to electrophoresis. The antibodies migrate according to their respective charge and mass. At the end of the electrophoresis, a narrow trough is cut in the gel, parallel to the path of migration of the sample components and the unknown sample is loaded into it and incubated, allowing the diffusion of the antigens in the sample and the electrophoresed antibodies to diffuse in the gel. If the unknown sample contains a specific antigen against any of the known electrophoresed antibodies, a precipitation reaction occurs and a characteristic arc (precipitation line or bow) appears in the gel between the trough and the specific antibody.

immunodiffusion Precipitation reaction that occurs between an antibody and an antigen in an agar gel medium.

immunofixation electrophoresis (IFE) An electrophoresis working on the antigen–antibody reaction principle. The proteins in an unknown sample are first separated by electrophoresis according to their charge. The gel is then overlaid with monospecific antisera reactive with a specific protein. If the specific antigen (protein) is present in the unknown sample, a characteristic immunoprecipitin band is formed. It is used to identify and characterize serum proteins. It is more sensitive than immunoelectrophoresis in detecting a monoclonal protein in the urine, especially in recognizing a monoclonal heavy chain fragment in the urine with myeloma or amyloidosis.

immunoglobulin A group of proteins (globulins) in the body that act as antibodies. It is produced by specialized cells (B lymphocytes) and are present in blood serum and other body fluids.

immunosensor A biosensor having an antibody as the biological part.

immunosuppression The suppression of immune response, necessary, following organ transplants in order to prevent the host rejecting the grafted organ.

immunosuppressor A substance, an agent or a condition that prevents or reduces the immune response.

immunotherapy The use of an antibody or a fusion protein containing the antigen-binding site of an antibody to cure a disease or increase the health of a patient.

immunotoxin Protein drugs consisting of an antibody joined to a toxin molecule. They are made by linking toxin and antibody molecules chemically, or by fusing the genes for the toxin and the antibody. The antibody portion of the molecule facilitates binding to a target molecule or cell, and the toxin inactivates or kills the target molecule or cell.

IMRAD or IMRaD The system of organizing a research paper (report, thesis, dissertation) into **I**ntroduction, **M**aterial and Methods, **R**esults **a**nd **D**iscussion, the acronym of these four sections.

***in situ* conservation** A conservation method that is followed to preserve the integrity of genetic resources by conserving them within the evolutionary dynamic ecosystems of their original habitat or natural environment. *In situ* conservation of farm animal genetic diversity refers to all measures to maintain live animal breeding populations, including those involved in active breeding programmes in the agro-ecosystem where they either developed or are now normally found, together with husbandry activities that are undertaken to ensure the continued contribution of these resources to sustainable food and agricultural production, now and in the future.

***in vivo* gene therapy** The delivery of a gene or genes to a tissue or organ of an individual to ease a genetic disorder.

inbred line The product of inbreeding; the result of mating of individuals that have ancestors in common. In plants and laboratory animals, it refers to populations resulting from at least 6 generations of selfing or 20 generations of brother–sister mating, that are for all practical purposes, completely homozygous; in farm animals, populations that have resulted from several generations of the mating of close relatives, without having reached complete homozygosity.

inbreeding Matings between individuals that have one or more ancestors in common; self-fertilization, which occurs in many plants and some primitive animals.

inbreeding depression Effect of increased level of inbreeding in a population; reduction in vigour, yield, etc. The greatest inbreeding depression are those that are most closely associated with viability and reproductive ability.

incineration A method of disposing of animal carcasses as well as anatomical and other laboratory waste, with or without prior decontamination, an alternative to autoclaving if used under laboratory control. It requires an efficient means of temperature control and a secondary burning chamber; the temperature in the primary chamber should be at least 800°C and that in the secondary chamber at least 1000°C. Materials for incineration, even with prior decontamination, should be transported under controlled conditions.

incubator An apparatus in which environmental conditions (light, photoperiod, temperature, humidity, etc.) are fully controlled, and used for hatching eggs, multiplying microorganisms, culturing plants, maintaining newborn babies, etc.

inoculation cabinet Small room or cabinet for inoculation of tissue or microorganism cultures and similar operations, often with a current of sterile air to carry contaminants away from the work area.

Intellectual Property Rights (IPR) *See* intellectual property.

intellectual property (IP) Refers to creations of the mind such as inventions (e.g. a software program) and artistic (e.g. a composition of music) and literary (e.g. a poem, novel, essay, etc.) creations. The legal right of the ownership of such intangible intellectual properties is the Intellectual Property Right (IPR).

introduction The first section of a research article proper. It aims to introduce the specific subject of research to the reader, to justify the choice of the topic and, if necessary, the specific methodology adopted, and to state clearly the objectives or hypotheses of the investigation.

ion-exchange chromatography A technique used in the separation and purification of biological molecules. Two types of ion exchange, viz., cation exchange and anion exchange, are used. In cation exchange, the stationary phase carries a negative charge, while in anion exchange, the stationary phase has positive charge. Charged molecules in the carrier liquid phase pass through the column and bind to cation or anion in the stationary phase. The bound molecules in the column are eluted by flushing the column with a solution of varying pH or ionic strength. The separation by ion-exchange process is highly selective.

ionizing radiation The portion of the electromagnetic spectrum that can produce positive and negative charges in molecules; electromagnetic radiation of very short wavelength and therefore of high frequency, which causes atoms to lose electrons, i.e., ionize. There are two forms: (i) X-rays, artificially produced and (ii) gamma rays, that are emitted during radioisotopic decay. They can cause mutations and even death, depending on the dose.

irradiation Refers to (i) exposure to any form of radiation; (ii) treatment with a ray, such as ultraviolet rays, etc.; (iii) exposure of food (in food technology) to controlled low levels of ionizing radiation to sterilize or disinfect it, without inducing radioactivity in the target.

isodensity centrifugation *See* isopycnic centrifugation.

isoelectric focusing (IEF) Electrophoretic separation of proteins in a sample on the basis of surface charge alone as a function of pH. A protein in a sample applied to a gel with pH gradient, will be either positively charged or negatively charged depending on the local pH at that point in the gel. When electric current is applied, the proteins in the sample will move towards either the anode or cathode, depending on its charge, until it encounters that part of gel which corresponds to its isoelectric point (pI). At this point a protein will have no net charge and will, therefore, stop migration. Proteins, which diffuse away from the isoelectric pH, become charged again and are electrophoresed back to their isoelectric pH, hence the name isoelectric focusing. An IEF gel is calibrated with proteins of known pI and a plot of position on the gel against pI allows an estimation of the pI of unknown proteins. It is used for the identification of the pI of a protein, as well as for separation and further purification of molecules.

isopycnic centrifugation A centrifugation used to separate particles only on the basis of their buoyant densities, instead of their sedimentation velocities. If the density of the particle and that of the medium are equal, then the sedimentation velocity will be zero. This is achieved in the isopycnic (isodensity) centrifugation, using a gradient in the centrifugal tube, such that at a certain distance from the top of the solution the fluid density increases to that of the particle, at which point the sedimentation velocity will disappear for a specific particle. During isopycnic centrifugation, a particle initially in a region where the fluid density is less than the particle density will migrate down the tube to the point of equal density, its isopycnic point. A particle in a region where the density is higher than its own density will migrate up (float up) to its isopycnic point. At equilibrium, the particles will be distributed along the length of the tube according to their density. It is also known as sedimentation equilibrium centrifugation.

isotope One of two or more forms of an element that have the same number of protons (atomic number) but differing numbers of neutrons (mass numbers). Radioactive isotopes are commonly used to make DNA probes and metabolic tracers.

IVEP *(In vitro* embryo production) The combination of ovum pickup, *in vitro* maturation of ova, and *in vitro* fertilization.

IVF *(In vitro* fertilization) A technique in human and animal science, whereby the egg is fertilized with sperm outside the body, before re-implantation into a female.

juvenile *in vitro* embryo technology (JIVT; JIVET) A method involving collection of immature ova from young animals, *in vitro* maturation of the ova, *in vitro* fertilization, and then transfer of the resultant embryos into recipient females; a technology for achieving rapid generation turnover.

key words A few words (about 3 to 4) used in a research article, useful for indexing services, given as key words just below the abstract of an article in a Journal. The choice of the words is such that they facilitate searching for the article in some database.

kilobase (kb) A length unit equal to 1000 base pairs of a double-stranded nucleic acid molecule. One kilobase of double-stranded DNA has a mass of about 660 kilodalton.

knock-out animal Animal with genome in which specific genes have been deleted; animal resulting from an embryonic stem cell in which a normal functional gene has been replaced by a non-functional form of the gene. This technique is used extensively in mice, to learn about the function of a gene by studying the phenotype of animals that lack the peptide product of that gene.

Köhler Illumination *See* optimum illumination.

label A tag, i.e., a compound or atom that is either attached to or incorporated into a macromolecule and is used to detect the presence of a compound, substance, or macromolecule in a sample.

labelling The method of substituting a stable atom in a compound with a radioactive isotope of the same element to enable it to be detected by autoradiography or other techniques. It is used to trace the path of the labelled compound through a biological or chemical system.

laboratory biosafety Containment principles, technologies and practices that are implemented to prevent unintentional exposure to pathogens and toxins, or their accidental release.

laboratory biosecurity Institutional and personal security measures designed to prevent the loss, theft, misuse, diversion or intentional release of biohazardous materials like pathogens and toxins.

laminar air-flow cabinet, laminar air-flow hood Cabinet for inoculation of cultures, in which the working area is kept sterile by a continuous, non-turbulent flow of sterilized air through a HEPA filter.

laser scanning confocal microscope (LSCM) A design of confocal microscope in which laser is employed to provide excitation light. The laser light, blue in colour and of very high intensity, is reflected off a dichroic mirror onto two other mirrors, one after the other. These mirrors are mounted on motors and they scan the laser across the sample, through the objective. The emitted light from the fluorescing element of the specimen travels back to the scanning mirrors, gets reflected to the dichroic mirror which allows it to go through to focus onto the pinhole. A detector, which is a photomultiplier tube (PMT), measures the light that passes through the pinhole and forms an image of the element on a high-resolution monitor.

Latin square design (LSD) A versatile experimental design involving ANOVA, useful for controlling the variations in an experiment that are related to rows and columns in a field. It can easily be adopted to many experimental situations in a greenhouse or a laboratory. In a field study, it basically consists of a row-and-column design for n treatments in n rows and n columns; each treatment occurs once in each row and once in each column.

LCR (Ligase Chain Reaction) A method for determining the presence or absence of a specific nucleotide pair within a target gene.

LD_{50} (Lethal dose$_{50\%}$) The amount of a substance required to kill 50% of the test population; the higher the LD_{50}, the lower the presumed toxicity of the substance.

liquid chromatography A column chromatography in which the mobile phase is liquid (e.g. methanol). More specifically, it may be a liquid–adsorption chromatography in which adsorbent stationary phase is used, or a liquid–liquid chromatography in which a liquid stationary phase is coated onto a solid support.

literature cited Section of the report that gives, in an organized way, the details of sources of the literature used while writing the different sections of a research report.

live vaccine A living, non-virulent form of a microorganism or virus that is used to produce an antibody response that will protect the inoculated organism (humans and other animals) against infection by a virulent form of the microorganism or virus.

longitudinal study An experimental investigation, in which the variables and factors in experimental units are manipulated and the effects observed; provides tangible evidence to demonstrate the causative agent of a particular event.

loop bioreactor A type of fermenter in which the fermenting material is cycled between a bulk tank and a smaller tank or loop of pipes. The circulation helps to mix the materials and to ensure that gas injected into the fermenter is well distributed in the liquid. It is useful for photosynthetic fermentations, where the photosynthesizing organisms are passed along a large number of small pipes such that they get light easily as against a single volume reactor in which only the organisms near the edges get much light.

macromolecules Molecules of large molecular weight, such as proteins, nucleic acids and polysaccharides.

macronutrient Essential component of growth media; an element normally required in concentrations >0.5 millimole/L.

macrophotography   The process of making photographs using a variety of cameras and lens attachments except a microscope; it may be still or movie photography, and can be underwater or aerial.

macropropagation   Production of plant clones from growing parts.

magnetic force microscope (MFM)   Scanning microscope which uses a magnetic tip to probe the magnetic stray field above the sample surface. The magnetic tip is mounted on a small cantilever that translates the force into a deflection that can be measured to create a map of the magnetic forces or force gradients above the surface.

major histocompatibility antigen   A cell-surface macromolecule that allows the immune system to distinguish foreign or "non-self" from "self"; histoglobulins; antigens that must be matched between donors and recipients during organ and tissue transplants to prevent rejection.

major histocompatibility complex   Large cluster of genes that encode the major histocompatibility antigens in mammals.

malignant   Possessing the properties of cancerous growth.

marker gene   A gene of known function and known location on the chromosome and used in the study of inheritance of a trait or a gene. It is used in mapping the order of genes along chromosomes and in following the inheritance of particular genes. Genes closely linked to the marker are generally inherited with it. It must be readily identifiable in the phenotype or at the molecular level.

marker peptide   A portion of fusion protein used in its identification or purification.

megabase (Mb)   A length of DNA consisting of 10^6 base pairs (if double-stranded) or 10^6 bases (if single-stranded); 1 Mb = 10^3 kb = 10^6 bp.

metabolic pathway engineering (MPE)   The application of molecular techniques to improve the efficiency of pathways that synthesize specific products.

metastasis   The transfer of a disease (e.g. cancer) from one organ to another not directly connected with it.

micellar electrokinetic capillary chromatography   An adaptation of capillary electrophoresis for the separation of not only the anionic and cationic molecules of the sample, but also the neutral molecules. Micelles which are amphiphilic monomeric surfactants, with hydrophilic head and a hydrophobic tail are used in the separation buffer. The hydrophobic tail may be a straight or branched chain of hydrocarbon, or steroidal skeleton. The hydrophilic head may be cationic, anionic, zwitterionic or nonionic; the micelles have a three-dimensional structure with the hydrophobic moieties in the interior and the charged hydrophilic moieties at the exterior. The separation of neutral molecules is based on the hydrophobic interaction of solutes with the micelles. The stronger the interaction, the longer the molecules migrate with the micelle; the selectivity is controlled by the choice of surfactant and also by the addition of modifiers to the buffer.

microarray   A two-dimensional array grid of spots on a glass, plastic, membrane filter, or silicon wafer, each spot having a unique DNA sequence, different from the sequence of the other spots around it. Each DNA fragment corresponds to a known gene or gene fragment. The fragments are deposited or synthesized in a predetermined spatial order allowing them to be used as probes for mRNAs coded by those genes. Each fragment in a spot will hybridize to its complementary DNA strand. Thus each spot functions as a probe to determine the levels of a specific mRNA produced by a collection of cells. When a solution of fluorescent-labelled DNA fragments is hybridized to the chip, spots to which hybridization occurs are visible as fluorescence. If the spots on the chip are genes (expressed sequence tags), hybridization with cDNA from a particular tissue shows which genes are expressed in that tissue. The number of DNA probes that is possible to place on a microarray is very large. A typical DNA chip would contain 10,000 discrete spots in an area of just a few square centimeters. There are microarrays with probes for every gene in yeast, and others with over 19,000 human cDNAs. Such an enormous number of probes permit the observation of the response of whole genomes to various stimuli instead of one gene at a time.

microelectrophoresis A free electrophoresis involving microscopic observation of movement of minute particles (e.g. blood cells, bacteria, etc.) in an electrical field. The apparatus includes microelectrophoretic cell (e.g. Abramson's microelectrophoretic cell), electrode compartments, ocular micrometer, and stop clock and is useful in measuring the zeta potential (electrokinetic potential, measurement of the velocity of the particle under an electrical field), blood cells and microorganisms (e.g. bacteria).

microenvironment The environment surrounding and close enough to the surface of a living or non-living thing.

microinjection The introduction of minute amounts of material (DNA, RNA, enzymes, cytotoxic agents) into a single eukaryotic cell with a fine, microscopic needle, penetrating the cell membrane.

micrometry The practice of measuring the dimensions of microscopic objects (e.g. diameter of RBC, thyroid follicles, ova, nuclei, etc., lengths of intestinal villi, thickness of connective tissue layers). The most common method of making such measurements is the use of microscopic ocular micrometer and stage micrometer.

micronutrient Essential component of growth media; an essential element normally required in concentrations < 0.5 millimole/L.

microphotography Photographic recording of images through a microscope. Wide-ranging subjects, from a whole-mount of a small insect to an organelle in a cell, can be photographed using an appropriate microscope, lens and lighting. The fields in which microphotography is used are cytology, histology, electron microscopy (TEM and SEM), autoradiography, etc.

micropropagation *In vitro* multiplication and/or regeneration of plants under aseptic and controlled environmental conditions on specially prepared media that contain substances necessary for growth. The three general types of tissue used are: excised embryos (embryo culture), shoot-tips (meristem culture or mericloning) and pieces of tissue that range from bits of stems to roots. The four stages involved in the process are: Stage I: establishment of an aseptic culture; Stage II: multiplication of propagules; Stage III: preparation of propagules for successful transfer to soil (rooting and hardening); Stage IV: establishment in soil.

minimum inoculum size   The critical volume of inoculum (material introduced into a culture medium) required to initiate culture growth, due to the diffusive loss of cell materials into the medium. The subsequent culture growth cycle is dependent on the inoculum size, which is determined by the volume of medium and size of the culture vessel.

minus strand RNA   *See* plus strand RNA.

mitochondrial DNA (mtDNA)   A circular ring of DNA found in mitochondria. In mammals, it constitutes less than 1% of the total cellular DNA; in plants the amount is higher and variable. It codes for ribosomal RNA and transfer RNA, but for only some mitochondrial proteins (up to 30 proteins in animals).

molality   The number of moles of solute per litre of solvent.

molar absorptivity   Defined as the absorbance of a 1 M solution measured over a path length of 1 cm, and is expressed as $cm^{-1} M^{-1}$; previously called as the absorption coefficient or extinction coefficient.

molarity   The number of moles of a substance contained in a kilogram of solution.

mole   Amount of substance that has a weight in grams numerically equal to the molecular weight of the substance; also called gram molecular weight (symbol, M). A mole contains 6.023×10^{23} molecules or atoms of a substance.

molecular cloning   The biological amplification of a specific DNA sequence through mitotic division of a host cell into which it has been transformed or transfected.

molecular exclusion chromatography   A chromatographic technique in which retention of sample components in the stationary phase is according to their molecular size. The stationary phase is a gel consisting of porous beads with strictly controlled pore size. If the sample components have molecular size smaller than that of the pore size, they get trapped in the pores and as a result migrate slowly. On the other hand, molecules larger than the pore dimension migrate faster without any hindrance. This technique is also useful in the determination of the relative molecular weight of a macromolecule. Also known as gel-filtration or gel-permeation chromatography.

monochromator   Device used in a spectrophotometer to resolve the polychromatic radiation into its component wavelengths, to isolate them into narrow bands and focus a specific band into the sample solution contained in the cuvette. It consists of an entrance slit through which a narrow beam of the polychromatic light from the radiation source is admitted, a collimating lens or mirror that directs the radiation as parallel beams onto the scattering device, a prism or a grating for scattering the radiation into its component wavelengths, a focusing lens to focus the required specific narrow band of wavelength and an exit slit through which the focused band is directed to pass through the sample solution.

monoclonal antibody (MAb) A single type of antibody that is directed against a specific epitope (antigen, antigenic determinant) and is produced by a single clone of a single hybridoma cell line, which is formed by the fusion of a lymphocyte cell with a myeloma cell. It has applications in (i) the typing of tissues, (ii) the identification and epidemiological study of infectious microorganisms, (iii) the identification of tumour and other surface antigens, (iv) the classification of leukemia and (v) the identification of functional populations of different types of T-cells. It has future applications such as (i) passive immunization against infectious diseases and toxic drugs, (ii) tissue or organ graft protection, (iii) stimulation of tumour rejection and elimination, (iv) manipulation of the immune response, (v) preparation of more specific and sensitive diagnostic procedures, and (vi) antitumour agents (immunotoxins) to tumour cells.

moving boundary electrophoresis A liquid–liquid electrophoresis, carried out in a U-tube (Tiselius tube, observation cell) fitted with electrodes at the two ends. A buffered solution containing macromolecules is introduced into the tube, and the same buffer but without the macromolecules is introduced into the two arms of the U-cell such that the sample solution is under a layer of pure buffer. The pH of the buffer is so adjusted that the net charge of all the molecules in the sample becomes negative, and when a constant electrical field is created in a vibration-free condition, all the molecules migrate towards the anode forming a moving boundary or front in the pure buffer. The rates of movement of the different molecules differ according to their respective electrophoretic mobility property. The observation cell would contain homogeneous boundaries containing specific macromolecules. The refractive indices of the solutions within each boundary will vary according to the physical properties of the molecules they contain and they are measured with appropriate optical system (e.g. Schleiren optics; Rayleigh interference optics). They are used in separating components of biological fluids (e.g. serum proteins) and also used in water purification.

native gel electrophoresis *See* nondissociating buffer system.

near-field scanning optical microscope (NSOM) Scanning microscope which uses a very small light source (quasipoint light source with a diameter less than the wavelength of light) close to the sample for scanning. Detection and digital interpretation of this light energy forms an image. The resolution is below that of any light microscope. The probe is very close to the surface, the gap between the tip and the surface being less than the wavelength of the light. This region is referred to as the "near-field'. Generally, laser light is used for illumination. It is fed to the aperture through an optical fibre. The aperture is a tapered fibre coated with aluminium, a microfabricated hollow AFM probe, or a tapered pipette.

negative films Photographic films used in macro-and microphotography. They are designed to produce negative images of the subject when exposed. The negative images are then printed with required enlargement on a print paper to get the positive image of the subject. They bear brand names such as Kodacolor, Fujicolor and Agfacolor. Slow speed films with fine grains are used in biological investigations, especially microphotography. They are also known as print films.

nitrification   A process in which nitrogen in plant and animal wastes and dead remains is oxidized, first to nitrites and then to nitrates.

nitrogen fixation   The conversion of atmospheric nitrogen into oxidized forms that can be assimilated by plants. Biological nitrogen fixation is catalysed by the enzyme nitrogenase, which is found only in prokaryotes; some blue-green algae and some genera of bacteria (e.g. *Rhizobium* spp.; *Azotobacter* spp.) can biochemically fix nitrogen; such bacteria are symbionts of plants growing in nitrogen-poor soils.

nondissociating buffer system   A type of buffer system used in PAGE. It is used especially for enzyme electrophoresis and is also referred to as native gel electrophoresis. It keeps intact the secondary and tertiary structure of the proteins in the sample. The gel contains only the native polyacrylamide without addition of any surfactants, and therefore the proteins in the sample carry their native charge at the pH of the electrophoresis buffer, and separate according to their respective electrophoretic mobility, and the sieving effect of the gel.

northern blotting   Transfer of electrophoresed RNA molecules to a cellulose or nylon membrane. The RNA molecules are attached by capillary action. The transferred RNA is hybridized to single-stranded DNA probes (DNA–RNA hybridization). The technique is used to measure expression (transcription) of a gene for which a specific cDNA is available for use as a probe. The transferred RNAs are often referred to as northern blot; the term "northern" is a jargon of Southern blotting.

northern hybridization   Hybridization of a labelled DNA probe to RNA fragments that have been transferred from an agarose gel to a nitrocellulose filter.

nuclear transfer   A technique to create animals by cloning a single diploid somatic cell. It involves isolating a single diploid cell from a culture of cells, and inserting it into an enucleated ovum (an ovum from which the haploid nucleus has been removed). The resultant diploid ovum when implanted into a female develops into an embryo and gives birth to a cloned animal. It is to be noted that the term "nuclear transfer" is a misnomer, since it is a whole cell that is transferred, not just the nucleus.

null-hypothesis   In statistical tests of significance, a hypothesis of no difference, hence the name null-hypothesis (H_o); it may state, for example, that the difference between observed sample mean and the population mean is zero; $H_o: \overline{X} - \mu = 0$

numerical aperture (NA)   An index of light-gathering capacity of a system of lenses (e.g. objective of a microscope) and measured in terms of the angle subtended by the optical axis and the outermost ray entering into the objective. The value is not expressed in degrees but as a sine value, hence the term "numerical" aperture. NA is given by the value $n \sin \theta$, where theta (θ) is ½ the angle of the cone of light entering the system.

ocular micrometer (OM) A glass disk with a diameter of 1 cm, engraved with an arbitrary scale of 100 divisions or less. It is also referred to as reticle, reticule or graticule. It is fitted into the eyepiece, hence called ocular micrometer. It is calibrated for a given microscope with specific objective and eyepiece, using a stage micrometer.

Okazaki fragment DNA is replicate in only one direction, i.e., nucleotides can be added only at the 3´ end. Therefore, only one of the two strands of a double helix is replicated continuously. The other strand is replicated in small segments called Okazaki fragments, which are subsequently joined together by DNA ligase.

oncogenes Genes that cause cancer. They are mutant forms of normal functional genes (proto-oncogenes) that play a role in normal cell proliferation. They are found in tumours and in retroviruses; once incorporated as part of the retroviral genome, they are activated at inappropriate times and places by retroviral promoters, thereby becoming oncogenes.

oncogenesis The progression of cytological, genetic and cellular changes that culminate in a cancerous growth.

open reading frame (ORF) A series of codons (base triplets) coding for amino acids without any termination codons. It starts with a start triplet (ATG), followed by a string of triplets each of which encodes an amino acid, and ends with a stop triplet (TAA, TAG or TGA). The term is used when, after the sequence of a DNA fragment has been determined, the function of the encoded protein is not known. The existence of open reading frames is usually inferred from the DNA (rather than the RNA) sequence.

optimum illumination The precise control of the light path before it reaches the specimen. It is essential for any sophisticated microscopic work such as photomicrography; given by the Köhler illumination (after August Köhler). It requires a low-voltage (6–10 volts) high intensity bulb with a small concentrated filament, fitted into a built-in illuminator or into an external lamp condenser, and an iris diaphragm in front of the lamp. It provides optimum and appropriate illumination of the field of observation.

outbreeding A mating system characterized by the breeding of genetically unrelated or dissimilar individuals. This process enhances the genetic diversity and increases the vigour or fitness of individuals. It is used to counter the detrimental effects of continuous inbreeding.

Ouchterloney technique *See* double diffusion agar assay.

packed bed column A type of stationary phase in a column chromatography. The granular material is packed tightly into the column as a homogeneous bed. The column is thus a cylinder with a solid core.

PAGE (Polyacrylamide gel electrophoresis) The most versatile of electrophoresis techniques used to separate proteins and nucleic acids, specifically in processes like protein characterization and purification, DNA sequencing, RNase protection assays and DNA shift assays. It uses a synthetic gel made by polymerization of the monomers of acrylamide (forming long chains) and bisacrylamide (N,N'-methylene bisacrylamide, forming cross links). The main advantage is the control of matrix pore size. The pore size increases with the increase in the percent composition of the acrylamide. It can be performed as a rod (tube) or slab, which may be vertical or horizontal.

palindrome sequence Literally palindrome refers to a word, sentence or verse that reads the same from left to right and as it does from right to left (e.g. Radar). In molecular biology, it refers to a segment of DNA in which the base-pair sequence reads the same in both directions from a central point of symmetry, that is, the sequence is the same when one strand is read left to right and the other strand is read right to left. Recognition sites for type II restriction endonucleases are palindromes.

paper electrophoresis A zone electrophoresis, one which uses a supporting medium, the paper, more specifically high quality filter paper (e.g. Whatman) or special electrophoresis paper, cut into desired dimension (e.g. 5-mm strips). It runs either horizontally or vertically. The apparatus includes a glass, plastic or Perspex tank with two buffer reservoirs separated by a platform to support the paper and electrodes, fixed to the reservoirs, and connected to a power pack.

parasexual hybridization *See* somatic cell hybridization.

particle radiation Gamma (γ) particles (positively charged) and beta (β) particles (negatively charged), electrons, protons and neutrons. Used to produce mutant cells or organisms.

partition chromatography Chromatographic technique in which separation of sample molecules is based on differences between the solubility of

these molecules in the stationary phase (gas chromatography), or on differences between the solubilities of the components in the mobile and stationary phases (liquid chromatography).

passive immunity Natural acquisition of antibodies by the foetus or neonate (newborn) from the mother; artificial introduction of specific antibodies by the injection of serum from an immune animal.

patent A right granted by the Government to the creator of an invention, conferring the exclusive right to make, use and sell that invention for a limited period of time, usually 20 years.

pathogen An organism that causes a disease in another organism.

pathotoxin Dilute substance synthesized and released by some pathogen and interacts with the host metabolism.

peer-reviewed journal One in which the manuscript submitted for publication is critically evaluated by peers, i.e., experienced scientists in the same field. If, in the opinion of the peers, the manuscript is not worthy of publication, it may be rejected.

pesticide A toxic substance that kills harmful organisms (e.g. insecticides, fungicides, weedicides, rodenticides).

pH Derived from the French word *pouvoir hydrogène* meaning "hydrogen power". It is a measure of acidity and alkalinity and is equal to the log of the reciprocal of the hydrogen ion concentration of a solution and expressed in grams per litre. The pH scale runs from 1 to 14. A reading of 7 is neutral (e.g. distilled water), whereas below 7 is acidic and above 7 is alkaline.

phase-contrast microscope A design of light microscope useful in observing live, transparent, unstained micro-organisms and cells. It uses a special condenser with annular stop and a special objective with phase analyser. It converts the different degrees of phase retardation of light rays passing through a translucent material into amplitude differences to create an image with increased contrast.

photomicrography See micro-photography.

photomultiplier (PM), photo-multiplier tube (PMT) Extremely sensitive detectors of light in the ultraviolet, visible and near infrared range. It multiplies the signal produced from the incident light from which single photons are detectable. It is a device for quantification of radiation (light) of very low intensity, used in spectrophotometers. A PMT consists of a photocathode and a series of dynodes in an evacuated glass tube. Photons that strike the photoemissive cathode emit electrons due to the photoelectric effect. Instead of being collected at an anode, these few electrons are accelerated towards a series of additional electrodes called dynodes, each of which is maintained at a more positive potential. Additional electrons are generated at each dynode. The cascading effect creates 10^5 to 10^7 electrons for each photon hitting the first cathode depending on the number of dynodes and the accelerating voltage. This amplified signal is finally collected at the anode where it can be measured.

photons Discrete particles that carry the energy of light waves; also known as quanta.

photoperiod The length of day or period of daily illumination provided or required by plants or animals for reaching the reproductive stage.

photoperiodism The response of an organism to changes in day length.

phototube A common detector used in spectrophotometers. It consists of two conducting electrodes, anode and cathode, that are enclosed in an evacuated glass tube. A source of potential difference (power supply) creates an electric field in the direction from anode to cathode. Light falling on the surface of the cathode causes a current (flow of emitted photoelectrons as a result of photoelectric effect) in the external circuit; the current is measured by a galvanometer; since the photocurrent is negligible, it is amplified by introducing high resistance in the circuit; the photocurrent varies with voltage and with the frequency and intensity of the light.

phylogeny An evolutionary history of populations of related organisms.

plagiarism It may refer to (i) reproducing materials from other workers' reports, verbatim, as one's own; (ii) paraphrasing materials from another report without giving due credit to the author of the original report, i.e., without citing the source or without obtaining permission.

planar chromatography Chromatography in which the geometry of the stationary phase is a thin two-dimensional matrix. For example, in thin layer chromatography the stationary phase is a thin film of solid particles bound together with a binder and coated on a glass plate or plastic or metal sheet.

plant cell immobilization Entrapment of plant cells in gel matrices. The cells are suspended in small drops of the material, which then is set or allowed to harden to make little carriers. Materials such as alginates, agar or polyacrylamide are used.

plant genetic resources (PGR) As defined in the International Undertaking on Plant Genetic Resources of FAO includes PGR refers to the reproductive or vegetative propagating material of the following categories of plants: (i) cultivated varieties (cultivars) in current use and newly developed varieties; (ii) obsolete cultivars; (iii) primitive cultivars (landraces); (iv) wild and weed species, near relatives of cultivated varieties; and (v) special genetic stocks (including elite and current breeder's lines and mutants).

plasmid An extrachromosomal, autonomous circular DNA molecule found in certain bacteria, capable of autonomous replication. It can transfer genes between bacteria and is an important tool of transformation in genetic engineering. It exists in an autonomous state and is transferred independently of chromosomes.

plates concept A useful concept in column chromatography based on the assumption that the column is separated into a number of segments that are identical and that one equilibrium distribution of a solute between the solid and mobile phases takes place in each segment. Each of these segments in a column is referred to as theoretical plate. The number of equilibrium distributions that takes place in a column is thus proportional to the number of plates in the column.

plus strand RNA Most RNA viruses have single-stranded RNA (ssRNA) as genetic material. The base sequence of the RNA may be identical to that of the viral mRNA, in which case the RNA strand is called the plus strand or positive strand (+ssRNA). On the other hand, if the viral genome RNA is complementary to the viral mRNA, it is called minus or negative strand. For example, polio virus and tobacco mosaic virus have positive strand, while rabies, mumps, measles and influenza viruses have negative RNA strand.

polarization microscope A microscope which uses plane-polarized (one that vibrates in only one plane, in contrast to the ordinary white light, which may vibrate in any plane with equal probability) light for illumination. Two polarizing filters, one placed between the light source and the specimen stage, polarizes the light and the second one, the analyser, located above the objective, are aligned so that a specimen appears brightest. If the two filters are not aligned, it will appear totally dark. In between these two extremes of maximal brightness and total darkness, there are infinite shades of brightness. Polarization microscopes are used to study minerals in thin sections of rocks. In biology it is used to study the molecular organization of crystalline or highly organized biological molecules such as DNA, starch, wood, urea, etc.

polaroid films Photographic films, black-and-white and colour, designed to produce instant prints of images of the subject and used in special Polaroid cameras. These films use silver halide emulsions and the processing chemicals that are combined in a self-contained paper envelope or within the print paper itself.

polyacrylamide gel electrophoresis *See* PAGE.

polyacrylamide gels Gels made by cross-linking acrylamide with *N,N'*-methylene-*bis*-acrylamide. They are used for the electrophoretic separation of proteins, DNA and RNA molecules. Polyacrylamide beads are also used as molecular sieves in gel chromatography.

polymerase An enzyme which catalyses the synthesis of polymeric molecules from monomers. A DNA polymerase synthesizes DNA from deoxynucleoside triphosphate using a complementary DNA strand and a primer. An RNA polymerase synthesizes RNA from monoribonucleoside triphosphate and a complementary DNA strand.

polymerase chain reaction (PCR) A method for amplification of a particular DNA sequence and involves multiple cycles of denaturation, annealing to oligonucleotide primers, and extension (polynucleotide synthesis), using a thermostable DNA polymerase, deoxyribonucleotides, and primer sequences in multiple cycles of denaturation–renaturation–DNA synthesis. It is carried out in a thermocycler.

polymerization Chemical union of two or more molecules of the same kind such as glucose or nucleotides to form a macromolecule (starch or nucleic acid) having the same elements in the same proportions but a higher molecular weight and different physical properties.

positive films Photographic films that produce positive images of the subject to which the film is exposed. The images can be viewed with the help of slide projector. They are useful in biological research for presentation of the results of an investigation to an audience; the brand names of these films end with a suffix chrome, such as Kodachrome, Ektachrome, Fujichrome and Agfachrome. They are also known as transparency films, slide films and reversal films.

primary culture A culture started from cells, tissues or organs taken directly from organisms. It is regarded as such until it is sub-cultured for the first time, after which it becomes a cell line.

primary immune response The immune response that occurs during the first encounter with a given antigen.

primary literature Research articles, which report the results of original research, published in peer-reviewed scientific journals.

primosome A protein-replication complex that catalyses the initiation of synthesis of Okazaki fragments during discontinuous replication of DNA. It involves DNA primase and DNA helicase activities.

prion Proteinaceous infectious particle; an abnormal form of a normal cell protein, with no detectable nucleic acid, found in the brain of mammals and considered to be the agent responsible for the class of diseases called spongiform encephalopathies, including scrapie in sheep and bovine spongiform encephalopathy (BSE; mad cow disease) in cattle.

prism An object made up of simple glass and used in the monochromator of a spectrophotometer for dispersion of light in the visible range. Fused silica or quartz or fluorite prism is used for dispersing radiation in the UV region. Ionic crystalline materials such as NaCl, KBr and CsBr are used for the dispersion of IR region.

probe It may be (i) the agent that is used to detect the presence of a molecule in a sample, as in a diagnostic test; (ii) a DNA or RNA sequence labelled or marked with a radioactive isotope or that is used to detect the presence of a complementary sequence by hybridization with a nucleic acid sample.

probe DNA A labelled DNA molecule used to detect complementary-sequence nucleic acid molecules by molecular hybridization. Auto-radiography or fluorescence is used to localize the probe DNA sequence and reveal the complementary hybridization sequence.

productivity The total amount of product that is produced within a given period of time from a specified quantity of resource.

promoter It may refer to (i) A nucleotide sequence of DNA to which RNA polymerase binds and initiates transcription; usually lies upstream of a coding sequence. A promoter sequence aligns the RNA polymerase so that transcription will initiate at a specific site. (ii) A chemical substance that enhances the transformation of benign cells into cancerous cells.

proofreading In molecular biology, the enzymatic scanning of newly-synthesized DNA for structural defects such as mis-matched base pairs. It is effected by DNA polymerase.

protein engineering Producing proteins with modified structures that confer properties such as higher catalytic specificity or thermal stability.

protein folding problem Proteins fold on a time scale. Starting from a random coil conformation, proteins find their stable fold quickly although the number of possible conformations is very high. The protein folding problem is to predict the folding and the final structure of a protein solely from its sequence.

protein sequencing The method of determining the amino-acid sequence of a protein or its component polypeptides.

protein structure prediction problem The combinatorial problem to calculate the three-dimensional structure of a protein from its sequence alone; one of the biggest challenges in structural bioinformatics.

proteome The protein complement expressed by a genome; while the genome is static, the proteome continually changes in response to external and internal events.

proteomics Study quantifying the expression levels of the complete protein complement (the proteome) in a cell at any given time. Initially focused on two-dimensional gel electrophoresis for protein separation and identification; now refers to any method that characterizes the function of large sets of proteins; used as a synonym for functional genomics.

protocol The step-by-step experiments proposed to describe or solve a scientific problem, or the defined steps of a specific procedure.

protoplast A bacterial or plant cell, the relatively rigid wall of which has been removed either chemically or enzymatically, leaving its cytoplasm enveloped by only a delicate peripheral membrane; spherical and smaller than the elongate, angular shaped and vacuolated cells from which they have been released.

protoplast culture   The isolation and culture of plant protoplasts by mechanical means or by enzymatic digestion of plant tissues or organs, or cultures derived from these; used for selection or hybridization at the cellular level.

protoplast fusion   The coalescence of the plasmalemma and cytoplasm of two or more protoplasts in contact with one another. The initial adhesion of two protoplasts is a random process but coalescence can be induced.

provirus   A retrovirus in which the single RNA strand has been converted into a double-stranded DNA, which has been integrated into a host genome.

pulsed-field gel electrophoresis (PFGE)   A method for separating very large DNA molecules by alternating the direction of electric current in a pulsed manner across a semisolid gel.

pure line   A strain in which all members have descended by self-fertilization or close inbreeding; genetically uniform.

pyrogen   Bacterial substance that causes fever in mammals.

quarantine   Keeping a person or living organism in isolation for a period (originally 40 days) after arrival at a country to allow disease symptoms to appear, if there was any disease present; regulations restricting the sale or shipment of living organisms, usually to prevent disease or pest invasion of an area.

race   A distinct group of organisms of a particular species, that are geographically, ecologically, physiologically, physically and/or chromosomically different from other members of the species.

radioactive isotope, radioisotope   An unstable isotope that emits ionizing radiation.

radioimmunoassay (RIA)   A very sensitive assay in which a purified radioisotope-labelled antigen and the same antigen in the sample fluid are allowed to compete for the binding sites of the specific, purified antibody against the antigen; in the process of competing, the quantity of each of the two antigens (radiolabelled and the unlabelled) that binds is proportional to its concentration. After binding has reached equilibrium, the antibody-antigen complex is separated and the quantity of radioactive antigen bound with antibody is measured by radioactive counting techniques. If a large amount of radioactive antigen has bound, it would indicate that there was only a small amount of natural antigen in the sample to compete with the radioactive antigen, and therefore the concentration of the antigen in the assayed sample is small. Conversely, if only a small amount of radioactive antigen has bound, it would indicate that there was a very large amount of natural antigen to compete for the binding sites. In order to make the assay quantitative, the radioimmunoassay procedure is performed also for "standard" solutions of the unlabelled antigen at different concentration levels, and a "standard curve" is plotted for comparing the radioactive counts recorded from the sample assay. The radioimmunoassay is so sensitive that as little as one trillionth of a gram of the antigen can be assayed.

randomization The process of ensuring that every experimental unit gets equal chance of being assigned to a treatment, in all aspects including space and time. Randomization begins from the collection of the sample and continues to the end of the experiment.

randomized block design (RBD) An experimental design involving ANOVA; also referred to as randomized complete block design (RCBD), used mainly to reduce the "error" due to natural variations. It involves the arrangement of the experiment in the form of blocks and the random assignment of the experimental units to each block.

rate-zonal centrifugation A centrifugation process effectively used for the separation of subcellular components that have similar buoyant density but different shapes or particle sizes (e.g. ribosomal subunits, different classes of polysomes, and various forms of DNA molecules). It is performed in either swinging-bucket rotors or zonal rotors. A shallow linear gradient is used to prevent convection in the centrifuge tubes or the chambers during centrifugation.

reading frame A series of triplets beginning from a specific nucleotide. Each triplet is represented by a single amino acid in the protein synthesized. The reading frame defines which sets of three nucleotides are read as triplets in the DNA, and hence as codons in the corresponding mRNA; this is determined by the initiation codon, AUG. For example, the sequence AUGGCAAAAUUUCCC would read as AUG/GCA/AAA/UUU/CCC/ and not as A/UGC/CAA/AAU/UUC/CC.

real-time PCR Device in which collection and analysis of data take place as the polymerase chain reaction proceeds in the thermocycler. All the reagents necessary for the entire reaction are added before amplification. There is no need for sample transfers, addition of reagents or electrophoresis and there is no risk of product contamination because the samples are never exposed during or in between steps. Detection, quantification and analysis of sample is rapid, about 20 minutes. It works on the principle that the fluorescence of the positive sample increases with the cycle number while that of the negative sample remains at base line. The variations in the intensity of the fluorescence is interpreted by a computer.

receptor-binding screening A method in biotechnology for discovering conventional drugs. It relies on the fact that many drugs act by binding to specific receptor proteins on or in cells. These proteins usually bind to hormones or to other cells, and control the cell's behaviour, although they may be enzymes or structural elements of the cell. The drug interferes with the normal function of the protein.

recombinant Refers to (i) an organism or cell that is the result of recombination (crossing-over); (ii) a molecule containing DNA from different sources. The word is used as an adjective, as in "recombinant DNA".

recombinant DNA technology A set of methods to manipulate DNA. One technique is DNA cloning, which can produce an unlimited number of copies of a particular DNA segment, and the result is sometimes called a DNA clone or gene clone (if the segment is a gene), or simply a clone. Any organism, animal or plant, manipulated by recombinant DNA technology is called a genetically modified organism (GMO). The recombinant DNA technology involves: (i) identifying desired genes; (ii) isolating and cloning genes; (iii) studying the expression of cloned genes; (iv) producing large quantities of the gene product.

refractive index (RI)   The ratio of the speed of light in vacuum (c) to the speed of light in a given medium (v). Thus,

$$RI = n = \frac{c}{v}.$$

The light travels more slowly in any medium than in vacuum. Therefore, the RI of any medium is always more than 1. The RI of a material and the speed of light in that material are inversely proportional. The greater the RI of a material, the slower the speed of light in that material. The refractive index of water is 1.33, and that of glass, about 1.52–1.80.

regulatory gene   A gene whose protein controls the activity of other genes or metabolic pathways.

relative centrifugal force (RCF) The centrifugal force F divided by 980 (earth's gravitational force); RCF means number of times g units ($\times g$). Thus,

$$RCF = \frac{\omega^2 r}{980}\, g \text{ units}$$

$= (1.119 \times 10^{-5}) \times (\text{rpm})^2 \times (r)\, g$ units. Since all factors other than rpm and r are constant, RCF is dependent on rpm and r, and since the radius of rotation r is also constant for a given centrifuge, rpm (rotations or revolutions/minute) is the only factor which determines the RCF.

relaxed plasmid   A plasmid that replicates independently of the main bacterial chromosome and is present in 10–500 copies per cell.

renaturation   It may refer to (i) the re-association of two nucleic acid strands after denaturation; (ii) the restoration of a molecule to its native form; (iii) the formation of a double-stranded helix from complementary single-stranded molecules.

repetitive DNA   DNA sequences which are present in a genome in multiple copies, sometimes more than a million times.

replacement therapy   The administration of metabolites, co-factors or hormones, which are lacking in the body as the result of a genetic disease.

replication   The experimental error that can be accounted for only if the treatment is replicated, i.e., the same treatment is administered to several different sampling units. The number of replicates, i.e., the size of the sample, depends on the expected variance of the data, before and after treatment; it also depends on the accuracy desired in the final prediction, higher the accuracy, larger is the size of the sample; if the number of replicates is very small the inferences will be inconclusive and invalid; on the other hand, too many replicates may yield diminishing returns.

reporter gene   A gene that encodes a product that can be easily assayed. It is used to determinate whether a particular DNA construct has been successfully introduced into a cell, organ or tissue.

repression   Inhibition of transcription by preventing RNA polymerase from binding to the transcription initiation site: a repressed gene is "turned off".

repressor   A protein which binds to a specific DNA sequence (the operator) upstream from the transcription initiation site of a gene or operon and prevents RNA polymerase from commencing mRNA synthesis. The C_1 protein of bacteriophage and the *lac*1 protein of the *lac* operon are examples of repressors.

research   A method to attain the goal of science, i.e., to expand the frontiers of knowledge, and involves the process of organized reflective thinking by the researcher. It may be carried out in any field of arts and science, wherever there is need and scope to expand the horizon of knowledge.

resolving power   The resolving power of the human eye is its ability to see two closely placed objects as two distinct objects, i.e., the minimum distance between the two objects so that they can be seen as separate objects. On the basis of the diameter of the cones in the retina and the distance between the eye lens and retina, human eye can resolve two objects separated by a distance of at least 20 μm. But practically, due to factors like illumination, contrast and object size, the resolving power of human eye is about 100 μm. The resolving power of a microscope is its ability to differentiate two objects (e.g. two adjacent chromatin granules, flagella, etc). It is determined by the optical system used in it and the wavelength of the light used to illuminate the object. The resolving power of a microscope is expressed as:

$$d = \frac{0.5\lambda}{NA}$$

where,

d = the minimum distance between two objects that reveals them as separate units or resolving power

0.5 = a constant which depends upon contrast

λ = wavelength of illumination

NA = the numerical aperture = $n \sin \theta$

The above relationship is also referred to as Abbè equation, after Ernst Abbè, the German Physicist.

restriction endonuclease   A class of endonucleases that cut DNA after recognizing a specific sequence, e.g. *Bam*H1 (5×GGATCC3×), *Eco* RI (5′GAATTC3′), and *Hind* III (5′AAGCTT3′). There are three types: Type I cleaves nonspecifically a distance greater than 1000 bp from its recognition sequence and contains both restriction and methylation activities. Type II cuts at or near a short, and often palindromic, recognition sequence. A separate enzyme methylates the same recognition sequence. They may make the cuts in the two DNA strands exactly opposite one another and generate blunt ends, or they may make staggered cuts to generate sticky ends. The type II restriction enzymes are the ones commonly exploited in recombinant DNA technology. Type III cuts 24–26 bp downstream from a short, asymmetrical recognition sequence. It requires ATP and contains both restriction and methylation activities.

restriction exonuclease   A class of nucleases that degrade DNA or RNA, starting from an end either 5′ or 3′.

restriction fragment   A fragment of DNA produced by cleaving (digesting, cutting) a DNA molecule with one or more restriction endonucleases.

restriction fragment length polymorphism (RFLP) The variation in the length of DNA fragments that are produced after cleavage with a type II restriction endonuclease. The differences in DNA lengths are due to the presence or absence of recognition site(s) for that particular restriction enzyme. It was initially detected using hybridization with DNA probes after separation of digested genomic DNA by gel electrophoresis (Southern analysis). It can also be detected by electrophoresis of digested PCR product.

restriction map The linear array of restriction endonuclease sites on a DNA molecule.

restriction nuclease A bacterial enzyme that cuts DNA at a specific site.

restriction site The specific nucleotide sequence in DNA that is recognized by a type II restriction endonuclease and within which it makes a double-stranded cut. Restriction sites usually comprise four or six base pairs that typically are palindromic; e.g. 5´GGCC3´ – 3´CCGG5´. The two strands may be cut either opposite to one another, to create blunt ends, or in a staggered manner, giving sticky ends, depending on the enzyme involved.

retardation factor value, retention factor, relative to front (*Rf*) A variable used to characterize the position of a substance in a chromatogram; an index of the degree of retention of a component. It is defined as the distance between the start line and the solvent front line divided by the distance between start line and solute front line. That is, the *Rf* of a solute (a component of sample mixture) is the ratio of distance migrated by solvent (eluent) to the distance migrated by the solute, in a chromatogram.

$$Rf = \frac{\text{Distance migrated by solvent}}{\text{Distance migrated by solute}}.$$

Rf values always lie between 0 and 1, depending on the specific chromatographic conditions, nature of the stationary phase, matrix of substances in the sample, etc. It is appropriate only to chromatograms that have been developed once. Good resolution results are obtained when *Rf* is about 0.4 to 0.8.

retro-poson/retro-transposon A transposable element that moves by reverse transcription (i.e., from DNA to RNA to DNA) but lacks the long terminal repeat sequences.

retroviral vectors Gene transfer systems based on viruses that have RNA as their genetic material.

retrovirus RNA viruses infecting eukaryotes. They are capable of forming double-stranded DNA copies of their genomes by using reverse transcription. The double-stranded forms integrate into chromosomes of an infected cell. Many naturally occurring cancers of vertebrates are caused by retroviruses. The AIDS virus is a retrovirus.

reverse phase chromatography A liquid column chromatography which employs a hydrophobic, low polarity stationary phase that is chemically bonded to an inert solid (e.g. silica). The separation of the sample components is by extraction process. It is used for the separation of non-volatile components of a sample.

reverse transcription The synthesis of DNA on a template of RNA, carried out by reverse transcriptase.

review of literature In research it refers to extensive, exhaustive, and systematic survey of publications relevant to the selected field of investigation. It refers to a thorough examination of the literature, i.e., the articles in science journals, review articles, books, monographs and other writings, which deal with a particular subject. The aim is to justify the selection of the research topic, methodology followed and discussion of the results and subsequent conclusions and generalizations made.

rho factor (ρ) A protein that causes RNA polymerase to detach from the terminator after it has stopped transcription.

sampling unit The experimental unit itself or a fraction of it. For example, in an experiment using a number of fish aquarium tanks as the experimental units, if the samples taken are water in each tank, the sampling units are these tanks; if the samples are fishes in these tanks, then the sampling units are the fishes.

SARS (Severe Acute Respiratory Syndrome) A disease caused by corona virus. The virus is about 120–150 nm large, roughly spherical, has membranous envelope, a helical nucleocapsid and a characteristic external layer or halo of spike like peplomers, which help to attach and enter into host cells, +ssRNA genetic material. It causes respiratory and enteric infections in birds and many mammals. The liver and nervous system may also be affected. SARS caused a worldwide outbreak in 2002–2003 infecting 8500 people and killing 800 in 27 countries, including 115 in USA. It started in China and spread to places like Hong-Kong, Singapore, Hanoi and Toronto.

satellite DNA That portion of the DNA in plant and animal cells consisting of highly repetitious sequences (millions of copies) typically in the range from 5 to 500 bases. Thousands of copies occur at each of the many sites. It can be isolated from the rest of the DNA by density gradient centrifugation.

satellite RNA A small, self-splicing RNA molecule that accompanies several plant viruses, including tobacco ringspot virus, also referred to as viroid.

scanning electron microscope (SEM) An electron microscope which creates a 3-D image from electrons emitted from the surface of the specimen. It is generally used to examine the surfaces of micro-organisms and those of eukaryotic cells.

scanning probe microscope Microscope which scans the surface of the sample with a probe-tip that is as small as a single atom. Because of the extremely fine nature of the scanning tip, it can achieve magnifications of 100 million times, which would enable one to view atoms on the surface of a solid sample.

scanning tunnelling microscope (STM) Microscope which works on the concept of quantum mechanics, viz., the tunnel effect of electrons; the electrons surrounding the surface atoms tunnel or project from the surface boundary a very short distance; a probe with a tip as thin as an atom is moved close to the sample surface until its electron cloud touches the electron cloud of the atoms of the surface. As the tip scans the surface contours, the variations in tunnelling current, which are proportional to the tip–surface distance, are recorded and analysed by a computer to produce an accurate three-dimensional image of surface atoms. The image is displayed on a computer monitor. It can also be plotted on a paper or photographed.

science Organized knowledge; involves observation, identification, description, experimental investigation, and theoretical explanation of any phenomenon that occurs in nature; the aim of science is to advance the frontiers of human knowledge, i.e., to add valid information to what is already known.

SDS (Sodium dodecyl sulphate) The most commonly used surfactant in protein electrophoresis. When treated, proteins in a sample become completely covered by negatively charged SDS ions, and unwind to assume an extended conformation. The number of SDS molecules covering the protein molecules is very large, as many as half the number of amino acid residues. Therefore, the intrinsic charge of the treated proteins becomes overpowered by the charge of the detergent. Even proteins that have widely different tertiary and quaternary structures, tend to have consistent and uniform charge-to-mass ratio. Electrophoretic separation of SDS-treated proteins is strictly on the basis of their molecular weights.

secondary immune response The rapid immune response that occurs during the second (and subsequent) encounters of the immune system of a mammal and many other vertebrates, with a specific antigen.

secondary literature Theses, project reports, proceedings of a conference, textbooks, unpublished data, etc. Citation of secondary literature may or may not be accepted by standard journals.

secondary messenger A chemical compound (e.g. cyclic AMP) within a cell that is responsible for initiating the response to a signal from a chemical messenger (e.g. a hormone) that cannot enter the target cell itself.

secondary metabolite A substance that is not necessary for growth or maintenance of cellular functions but is synthesized, generally, for the protection of a cell or microorganism, during the stationary phase of the growth cycle.

sedimentation Deposition of particles (molecules, cell organelles, microorganisms, etc.) suspended in a liquid medium, to the bottom of the container, according to their density and size, due to the force of gravity.

sedimentation coefficient The sedimentation rate of a particle expressed per unit of centrifugal field is called sedimentation coefficient i.e., the rate of descent per unit of centrifugal field per second. It is expressed in Svedberg unit (S) as follows.

$$S = \frac{\nu}{\omega^2 r}$$

where,

S = sedimentation coefficient
ν = sedimentation rate
ω = angular velocity and
r = radius of rotation.

sedimentation constant The sedimentation coefficient standardized for a medium of water at a temperature of 20°C. It is also referred to as standard sedimentation coefficient ($S_{20.w}$).

$$S_{20.w} = S_{obs} \times \frac{\eta_T}{\eta_{20}} \times \frac{\eta_C}{\eta_0} \times \frac{(1 - \bar{V}\rho_{20.w})}{(1 - \bar{V}\rho_T)}$$

where,

$S_{20.w}$ = standard sedimentation co-efficient or sedimentation constant,

S_{obs} = experimentally observed sedimentation coefficient,

η_T = viscosity of the water at temperature T,

η_{20} = viscosity of water at 20°C,

η_c = viscosity of the solvent at the given temperature,

η_o = viscosity of water at the given temperature,

$\overline{v}$ = partial specific volume of the solute,

$\rho_{20.w}$ = density of water at temperature 20°C,

ρ_T = density of water at the given temperature.

seed storage proteins   Proteins that are accumulated in large quantities in seeds not because of their enzymatic or structural properties but simply as a convenient source of amino acid for use when the seed germinates i.e., as a source of protein. Much of the world's food comes from plant seeds or fruits, and much of the protein in those seeds is storage protein; improvement of the nutritional content of these proteins could improve human diet. These proteins are also useful as expression systems. Storage proteins are produced in very large amounts relative to other proteins, and are stored in stable, compact bodies in the plant seed. Attempts are on to make the plants produce other proteins in similarly large amounts and in as convenient a form, by splicing the gene for a desired protein into the middle of a plant storage protein gene.

segmented genomes   Many of the viral RNA genomes are divided into separate parts, and hence called segmented genomes. Each segment is assumed to code for one protein.

selectable marker   A gene the expression of which allows the identification of: (i) A specific trait or gene in an organism; (ii) Cells that have been transformed or transfected with a vector containing the marker gene.

selection   It may refer to (i) differential survival and reproduction of phenotypes; (ii) a system for either isolating or identifying specific organisms in a mixed culture.

senescence   The final stage in the post-embryonic development of multicellular organisms, during which loss of functions and degradation of biological components occur; a physiological ageing process in which cells and tissues deteriorate and finally die.

sepsis   Destruction of tissue by patho-genic microorganisms or their toxins, especially through infection of a wound.

sequence hypothesis   Francis Crick's decisive concept that genetic information exists as a linear DNA code; DNA and protein sequence are collinear.

sequencing   The determination of the order of nucleotides in a DNA or RNA molecule, or that of amino acids in a polypeptide chain.

serology   The study of antigen–antibody reactions in serum. It is used to identify and distinguish between antigens, such as those specific to microorganisms or viruses. It is also used as an indicator technique to assay plants suspected of being virus-infected.

serotyping A serological procedure used to differentiate between strains (serovars or serotypes) of microorganisms that have differences in antigenic composition of structure or product.

sewage treatment Treatment, on biological basis, of the huge amounts of human and animal waste to break down the organic material and convert it into something that can be safely discharged into the environment, usually rivers or seas.

shear It may refer to (i) the sliding of one layer across another, with deformation and fracturing in the direction parallel to the movement; (ii) the forces that cells are subjected to in a bioreactor or a mechanical device used for cell breakage; (iii) the fragmentation of DNA molecules into smaller pieces. DNA, as a very long and fairly stiff molecule, is very susceptible to hydrodynamic shear forces. Forcing a DNA solution through a hypodermic needle will fragment it into small pieces. The size of the fragments obtained is inversely proportional to the diameter of the needle's bore. The actual sites at which the shear force breaks a DNA molecule are approximately random. It is useful for cloning and to create a complete gene library of an organism.

shuttle vector A plasmid which can replicate in two different host organisms because it has two different origins of replication. It is used to 'shuttle' genes from one to the other; for instance, the YEp, pJDB219, is a shuttle vector able to replicate in *E. coli* from its pMB9 origin and in *Saccharomyces cerevisiae* from its 2 μm-plasmid origin.

sigma factor The sub-unit of prokaryotic RNA polymerases which is responsible for the initiation of transcription at specific initiation sequences.

signal sequence A section of about 15 to 30 amino acids at the N terminus of a protein, that makes the protein to be secreted, i.e., pass through a cell membrane. The signal sequence is removed as the protein is secreted. It is also called signal peptide or leader peptide.

signal transduction The biochemical process which conducts the signal of a hormone or growth factor from the cell exterior, through the cell membrane, and into the cytoplasm. This involves a number of molecules, including receptors, ligands and messengers.

single radial immunodiffusion assay (RID) Mancini's technique for quantifying antigens. Agar mixed with monospecific antibody is poured on glass slides and set. Wells are cut in the gel and known amounts of standard antigen are added to the wells. The unknown test antigen is added to a separate well. The gel is incubated for about 24 hours to attain equilibrium. During this period the antigens in the wells diffuse out of the wells and form insoluble complexes in the form of precipitation rings around the wells. The size of the ring is proportional to the amount of antigen in the well, i.e., wider the ring, the greater the concentration. A standard graph is constructed using the square of the diameters of the standard rings along the X-axis and the corresponding known concentrations along the Y-axis, and this graph is used to read the concentration of the antigen in the sample.

silencer   A DNA sequence that helps to reduce or shut off the expression of a nearby gene.

single-beam spectrophotometer The simplest design of spectrophotometer, in which the absorbance of the blank is measured first and then the absorbance of the sample, using the same cuvette. Since a significant amount of time elapses between the first reading (blank or reference) and the second reading (sample), there can be problems with drift (variations in the intensity of the radiation).

single nucleotide polymorphism (SNP)   SNPs are single base pair positions in genomic DNA at which normal individuals in a given population show different sequence alternatives (alleles) with the least frequent allele having an abundance of 1% or greater. They occur once every 100 to 300 bases and are hence the most common genetic variations. SNP is pronounced as "snip".

single-cell line/cell strain   A culture initiated from a single cell, usually from suspension cultures of single cells or small aggregates plated on solidified medium. The latter may incorporate a selective agent, from which tolerant or resistant individual cell lines or cell clones can be selected.

single-cell protein (SCP)   Microbial biomass or protein extract used as food or feed additive. Protein extracted from pure microbial cell culture (monoculture) which can be used as protein supplement for humans (food grade) and animals (feed grade); e.g. *Pseudomonas* sp., *Saccharomycopsis lipolytica, Spirulina maxima*.

size exclusion chromatography   A column chromatography in which the column packing is a porous gel that can separate molecules on the basis of their sizes. Larger molecules elute first because they cannot penetrate the pores of the gel. The porosity of the gel can be adjusted to exclude all molecules above a certain size. It is a technique commonly used in protein separation and purification and is also known as gel permeation or filtration chromatography.

SLR camera (single-lens-reflex camera)   A type of camera in which there is only one lens for both viewing and framing the subject and for exposure setting. Light rays from the subject come through the lens and fall onto a mirror, which also acts as a shutter. The mirror reflects the light rays into the viewfinder through a five-sided prism. When clicked for exposure, the mirror is lifted out of the path between the lens and film, thus opening the way for the light rays from the subject to strike the film. During the exposure period, the viewfinder is blocked. The use of single lens for viewing as well as exposing eliminates parallax error. Because of its versatility, it is most suitable for biological investigations including close-up photography and microphotography and is available in a vast array of interchangeable lenses, macrolenses, adapters, etc.

small nuclear RNA (snRNA)   Short RNA transcripts of 100–300 bp that associate with proteins to form small nuclear ribonucleoprotein particles (snRNPs). Most snRNAs are components of the spliceosomes that excise introns from pre-mRNAs in RNA processing.

soil amelioration The development of poor soils, using bacteria or fungi. It includes breaking down organic matter, forming humus by solubilizing them and making minerals like phosphates, in the soil available to plants. It also includes nitrogen fixation and sometimes an element of bioremediation.

somatic cell embryogenesis Production of embryos either from somatic cells of explants (direct embryogenesis) or by induction on callus formed by explants (indirect embryogenesis). It is also referred to as asexual embryogenesis.

somatic cell gene therapy The delivery of a gene or genes to a tissue other than reproductive cells of an individual, with the objective of correcting a genetic defect.

somatic embryo/somatic embryoid Embryonic structure morphologically similar to a zygotic embryo but created from somatic cells. It can go through developmental processes under *in vitro* conditions.

somatic hybridization It may refer to (i) asexual fusion of protoplasts from somatic cells of genetically different parents; (ii) hybridization by induced fusion of cells (protoplasts) from two divergent genotypes for production of hybrids or cybrids which contain various mixtures of nuclear and/or cytoplasmic genomes. It is also referred to as parasexual hybridization.

Southern blotting Named after Ed Southern, it is a method for transferring denatured DNA molecules that have been separated electrophoretically, from a gel to a matrix (e.g. a nitrocellulose membrane) on which a hybridization assay can be performed. The cellulose or nylon membrane to which DNA fragments have been transferred by capillary action is called Southern blot.

Southern hybridization A method in which a cloned, labelled segment of DNA is hybridized to DNA restriction fragments on a Southern blot.

Southern transfer *See* Southern blotting.

spacing Refers to vertical and horizontal spacing of typed text in a report. Vertical spacing pertains to the number of typed lines in a page. The space between lines (single, one-and-a-half and double spacing) makes a manuscript appear either cluttered or uncluttered. The horizontal spacing determines the space between letters.

species A group of potentially interbreeding individuals that are reproductively isolated from other such groups having many characteristics in common.

sperm sexing The sorting of sperm into those carrying an X chromosome and those with a Y chromosome, to produce, via artificial insemination or *in vitro* fertilization, animals of a specified sex. It is achieved by inactivation of X-bearing or Y-bearing sperm by antibodies directed against sex-specific peptides on the surface of sperm cells, or fluorescence-activated cell sorting.

spherical aberration An aberration in the image of the specimen observed in a light microscope; caused by spherical lenses because of their inability to focus light waves passing through different regions, i.e., the thicker central and thinner peripheral regions, in the same plane. The result is the absence of a well-defined image plane, the image appearing blurred in the periphery when the central portion is focused and vise versa. It can be rectified in a microscope by the use of lens group which is a combination of positive and negative lens elements with different thickness, all cemented together into a single group. It can be reduced by preventing light from striking the periphery of a spherical lens by closing the diaphragms to the necessary extent. Modern microscopes have objectives that are highly corrected for spherical aberration, using special glass-grinding techniques, improved glass formulations and superior control of optical pathways.

spheroplast   A microbial or plant cell from which most of the cell wall has been removed, usually by enzyme treatment. It differs from protoplast in the sense that in a spheroplast, some of the wall remains, while in a protoplast the wall has been completely removed. The two words are often used as synonyms.

splicing   A process in the maturation of eukaryotic mRNA, which eliminates intervening intron sequences and covalently joins exon sequences of RNA. In recombinant DNA technology, it may refer to joining fragments of DNA together.

split-beam spectrophotometer   A design of the spectrophotometer in which a beam splitter is used to direct the radiation along the reference (blank) and the sample paths. Thus, the measurements of both the blank and the sample takes place at the same moment in time; the intensities of the transmitted radiation through the reference solution and the sample solution are measured by two different phototubes.

split genes   Structural genes in eukaryotes that are typically split by a number of non-coding regions called introns.

split-plot design   An experimental design involving ANOVA. It is useful for two factors, one of which may have to be applied to larger experimental units and the other factor, to smaller areas within the larger units. The design can be adopted for a variety of experimental situations such as effect of environmental conditions, disease trials, nutrient trials, etc. on different plants.

sporocide   A substance used to kill microorganisms and spores.

stage micrometer   A standard microscope slide having a scale of defined length, etched; the scale is 1 mm ($1000\,\mu$m) divided into 100 divisions, so that one stage micrometer division = $10\,\mu$m. It is used for calibrating an ocular micrometer for use in *micrometry*.

starch   A gel supporting medium of electrophoresis. It is obtained from potato, the hydrolysed starch is dissolved in a buffer, heated and cooled to set as a gel by intertwining of amylopectin chains. Gels prepared using 2% (w/v) starch have high porosity, while those prepared with higher concentration (10–15%) of starch have low porosity. However, the pore size of the starch gel cannot be controlled by varying the concentration of the starch.

stem cell   An undifferentiated, pluripotent, active somatic cell that undergoes division and gives rise to other stem cells or to cells that differentiate to form specialized cells.

stereomicroscope   A microscope in which twin images are transmitted to the observer's eyes from two slightly different angles, differing by about 10–12 degrees, to produce a true stereoscopic effect. In a simple design (Greenough design), stereomicroscopes have two separate compound optical trains. The common main objective (CMO) stereomicroscopic design has a single objective but two optical channels for binocular vision. Both designs have a pair of prisms located before each eyepiece, to invert the image and give the correct orientation of the specimen. Modern stereomicroscopes have zoom lens system or a rotating drum with telescopic optics that can be utilized to increase or decrease the overall magnification.

sterile It may refer to (i) a medium or object with no perceptible or viable microorganisms; (ii) incapability of fertilizing or being infertile.

sterile room Room for performing inoculations under aseptic conditions. Now replaced by laminar air-flow cabinets, in which filtered air is blown from the inside to the outside.

sterilization It may refer to (i) A process that kills and/or removes all classes of microorganisms and spores; (ii) Complete destruction or elimination of the pathogenic, reproductive or infective potential of a biological agent. It is the best method of eliminating the risks associated with biohazardous wastes. Autoclaving (use of saturated steam under pressure) and dry heat are the sterilization methods commonly used in research laboratories and hospitals.

stop codon/termination codon A codon for which there is no corresponding tRNA molecule to insert an amino acid into the polypeptide chain. Therefore protein synthesis is terminated and the completed polypeptide is released from the ribosome. Three stop codons are found: UAA (ochre), UAG (amber) and UGA (opal). Mutations which generate any of these three codons in a position which normally contains a codon specifying an amino acid are known as *nonsense mutations*. Stop codons can also be called nonsense codons.

stringent plasmid A plasmid that replicates only along with the main bacterial chromosome and is present as a single copy, or at most several copies, per cell.

structural bioinformatics The branch of bioinformatics concerned with computational approaches to predict and analyse the spatial structure of proteins and nucleic acids.

substage condenser The condenser of a light microscope, fitted just below the specimen stage, and functions to condense the scattering light rays from the source (sunlight or external lamp through mirror or built-in lamp) into a cone of bright light, align and focus onto a tiny spot of the specimen.

summary (and conclusions) Term used in addition to the term "abstract". In research articles published in journals and in other research documents such as theses there is not much difference in the meanings. The abstract of a thesis is usually called summary. It is usually given at the end, after the Discussion of the results. It is much more elaborate than an abstract in an article of a journal. It deals with more than one chapter of a thesis, hence has numbered paragraphs, sections, etc., running into more than one page. It gives the gist of the entire report including the materials used, methods adopted, observations made, the inferences drawn and the conclusions arrived at.

superbug   Name of the bacterial strain of *Pseudomonas* developed by combining hydrocarbon-degrading genes carried on different plasmids into one organism. This genetically engineered microorganism is neither "super" nor a "bug"; it is an example of how genetically modified microbial strains could be used in a novel way because it was the basis for the legal decision that declared that genetically engineered organisms were patentable.

supercoil   A DNA molecule that has extra twists as a result of overwinding (positive supercoils) or underwinding (negative supercoils).

supercoiled plasmid   The common *in vivo* form of plasmid in which the plasmid is coiled around histone-like proteins. When supporting proteins are removed during extraction from the bacterial cell, the plasmid molecule supercoils around itself *in vitro*.

supercritical fluid chromatography (SFC)   A column chromatography using supercritical carrier fluid (e.g. carbon dioxide mixed with a modifier) as mobile phase, which when compared to the usual mobile phase liquids, has higher solubility, density, and diffusivity.

supergene   A group of adjacent genes on a chromosome that tend to be inherited together and are sometimes, functionally related.

surfactants/solubilizers   A surface-active agent (e.g. a detergent) that affects, especially reduces, the surface tension of liquids containing it. It is used in electrophoresis as agents to promote solubility of the sample molecules and for separation of the subunits of a molecule. Many biomolecules have more than one unit and have extensive nonpolar interactions between the units. Nonionic (e.g. Tween-20 and Triton X-100), anionic and cationic detergents are commonly used.

Svedberg unit (S)   The unit of expressing the sedimentation coefficient. The greater a particle's Svedberg value, the faster it gets deposited during centrifugation.

synopsis   A concise outline survey of the contents of a thesis or a monograph or a review paper.

T cell receptor   An antigen-binding protein that is located on the surfaces of killer T cells and mediates the cellular immune response of mammals.

T cells/T lymphocytes   Lymphocytes that mature in the thymus gland. They are primarily responsible for the cellular immune response.

***Taq* polymerase**   A heat-stable DNA polymerase isolated from the thermophilic bacterium *Thermus aquaticus* and used in PCR.

targeted drug delivery   A technique of delivering a drug to the site in the body where it is needed, rather than allowing it to diffuse into many sites.

TATA box   A nucleotide sequence located at the 5× end of eukaryotic mRNA encoding genes about 25–30 bp upstream of a gene, which is the binding site of RNA polymerase and thus plays a critical role in the initiation of transcription. It is analogous to Prinbow box of prokaryotic promoters. It is also known as Hogness box or Hogness-Goldberg box.

T-cell-mediated immune response Cellular immune response; the synthesis of antigen-specific T cell receptors and the development of killer T cells in response to an interaction of immune cells with an immunogen.

temperature gradient gel electrophoresis (TGGE) A type of electrophoresis that studies the behaviour of molecules under different temperatures by having a temperature gradient in the gel. It exploits the temperature-dependent change of conformation for separating molecules. It starts with double-stranded DNA molecules, and at a specific temperature the DNAs begin to melt, resulting in fork-like structures. In this conformation, the migration is retarded compared to double-stranded fragments. Since the melting temperature is highly dependent on the base sequence, DNA fragments of the same size but having different sequence can be resolved. Thus, TGGE not only resolves molecules but also provides additonal information about melting behaviour and stability. It is a new and powerful technique for the separation of nucleic acids like DNA or RNA and for the analysis of proteins.

terminator A DNA sequence that terminates transcription. It may refer to (i) a sequence just downstream of the coding segment of a gene, which is recognized by RNA polymerase as a signal to stop synthesizing mRNA. (ii) A term indicating antisense DNA inserted in plants to make impossible the use of a second generation of seed by a farmer.

thermal radiation Emission of electromagnetic radiation due to thermal motion of the molecules of an object.

thermophile An organism that grows at a higher temperature than most other organisms. Organisms which can grow at above 50°C as against a wide range of bacteria, fungi and simple plants and animals can grow at temperatures up to 50°C. They are classified according to their optimal growth temperature, into simple thermophiles (50–65°C), thermophiles (65–85°C), and extreme thermophiles (>85°C). Thermophiles and extreme thermophiles are usually found growing in very hot places, such as hot springs and geysers, smoker vents on the sea floor, and domestic hot water pipes.

thermotherapy Procedure used for virus or mycoplasma eradication. Plants are exposed to elevated temperatures as a treatment.

threading A technique to match a target sequence on a library of known three-dimensional structures by "threading" the target sequence over the known coordinates. It attempts to predict the three-dimensional structure starting from a given protein sequence. It is successful when comparisons based on sequences or sequence profiles alone fail due to a too low sequence similarity.

Ti plasmid (Tumour-inducing plasmid) A giant plasmid of *Agrobacterium tumefaciens* which is responsible for the induction of tumours in infected plants. It is used as a vector to introduce foreign DNA into plant cells.

tonoplast The cytoplasmic membrane bordering the vacuole, with a role in regulating the pressure exerted by the cell sap.

toxicity Harmful effect of a substance, as expressed by altered morphology or physiology. It is meaningful only when the effect is observable, such as changes in the rate of cell growth, cell death, etc.

toxin A substance produced by an organism and poisonous to plants or animals.

tracer An added or injected compound that can be followed within a reaction or an organism; e.g. radioactive isotopes, dyes.

trade secret Any information that allows the person who holds the secret to make money because others do not know the information. It could be a formula, computer program, process, method, device, technique, customer list or other non-public information.

trademark (TM) A word, name, symbol, logo, slogan or other distinctive sign or a combination thereof identifying and distinguishing a product or services from other similar products. It is officially registered at the trademark office, and legally restricted to the use of the owner or manufacturer of the product. It gives protection to the owner of the mark by ensuring exclusive right to use it to identify goods or service. The owner of a trademark can authorize another person to use it in return for money. Trademarks for services such as banking and insurance are called service marks (SM). The abbreviations "TM" or "SM" are used as notations to inform third parties that the person claims trademark or service mark rights to the word, slogan and phrases; after the official grant of the trademark right, the symbol ® is used.

traditional knowledge (TK) The knowledge that has been continually developed, acquired, used, practised, transmitted and sustained by the communities and or individuals through generations and includes artistic expressions, agricultural methods, and knowledge about the medicinal properties of plants and animals.

transcript An RNA molecule that has been synthesized from a specific DNA template. In eukaryotes, the primary transcript produced by RNA polymerase must be processed or modified to become mature, functional mRNA, rRNA or tRNA.

transcription Mechanism by which RNA is synthesized along a DNA template; the enzyme RNA polymerase catalyses the formation of RNA from ribonucleoside triphosphates.

transduction The transfer of DNA sequences from one bacterium to another via lysogenic infection by a bacteriophage (transducing phage); genetic recombination in bacteria mediated by bacteriophage. Abortive transduction is the inability of a bacterial DNA that was injected into a bacterium by a phage, to replicate.

transfection The transfer of DNA to a eukaryotic cell.

transformation It may refer to (i) the uptake and establishment of DNA in a bacterium or yeast cell, in which the introduced DNA often changes the phenotype of the recipient organism; (ii) conversion of animal cells in tissue culture from controlled to uncontrolled cell growth, through infection by a tumour virus or transfection with an oncogene.

transgene A gene from one genome that has been transferred to the genome of another organism; gene that has been introduced into a multicellular organism.

transgenesis The process of introducing a foreign gene or genes into animal or plant cells, which leads to the transmission of the introduced gene (transgene) to successive generations.

transgenic An organism in which a foreign gene (a transgene) is incorporated into its genome. The transgene is present in both somatic and germ cells, is expressed in one or more tissues, and is inherited by the offspring in a Mendelian fashion.

transgenic animals Animals carrying foreign genetic material.

translation The process of protein synthesis in which the amino acid sequence is determined by mRNA, mediated by tRNA molecules, and carried out on ribosomes.

transmission electron microscope (TEM) A modern TEM consists of an electron source, electromagnets to function as "lenses", and a system that projects an image onto a fluorescent screen or photographic film. It forms an image from radiation that is transmitted through the specimen.

transposable genetic element A DNA element that can move from one location in the genome to another.

transposon A transposable or movable genetic element; a relatively small DNA segment that can move from one chromosomal position to another, hence also referred to as mobile genetic element; e.g. Tn5 is a bacterial transposon that carries the genes for resistance to the antibiotics neomycin and kanamycin and the genetic information for insertion and excision of the transposon.

turbidimetry, nephelometry A method for quantitative estimation of the number of bacteria or any other microorganism or any other particulate material in a liquid medium using spectrophotometers. When viable bacteria or other cellular or particulate matter is suspended in a liquid, the liquid becomes turbid. The turbidity of the liquid is proportional to the number of the bacteria or other material suspended. When monochromatic light passes through such a turbid medium, certain amount of light is absorbed by the liquid medium, some light is scattered by the suspended materials, and the remaining light is transmitted. The intensity of the transmitted light is proportional to the degree of turbidity, which in turn is proportional to the number of bacteria or other materials suspended. Since bacteria do not absorb radiation but only scatter it, the term optical density (OD) is more appropriate to describe the measurement of turbidity due to bacterial suspension. The observed OD is interpreted with reference to the OD of a known bacterial concentration. When the turbidity of a solution is measured in terms of the intensity of the transmitted light, it is called turbidimetry. If the measurement is in terms of the intensity of the scattered light, it is referred to as the nephelometry.

two-dimensional chromatography An adaptation of the paper chromatography in which the paper is developed in two dimensions, sequentially using two different solvents. In one-dimensional chromatography the sample is developed only once in one direction and using one solvent, with limited resolution of the sample molecules in the chromatogram. In order to increase the resolution, the same paper is developed after turning it 90° clockwise, and using a different solvent.

After a second run, the various sample components will be spread out at distinct spots across the sheet.

two-way classification design   An experimental design involving ANOVA. When two factors are involved in an experiment, the design is referred to as two-way classification, completely random design with fixed effects or factorial experiment. The effect of two or more factors operating on a third factor is investigated. The first two factors may be in two or more levels. Also referred to as factorial design.

ultraviolet light (UV)/ultraviolet radiation   The portion of the electromagnetic spectrum with wavelengths from about 100 to 400 nm, between ionizing radiation (X-rays) and visible light. It is absorbed by DNA and therefore highly mutagenic to unicellular organisms and to the epidermal cells of multicellular organisms. It is used in tissue culture for its mutagenic and bactericidal properties.

upstream processing   In molecular biology it refers to the stretch of DNA base pairs that lie in the 5´ direction from the site of initiation of transcription; the first transcribed base is designated $+1$ and the upstream nucleotides are marked with minus signs, e.g. -1, -10; also, to the 5´ side of a particular gene or sequence of nucleotides. In chemical engineering, biotechnology and similar technologies (e.g. fermentation technology), it refers to those phases of a manufacturing process that precede the bio-transformation step. It refers to the preparation of raw materials for a fermentation process.

useful magnification   A magnification of a microscope (magnification of objective × magnification of eyepiece) that increases the size of the smallest resolvable object sufficient to be clearly visible; in most of the light microscopes the largest useful magnification is 1000× when a 10× eyepiece is used and 1500× when 15× eyepiece is used along with 100× oil immersion objective.

vaccine   A preparation of dead or weakened pathogens, or of derived antigenic determinant that is used to induce synthesis of antibodies or immunity against the pathogen.

vector   It may refer to (i) an organism (e.g. an insect) that carries and transmits disease-causing organisms between hosts or between reservoir and host; (ii) a plasmid or phage that is used to deliver selected foreign DNA for cloning and in gene transfer.

vehicle   The host organism used for the replication or expression of a cloned gene or other sequence.

viral vaccines   Vaccines consisting of live viruses instead of dead ones or separated parts of viruses. Since the virus itself cannot be used, as that would simply cause the disease, the virus is genetically engineered so that it elicits the immune response to the viral pathogen without causing the disease itself. Two genetic engineering methods are used: (i) by making the disease causing virus harmless, but still able to replicate in cultured animal cells. This is similar to producing an 'attenuated' virus, i.e., one which has been grown in the laboratory until it loses its ability to cause disease; it is made sure that the attenuated virus has no chance of mutating back to a wild type pathogenic virus, by deleting whole genes or replacing key regions of genes with completely different genetic material; (ii) by cloning the gene for a protein from the pathogenic virus into another harmless virus, so that the result 'looks' like the pathogenic virus but does not cause disease.

virion An infectious virus particle; a plant pathogen that consists of a naked RNA molecule of approximately 250–350 nucleotides, the extensive base pairing of which results in an almost correct double helix. Virion may also refer to a virus particle existing freely outside a host cell.

viroid An infectious element similar to a virus but smaller, consisting only of a strand of nucleic acid without the protein coat characteristic of a virus.

virus An infectious particle composed of a protein capsule and a nucleic acid core (DNA or RNA). It is dependent on a host organism for replication. The DNA or a double-stranded DNA copy of an RNA virus genome is incorporated into the host chromosome during lysogenic infection or replicated during the cystic cycle.

visible light The part of the electromagnetic spectrum with wavelengths between 380 nm and 750 nm and visible to the human eye.

western blotting A process in which electrophoresed protein is transferred from the gel to a cellulose or nylon support membrane. A particular blot of a protein molecule can then be identified by probing with a radiolabelled antibody which binds only the specific protein to which the antibody was prepared. It is useful, for example, for measurement of levels of production of a specific protein in a

particular tissue or at particular developmental stage.

widow and orphan lines Single lines of a paragraph at the top (widow line) and bottom (orphan line) of a page; should be avoided while typing a report.

wobble hypothesis An explanation of how one tRNA may recognize more than one codon. The first two bases of the mRNA codon and anticodon pair properly, but the third base in the anticodon has some wobble that permits it to pair with more than one base.

working bibliography A wide range of works pertaining not only to the specific study but also those required to have a strong grasp of broad background information.

working distance The distance in a microscope between the objective lens and the specimen slide when a specimen on glass slide is focused under each of the objectives. Higher the magnification, shorter is the distance. Specifically it is the distance between the surface of the objective lens and the coverslip or the specimen itself, when the specimen is in sharp focus. Objectives with larger NA and higher resolving power have shorter working distance.

xenobiotic A chemical substance that is not produced by living organisms; a manufactured chemical compound.

xenotransplantation The transplantation of tissue from one species to another species, typically from non-human mammals to humans. It is important because of a worldwide shortage of human organs for humans requiring a new organ. The most popular non-human species involved in this technology is the pig.

X-ray crystallography   The deduction of crystal structure from analysis of the diffraction pattern of X-rays passing through a pure crystal of a substance.

yeast cloning vectors   The yeasts, and especially *Saccharomyces cerevisiae*, that are suitable for cloning and expressing DNA; since they are eukaryotes, they can splice out introns, the non-coding sequences in the middle of many eukaryotic genes.

Z-DNA (Zig-zag DNA)   A form of DNA duplex in which the double helix is wound in a left-handed, instead of a right-handed manner. It adopts the Z configuration when purines and pyrimidines alternate on a single strand, as in

    5´CGCGCGCG3´
    3´GCGCGCGC5'

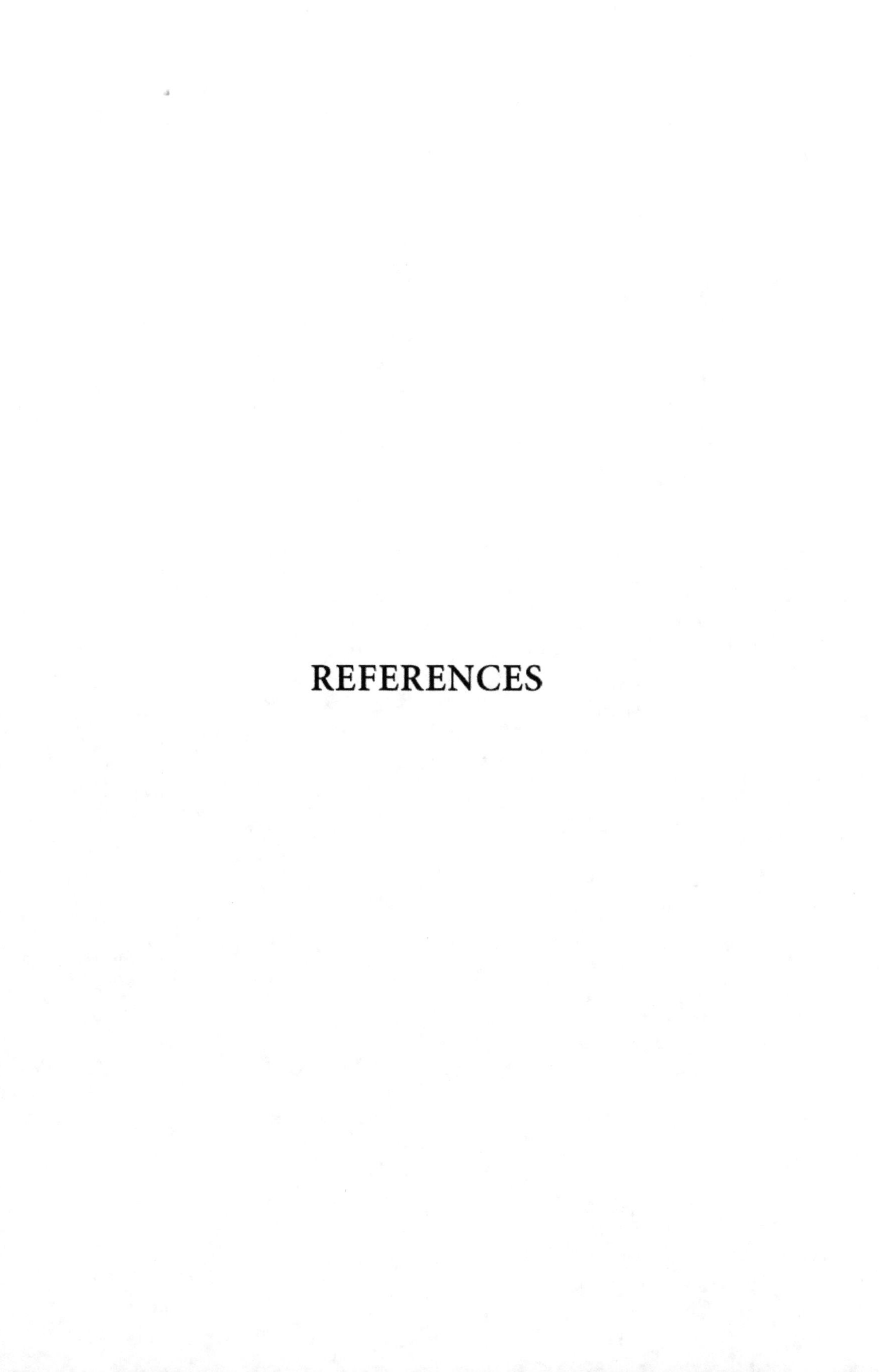

REFERENCES

Anderson, J., B. H. Durston, and M. Poole, *Thesis and Assignment Writing*. Wiley Eastern Private Limited, New Delhi, 1970.

Atwood, T. K. and D. J. Perry-Smith, *Introduction to Bioinformatics*. Pearson Education (Singapore) Pte Ltd., Indian Branch, Delhi, 1999.

Bidwai, P., *The Case for Animals in Research*. Frontline, 19 (23): 9. 2002.

Casey, E. J., *Biophysics: Concepts and Mechanisms*. Van Nostrand Reinhold Co., NY, and Affiliated East-West Press Pvt. Ltd., New Delhi, 1962.

CBE Style Manual Committee, *CBE Style Manual* (5th edition). Council of Biology Editors, Inc., Bethesda, Md., 1983.

Cromwell, L., F. J. Weibell, and E. A. Pfeiffer, *Biomedical Instrumentation and Measurements* (Second edition). Prentice-Hall of India Private Limited, New Delhi, 1980.

Day, R. A., *How to Write and Publish a Scientific Paper*, 3rd edition. Oryx Press, Phoenix, Arizona, 1988.

Dubey, R., *India: An Insight On Copyright, Geographical Indication and Confidential Information Laws*. http://www.musicjournal.org/home.html, 2005.

Gerhardt, P., R.G.E. Murray, W. A. Wood, and N. R. Krieg, (editors), *Methods for General and Molecular Bacteriology*. American Society for Microbiology, Washington DC, 1994.

Gurumani, N., *An Introduction to Biostatistics* (Second edition). MJP Publishers, Chennai, 2005.

INSA, *Guidelines for Care and Use of Animals in Scientific Research*. Indian National Science Academy, Bahadur Shah Zafar Marg, New Delhi 110 002, 2000.

Jayaraman, J., *Laboratory Manual in Biochemistry*. Wiley Eastern Limited, New Delhi, 1981.

Misener, S., and A. Krawetz (editors), *Bioinformatics: Methods and Protocols. Methods in Molecular Biology Series*. Humana Press, Totowa, New Jersey, 2000.

Murthy, K. R, *Animal experimentation and research in India*. JPGM, 46 (4): 251–252, 2000.

Pereira, S., P. Veeraraghavan, S. Ghosh and Maneka Gandhi, *Animal experimentation and ethics in India: The CPCSEA makes a difference*. ATLA 32, Supplement 1, 411–415, 2004.

Persing, D. H., F. C. Tenover, J. Versolovic, Yi-Wei Tang, E. R. Unger, D. A. Relman and T. J. White (editors), *Molecular Microbiology: Diagnostic Principles and Practice*. American Society for Microbiology, Washington DC, 2004.

Prescott, L.M., J. P. Harley, and D. A. Klein, *Microbiology* (Sixth edition). McGraw-Hill, New York, 2005.

Rastogi, S. C., *Biochemistry*. Tata McGraw-Hill Publishing Company Limited, New Delhi, 1993.

Ray, K., *And the dispute goes on...* Deccan Herald, Tuesday, January 13, DH News Service, New Delhi, 2004.

Rieger, R, A. Michaelis and M. M. Green, *Glossary of Genetics: Classical and Molecular* (Fifth edition). Narosa Publishing House, New Delhi, 1991.

Saravanan, P., *Virology*, MJP Publishers, Chennai, 2006.

Schefler, W. C., *Statistics for the Biological Sciences.* Addison-Wesley Publishing Company. Reading, Massachusetts, 1969.

Sen, N., *Animals in research.* Current Science, 84(3):273–276, 2003.

Sharma, K. R., *Research Methodology.* National Publishing House, Jaipur and New Delhi, 2002.

Shaw, V., *Reporting Research. Papers on Survey Research Methodology Series.* The Agricultural Development Council, New York, 1977.

Sokal, R. R. and F. J. Rohlf, *Biometry: The Principles and Practice of Statistics in Biological Research.* W. H. Freeman and Company, San Francisco, 1969.

Subramanian, M. A., *Biophysics: Principles and Techniques.* MJP Publishers, Chennai, 2005.

Sullia, S.B., *Biophysical chemistry: Principles and Techniques.* Himalaya Publishing House, New Delhi, 2002.

Upadhyay, A., K. Upadhyay and N. Nath, *Biophysical chemistry: Principles and Techniques.* Himalaya Publishing House, New Delhi, 2002.

Vandana Shiva, *Biopiracy: need to change Western IPR systems.* The Hindu, Wednesday, July 28, 2002.

Veerakumari, L., *Biochemistry.* MJP Publishers, Chennai, 2004.

Webster, J. G. (editor), *Bioinstrumentation.* John Wiley & Sons (Asia) Pte. Ltd., Singapore, 2004.

Whitney, F. L., *The Elements of Research.* Prentice-Hall, Inc, Englewood, N. J, 2004.

Young, H. D., and R. A. Freedman, *University Physics* (Ninth edition). Addison-Wesley Reading, Massachusetts, 1996.

Consulted and useful websites

http://www.4colorvision.com

http://www.advantecmfs.com/labo/electrophoresis

http://www.agbioworld.org/about/index.html

http://www.american.edu/initeb/mk5916a/india.htm

http://www.answers.com

http://www.apelex.fr/anglais/applications/sommaire1/apparatus.htm

http://www.atech.wku.edu/cgi-bin/proofreading.cgi

http://www.bedfordsmartins.com

http://www.beloit.edu/~biology/HHMIsumwork98/lessonplans.html

http://www.bio.org/index.asp?

http://www.biology.clc.uc.edu

http://www.biology.kenyon.edu/Bio_InfoLit/methods/index.html

http://www.biospectrumindia.com

http://www.bluecross.org.in/CPCSEA

http://www.bluecross.org.in/PCAact.html

http://www.britannica.com

http://www.cameraguild.com

http://www.cam-info.net/enc.html

http://www.ccc.comment.edu

http://www.chemistry.nmsu.edu

http://www.chm.davidson.edu/ChemistryApplets/spectrophotometry/index.html

http://www.colorado.edu/Publications/styleguide/symbols.html

http://www.csicop.org

http://www.depthome.brooklyn.cuny.edu

http://www.dl.clackamas.cc.or.us

http://www.ec.njit.edu

http://www.education.indiana.edu

http://www.elchem.kaist.ac.kr

http://www.electrontubes.com

http://www.em-outreach.ucsd.edu

http://www.en.wikipedia.org

http://www.exportmichigan.com/main.htm

http://www.factmonster.com

http://www.fao.org

http://www.geodata.soton.ac.uk/biology/lexstats.html

http://www.gslc.genetics.utah.edu/

http://www.healthtech.com

http://www.helios.bto.ed.ac.uk/bto/statistics

http://www.home.agilent.com

http://www.howstruffworks.com

http://www.hyphenologist.co.uk

http://www.icnet.ic.gc.ca/publication/english/index_e.html

http://www.indiaonestop.com/markets-intellectual-prop-rights.htm

http://www.ipindia.nic.in/ipr/patent/patents.htm

http://www.ipr-helpdesk.org

http://www.iprindia.net/newiprindia/faq/faq.htm

http://www.iprindia.net/newiprindia/index.htm

http://www.isntrumentsystems.de

http://www.iuj.ac.jp/gsir/thesis/writing.html#toc

http://www.iws.ohiolink.edu/chemistry

http://www.libraries.mit.edu

http://www.library.csusm.edu/subject_guides/biology

http://www.lifesciences.napier.ac.uk

http://www.math.liu.se/~behal

http://www.medbioworld.com

http://www.metrolux.de

http://www.micr.magnet.fsu.edu

http://www.microscopyu.com

http://www.microvet.arizona.edu

http://www.monroecc.edu./depts./library/cbe.htm

http://www.nationaldiagnostics.com

http://www.navdanya.org/about/index.htm

http://www.ncbi.nlm.nih.gov/entrez/query.fcgi?db=PubMed

http://www.newscientist.com

http://www.nipoonline.org/index.asp

http://www.nist.gov

http://www.nlm.nih.gov

http://www.onthehill.hamilton.edu/writing/style/

http://www.organicconsumers.org/

http://www.owl.english.purdue.edu

http://www.patents.com/copyrigh.htm

http://www.pathomicro.med.sc.edu

http://www.people.eku.edu

http://www.photo.net

http://www.photozone.de

http://www.plos.org/journals

http://www.sciencedirect.com

http://www.scipeeps.com

http://www.shu.ac.uk

http://www.shutterbutton.com

http://www.siu.edu/~tw3a/index.html

http://www.sja.ucdavis.eduwww.grad-college.iastate.edu

http://www.statistics.com/content/teaching/biology.html

http://www.thehindu.com/thehindu/mp/2003/05/22

http://www.ucar.edu/

http://www.uwlax.edu/murphylibrary/

http://www.vlib.org/Biosciences

http://www.voicesforall.org/

http://www.weather.nmsu.edu/Teaching_Material/SOIL698/electrophoresis/acrylami.html

http://www.weather.nmsu.edu/Teaching_Material/SOIL698/electrophoresis/index.html

http://www.webster.commnet.edu/mla/index.shtml

http://www.webster.commnet.edu/writing/symbols.htm

http://www.wikimediafoundation.org

http://www.wipo.int/treaties/en/

http://www.wku.edu/~sally.kuhlenschmidt/proofprc.htm

http://www.wku.edu/~sally.kuhlenschmidt/proofrd2.htm

http://www.wordnet.princeton.edu

Index

A

A4 bond 113
Abbe condenser 207
Abbè equation 193, 708
abbreviation 607
absorbance 377, 655
absorbance spectrum 390, 655
absorption 278
absorption coefficient 381
absorption maxima 391, 655
absorption spectrum 390, 655
abstract 58, 60, 655
abstract page 125
accession number 655
acclimatization 655
accuracy 159
acid dissociation constant 332
acidity 267
acknowledgement page 123
acknowledgements 80
acoustic optical deflector (AOD) 223
acridine-orange 215, 218
acronyms 607
acrylamide 328
active site 655
activity 524
address 58
addresses 56
adenovirus 655
ADEPT (antibody-directed enzyme
    pro-drug therapy) 655
adsorbate 278
adsorbed phase 278
adsorbent 278, 279
adsorption 278
    chromatography 280, 655
    coefficient 278
aerial photography 406
aerial survey 406
aerosol 464
affinity chromatography 282, 315, 655
affinity matrix 282, 655
affinity tag 656

affinity tails 282
aflatoxin 656
AFLP (amplified fragment length
    polymorphism) 656
agar 327, 657
agaropectin 327, 657
agarose 327, 657
agarose gel electrophoresis 657
AIDS (acquired immunodeficiency
    syndrome) 657
aim 153
algorithm 657
alignment 115, 657
alkalinity 267
alphabet–number system 20
amber 667
ampere 594
ampholytes 364, 372
amplitude 182, 183
amplitude objects 210
anaesthesia 531, 532
analyser 220
analyser-polarizing filter 214
analysis of covariance 169
analytical centrifugation 657
analytical ultracentrifuges 260
angle of incidence 186
angle of reflection 186
angular velocity 252, 667
anilinonaphthalene sulphonate (ANS) 339
animal cell immobilization 658
animal husbandry 522
animal model systems 658
animal procurement 514
animal room 518
animal welfare 480
Animal Welfare Board of India 485
anion 322
anisotropic 221
annotated bibliography 24
annular stop 210
anode 322
ANOVA 169

F

G

H

Q

quanta 181
quarantine 514, 705
quotations 138

R

race 705
radial chromatography 289
radiation hazard 475
radioactivity 598
radioactive detector 314
radioactive isotope, radioisotope 705
radioimmunoassay (RIA) 705
random 158, 175
random effects 172
randomization 158, 706
randomized block design (RBD) 169,
    175, 706
randomized complete block design
    (RCBD) 175
rangefinder camera 420
rangefinders 426
rate of the migration 277
rate-zonal centrifugation 260, 706
rate-zonal centrifugation 262
ray 186
reading frame 706
real-time PCR 706
receptor-binding screening 706
recombinant 706
recombinant DNA technology 706
red eye 424
redundancy 103
reference 24
    beam 214
    card 15
    citation 84
    cited 84
    electrode 270
references 83, 141
reflected ray 186
reflection 3, 186

reflective thinking 3
refraction 186
refractive index (RI) 187, 707
regulatory gene 707
relative centrifugal force (RCF) 253, 707
relative to front (Rf) 709
relaxed plasmid 707
renaturation 707
repetitive DNA 707
replacement therapy 707
replicates 169
replication 154, 707
reporter gene 707
repression 707
repressor 708
reputed journal 12
research 4, 708
    reading 10
    report 53
        formatting and typing 113
    scholar 57
residual variance 154
resistance thermometers 389
resolution 193, 237, 431
resolving power 192, 237, 708
restriction
    endonuclease 708
    exonuclease 708
    fragment 708
    map 709
    nuclease 709
    site 709
restriction fragment length polymorphism
    (RFLP) 709
results 72
retardation factor 291
retardation factor value (RF) 291, 709
retention
    factor 709
    time 303
    volume 303
reticle 225
reticule 199, 225

retro-poson 709
retro-transposon 709
retroviral vectors 709
retrovirus 709
reverse phase chromatography 310, 709
reverse transcription 710
review of literature 63, 710
Rf value 279, 291
rho factor 710
risk groups 460
rotor 256
running head 55
running title 55

S

S phase 666
safety
    in genetic engineering 477
    of laboratory animals 479
    of transgenic animals 479
    of transgenic plants 479
    of viral vectors 478
sample 155
sampling unit 153, 710
sanitation and cleanliness 526
SARS 710
satellite DNA 710
satellite RNA 710
scanning electron microscope (SEM) 234,
    237, 710
scanning probe microscope 239, 710
scanning tunnelling microscope (STM)
    241, 710
SCI (Science Citation Index) 668
science 4, 711
SDS (sodium dodecyl sulphate) 711
SDS-PAGE 362
secondary emission 388
secondary immune response 711
secondary literature 12, 84, 711
secondary messenger 711
secondary metabolite 711
sectoral chimera 667

sedimentation 251, 711
    coefficient 255, 711
    constant 255, 711
    rate 255
sedimentation equilibrium
    centrifugation 262
seed storage proteins 712
segmented genomes 712
selectable marker 712
self-luminous 181
senescence 712
senior author 57
separation 514
sepsis 712
sequence hypothesis 712
sequencing 712
series title 55
serology 712
serotyping 713
service marks (SM) 446
sewage treatment 713
shadow casting 237
shadowing 236
shallow zone of focus 428
shear 713
sheltered or outdoor housing 523
short communication 54
short note 54
shutter 428
    release cord or cable 424
    speed 428
shuttle vector 713
SI (Système Internationale) 593
Siemens 597
sigma factor 713
signal
    peptide 713
    sequence 713
    transduction 713
silencer 714
simple microscope 190
single nucleotide polymorphism
    (SNP) 714

www.ingramcontent.com/pod-product-compliance
Lightning Source LLC
LaVergne TN
LVHW060458070726
842759LV00029BA/854